Transactions on Engineering Technologies

Gi-Chul Yang · Sio-Iong Ao
Xu Huang · Oscar Castillo
Editors

Transactions on Engineering Technologies

International MultiConference of Engineers
and Computer Scientists 2014

 Springer

Editors
Gi-Chul Yang
Department of Multimedia Engineering,
  College of Engineering
Mokpo National University
Chonnam
Korea, Republic of (South Korea)

Sio-Iong Ao
Unit 1, 1/F, Hung To Road, IAENG
  Secretariat
International Association of Engineers
Hong Kong
Hong Kong SAR

Xu Huang
Faculty of Information Sciences
  and Engineering
University of Canberra
Canberra, ACT
Australia

Oscar Castillo
Calzada Tecnologico s/n
Instituto Tecnologico de Tijuana
Tijuana, Baja California
Mexico

ISBN 978-94-017-9587-6   ISBN 978-94-017-9588-3   (eBook)
DOI 10.1007/978-94-017-9588-3

Library of Congress Control Number: 2013953195

Springer Dordrecht Heidelberg New York London

Printed on acid-free paper

Springer Science+Business Media B.V. Dordrecht is part of Springer Science+Business Media
(www.springer.com)

Preface

A large international conference on Advances in Engineering Technologies and Physical Science was held in Hong Kong during 12–14 March 2014, under the International MultiConference of Engineers and Computer Scientists 2014 (IMECS 2014). The IMECS 2014 is organized by the International Association of Engineers (IAENG). IAENG is a non-profit international association for engineers and computer scientists, which was founded originally in 1968 and has been undergoing rapid expansions in recent few years. The IMECS conference serves as a good platform for the engineering community to meet with each other and to exchange ideas. The conference has also struck a balance between theoretical and application development. The conference committees have been formed with over 300 committee members who are mainly research center heads, faculty deans, department heads, professors, and research scientists from over 30 countries with the full committee list available at our conference website (http://www.iaeng.org/IMECS2014/committee.html). The conference is truly international meeting with a high level of participation from many countries. The response that we have received for the conference is excellent. There have been more than 600 manuscript submissions for IMECS 2014. All submitted papers have gone through the peer-review process and the overall acceptance rate is 51.24 %.

This volume contains 33 revised and extended research articles written by prominent researchers participating in the conference. Topics covered include engineering physics, engineering mathematics, scientific computing, control theory, artificial intelligence, electrical engineering, communications systems, and industrial applications. The book offers the state of art of tremendous advances in engineering technologies and physical science and applications, and also serves as an excellent reference work for researchers and graduate students working with/on engineering technologies and physical science and applications.

Gi-Chul Yang
Sio-Iong Ao
Xu Huang
Oscar Castillo

Contents

Active-Force Control on Vibration of a Flexible Single-Link Manipulator Using a Piezoelectric Actuator

Abdul Kadir Muhammad, Shingo Okamoto and Jae Hoon Lee

Abstract The purposes of this research are to formulate the equations of motion of the system, to develop computational codes by a finite-element method in order to perform dynamics simulation with vibration control, to propose an effective control scheme using three control strategies, namely active-force (AF) proportional (P), and proportional-derivative (PD) controls and to confirm the calculated results by experiments of a flexible single-link manipulator. The system used in this paper consists of an aluminum beam as a flexible link, a clamp-part, a servo motor to rotate the link and a piezoelectric actuator to control vibration. Computational codes on time history responses, Fast Fourier Transform (FFT) processing and eigen-values–eigenvectors analysis were developed to calculate the dynamic behavior of the link. Furthermore, the AF, P, and PD controls strategies were designed and compared their performances through calculations and experiments. The calculated and experimental results showed the superiority of the proposed AF control compared to the P and PD ones to suppress the vibration of the flexible link manipulator.

Keywords Active-force control · Finite-element method · Flexible manipulator · Piezoelectric actuator · Proportional control · Proportional-derivative control · Vibration control

A.K. Muhammad (✉) · S. Okamoto · J.H. Lee
Graduate School of Science and Engineering, Ehime University, 3 Bunkyo-cho, Matsuyama 790-8577, Japan
e-mail: y861008b@mails.cc.ehime-u.ac.jp; kadir_muhammad@yahoo.co.id

S. Okamoto
e-mail: okamoto.shingo.mh@ehime-u.ac.jp

J.H. Lee
e-mail: jhlee@ehime-u.ac.jp

A.K. Muhammad
Center for Mechatronics and Control Systems, Mechanical Engineering Department, State Polytechnic of Ujung Pandang, Jl. Perintis Kemerdekaan KM 10, Makassar 90-245, Indonesia

1

1 Introduction

Employment of flexible link manipulator is recommended in the space and industrial applications in order to accomplish high performance requirements such as high-speed besides safe operation, increasing of positioning accuracy and lower energy consumption, namely less weight. However, it is not usually easy to control a flexible manipulator because of its inheriting flexibility. Deformation of the flexible manipulator when it is operated must be considered by any control. Its controller system should be dealt with not only its motion but also vibration due to the flexibility of the link.

In the past few decades, a number of modeling methods and control strategies using piezoelectric actuators to deal with the vibration problem have been investigated by researchers [1–3]. Nishidome and Kajiwara [1] investigated a way to enhance performances of motion and vibration of a flexible-link mechanism. They used a modeling method based on modal analysis using the finite-element method. The model was described as a state space form. Their control system was constructed with a designed dynamic compensator based on the mixed of H_2/H_∞. They recommended separating the motion and vibration controls of the system. Zhang et al. [2] has studied a flexible piezoelectric cantilever beam. The model of the beam using finite-elements was built by ANSYS application. Based on the Linear Quadratic Gauss (LQG) control method, they introduced a procedure to suppress the vibration of the beam with the piezoelectric sensors and actuators were symmetrically collocated on both sides of the beam. Their simulation results showed the effectiveness of the method. Gurses et al. [3] investigated vibration control of a flexible single-link manipulator using three piezoelectric actuators. The dynamic modeling of the link had been presented using Euler-Bernoulli beam theory. Composite linear and angular velocity feedback controls were introduced to suppress the vibration. Their simulation and experimental results showed the effectiveness of the controllers.

Furthermore, applications of the AF control strategy to suppress vibration of a flexible system were done by some researchers [4–6]. Hewit et al. [4] used the AF control for deformation and disturbance attenuation of a flexible manipulator. Then, a PD control was used for trajectory tracking of the flexible manipulator. They used a motor as an actuator. Modeling of the manipulator was done using virtual link coordinate system (VLCS). Their simulation results had shown that the proposed control could cancel the disturbance satisfactorily. Tavakolvour et al. [5] investigated the AF control application for a flexible thin plate. Modeling of their system was done using finite-difference method. Their calculated results showed the effectiveness of the proposed controller to reduce vibration of the plate. Tavakolvour and Mailah [6] studied the AF control application for a flexible beam with an electromagnetic actuator. Modeling of the beam was done using finite-difference method. The effectiveness of the proposed controller was confirmed through simulation and experiment.

The purposes of this research are to derive the equations of motion of a flexible single-link system by a finite-element method, to develop the computational codes in order to perform dynamics simulations with vibration control, to propose an effective control scheme of a flexible single-link manipulator using three control strategies namely active-force (AF), proportional (P) and proportional-derivative (PD) controls and to confirm the calculated results by experiments of the flexible single-link manipulator.

The flexible manipulator used in this paper consists of an aluminum beam as a flexible link, a clamp-part, a servo motor to rotate the link and a piezoelectric actuator to control vibration. Computational codes on time history responses, Fast Fourier Transform (FFT) processing and eigenvalues–eigenvectors analysis were developed to calculate the dynamic behavior of the link and validated by the experimental one. Furthermore, the AF, P, and PD controls strategies were designed to suppress the vibration. It was done by adding bending moments generated by the piezoelectric actuator to the single-link. Finally, their performances were compared through calculations and experiments.

2 Formulation by Finite-Element Method

The link has been discretized by finite-elements [7, 8]. The finite-element has two degrees of freedom, namely the lateral deformation $v(x, t)$, and the rotational angle $\psi(x, t)$. The length, the cross-sectional area and the area moment of inertia around z-axis of every element are denoted by l_i, S_i and I_{zi} respectively. Mechanical properties of every element are denoted as Young's modulus E_i and mass density ρ_i.

2.1 Kinematics

Figure 1 shows the position vector of an arbitrary point P in the link in the global and rotating coordinate frames. Let the link as a flexible beam has a motion that is confined in the horizontal plane as shown in Fig. 1. The O–XY frame is the global coordinate frame while O–xy is the rotating coordinate frame fixed to the root of the link. A motor is installed on the root of the link. The rotational angle of the motor when the link rotates is denoted by $\theta(t)$.

The position vector $\boldsymbol{r}$ (x, t) of the arbitrary point P in the link at time $t = t$, measured in the O–XY frame shown in Fig. 1 is expressed by

$$r(x,t) = X(x,t)\boldsymbol{I} + Y(x,t)\boldsymbol{J} \tag{1}$$

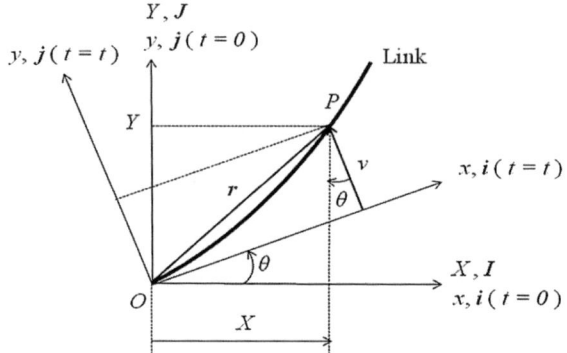

Fig. 1 Position vector of an arbitrary point P in the link in the global and rotating coordinate frames. O–XY Global coordinate frame. O–xy Rotating coordinate frame

where

$$X(x,t) = x \cos \theta(t) - v(x,t) \sin \theta(t) \tag{2}$$

$$Y(x,t) = x \sin \theta(t) + v(x,t) \cos \theta(t) \tag{3}$$

The velocity of P is given by

$$\dot{r}(x,t) = \dot{X}(x,t)\boldsymbol{I} + \dot{Y}(x,t)\boldsymbol{J} \tag{4}$$

2.2 Finite-Element Method

Figure 2 shows the rotating coordinate frame and the link divided by one-dimensional and two-node elements. Then, Fig. 3 shows the element coordinate frame of the i-th element. Here, there are four boundary conditions together at nodes i and $(i + 1)$ when the one-dimensional and two-node element is used. The four boundary conditions are expressed as nodal vector as follow

$$\boldsymbol{\delta}_i = \left\{ v_i \quad \psi_i \quad v_{i+1} \quad \psi_{i+1} \right\}^T \tag{5}$$

Then, the hypothesized deformation has four constants as follows [9]

$$v_i = a_1 + a_2 x_i + a_3 x_i^2 + a_4 x_i^3 \tag{6}$$

Fig. 2 Rotating coordinate frame and the link divided by the one-dimensional and two-node elements. o–xy Rotating coordinate frame

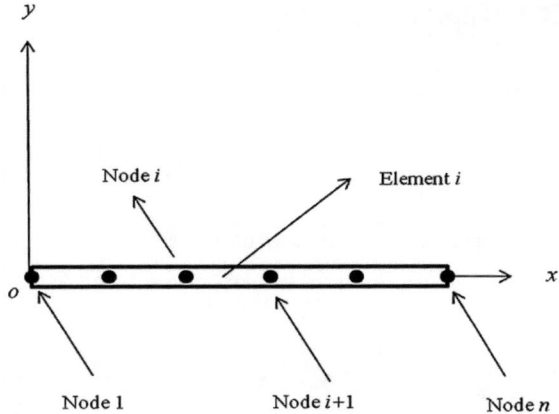

Fig. 3 Element coordinate frame of the i-th element. o_i–x_i y_i Element coordinate frame of the i-th element

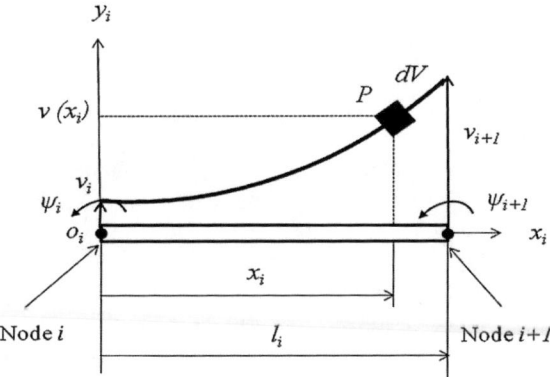

The relation between the lateral deformation v_i and the rotational angle ψ_i of the node i is given by

$$\psi_i = \frac{\partial v_i}{\partial x_i}.$$

(7)

Furthermore, from mechanics of materials, the strain of node i can be defined by

$$\varepsilon_i = \varepsilon_{x_i} = -y_i \frac{\partial^2 v_i}{\partial x_i^2}$$

(8)

2.3 Equations of Motion

Equation of motion of the i-th element is given by

$$M_i\ddot{\delta}_i + C_i\dot{\delta}_i + \left[K_i - \dot{\theta}^2(t)M_i\right]\delta_i = \ddot{\theta}(t)f_i \qquad (9)$$

where M_i, C_i, K_i, $\ddot{\theta}(t)f_i$ are the mass matrix, damping matrix, stiffness matrix and the excitation force generated by the rotation of the motor respectively. The representation of the matrices and vector in Eq. (9) can be found in [7]. Finally, the equation of motion of the system with n elements considering the boundary conditions is given by

$$M_n\ddot{\delta}_n + C_n\dot{\delta}_n + \left[K_n - \dot{\theta}^2(t)M_n\right]\delta_n = \ddot{\theta}(t)f_n \qquad (10)$$

3 Modeling

Figure 4 shows a model of the single-link manipulator, the clamp-part and the piezoelectric actuator. The link including the clamp-part and actuator were discretized by 35 elements. The clamp-part is more rigid than the link. Therefore Young's modulus of the clamp-part was set in 1,000 times of the link's. The piezoelectric actuator was bonded to a one-side surface of Element 4. A schematic representation on modeling of the piezoelectric actuator is shown in Fig. 5. Furthermore, a strain gage was bonded to the position of Node 6 of the single-link (0.11 m from the origin). Physical parameters of the single-link model and the piezoelectric actuator are shown in Table 1 [8].

The piezoelectric actuator suppressed the vibration of the flexible link manipulator by adding bending moments at Nodes 3 and 6, M_3 and M_6 to the flexible link. The bending moments are generated by applying voltages $+E$ to the piezoelectric actuator as shown in Fig. 5. The relation between the bending moments and the voltages are related by

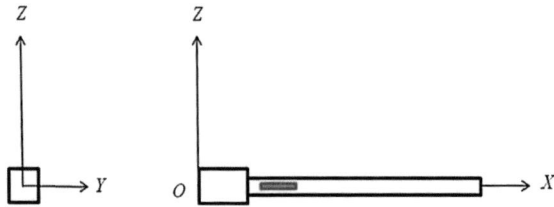

Fig. 4 Computational model of the flexible single-link manipulator

Fig. 5 Modeling of the
piezoelectric actuator

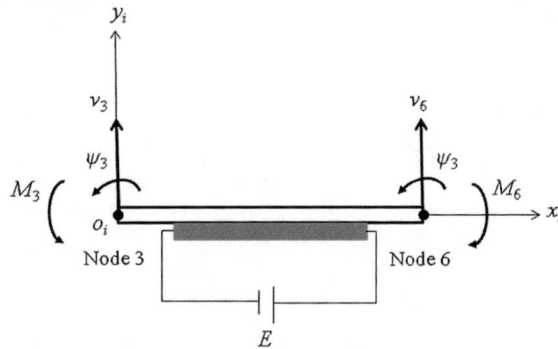

Node 3　　　　　　　Node 6

E

Table 1 Physical parameters of the flexible link and the piezoelectric actuator [10]

l	Total length	m	3.91×10^{-1}
l_l	Length of the link	m	3.50×10^{-1}
l_c	Length of the clamp-part	m	4.10×10^{-2}
l_a	Length of the actuator	m	2.00×10^{-2}
S_l	Cross section area of the link	m^2	1.95×10^{-5}
S_c	Cross section area of the clamp-part	m^2	8.09×10^{-4}
S_a	Cross section area of the actuator	m^2	1.58×10^{-5}
I_{zl}	Cross section area moment of inertia around z-axis of the link	m^4	2.75×10^{-12}
I_{zc}	Cross section area moment of inertia around z-axis of the clamp-part	m^4	3.06×10^{-8}
I_{za}	Cross section area moment of inertia around z-axis of the actuator	m^4	1.61×10^{-11}
E_l	Young's Modulus of the link	GPa	7.03×10^{1}
E_c	Young's Modulus of the clamp-part	GPa	7.00×10^{4}
E_a	Young's Modulus of the actuator	GPa	4.40×10^{1}
ρ_l	Density of the link	kg/m^3	2.68×10^{3}
ρ_c	Density of the clamp-part	kg/m^3	9.50×10^{2}
ρ_a	Density of the actuator	kg/m^3	3.33×10^{3}
α	Damping factor of the link	–	2.50×10^{-4}

$$M_3 = -M_6 = d_1 E \tag{11}$$

here d_1 is a constant quantity.

Furthermore, the voltage to generate the bending moments is proportional to the strain ε of the single-link due to the vibration. The relation can be expressed as follows

$$E = \pm \frac{1}{d_2} \varepsilon \qquad (12)$$

here d_2 is a constant quantity. Then, d_1 and d_2 will be determined by comparing the calculated results and experimental ones.

Computational codes were developed to perform dynamics simulation of the system based on the formulation that explained above. The validation was done using time history responses analysis of free vibration, natural frequencies using Fast Fourier Transform (FFT) processing, vibration modes and natural frequencies using eigenvalues–eigenvectors analysis and time history responses analysis due to the base excitation [8].

4 Control Scheme and Strategies

A control scheme to suppress the vibration of the single-link was designed using the piezoelectric actuator. It was done by adding bending moments generated by the piezoelectric actuator to the single-link. Therefore, the equation of motion of the system become

$$M_n \ddot{\delta}_n + C_n \dot{\delta}_n + \left[K_n - \dot{\theta}^2(t) M_n \right] \delta_n = \ddot{\theta}(t) f_n + u_n(t) \qquad (13)$$

where the vector $u_n(t)$ containing M_3 and M_6 is the control force generated by the actuator to the single-link.

To drive the actuator, three different control strategies namely AF, P and PD controls have been designed and examined. Their performances were compared through calculations and experiments.

4.1 Active-Force Control

Figure 6 shows the block diagram of the AF control that is proposed in this research. In this strategy, vibration of the system is controlled by canceling bending moments acting at Nodes 3 and 6 due to the base excitation (excitation bending moments). The following steps are the way to estimate and cancel the excitation bending moments.

Firstly, the strain, ε_6 at Node 6 is measured to estimate the lateral deformation, v_6 at the Node 6. Substituting Eq. (6) to Eq. (8) considering the boundary conditions then the relation between the strain and the lateral deformation can be defined as follows

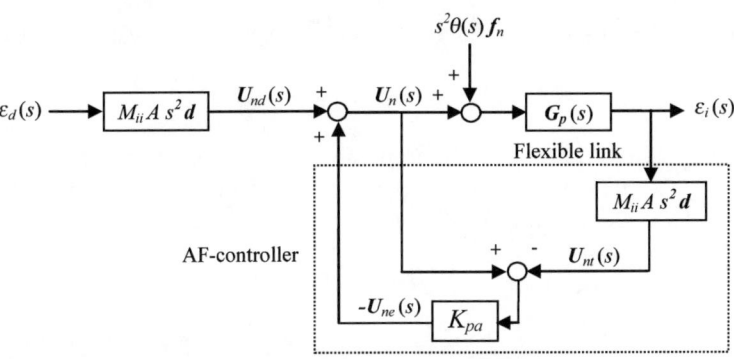

ε_d : Desired strain　　　　　　　　　　ε_i : Measured strains at Node i

θ : Rotation angle of the motor　　　　　M_{ii} : Component of mass matrix

A : Conversion from ε_i to v_i　　　　　d : Position vector

U_{nd} : Desired bending moments　　　U_n : Applied bending moments

U_{ne} : Excitation bending moments　　　U_{nt} : Bending moments

Fig. 6 Block diagram of active-force control of the flexible link manipulator. ε_d Desired strain. ε_i Measured strains at Node i. θ Rotation angle of the motor. M_{ii} Component of mass matrix. A Conversion from ε_i to v_i. d Position vector. U_{nd} Desired bending moments. U_n Applied bending moments. U_{ne} Excitation bending moments. U_{nt} Bending moments

$$\frac{v_6}{\varepsilon_6} = -\frac{x^2(x-3l)}{6y(x-l)} = A \tag{14}$$

where l, x and y are the length of the link, the position of Node 6 in x and y directions, respectively.

Secondly, the actual force in the s-domain acting at Node 6 can be defined in the form of the Newton's equation of motion as follows

$$F_6(s) = M_{ii(i=11)}\, s^2\, v_6 \tag{15}$$

where $M_{ii(i=11)}$ is the component of the mass matrix corresponding to v_6.

Thirdly, the bending moments acting at Nodes 3 and 6 are estimated using the following equation

$$U_{nt}(s) = \pm F_6(s)\, d \tag{16}$$

The vector d that represents the position vector from the reference point to the position where the excitation force acting can be written as follows

$$d = \{ 0 \quad 0 \quad 0 \quad l_2 \quad 0 \quad l_2 \quad 0 \quad \ldots \quad 0 \}^T \tag{17}$$

Fourthly, based on Fig. 6, the excitation bending moments can be calculated as

$$U_{ne}(s) = K_{pa}\{U_{nt}(s) - U_n(s)\} \tag{18}$$

where K_{pa} is the non-dimensional proportional gain of the proposed AF control.

Finally, the bending moments applying as a control force to control the vibration of the system can be calculated as follows

$$U_n(s) = -U_{ne}(s) + U_{nd}(s) \tag{19}$$

where $U_{nd}(s)$ is the desired bending moments which is zero. The negative of $U_{ne}(s)$ indicates that the bending moments used to cancel the vibration of the system.

4.2 Proportional and Proportional-Derivative Controls

Substituting Eq. (12) to Eq. (11) gives

$$M_{3,6} = \pm\frac{d_1}{d_2}\varepsilon \tag{20}$$

Based on Eq. (20), the bending moments for P and PD controllers can be defined in s-domain as follows

$$U_n(s) = G_{C1,2}(s)(\varepsilon_d(s) - \varepsilon_6(s)) \tag{21}$$

where ε_d and ε_6 denote the desired and measured strains at Node 6, respectively. The gain of P and PD controllers can be written by a vector in s-domain respectively as follows

$$G_{C1}(s) = \{0 \quad 0 \quad 0 \quad K_p \quad 0 \quad -K_p \quad 0 \quad \dots \quad 0\}^T \tag{22}$$

and

$$G_{C2}(s) = \{0 \quad 0 \quad 0 \quad K_p + K_d s \quad 0 \quad -(K_p + K_d s) \quad 0 \quad \dots \quad 0\}^T \tag{23}$$

A block diagram of the proportional and proportional-derivative controls strategies for the single-link system is shown in Fig. 7.

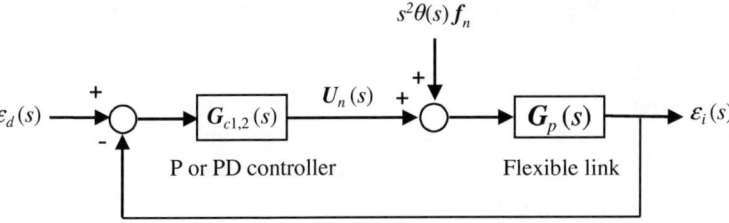

Fig. 7 Block diagram of P and PD controls of the flexible link manipulator. ε_d Desired strain. ε_i Measured strains at Node i. θ Rotation angle of the motor. U_n Applied bending moments

5 Experiment

5.1 Experimental Set-up

In order to investigate the validity of the proposed control strategies, an experimental set-up was designed. The set-up is shown in Fig. 8. The flexible link manipulator consists of the flexible aluminum beam, the clamp-part, the servo motor and the base. The flexible link was attached to the motor through the clamp-part. In the experiments, the motor was operated by an independent motion controller. A strain gage was bonded to the position of 0.11 [m] from the origin of the link.

The piezoelectric actuator was attached on one side of the flexible manipulator to provide the blocking force against vibrations. A Wheatstone bridge circuit was developed to measure the changes in resistance of the strain gage in the form of

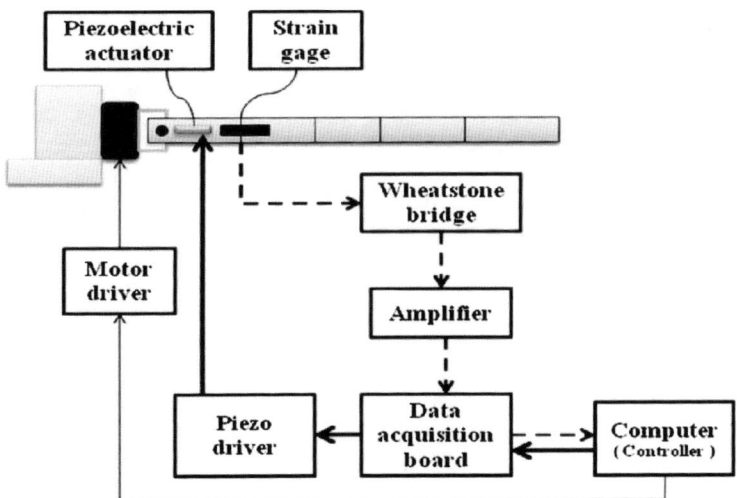

Fig. 8 Schematics of measurement and control system [8]. *Dashed arrow* Measurement of strains. *Thick arrow* Vibration control. *Thin arrow* Motion control

voltages. An amplifier circuit was designed to amplify the small output signal of the Wheatstone bridge.

Furthermore, a data acquisition board and a computer that have functionality of analog to digital (A/D) conversion, signal processing, control process and digital to analog (D/A) conversion were used. The data acquisition board connected to the computer through USB port. Finally, the controlled signals sent to a piezo driver to drive the piezoelectric actuator in its voltage range.

5.2 Experimental Method

The rotation of the motor was set from 0 to $\pi/2$ radians (90°) within 0.68 [s]. The outputs of strain gage were converted to voltages by the Wheatstone bridge and magnified by the amplifier. The noises that occur in the experiment were reduced by a 100 [µF] capacitor attached to the amplifier. The output voltages of the amplifier sent to the data acquisition board and the computer for control process.

The control strategies were implemented in the computer using the visual C++ program. The analog output voltages of the data acquisition board sent to the input channel of the piezo driver to generate the actuated signals for the piezoelectric actuator.

6 Calculated and Experimental Results

6.1 Calculated Results

Time history responses of strains on the uncontrolled and controlled systems were calculated when the motor rotated by the angle of $\pi/2$ radians (90°) within 0.68 [s]. Time history responses of strains on the controlled system were calculated for the model under three control strategies as shown in Figs. 6 and 7.

Examining several gains of the AF, P and PD controllers leaded to $K_{pa} = 0.83$ [−], $K_p = 30$ [Nm], $K_d = 0.02$ [Nms] as the better ones. Figure 9 shows the uncontrolled and controlled time history responses of strains at Node 6. The maximum and minimum strains of uncontrolled system in positive and negative sides were 387.00×10^{-6} and -435.00×10^{-6}, as shown in Fig. 9a. By using AF-controller they became 66.50×10^{-6} and -88.60×10^{-6}, as shown in Fig. 9b. Moreover, by using P-controller they became 127.00×10^{-6} and -145.00×10^{-6}, as shown in Fig. 9c. By adding D-gain they became -124.00×10^{-6} and -143.00×10^{-6}, as shown in Fig. 9d.

Based on the calculated results, the effect of D-controller was very small compare to P-controller, therefore using a P-controller will be sufficient for experiment [7].

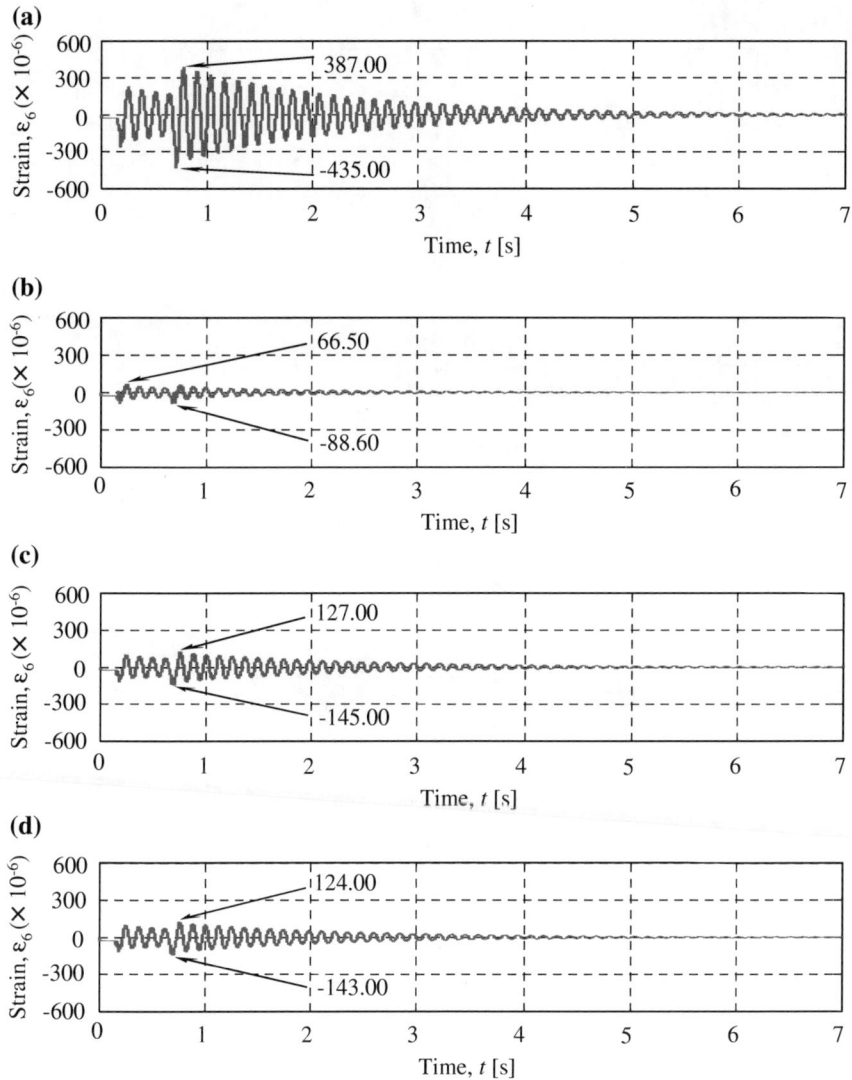

Fig. 9 Calculated time history responses of strains at Node 6 for uncontrolled and controlled systems due to the base excitation. **a** Uncontrolled system. **b** Controlled by AF-controller, $K_{pa} = 0.4$ [−]. **c** Controlled by P-controller, $K_p = 30$ [Nm]. **d** Controlled by PD-controller, $K_p = 30$ [Nm], $K_d = 0.02$ [Nms]

(a)

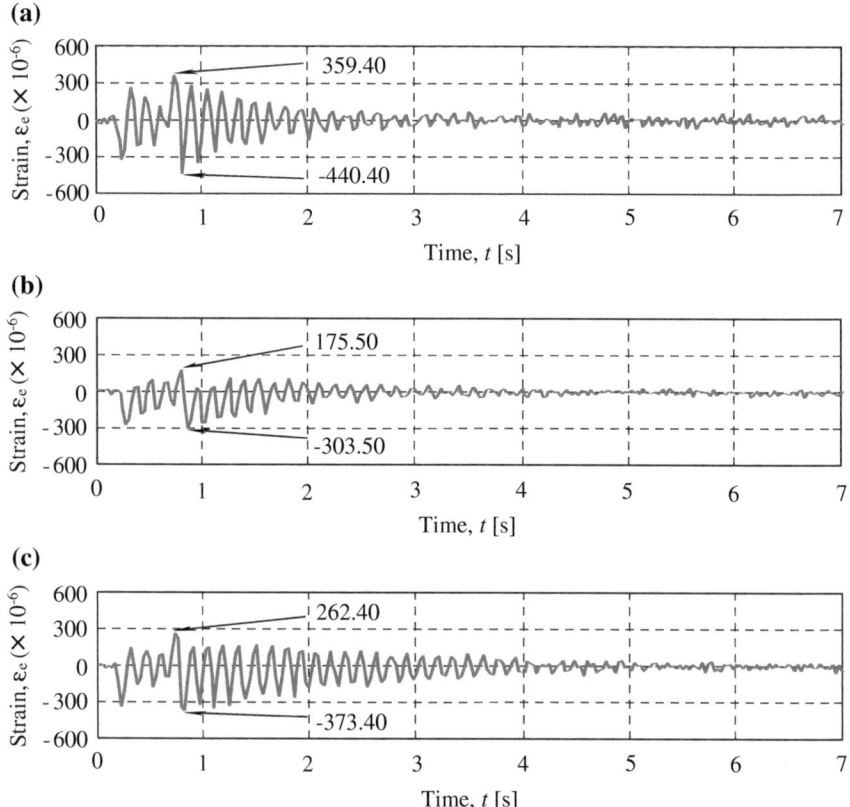

Fig. 10 Experimental time history responses of strains at 0.11 (m) from the link's origin for uncontrolled and controlled systems due to the base excitation. **a** Uncontrolled system. **b** Controlled by AF-controller, $K_{pa}' = 125$ [−]. **c** Controlled by P-controller, $K_p' = 600$ [−]

6.2 Experimental Results

Experimental time history responses of the strains on the uncontrolled and controlled systems were measured when the motor rotated by the angle of $\pi/2$ radians (90°) within 0.68 [s]. Experimental time history responses on the controlled system were measured under the control strategies as shown in Figs. 6 and 7.

Furthermore, the experimental active-force and proportional gains that are non-dimensional gains, K_{pa} and K_p' were examined. The examination of gains leaded to $K_p' = 600$ [−] and $K_{pa}' = 125$ [−], as the better ones. Figure 10 shows the experimental uncontrolled and controlled time history responses of strains at the same position in the calculations. The maximum and minimum strains of uncontrolled system in positive and negative sides were 359.40×10^{-6} and -440.40×10^{-6}, as shown in Fig. 10a. By using AF-controller they became 175.50×10^{-6} and -303.50×10^{-6}, as shown in Fig. 10b. Moreover, by using P-controller they became 262.40×10^{-6} and -373.40×10^{-6}, as shown in Fig. 10c.

It was verified from these results that the vibration of the flexible link manipulator can be more effectively suppressed using the proposed AF-control compared to the P and PD ones.

7 Conclusion and Future Work

The equations of motion for the flexible link manipulator had been derived using the finite-element method. Computational codes had been developed in order to perform dynamic simulations of the system. Experimental and calculated results on time history responses, natural frequencies and vibration modes show the validities of the formulation, computational codes and modeling of the system. The active-force (AF), proportional (P) and proportional-derivative (PD) controls strategies were designed to suppress the vibration of the system. Their performances were compared through the calculations and experiments. The calculated and experimental results show the superiority of the proposed AF control compared to the P and PD ones to suppress the vibration of the flexible single-link manipulator.

A flexible two-link manipulator is being prepared. The control scheme and strategies presented in this paper will be applied to the flexible two-link system.

References

1. C. Nishidome, I. Kajiwara, Motion and vibration control of flexible-link mechanism with smart structure. JSME Int. J. **46**(2), 565–571 (2003)
2. J. Zhang et al., Active vibration control of piezoelectric intelligent structures. J. Comput. **5**(3), 401–409 (2010)
3. K. Gurses et al., Vibration control of a single-link flexible manipulator using an array of fiber optic curvature sensors and PZT actuators. Mechatronics **19**, 167–177 (2009)
4. J.R. Hewit et al., Active force control of a flexible manipulator by distal feedback. Mech. Mach. Theory **32**(5), 583–596 (1997)
5. A.R. Tavakolpour et al., Modeling and Simulation of a novel active vibration control system for a flexible structures. WSEAS Trans. Syst. Control **6**(5), 184–195 (2011)
6. A.R. Tavakolpour, M. Mailah, Control of resonance phenomenon in flexible structures via active support. J. Sound Vib. **331**, 3451–3465 (2012)
7. A.K. Muhammad et al., in Computer simulations on vibration control of a flexible single-link manipulator using finite-element method. *Proceeding of 19th International Symposium of Artificial Life and Robotics*, 22–24 Jan 2014, Beppu, Japan, pp. 381–386
8. A.K. Muhammad et al., in Computer simulations and experiments on vibration control of a flexible link manipulator using a piezoelectric actuator, lecture notes in engineering and computer science. *Proceeding of The International MultiConference of Engineers and Computer Scientists* 2014, IMECS 2014, 12–14 Mar 2014, Hong Kong, pp. 262–267
9. M. Lalanne et al., *Mechanical vibration for engineers* (Wiley, New York, 1983), pp. 262–267
10. Resin Coated Multilayer Piezoelectric Actuators [Online]. http://www.mmech.com

Numerically Stable Computer Simulation of Solidification: Association Between Eigenvalues of Amplification Matrix and Size of Time Step

Elzbieta Gawronska and Norbert Sczygiol

Abstract The constantly increasing demand for efficient and precise computational solvers becomes the crucial factor deciding about usability of a given domain specific simulation software. The main idea of this article is the use of eigenvalues of amplification matrices to determine the size of time step in modeling of solidification. As far as numerical simulations are concerned it is very important to obtain solutions which are stable and physically correct. It is acquired by fulfilling many assumptions and conditions during the construction a numerical model and carrying out computer simulations. One of the conditions is a proper selection of time step. The size of time step has a great impact on the stability of used time integration schemes (e.g. explicit scheme), or on a proper image of physical phenomena occurring during the simulation (e.g. implicit scheme). The eigenvalues of amplification matrix in governing equations influence on the appropriate selection of size of time step in computer simulations. Hence, it allows to better fit the size of time step and time integration scheme for modeled structure.

Keywords Amplification matrix · Computer simulation · Computer modeling · Eigenvalues · Explicit scheme · Implicit scheme · Solidification process · Stability · Time step

1 Introduction

Modeling and computer simulation is one of the most effective methods of studying of difficult problems in foundry and metallurgical manufacture. Numerical simulations are use for optimization of casting production. In many cases they are a

E. Gawronska · N. Sczygiol (✉)
Czestochowa University of Technology, Dabrowskiego 69, 42201 Czestochowa, Poland
e-mail: norbert.sczygiol@icis.pcz.pl

E. Gawronska
e-mail: elzbieta.gawronska@icis.pcz.pl

© Springer Science+Business Media Dordrecht 2015
G.-C. Yang et al. (eds.), *Transactions on Engineering Technologies*,
DOI 10.1007/978-94-017-9588-3_2

unique possible technique for carrying out of the experiments which real statement is complicated. Computer modeling allows defining the major factors for a quality estimation of alloy castings. Simulations help to investigate interaction between solidifying casting and changes of its parameters or initial conditions.

A numerical modeling of solidification is known to be a very time consuming task. The constantly increasing demand for efficient and precise computational solvers becomes the factor that decides about usability of a given solidification simulation software. In many cases practitioners require multiple scenarios to be tested, e.g. for different input parameters, before they make a final decision about the setup of a given technological process. At the same time increasing size of computer memory makes it possible to consider problems with increasing size, which in turn results in increased precision of simulations. There are several possible ways to tackle this kind of problems. For instance, one can use parallel computers or accelerated architectures such as GPUs or FPGAs [1]. However, these solutions require another level of expertise in both, parallel hardware and software, which very often is not easily available. In papers [2–4] we proposed new method, which relies on the application of the technique called mixed time partitioning. Our approach exploits the fact that physical processes inside a mould are of different nature than those in a solidifying casting. As a result different time steps can be used to run computations within both sub-domains. Because processes that are modeled in the casting sub-domain are more dynamic they require very fine-grained time step. On the other hand a heat transfer within the mould sub-domain is less intense, and thus coarse-grained step is sufficient to guarantee desired precision of computations. Obviously, increasing length of a single time step results in decreased computational load, which in turn greatly improves performance of our approach. In this paper we put emphasis on determination of stability criteria for the selected integration method. Mathematical apparatus of the chosen stability analysis method is applied for the homogeneous form of the semi-discretized (after spatial discretization) equation of solidification, as the stability is independent of the inhomogeneous part. The analysis of numerical stability of mixed time partitioning methods for the structural dynamics [5] and for heat conduction problem was adapted to the solidification problem with temperature-dependent material properties.

2 Solidification in Terms of the Finite Element Method

The finite element method facilitates the modeling of many complex problems. Its wide application for founding comes from the fact that it permits an easy adaptation of many existing solutions and technique of solidification modeling.

Computer calculations need to use discrete models, which means problems must be formulated by introducing time-space mesh. These methods convert given physical equations into matrix equations (algebraic equations). This system of algebraic equations usually contain many thousands of unknowns, that is why the efficiency of method applied to solve them is crucial.

After essential transformations we obtain an ordinary differential equation containing only the time derivative [6] as following:

$$\mathbf{M}(T)\dot{\mathbf{T}} + \mathbf{K}(T)\mathbf{T} = \mathbf{b}(T), \tag{1}$$

where $\mathbf{M}$ is the capacity matrix, $\mathbf{K}$ is the conductivity matrix, $\mathbf{T}$ is temperature vector and $\mathbf{b}$ is right-hand side vector values of which are calculated using boundary conditions. The global form of these matrices is obtained by summing of coefficients for all the finite elements. The matrix components are defined for a single finite element as follows:

$$\mathbf{M} = \sum_e \int_\Omega c^* \mathbf{N}^T \mathbf{N} \, d\Omega, \tag{2}$$

$$\mathbf{K} = \sum_e \int_\Omega \lambda \nabla^T \mathbf{N} \cdot \nabla \mathbf{N} d\Omega, \tag{3}$$

$$\mathbf{b} = \sum_e \int_\Gamma \mathbf{N}_\Gamma^T \mathbf{q}^T \mathbf{n} \, d\Gamma, \tag{4}$$

where $\mathbf{N}$ is a shape vector in the area Ω, $\mathbf{N}_\Gamma$ is a shape vector on the boundary Γ, $\mathbf{n}$ is an ordinary vector towards the boundary Γ, and $\mathbf{q}$ is vector of nodal fluxes.

The system of ordinary differential equations (1) containing time derivative was obtained as a result of spatial integration and it may only be solved in approximation. In order to receive an approximate solution it is needed to use the division of time interval $(0, t_{max})$ into sub-intervals (t_k, t_{k+1}) with the length of $\Delta t_k = t_{k+1} - t_k$ and time integration is performed by the use of one step or multisteps methods [7]. We used the one step schemes, so-called Θ schemes, in the following form:

$$\begin{aligned} \mathbf{T}^{n+1} = \mathbf{T}^n &+ (-(\mathbf{M}^{n+\Theta})^{-1}\mathbf{K}^{n+\Theta}\mathbf{T}^n \\ &+ (\mathbf{M}^{n+\Theta})^{-1}\mathbf{b}^n)(1 - \Theta)\Delta t \\ &+ (-(\mathbf{M}^{n+\Theta})^{-1}\mathbf{K}^{n+\Theta}\mathbf{T}^{n+1} \\ &+ (\mathbf{M}^{n+\Theta})^{-1}\mathbf{b}^{n+1})\Theta\Delta t. \end{aligned} \tag{5}$$

Due to a possible dependence of materials properties from the temperature, namely $\mathbf{M}$, $\mathbf{K}$ and $\mathbf{b}$ for $\Theta \neq 0$ it is a system of nonlinear equations. To solve this system iterative methods must be used.

The forward Euler scheme:

$$\mathbf{M}^n\mathbf{T}^{n+1} = \mathbf{M}^n\mathbf{T}^n - \Delta t\mathbf{K}^n\mathbf{T}^n + \Delta t\mathbf{b}^n, \tag{6}$$

is obtained for one extreme value $\Theta = 0$ and the backward Euler scheme:

$$\left(\mathbf{M}^{n+1} + \Delta t\mathbf{K}^{n+1}\right)\mathbf{T}^{n+1} = \mathbf{M}^{n+1}\mathbf{T}^n + \Delta t\mathbf{b}^{n+1}, \tag{7}$$

is obtained for the other extreme value $\Theta = 0$. And if the values of matrices coefficients $\mathbf{M}$ i $\mathbf{K}$ in the Eq. (7) are evaluated on the level of previous time step then a modified backward Euler scheme is obtained as follows:

$$(\mathbf{M}^n + \Delta t\mathbf{K}^n)\mathbf{T}^{n+1} = \mathbf{M}^n\mathbf{T}^n + \Delta t\mathbf{b}^{n+1}. \tag{8}$$

3 Basic Equations

In the computer simulations apparent heat formulation (AHC) of solidification is used [8]:

$$\nabla \cdot (\lambda \nabla T) = c^*(T)\frac{\partial T}{\partial t}. \tag{9}$$

The Eq. (9) is solved by mixed time partitioning method considering:

1. semi-discretization,
2. initial-value problem which consists of given functions $\mathbf{T} = \mathbf{T}(t)$ satisfying the Eq. (9) and being the part of initial conditions $\mathbf{T}(t = 0) = \mathbf{T}_0$ for $t \in \langle 0, T \rangle$, $T > 0$,
3. one step Θ time integration scheme.

The finite elements mesh consists of two groups elements (e): connected with a mould (A), where $e \in A$ and connected with a casting (B), where $e \in B$. Each of these can be integrated with the use of different schemes of time integration.

This fact simplifies finding of the critical time step and the stability analysis. If this division is assumed then it may be written as:

$$\mathbf{M}_A = \sum_e \mathbf{M}_e, \mathbf{K}_A = \sum_e \mathbf{K}_e, \mathbf{b}_A = \sum_e \mathbf{b}_e,$$
$$\mathbf{M}_B = \sum_e \mathbf{M}_e, \mathbf{K}_B = \sum_e \mathbf{K}_e, \mathbf{b}_B = \sum_e \mathbf{b}_e. \tag{10}$$

All vectors are also divided into parts according to finite elements mesh division $\mathbf{T} = (\mathbf{T}_A\mathbf{T}_B)^T$, $\dot{\mathbf{T}} = (\dot{\mathbf{T}}_A\dot{\mathbf{T}}_B)^T$, the upper index T represents transportation. As above, vector $\dot{\mathbf{T}}$ may be written as:

$$\dot{\mathbf{T}} = \mathbf{v}_A + \mathbf{v}_B,$$
$$\mathbf{v}_A = \mathbf{M}^{-1}(\mathbf{b}_A - \mathbf{K}_A \mathbf{T}), \tag{11}$$
$$\mathbf{v}_B = \mathbf{M}^{-1}(\mathbf{b}_B - \mathbf{K}_B \mathbf{T}).$$

In domain connected with a mould the integration is carried out with a bigger time step ($m\varDelta t$, where m is positive integer) whereas in domain connected with a casting with a smaller time step ($\varDelta t$). This allows to build a system of equations on the basis of Eq. (1) separately for the sub-domain B elements and to carry out calculations more often for it than for the whole mesh with maintaining condition of stability.

4 General Outline of Numerical Stability

Numerical method is stable when a little error in any solution stage moves further with a decreasing amplitude. An error appearing on time level n may be defined as ε^n, on time level $n + 1$ as ε^{n+1}, whereas values of this error may be determined with equation:

$$\varepsilon^{n+1} = g\varepsilon^n, \tag{12}$$

where g is amplification factor connected with integral operator $\mathscr{T}(\varDelta t, \varDelta)$. The amplification factor refers to a method error and is connected with time integration scheme. That is why it is necessary fulfilling one of conditions for stability of the method: the value of an error on time level $n + 1$ must not be bigger than value of an error on time level n. That may be written in this formula:

$$|\varepsilon^{n+1}| \leq |\varepsilon^n|, \tag{13}$$

and using the definition of amplification factor (12):

$$|g\varepsilon^n| \leq |\varepsilon^n|. \tag{14}$$

It follows that numerical stability may be achieved if condition:

$$|g| \leq 1 \tag{15}$$

is fulfilled. This condition is limited to issues leading to finite solutions.

For the system N of ordinary differential first-order equations an error vector is defined as ε^n. Each coordinate of this vector is an error connected with an appropriate dependent variable of the system. For each time step an error is multiplied by *amplification matrix* $\mathbf{G}$ in order to obtain an error vector in a new time step:

$$\varepsilon^{n+1} = \mathbf{G}\varepsilon^n. \tag{16}$$

Amplification matrix is connected to an integral operator which couples solutions in consecutive time steps. It means that if an error ε^n appeared in a solution $\mathbf{T}^n$ on time level n then after some necessary transformations is obtained:

$$\mathbf{T}^{n+1} + \varepsilon^{n+1} = \mathcal{T}(\mathbf{T}^n + \varepsilon^n). \tag{17}$$

Assuming that an error vector has a small amplitude, the Eq. (17) may be expanded into the Taylor series, taking into account only its two first terms. After some transformations an expression joining together two time levels is obtained:

$$\varepsilon^{n+1} = \left\{ \frac{\partial}{\partial \mathbf{T}} (\mathcal{T}\mathbf{T}) \right\}^n \varepsilon^n. \tag{18}$$

This expression also defines the amplification matrix in the Eq. (16). The operator $\mathcal{T}$ on the right-hand side of this equation is a linear matrix operator. Using the given integration scheme it is possible to determine an amplification matrix for it. An error vector on a new time level connects with an error vector in a previous step. If in a amplification Eq. (16) a matrix $\mathbf{G}$ is diagonal then the amplitudes of each of the error eigenvectors ε_i connected to each other by appropriate eigenvalues g_i of amplification matrix may be written as:

$$\varepsilon_i^{n+1} = g_i \varepsilon_i^n. \tag{19}$$

Stability condition must be used separately for amplitudes of each error eigenvectors:

$$|\varepsilon_i^{n+1}| \leq |\varepsilon_i^n|, \tag{20}$$

for all i, that is:

$$|g_i| \leq 1. \tag{21}$$

Stability criterion defined in a given way is limited to a demand that each eigenvalue g_i of an amplification matrix $\mathbf{G}$ was smaller or equal to a unit. In the paper this condition is used for the stability analysis of the mixed time partitioning method of solidification issues.

5 Association Between the Eigenvalues of Amplification Matrix and Size of Time Step

It is essential to find the criterion to determine the size of time step for the explicit scheme [2]. If we assume that $\theta = 0$ the Eq. (33) is reduced to the form:

$$\mathbf{T}^{n+1} = (\mathbf{I} - \varDelta t \mathbf{M}^{-1} \mathbf{K}) \mathbf{T}^n. \tag{22}$$

The Eq. (22) is called the evolution equation, because it gives the possibility to obtain the value of searched size $\mathbf{T}$ at the time level $n + 1$ from appropriate values of nodal quantities at the time level n.

In the evolution equation the capacity matrix $\mathbf{M}$ can be full or diagonal. Depending on the type of matrix in the equation of evolution, numerical stability analysis is combined with carrying out various algebraic operations.

In case of the capacity matrix is diagonal matrix, the calculation of the inverse matrix, namely $\mathbf{M}^{-1}$, is very simple and then finding its eigenvalues, necessary to determine the critical value of time step, is not difficult. However, in case of full capacity matrix which is symmetric and positively definite, in order to determine the inverse matrix we need to use the distribution $\mathbf{M} = \mathbf{L}\mathbf{L}^T$ or other transformations which keep the eigenvalues of full matrix. In case of diagonal matrix the inversion process and searching eigenvalue, which decides about the maximum, acceptable value of time step, is less complicated. The evolution Eq. (22) after converting can be written as follows:

$$\mathbf{T}^{u+1} = \mathbf{G}\mathbf{T}^n, \tag{23}$$

where amplification matrix $\mathbf{G}$ is given as:

$$\mathbf{G} = \mathbf{I} - \varDelta t \mathbf{M}^{-1} \mathbf{K}. \tag{24}$$

The scheme is explicit if the size $\mathbf{T}^{n+1}$ can be received from the Eq. (23), without solving the system of algebraic equations and if updates of searched quantities can be repeated m—times according to the formula [9]:

$$\mathbf{T}^{n+m} = \mathbf{G}^m \mathbf{T}^n. \tag{25}$$

Finding the maximum eigenvalues of the amplification matrix is a sufficient condition for the numerical stability:

$$\mathbf{G}\mathbf{x} = \lambda \mathbf{x}, \tag{26}$$

where $\mathbf{G}$ is the matrix of N degree, and N is the number of nodes of sub-domain connected with casting or mould domain. The analysis of numerical stability is conducted separately for each sub-domain on the basis of finite elements inside the sub-domain.

Using the theory of eigenvalues, eigenvectors and algebraic operations on matrices it is known that the size $\mathbf{G}^m \mathbf{T}^n \to \mathbf{0}$, if $m \to \infty$ for any $\mathbf{T}^n \in R^N$, if $|\lambda_i| < 1$ for $i = 1 \ldots N$, moreover, the size $\mathbf{G}^m \mathbf{T}^n$ is limited, if $m \to \infty$, if $|\lambda_i| \leq 1$ for $i = 1 \ldots N$, if there are linearly independent eigenvectors $\mathbf{x}_i$ for each $|\lambda_i| = 1$, which is satisfied because of symmetry of matrices $\mathbf{G}$.

After substituting $\mathbf{G}$ from the Eq. (24) to the Eq. (26), multiplying this equation by $\mathbf{M}$ and doing transformations the formula for generalized problem of eigenvalues is received:

$$\mathbf{Kx} = \frac{1-\lambda}{\Delta t} \mathbf{Mx}, \tag{27}$$

where $(1-\lambda)/\Delta t$ is an eigenvalue of couple of matrices $\mathbf{K}$ and $\mathbf{M}$. From the Eq. (27) it is known that if λ is equal to the unity then $\mathbf{Kx} = \mathbf{0}$ only if $\mathbf{x} = \mathbf{0}$:

$$\lambda_i = 1 - \Delta t \mu_i. \tag{28}$$

As $|\lambda_i| \leq 1$, the size of time step, which can be used to solve the system of Eq. (33), to be numerical stable and is limited by the inequality:

$$\Delta t \leq \frac{2}{\mu_i}, \tag{29}$$

The most restrictive limitation of the size of time step, which assures the stability is the case in which μ_i is the maximum eigenvalue $\mu_{\max}$ of the matrix of Eq. (37). Taking into account the way of assembly of capacity and conductivity matrices, the Eq. (37) may be written for a definite element e of the given domain:

$$\mathbf{K}^{(e)} \mathbf{x}^{(e)} = \mu^{(e)} \mathbf{M}^{(e)} \mathbf{x}^{(e)}, \tag{30}$$

whereas the limitation of a size of time step may be written as follows:

$$\Delta t \leq \frac{2}{\mu_i^{(e)}}. \tag{31}$$

In order to find a maximum acceptable size of time step for the casting and mould domains it is necessary to determine, for all the elements, their biggest eigenvalues and create from them double inequality. This inequality is limited from the smallest value to the biggest one:

$$\mu_{min}^{(e)} \leq \mu \leq \mu_{max}^{(e)}, \tag{32}$$

where $e = 1 \ldots ne$, and ne is the number of elements in the considered domain.

6 The Criterion of Determination of the Critical Time Step

In order to determine the criterion of numerical stability of chosen method, operations converting this equation into general problem of the eigenvalues are conducted. The analysis of stability is carried out to determine the maximum size of time step, which exceeding may be cause of unsteady solutions.

The one step time integration scheme of the equation obtained after spatial discretization is presented by the formula:

$$(\mathbf{M} + \Theta \Delta t \mathbf{K})\mathbf{T}^{n+1} = (\mathbf{M} - (1 - \Theta)\Delta t \mathbf{K})\mathbf{T}^n. \tag{33}$$

The right-hand side vector is not taken into consideration because the homogeneous equation is only essential for the numerical stability. If the homogeneous expression is stable so the inhomogeneous one is also stable [7].

The generalised problem of the eigenvalues is connected with casting domain B in sub-cycle and with mould domain A in total cycle [3, 4] and can be written in the universal form:

$$\mathbf{A}\mathbf{x}_i = \lambda_i \mathbf{B}\mathbf{x}_i, \quad i = 1, \ldots, N, \tag{34}$$

where N is a grade of the matrix $\mathbf{A}$ i $\mathbf{B}$ expressed by the formulas:

$$\mathbf{A} = \mathbf{M} - (1 - \Theta^{(e)})\Delta t \mathbf{K}, \tag{35}$$

$$\mathbf{B} = \mathbf{M} + \Theta^{(e)}\Delta t \mathbf{K}, \tag{36}$$

for

$$\Theta^{(e)} = \Theta_A \text{ for } e \in A,$$
$$\Theta^{(e)} = \Theta_B \text{ for } e \in B.$$

After substituting the Eqs. (35) and (36) into the formula (34) and doing the transformations the expression is received:

$$\mathbf{K}\mathbf{x}_i = \mu_i \mathbf{M}\mathbf{x}_i, \tag{37}$$

where μ_i is the eigenvalue of couple matrices of $\mathbf{M}$ i $\mathbf{K}$ form:

$$\mu_i = \frac{1 - \lambda_i}{(1 - \Theta + \Theta\lambda_i)\Delta t}. \tag{38}$$

After transformation the homogeneous Eq. (9) can be written as follows:

$$\dot{\mathbf{T}} + \mathbf{B}\mathbf{T} = \mathbf{0}, \tag{39}$$

where $\mathbf{B} = \mathbf{M}^{-1}\mathbf{K}$. Naturally, such inversion of the matrix $\mathbf{M}$ would cause its asymmetry, therefore Cholesky decomposition is used in this purpose instead of explicit inversions.

The one step Θ method is used in a scalar equation, which comes from the modal decomposition of system of Eq. (39), gives:

$$T^{n+1} = \lambda T^n, \tag{40}$$

where the eigenvalue λ is expressed by the formula:

$$\lambda = \frac{1 - (1 - \Theta)\mu\Delta t}{1 + \Theta\mu\Delta t}. \tag{41}$$

As far as the Eq. (41) and the inequality $|\lambda| \leq 1$ are concerned, the stability of the method is obtained if the following condition is satisfied:

$$2 + (2\Theta - 1)\mu\Delta t \geq 0. \tag{42}$$

It arises from (42) that for $\Theta \geq 1/2$ the condition of the inequality is always satisfied, so the method is stable. Moreover, for $\Theta < 1/2$ the stability of the method depends on the size of quotient $\mu\Delta t$, because of that for the explicit scheme ($\Theta = 0$) the size of maximum and accessible time step is strictly connected with the maximum eigenvalue in a given domain (the casting, the mould).

7 Restrictions Imposed on the Eigenvalues

The solution of N system of Eq. (9) consists of particular integral and complementary function of the solution of the homogeneous equation [9–11]:

$$\mathbf{M}\dot{\mathbf{T}} + \mathbf{K}\mathbf{T} = \mathbf{0}. \tag{43}$$

Substituting $\mathbf{T} = e^{-\lambda t}\mathbf{v}$ to the Eq. (43) is obtained an equivalent system of equations:

$$\lambda\mathbf{M}\mathbf{v} = \mathbf{K}\mathbf{v}. \tag{44}$$

Because of the semi-discretization the Eq. (44) is satisfied for $\lambda = \lambda_i$ and $\mathbf{v} = \mathbf{v}_i$.

The mass matrix $\mathbf{M}$ is diagonal and it helps to reduce the analysis of stability. If this matrix is the full symmetric matrix, the analysis of stability of the equation is conducted in a different way, however, the effect of both operations is the same as the criterion limiting the size of time step in the explicit scheme of the integration.

If the matrix $\mathbf{M}$ is positively definite, Cholesky decomposition can be executed, namely $\mathbf{M} = \mathbf{L}\mathbf{L}^T$, where $\mathbf{L}$ is lower triangular and non singular matrix. Using such distribution in the Eq. (44) and multiplying both sides of the equation by $\mathbf{L}^{-1}$, it is obtained:

$$\lambda_i \mathbf{L}^T \mathbf{v}_i = \mathbf{L}^{-1} \mathbf{K} (\mathbf{L}^{-1})^T \mathbf{L}^T \mathbf{v}_i, \tag{45}$$

where $\mathbf{L}^T \mathbf{v}_i$ is the eigenvector, and λ_i is the eigenvalue of the symmetric matrix $\mathbf{P} = \mathbf{L}^{-1} \mathbf{K} (\mathbf{L}^{-1})^T$. The matrix $\mathbf{P}$ has the set of linearly independent eigenvalues $\mathbf{v}_i$.

If the matrix $\mathbf{V}$ is composed of $\mathbf{v}_i$, which are the columns of such a matrix and $\mathbf{L}^T \mathbf{V}$ is orthogonal, it can be written:

$$\mathbf{V}^T \mathbf{L} \mathbf{L}^T \mathbf{V} = \mathbf{V}^T \mathbf{M} \mathbf{V} = \mathbf{I}. \tag{46}$$

Moreover, on the basis of the Eq. (44) and the Cholesky decomposition process it can be written:

$$\lambda_i = \lambda_i \mathbf{v}_i^T \mathbf{M} \mathbf{v}_i = \mathbf{v}_i^T \mathbf{K} \mathbf{v}_i. \tag{47}$$

Substituting $\mathbf{T} = \mathbf{V}\mathbf{x}$ into the Eq. (43) and left-multiplying both sides by $\mathbf{V}^T$ it is obtained:

$$\mathbf{V}^T \mathbf{M} \dot{\mathbf{x}} + \mathbf{V}^T \mathbf{K} \mathbf{V} \mathbf{x} = \mathbf{0}, \tag{48}$$

and then:

$$\mathbf{I} \dot{\mathbf{x}} + \Lambda \mathbf{x} = \mathbf{0}, \tag{49}$$

where $\Lambda = diag(\lambda_i)$. Such distribution is known as *modal decomposition* and allows to write the system of equations in the scalar form:

$$\dot{x}_i + \lambda_i x_i = 0. \tag{50}$$

The problem of the stability is connected with some restrictions of the eigenvalues. For the problems described by the prime row equations, from the Eq. (44) the eigenvalues and the eigenvectors can be designated. However, the restrictions imposed on the eigenvalues in the Eq. (47) can be derived from Rayleigh quotient:

$$\lambda = \frac{\mathbf{v}^T \mathbf{K} \mathbf{v}}{\mathbf{v}^T \mathbf{M} \mathbf{v}}. \tag{51}$$

Taking into consideration the way of matrix assembling $\mathbf{K}$ i $\mathbf{M}$:

$$\lambda = \frac{\sum_i (\mathbf{v}_i^T \mathbf{K}_i^{(e)} \mathbf{v}_i)}{\sum_i (\mathbf{v}_i^T \mathbf{M}_i^{(e)} \mathbf{v}_i)}, \tag{52}$$

where e is an element, and $\mathbf{v}_i$ is appropriate component of the eigenvector, Rayleigh quotient for an element can be written as follows:

$$\lambda_i^{(e)} = \frac{\mathbf{v}_i^T \mathbf{K}_i^{(e)} \mathbf{v}_i}{\mathbf{v}_i^T \mathbf{M}_i^{(e)} \mathbf{v}_i}.$$ (53)

Inserting (53) into (52) and doing certain transformations it is obtained:

$$\lambda = \frac{\sum_i \alpha_i \lambda_i^{(e)}}{\sum_i \alpha_i},$$ (54)

where $\alpha_i = \mathbf{v}_i^T \mathbf{M}_i^{(e)} \mathbf{v}_i > 0$, because the capacity matrix is positively definite. It is resulted from the Eq. (54) that λ is determined as the weighted average from $\lambda_i^{(e)}$ with positively weight, so the restrictions resulting from Rayleigh quotient can be written as follows:

$$\lambda_{\min}^{(e)} \leq \lambda \leq \lambda_{\max}^{(e)}.$$ (55)

Estimation of the extreme values is received from the formulas:

$$\lambda_{\min}^{(e)} \leq \frac{\min\{\mathbf{v}_i^T \mathbf{K}_i^{(e)} \mathbf{v}_i\}}{\max\{\mathbf{v}_i^T \mathbf{M}_i^{(e)} \mathbf{v}_i\}},$$ (56)

$$\lambda_{\max}^{(e)} \leq \frac{\max\{\mathbf{v}_i^T \mathbf{K}_i^{(e)} \mathbf{v}_i\}}{\min\{\mathbf{v}_i^T \mathbf{M}_i^{(e)} \mathbf{v}_i\}}.$$ (57)

8 Remarks and Conclusion

There are many types of methods used for the integration with respect to time, but two of them are basic: explicit and implicit. Explicit methods usually need few computations per time step, but numerical stability requires small the size of time step. In practice, a time step which is too small results in any unnecessarily long entire simulation time. Implicit methods, on the other hand, need many computations per time step, but allows to use larger the size of time step.

With respect to this problems we proposed to use both explicit and implicit schemes in this study. By using mixed time partitioning methods, computations in different parts of modelled domain were carried out by different integration schemes (E—explicit, I—implicit) and different the size of time step.

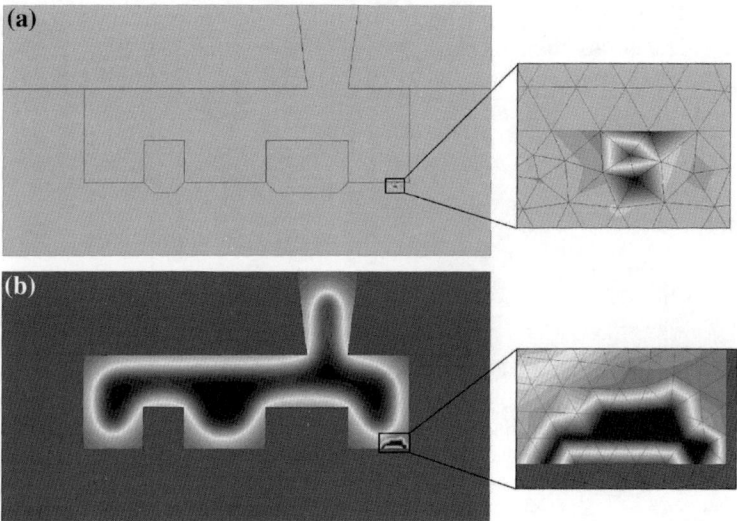

Fig. 1 *Erroneous results.* Errors appearing in computational results for **a** temperature and **b** solid phase fraction are caused by too large time step (e.g. in explicit scheme or when multiplication factor $m > 15$)

If a conditionally stable time integration scheme is used, numerical stability requires less the size of time step than its critical value (Δt_{critic}) calculated on the basis of stability analysis.

Numerical simulation of solidification was carried out for Al–2%Cu alloy casting, solidifying in the metal form [6, 12]. Computations were carried out by using mixed time partitioning methods on the basis of eigenvalues, where $m = 15$ was the largest acceptable value with respect to the numerical stability criterion. The time step equal to 0.0035 s was used in the simulations.

We focused on computational framework to simulate solidification of binary system with casting and mould considered. In our approach, we used a fixed time step in a casting domain and much larger time steps in other parts of mould, while maintaining high accuracy (comparable with case when small time step is used for all domains). We performed series of numerical experiments and noticed the eigenvalues of amplification matrix strongly affect the size of time step. The proper selection of the size of the time step is important for the stability of the method and the accuracy of the results. The simulation results are erroneous and inconsistent with the physics of the phenomenon after crossing the critical time step (see Fig. 1).

The eigenvalues remain with close relation to the stability of numerical method and hence with the size of the time step. For explicit schemes of time integration such a step cannot exceed a certain critical value. For implicit schemes of time integration the size of the time step cannot be unlimited because exceeding certain limit can result in omission of important physical phenomena. The use of the analysis of the relation between the eigenvalues and the size of time step allows to designate the maximum permissible size of the time step and to conduct the computer simulations correctly. The problem of the eigenvalues of the matrices is a very extensive issue and the works have very deep scientific and practical justification.

References

1. G. Michalski, N. Sczygiol, Using CUDA architecture for the computer simulation of the casting solidification process, in *Proceedings of The International MultiConference of Engineers and Computer Scientists, IMECS 2014*, vol. 2, Hong Kong, 12–14 Mar 2014. Lecture Notes in Engineering and Computer Science, pp. 933–937
2. E. Gawronska, N. Sczygiol, Relationship between eigenvalues and size of time step in computer simulation of thermomechanics phenomena, in *Proceedings of The International MultiConference of Engineers and Computer Scientists 2014, IMECS 2014*, vol. 2, Hong Kong, 12–14 Mar 2014. Lecture Notes in Engineering and Computer Science, pp. 881–885
3. E. Gawronska, N. Sczygiol, Stability of the mixed time partitioning methods in relation to the size of time step, in *Proceedings of the Annual Meeting of the International Association of Applied Mathematics and Mechanics*, GAMM, pp. 467–468 (2011)
4. E. Gawronska, N. Sczygiol, Application of mixed time partitioning methods to raise the efficiency of solidification modeling, in *Proceedings of the 12th International Symposium on Symbolic and Numeric Algorithms for Scientific Computing*, SYNASC, pp. 99–103 (2010)
5. A. Gravouil, A. Combescure, Multi-time-step explicit–implicit method for non-linear structural dynamics. Int. J. Numer. Meth. Eng. **50**, 199–225 (2001)
6. N. Sczygiol, G. Szwarc, Application of enthalpy formulation for numerical simulation of castings solidification. Comput. Assist. Mech. Eng. Sci. **8**, 99–120 (2001)
7. L.W. Wood, *Practical Time-stepping Schemes* (Clarendon Press, Oxford, 1990)
8. N. Sczygiol, Approaches to enthalpy approximation in numerical simulation of two-component alloy solidification. Comput. Assist. Mech. Eng. Sci. **7**, 717–734 (2000)
9. P. Smolinski, Y.S. Wu, Stability of explicit subcycling time integration with linear interpolation for first-order finite element semidiscretizations. Comput. Method Appl. Mech. Eng. **151**, 311–324 (1998)
10. T. Belytschko, P. Smolinski, W.K. Liu, Stability of multi-time step partitioned integrators for first-order finite element systems. Comput. Method Appl. Mech. Eng. **49**, 281–297 (1985)
11. P. Smolinski, T. Belytschko, M. Neal, Multi-time-step integration using nodal partitionig. Int. J. Numer. Meth. Eng. **26**, 349–359 (1988)
12. N. Sczygiol, Numerical modelling of thermo-mechanical phenomena in solidifying cast and casting mould. Czestochowa University of Technology, Monographs No. 71, Czestochowa (in polish) (2000)

Modelling of Process Parameters Influence on Degree of Porosity in Laser Metal Deposition Process

Rasheedat Modupe Mahamood and Esther Titilayo Akinlabi

Abstract Additive manufacturing process is an advanced manufacturing process that fabricates component directly from the three dimensional (3D) image of the part being produced by adding materials layer by layer until the part is completed. Laser Metal Deposition (LMD) process is an important additive manufacturing technique that is capable of producing complex parts in a single manufacturing run. A difficult to manufacture material such as Titanium and its alloys can readily be manufactured using the LMD process. Titanium and its alloys possess excellent corrosion properties that made them to find applications in many industries including biomedical. The biocompatibility of Ti6Al4V made then to be used as implants. Porous implants are desirable in some applications so as to reduce the weight as well as to aid the healing and proper integration of the implant with the body tissue. In this chapter, the effect of laser power and scanning speed on the degree of porosity was investigated and empirically modelled in laser metal deposition of Ti6Al4V. The model was developed using full factorial design of experiment and the results were analyzed using Design Expert software. The model was validated and was found to be in good agreement with the experimental data.

Keywords Design of experiment · Laser metal deposition · Medical implants · Model · Porosity · Processing parameters · Titanium alloy

R.M. Mahamood (✉) · E.T. Akinlabi
Department of Mechanical Engineering Science, University of Johannesburg,
Auckland Park Kingsway Campus, Johannesburg 2006, South Africa
e-mail: mahamoodmr@unilorin.edu.ng; mahamoodmr2009@gmail.com

E.T. Akinlabi
e-mail: etakinlabi@uj.ac.za

R.M. Mahamood
Department of Mechanical Engineering, University of Ilorin, Ilorin, Nigeria

© Springer Science+Business Media Dordrecht 2015
G.-C. Yang et al. (eds.), *Transactions on Engineering Technologies*,
DOI 10.1007/978-94-017-9588-3_3

31

1 Introduction

Additive manufacturing (AM) also known as 3D printing [1] is a revolutionary advanced manufacturing process [2] that produces part in a layer wise manner from the 3D Computer Aided Design (CAD) model of the part [3]. Laser Material Deposition (LMD) process belongs to a class of Directed Energy Deposition (DED) process, as it was classified by the F42 committee on additive manufacturing standards [1]. Laser Material Deposition, like any other additive manufacturing technique, can produce a very complex part at no extra cost in a single step. Also, parts are made by adding materials layer by layer as against the traditional manufacturing processes which remove material in a subtractive manner in order to shape parts. Complex parts need to be broken down into various parts which are later assembled using traditional manufacturing process and they generate lots of scrap. Of all the family of AM technologies, Directed Energy Deposition (DED), of which LMD belongs, is the only class that can be used to repair worn out components [1]. LMD technologies are advantageous in producing customized part such as medical implants and functionally graded materials [4].

Ti6Al4V is an important Titanium alloy and it is the most widely used titanium alloy in the aerospace industry and as medical implants because of its high specific strength to weight ratio and good corrosion resistance [5]. The biocompatibility of Ti6Al4V is responsible for its wide use as medical implant [6]. Despite all these good properties, titanium and its alloys are generally referred to as difficult to machine materials [7]. It becomes uneconomical to produce highly customized part such as in the case of medical implants using traditional manufacturing processes. The chemical properties of Titanium make it difficult to machine and it makes the tool to fail prematurely during cutting operation as a result of galling, which is the tearing of cutting tool as a result of friction and adhesion of the cutting tool to the material being cut. LMD offers advantage in producing this important material especially for medical implants because there is no contact between the material being processed and the LMD process because it is a tool-less process.

Porosity in some laser metal deposited parts is seen as a defect in most structural engineering applications [8]. Porosity is of great importance in biomedical applications, because it aids healing process as well as proper integration of medical implants and body tissues in case of implants [9, 10]. Porosity is desirable to further improve compatibility of Titanium alloy implant and human bone. This is because the modulus of elasticity of titanium is higher than that of the human bone and causes mismatch in titanium implants and the host bone. By introducing porosity in the implants, the modulus of elasticity of titanium can be reduced and made to be as close to the human bone as possible [11–13]. Different methods have been used to produce various types of porous titanium alloy implants and have been reported in the literature [14–17]. A number of porous titanium alloys has also been reported in the literature using additive manufacturing method [18, 19]. Xiong et al. [19] studied the feasibility of using 3D-printing to fabricate porous titanium implants. The study revealed that 3D-printing can be used to produce a porous implant with properties close to those of human bones. Most of the porous medical implants produced in the literature have designed porosity [14–19]. That is, they are produced in such a way that the part is porous. For example the way the porous part was produced was such that the layers are arranged in such a way that porosity is left on the part according to the 3D CAD profile of the part.

Processing parameters in LMD process has been shown to affect the evolving properties as well as the porosity of deposited part [20]. In this chapter, the effect of laser power and scanning speed on the degree of porosity of laser deposited Ti6Al4V was investigated. An empirical model was developed using full factorial design of experiment. It was found that increasing the laser power results in a decrease in the degree of porosity. The developed model was validated by producing porous samples outside the process parameters studied. The model was found to be in good agreement with the experimental data. The detailed results are presented and explained fully in this chapter.

2 Materials and Methods

2.1 Materials and Substrate Preparation

The materials used in this study comprise of gas atomized Ti6Al4V powder and 5 mm thick 72 mm × 72 mm hot rolled Ti6Al4V sheet supplied by VSMPO-AVISMA Corporation, Russia. The powder particle size distribution ranged between 150–200 µm. Before the deposition process, the substrate was sandblasted, washed and degreased with acetone in order to aid the absorption of the laser beam. The sandblasting makes the surface of the substrate to become rough. Rough surface aids laser energy absorption as against shinning surface that reflects most of the laser beam. Argon gas was used as the carrier gas for the powder as well as for shielding of the deposit in order to prevent environmental contamination of the powder through oxidation. The shield gas also protects the hot deposited part from oxidation because, at high temperature, titanium and its alloys have a high affinity for oxygen.

2.2 Laser Metal Deposition Process

The experimental set-up of the LMD process was achieved using a Kuka robot carrying in its end effector, a 4 kW Nd-YAG laser and coaxial powder nozzle which is available at the Council of Scientific and Industrial Research (CSIR) National Laser Centre (NLC), Pretoria, South Africa. The laser beam diameter was maintained at 2 mm at a focal distance of 195 mm. To prevent oxidation during the deposition process, a shielding mechanism was provided using plastic wrapping as shown in Fig. 1a. The glove box (shielding mechanism) was filled with argon gas to maintain the oxygen level below 10 ppm. During the LMD process, the laser beam was

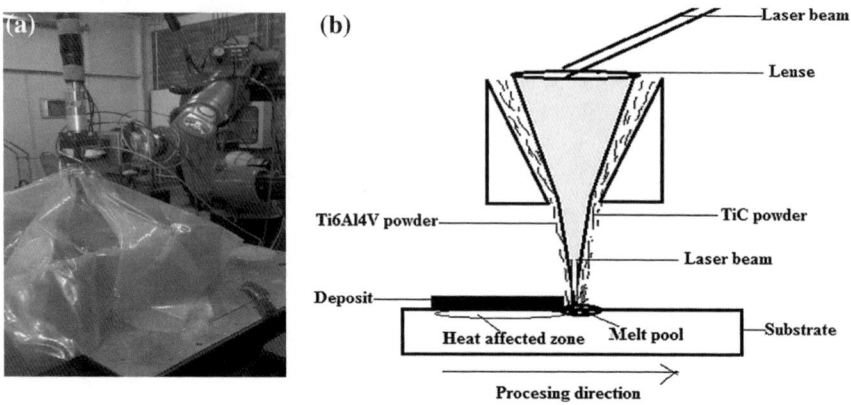

Fig. 1 **a** Experimental set-up. **b** Schematic of the laser metal deposition process (adapted from [21])

focused on the surface of the substrate. The laser beam generated a lot of heat thereby creating a melt pool on the substrate. The powder is then focused and delivered into the melt pool through the powder delivery system and melted. Upon solidification of the melt pool, a solid track of the metal is left on the path of the laser beam. The schematic diagram of the laser metal deposition process is shown in Fig. 1b.

The processing parameters in coded form as generated by the Design Expert software are presented in Table 1. A single track was produced at each processing parameter. The laser power was set at a low level of 0.4 kW and a high level of 0.8 kW. The scanning speed was also set at a lower level of 0.1 m/s and at a higher level of 0.2 m/s. The powder flow rate and the gas flow rate were maintained at constant values of 1.44 g/min and 2 l/min respectively. The settings were entered in the Design expert software to generate the experimental matrix with two (2) replicates. The lower level setting is coded as −1 and the upper level is coded as 1.

Table 1 Processing parameters settings

		Factor 1	Factor 2
Std.	Run	A: laser power	B: scanning speed
1	1	−1	−1
2	2	−1	−1
7	3	1	1
4	4	1	−1
5	5	−1	1
3	6	1	−1
6	7	−1	1
8	8	1	1

2.3 Material Characterization

After the deposition process, the samples were sectioned along the transverse direction for porosity analysis. The cut samples were mounted in hot resin, ground and polished according to the standard metallographic sample preparation of titanium alloys. The porosity analysis was performed using the BMX optical microscope equipped with ANALYSIS Docu image processing software to determine the percentage of porosities.

3 Results and Discussion

The morphology of the Ti6Al4V powder is shown in Fig. 2. The powder is characterized by spherically shaped gas atomized powder. Spherically shaped powder is preferred more in laser processing because of their better laser absorption property.

The percentages of porosity in the deposited samples were measured at three different points on each sample and the averages of the percentage porosity are presented in Table 2.

The result was analyzed in Design Expert and the analysis of variance (ANOVA) of the selected model is presented in Table 3.

From the table, the Model F-value of 14062.99 implies that the model is significant and that there is only a 0.01 % chance that an F-value this large could occur due to noise. Values of "Prob > F" less than 0.0500 indicate the model terms are significant and in this case A, B, AB are significant model terms. The analysis of the coefficient of determinates is presented in Table 4.

From the table, the "Predicted R-Squared" of 0.9996 is in reasonable agreement with the "Adjusted R-Squared" of 0.9998 because the difference between them is less than 0.2. The "Adequate Precision" measures the signal to noise ratio. A ratio greater than 4 is desirable. The ratio of 269.224 for this model indicates an adequate

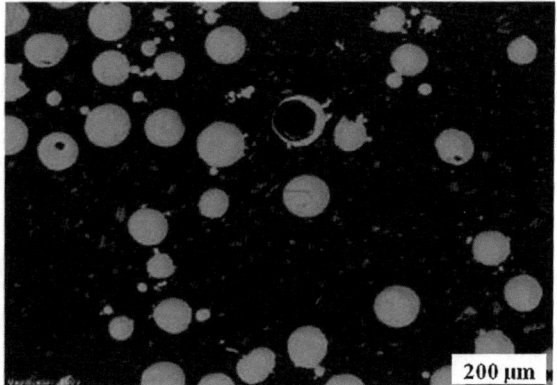

Fig. 2 Morphology of Ti6Al4V powder [22]

200 μm

Table 2 Average percentage porosities

Std.	Run order	Factor 1 A: laser power	Factor 2 B: scanning speed	Response 1 average porosity (%)
1	1	−1	−1	4.12
2	2	−1	−1	4.05
7	3	1	1	3.98
4	4	1	−1	2.07
5	5	−1	1	11.15
3	6	1	−1	2.12
6	7	−1	1	11.25
8	8	1	1	4.01

Table 3 Analysis of variance (ANOVA) Table

Source	Sum of squares	df	Mean square	F-value	p-value Prob > F	
Model	96.51	3	32.17	14062.99	<0.0001	Significant
A-laser power	42.27	1	42.27	18480.44	<0.0001	
B-scanning speed	40.64	1	40.64	17763.98	<0.0001	
AB	13.60	1	13.60	5944.53	<0.0001	
Pure error	9.150E-003	4	2.287E-003			
Cor total	96.52	7				

Table 4 Coefficient of determinates R-square

Std. Dev.	0.048	R-squared	0.9999
Mean	5.34	Adjusted R-squared	0.9998
C.V. %	0.90	Predicted R-squared	0.9996
PRESS	0.037	Adequate precision	269.224

signal and this model can be used to navigate the design space. The estimated model coefficients at 95 % confidence level and their analysis are presented in Table 5. The Final Equation in Terms of the actual Factors is presented in Eq. (1).

$$\text{Average porosity }(\%)$$
$$= 5.34 - 2.30 \times \text{Laser Power} + 2.25 \times \text{Scanning Speed} \tag{1}$$
$$- 1.30 \times \text{Laser Power} \times \text{Scanning Speed}$$

Table 5 Analysis of the model coefficients terms

	Coefficient			95 % CI	95 % CI	
Factor	Estimate	df	Standard Error	Low	High	VIF
Intercept	5.34	1	0.017	5.30	5.39	
A-laser power	−2.30	1	0.017	−2.35	−2.25	1.00
B-scanning speed	2.25	1	0.017	2.21	2.30	1.00
AB	−1.30	1	0.017	−1.35	−1.26	1.00

Normal plot of the residuals is shown in Fig. 3. The residuals are seen to be randomly distributed. The graph of predicted versus the actual experimental data is shown in Fig. 4. It is seen in this graph that the model is a true representation of the experimental data.

The main effect plot of Laser power on the percentage porosity is shown in Fig. 5a. The degree of porosity is seen to reduce as the laser power is increased. The porosities in the samples are produced as a result of unmelted powder particles or as a result of gas entrapment in the melt pool that results in a blow-hole kind of porosity. As the laser power was increased, the laser-material interaction time also increased, hence there is more melting of the powder taking place thereby reducing the degree of porosity. The main effect plot of the scanning speed is shown in

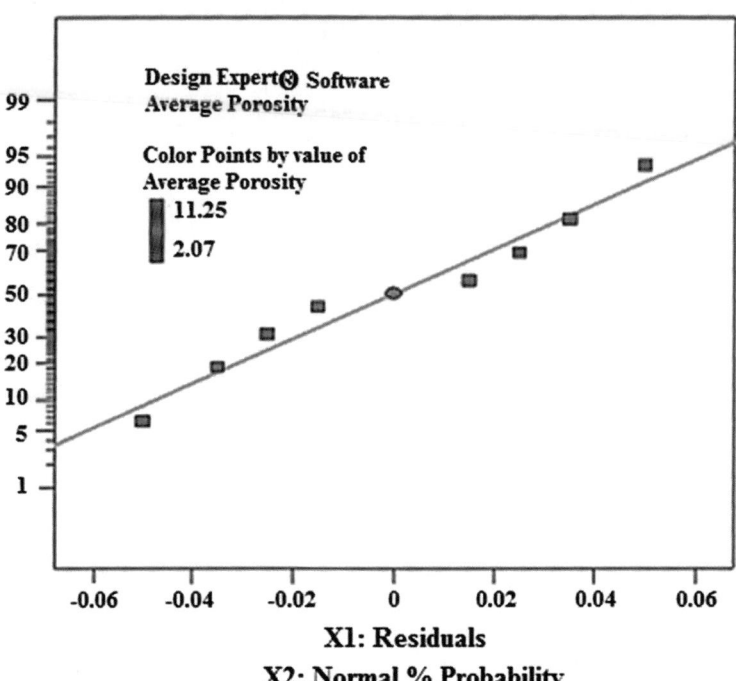

Fig. 3 Normal plot of residual

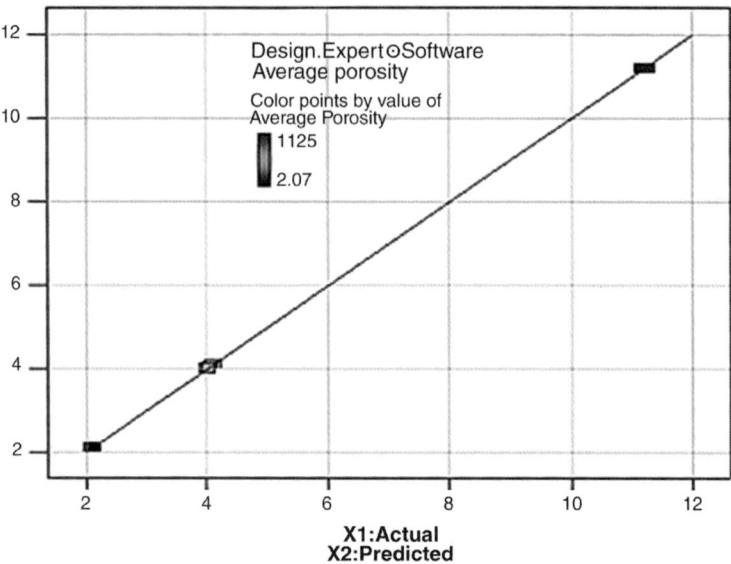

Fig. 4 Graph of predicted versus actual

Fig. 5b. The higher the scanning speed, the higher the average porosity. This is because as the scanning speed is increased; the laser material interaction time is reduced thereby causing the quantity of unmelted powder particles to increase hence, increasing the average porosity. The combine effect of the laser power and the scanning speed is shown in Fig. 6. The surface plot of the average porosity against the laser power and the scanning speed shows that the lowest porosity occurs at the lowest scanning speed and the highest laser power. The highest porosity is seen at the lowest laser power and the highest scanning speed.

The micrograph of the sample at the lowest laser power of 0.4 kW and the lowest scanning speed of 0.1 m/s showing the porosities is shown in Fig. 7a and that of the sample at the highest laser power of 0.8 kW and lowest scanning speed of 0.1 m/s is shown in Fig. 7b. It can be seen from the Figure that the porosities at low laser power are smaller in size (see Fig. 7a) while those seen at the highest laser power are bigger. This can be attributed to the fact that the porosities at high laser power could be as a result of gas trapped in the belt pool during the deposition process.

4 Model Validation

In order to validate the developed model, the model was used to predict the average porosity based on the constraints dictated and the results are compared with the experimental data. Table 6 shows the predicted and actual experimental data. The experiments are repeated twice.

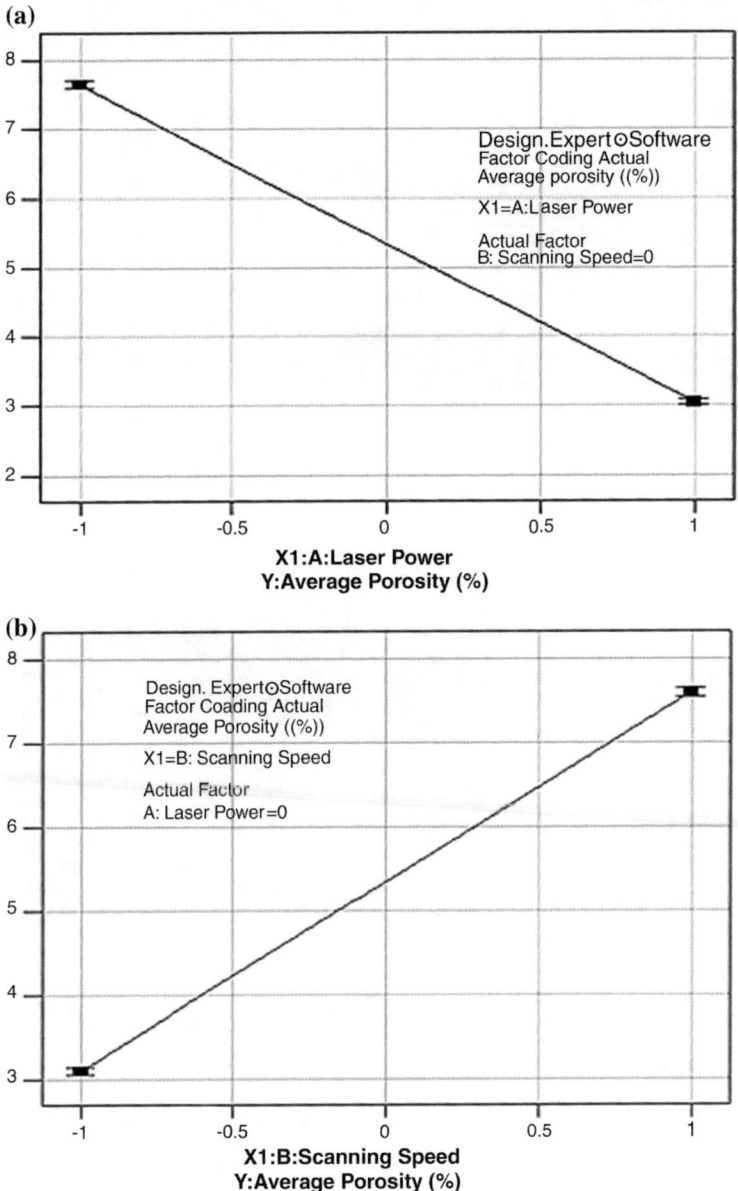

Fig. 5 Main effect plot of **a** Laser power, **b** Scanning speed

From the Tables 1, 2, 3, 4, 5 and 6, it can be observed that there is good agreement between the predicted data and that of the actual experimental data. This shows that the model is a good representation of the experimental data.

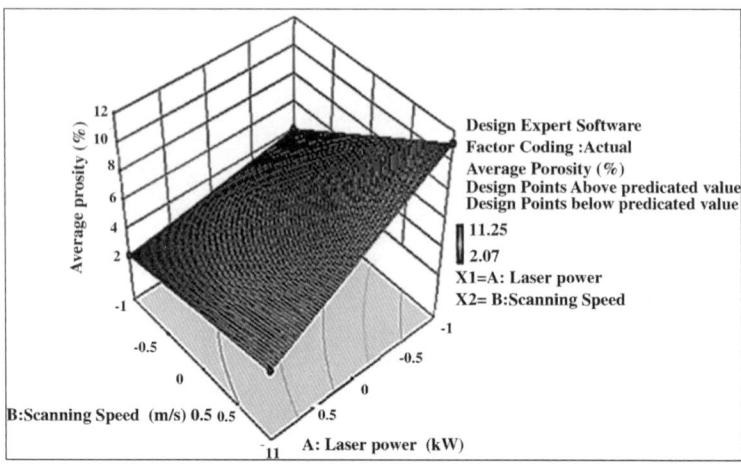

Fig. 6 Surface plot of average porosity against laser power and scanning

Fig. 7 Micrograph of samples at **a** 0.4 kW laser power and 0.1 m/s scanning speed [sample 1 (run order)] and **b** 0.8 kW laser power and 0.1 m/s scanning speed (Sample 6) [23]

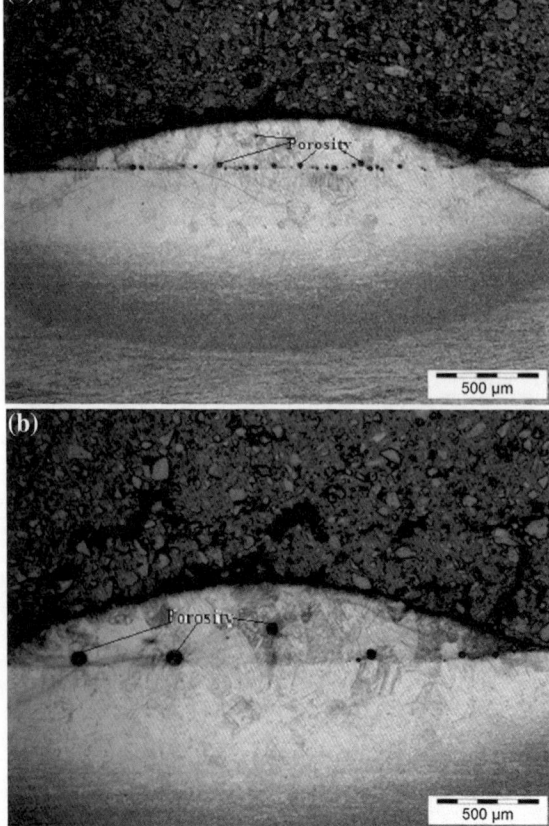

Table 6 Predicted versus actual experimental data

Laser power (kW)	Scanning speed (m/s)	Predicted average porosity (%)	Actual average porosity (%)
0.6	0.1	3.09	3.12
0.7	0.1	2.592	2.7
0.8	0.15	3.045	2.98
0.8	0.175	3.520	3.49
0.85	0.075	2.200	2.08

5 Conclusions

Porosity is of great importance in medical implants. The effect of laser power and scanning speed on the degree of porosity and size of porosity was investigated in this chapter. Design Expert, a statistical software, was used to analyze the results and develop a model. The study revealed that, the lower the scanning speed, the lower is the degree of porosity Also, the higher the laser power, the lower the degree of porosity. The developed model was validated by performing experiments at processing parameters that are outside those parameters used in the development of the model. The results are compared to those predicted from the model and it was found to be in good agreement with one another. It can be concluded that the model is a good representation of the experimental data and it can be used to design and produce porous part implants depending on the degree of porosity.

6 Future Work

The future research work is to develop a model for the influence of processing parameters on the average size of the pore produced in laser metal deposited porous parts. This will be useful in application requiring specific pore size in porous part or implant.

Acknowledgments This work is supported by the Rental Pool Grant of the National Laser Centre–Council of Scientific and Industrial Research (NLC–CSIR), Pretoria South Africa.

References

1. J. Scott, N. Gupta, C. Wember, S. Newsom, T. Wohlers, T. Caffrey, Additive manufacturing: status and opportunities, Science and Technology Policy Institute (2012), https://www.ida.org/stpi/occasionalpapers/papers/AM3D_33012_Final.pdf. Accessed 11 July 2012
2. R.M. Mahamood, E.T. Akinlabi, M. Shukla, S. Pityana, Evolutionary additive manufacturing: an overview. Lasers Eng. **27**, 161–178 (2013)

3. X.H. Wu, L. Jing, J.F. Mei, C. Mitchell, P.S. Goodwin, W. Voice, Microstructures of laser-deposited Ti-6Al-4V. Mater. Des. **25**, 137–144 (2004)
4. R.M. Mahamood, E.T. Akinlabi, M. Shukla, S. Pityana, Functionally graded material: an overview. Lect. Eng. **3**, 1593–1597 (2012)
5. Y. Lu, H.B. Tang, Y.L. Fang, D. Liu, H.M. Wang, Microstructure evolution of sub-critical annealed laser deposited Ti–6Al–4V alloy. Mater. Des. **37**, 56–63 (2012)
6. H. Schiefer, M. Bram, H.P. Buchkremer, D. Stöver, Mechanical examinations on dental implants with porous titanium coating. J. Mater. Sci. Mater. **20**, 1763–1770 (2009)
7. A.R. Machado, J. Wallbank, Machining of titanium and its alloys: A review. Proc. Inst. Mech. Eng. B Manage. Eng. Manuf. **204**(11), 53–60 (2005)
8. G.K.L. Ng, A.E.W. Jarfors, G. Bi, H.Y. Zheng, Porosity formation and gas bubble retention in laser metal deposition. Appl. Phys. A **97**, 641–649 (2009)
9. A. Naumann, S. Ehrmantraut, V. Willnecker, M.D. Menger, B. Schick, M.W. Laschke, Ear reconstruction using porous polyethylene implants. Effect of cortisone on edema reduction and healing process]. HNO **59**(3), 268–273 (2011)
10. M. Motomiya, M. Ito, M. Takahata, K. Kadoya, K. Irie, K. Abumi, A. Minami, Effect of Hydroxyapatite porous characteristics on healing outcomes in rabbit poster lateral spinal fusion model. Eur. Spine J. **12**, 2215–2224 (2007)
11. H. Schiefer, M. Bram, H.P. Buchkremer, D. Stöver, Mechanical examinations on dental implants with porous titanium coating. J Mater. Sci. Mater. **20**, 1763–1770 (2009)
12. B.V. Krishna, S. Bose, A. Bandyopadhyay, Low stiffness porous Ti structures for load-bearing implants. Acta Biomater. **3**, 997–1006 (2007)
13. Y.J. Chen, B. Feng, Y.P. Zhu, J. Weng, J.X. Wang, X. Lu, Fabrication of porous titanium implants with biomechanical compatibility. Mater. Lett. **63**, 2659–2661 (2009)
14. J.P. Li, S.H. Li, CA Van Blitterswijk, Cancellous bone from porous TI6Al4V by multiple coating technique. J. Mater. Sci. Mater. Med. **17**, 179–185
15. B.V. Krishna, W. Xue, S. Bose, A. Bandyopadhyay, Engineered porous metals for implants. J. Mater.
16. C.E. Wen, M. Mabuchi, Y. Yamada, K. Shimojima, Y. Chino, T. Asahina, Processing of biocompatible porous Ti and Mg. Scripta Mater. **45**, 1147–1153 (2001)
17. I.H. Oh, N. Nomura, N. Masahashi, S. Hanada, Mechanical properties of porous titanium compacts prepared by powder sintering. Scripta Mater. **49**, 1197–1202 (2003)
18. D.F. Justin, B.E. Stucker, Laser based metal deposition (LBMD) of implant structures. US patent US7632575 B2
19. Y. Xiong, C. Qian, J. Sun, Fabrication of porous titanium implants by three-dimensional printing and sintering at different temperatures. Dent. Mater. J. **31**(5), 815–820 (2012)
20. M.K. Imran, S. Masood, M. Brandt, S. Bhattacharya, J. Mazumder, Parametric Investigation of Diode and CO_2 Laser in Direct Metal Deposition of H13 Tool Steel on Copper Substrate. World Acad. Sci. Eng. Technol. **55**, 437–442 (2011)
21. R.M. Mahamood, E.T. Akinlabi, M. Shukla, S. Pityana, Characterization of laser deposited Ti6A4V/TiC Composite. Lasers Eng. **29**(2–4), 197–213 (2014)
22. R.M. Mahamood, E.T. Akinlabi, M. Shukla, S. Pityana, Scanning velocity influence on microstructure, microhardness and wear resistance performance on laser deposited Ti6Al4V/TiC composite. Mater. Des. **50**, 656–666 (2013)
23. R.M. Mahamood, E.T. Akinlabi, M. Shukla, S. Pityana, in Characterizing the effect of processing parameters on the porosity of laser deposited titanium alloy powder. *Lecture Notes in Engineering and Computer Science: Proceedings of The International Multiconference of Engineers and Computer Scientists 2014*, IMECS 2014, 12–14 Mar 2014, Hong Kong, pp. 904–908 (2014)

Statistics of End-to-End Distance of a Linear Chain Trapped in a Cubic Lattice of Binding Centers

Zbigniew Domański and Norbert Sczygiol

Abstract Two and three dimensional nanostructured substrates are widely employed in a variety of biomedically-oriented nanodevices as well as in functional devices created with the use of DNA scaffolding. In this context spatial arrangements of binding centers influence the efficiency of these substrates. Here, we concentrate on 3D substrates and we compute and analyze the distribution of distances (q) between binding centers in the case where the centers are localized in nodes of a cubic lattice. We find that for this particular lattice the exact node-to-node probability distribution is a fifth-degree polynomial in q. We merge this polynomial-shaped distribution with an end-to-end distance distribution of a linear chain and we find an excellent agreement between it and the corresponding distribution for a self-avoiding walk in 3D.

Keywords Distance distribution · DNA scaffolding · Micropatterned substrates · Polymer adhesion · Self-avoiding walks · Zigzag path statistics

1 Introduction

Macromolecules are present anywhere in life processes. They are fundamental constituents of living organisms and plants. Among them polymers are the objects of long-term, intensive scientific works. Biologically oriented physicists are

Z. Domański (✉)
Institute of Mathematics, Częstochowa University of Technology, Dąbrowskiego 69, 42201 Częstochowa, Poland
e-mail: zbigniew.domanski@im.pcz.pl

N. Sczygiol
Institute of Computer and Information Sciences, Częstochowa University of Technology, Dąbrowskiego 69, 42201 Częstochowa, Poland
e-mail: norbert.sczygiol@icis.pcz.pl

© Springer Science+Business Media Dordrecht 2015
G.-C. Yang et al. (eds.), *Transactions on Engineering Technologies*,
DOI 10.1007/978-94-017-9588-3_4

43

studying polymers theoretically and experimentally and they are trying to find these polymer's underlying properties which are valuable for biomedical and technological purposes [1].

From the engineering sciences point of view numerous polymer-involved approaches have been elaborated and then implemented in factory processes. One of such recent advancement in the field of nanotechnology allows 6-nm-resolution pattern of biding sites [2, 3]. This spectacular resolution is due to the so-called DNA origami technique [4, 5]. A particularly appealing feature of DNA origami comes from the precise location of the ends of DNA strand on a given substrate. In this way, functional devices created via DNA scaffolding can be sized down to reach sizes of the order of 10^{-8} m or even less.

In order to achieve a few nanometer resolution the binding centers have to be accurately positioned in a given volume. Then, a functionalized polymer trapped by a pair of these centers is used as a piece of scaffolding [4]. It is worth to mention that a successful-polymer-capture takes place if the end-polymer molecules are sufficiently sensitive to the binding centers. Such an attractive interaction between the polymer ends and the binding centers creates an additional tension along the polymer backbone which in turn may result in a formation of knots or a polymer-strand entanglement. Thus, the polymer-functionalization process, i.e. the attachment of appropriate molecules to polymer's ends is a subtle process and one should take care of the resulting attractive forces between the polymer segments and the armed-polymer-ends.

A biomedical example of a three-dimensional substrate is a silicon-nanopillar array which allows one to create enhanced-local-interactions between the substrate and other macromolecules [6]. This nanostructured substrate yields a high capability to capture cancer cells detached from the solid primary tumor and thus enables one to isolate these circulating cells from the blood. In this spirit, another recently reported example of functionalized substrate, the functionalized graphene oxide nanosheets [7], clearly shows that 3D substrates can operate as valuable bio-markers for disease diagnosis.

The DNA origami technique and the medical bio-markers employ 3D functionalized substrates [6, 8]. The spatial arrangement of binding centers of these substrates plays an important role in the capture yield. However, a considerable scientific activity is concentrated mainly on the biochemical and physical properties of adhesion process and less attention is paid to the geometry-induced characteristics of the substrate itself and to the resulting impact on the binding efficiency.

In this work we analyze how the polymer-chain-capture phenomena is influenced by the spacial arrangement of binding centers of a given substrate. For this purpose we use a simple model of the 3D substrate, i.e. we assume that the binding centers are periodically arranged in a limited volume of a three-dimensional cubic lattice. Because of the attractive force between the functionalized-polymer ends and the binding centers, the polymer feels an effective non-homogeneous electrochemical potential. This potential is modulated by the relative positions of the substrate's uptake centers and, in consequence, the polymer trajectories resemble zigzag lines. In such circumstances the Euclidean norm is not adequate to measure the distances traversed by the polymer. The lengths of zigzag-like trajectories should be measured in terms of the taxicab metric in which the distance q between points $\mathbf{x}(x_1, x_2, x_3)$ and $\mathbf{y}(y_1, y_2, y_3)$ is given by

$$q(\mathbf{x}, \mathbf{y}) = \Sigma_{i=1,2,3} |x_i - y_i|. \tag{1}$$

Below we analyze the distributions of such a distance between the functional-ized-polymer-end points.

2 Distinct Distances in a Simple Cubic Lattice

Consider a set of $(L+1)^3$ binding centers confined in a simple cubic lattice (SC) in such a way that the biding centers are represented by the nodes of the SC and the edges of equal length $(a = 1)$ measure the distance between any pair of centers. In this scenario, the SC is seen as a unit distance grid graph. For a given pair of binding centers the distance between them is the length of the shortest path between the corresponding nodes, i.e. the number of edges in such path.

A polymer model can be chosen in a way related to the studied problem. For the purpose of this work we choose a self-avoiding walk (SAW) [9–11] and we rep-resent the long polymer body by the path on a lattice. Due to the excluded volume effect two monomers cannot be closer than their diameter and thus, the SAW is a path without self-intersections [11]. An example of the SAW is presented in Fig. 1. The SAW's paths have been extensively studied from the statistical physics per-spective and the enumeration of these paths still is the center of interest [11]. Here, another point of view is taken into account. We are interested in the statistics of

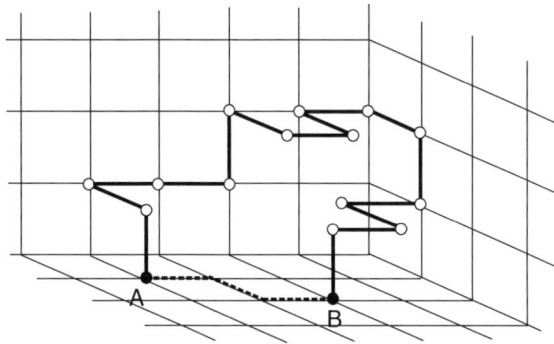

Fig. 1 Schematic illustration of the m-segment chain embedded in cubic lattice. *Black-filled circles* mark the chain's terminals: A, B. Here, $m = 15$ and the end-to-end distance $q(A, B)$ equals to 3 lattice spacing, i.e. $q = 3$

distances between nodes of the finite 3D SC lattice, under assumption that the distances are measured according to the metric (1). In the literature different wordings are used for such a distance. Here, we express it as the Manhattan distance.

In a case of the SC lattice all node-to-node Manhattan distances can be easily computed and sorted. Since the shortest path between two nodes in this lattice is at most three segments zigzag line, then a straightforward listing of all different non-ordered pairs of nodes enables one to assign the Manhattan distance to each pair of nodes and then to form an appropriate distribution. However, such an approach is justified if one deals with a primitive Bravais lattice [12], i.e. the lattice which has only one node in its unit cell. In a case of non-Bravais lattice or a decorated lattice the shape of the shortest path is not obvious and one has to either rely on the graph-theory-based tools or use some dedicated algorithms [13].

The graph-theory makes it possible to represent a lattice by the so-called adjacency matrix, also termed the connectivity matrix. For a given lattice with n nodes its adjacency matrix is the $n \times n$ matrix $\mathbf{A}$ with entries $A_{ij} = 1$ only if i and j nodes share the same edge. If such an edge does not exist the corresponding entry equals to zero. A useful property of the adjacency matrix consists in a direct relation between consecutive powers $\mathbf{A}^k$, $k \in (1, 2, \ldots, n)$ and the number of distinct paths in a graph. More precisely, an entry $\mathbf{A}_{ij}^k$ is the number of paths of the length k from the node i to the node j. It is this property that we use to sum up the number of Manhattan distance in the SC lattice, i.e. (i) to each pair of nodes we assign the smallest value of k for which $A_{ij}^k \neq 0$ and then (ii) for each value of k we count the number of pairs of nodes related to this value. Since $n < \infty$ due to (i) and (ii) we get the required distribution of Manhattan distances [14, 15].

This is quit a general procedure and we employ it here despite the relative simplicity of the SC lattice.

3 Results and Discussion

Using the method described in the previous section we compute the number $N(L, q)$ of pairs of nodes separated by the distance $1 \leq q \leq 3L$ in the SC with $(L+1)^3$ nodes. Note that the maximum value of the node-to-node distance $q_{max} = 3L$ corresponds to four pairs of nodes located in the opposite corners of the cube, whereas $q_{min} = 1$ corresponds to all cube's edges. As a result we get $N(L, q)$, a fifth-degree polynomial in q

$$
N(L, q) = \begin{cases} \sum_{n=0}^{n=5} a_n(L) q^n; & 1 \leq q \leq L \\ \sum_{n=0}^{n=5} b_n(L) q^n; & L+1 \leq q \leq 2L \\ \sum_{n=0}^{n=5} c_n(L) q^n; & 2L+1 \leq q \leq 3L \end{cases} \tag{2}
$$

where, the coefficients $a_n(L), b_n(L)$ and $c_n(L)$ depend on L and they are univariate polynomials of the degree ≤ 5. Specifically, the coefficients $a_n(L)$ are of the degree ≤ 3 and they read [16]

$$a_0(L) = (L+1)^3$$

$$a_1(L) = -(L+1)^2 - \frac{2}{15}$$

$$a_2(L) = (L+1)\left[2(L+1)^2 - \frac{1}{2}\right]$$

$$a_3(L) = -2(L+1)^2 + \frac{1}{6}$$

$$a_4(L) = \frac{1}{2}(L+1)$$

$$a_5(L) = -\frac{1}{30}$$

(3)

whereas the fifth degree $b_n(L)$ and $c_n(L)$ are of the form

$$b_0(L) = -\frac{1}{10}(L+1)\left[31(L+1)^4 - 15(L+1)^2 + 4\right]$$

$$b_1(L) = \frac{7}{2}(L+1)^2\left[3(L+1)^2 - 1\right] + \frac{4}{15}$$

$$b_2(L) = -(L+1)\left[11(L+1)^2 - 2\right]$$

$$b_3(L) = 5(L+1)^2 - \frac{1}{3}$$

$$b_4(L) = -(L+1)$$

$$b_5(L) = \frac{1}{15}$$

(4)

and

$$c_0(L) = \frac{9}{2}L^2(15 + 17L + 9L^2 + \frac{9}{5}L^3) + \frac{137}{5}L + 4$$

$$c_1(L) = -9(1 + 5L + \frac{17}{2}L^2 + \frac{3}{2}L^4) - \frac{2}{15}$$

$$c_2(L) = 3(L+1)\left[3(L+1)^2 - \frac{1}{2}\right]$$

$$c_3(L) = -3(L+1)^2 + \frac{1}{6}$$

$$c_4(L) = \frac{1}{2}(L+1)$$

$$c_5(L) = -\frac{1}{30}$$

(5)

Equation (2) can be written in the form of the probability distribution for q. To do this we divide (2) by a function

$$\Sigma(L) = \frac{1}{2}(L+1)^3\left[(L+1)^3 - 1\right],\tag{6}$$

which is the total number of pairs of nodes in the cube. After such a normalization we obtain

$$P(L,q) = \frac{2}{(L+1)^3\left[(L+1)^3 - 1\right]}N(L,q).\tag{7}$$

Figure 2 depicts the distribution (7) for values of $L = 10$, 11, 13 and 14.

With the help of Eq. (7) we can compute the mean value $\langle q \rangle$ of the node-to-node distance. It appears that $\langle q \rangle$ scales with the length N of the functionalized SAW as

$$\langle q \rangle = \frac{(L+1)^2(L+2)}{(L+1)^2 + L + 2}\tag{8}$$

and $\lim_{L\to\infty}\frac{\langle q \rangle}{L} = 1$.

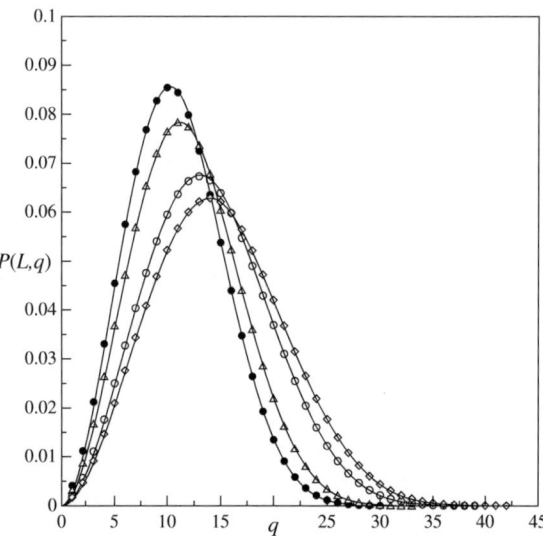

Fig. 2 Probability distribution functions p.d. of end-to-end distance q of polymer chain with ends binded to the nodes of a simple cubic lattice with $(L+1)^3$ nodes: *Dark circle, $L = 10$; Triangle, $L = 11$; Circle, $L = 13$ and Diamond, $L = 14$. The distance q between any pair of nodes is given by the smallest number of edges connecting these nodes. The lines are drawn using Eq. (7) and they are only visual guides

3.1 The Large L Limit

When one deals with a large number of binding centers, i.e. when L grows significantly, instead of distribution (7), a density probability function is more suitable for characterizing the node-to-node distance distribution. Such a density formulation can be introduced in a straightforward manner. Since $N(L, q)$, Eq. (2), is the fifth-degree polynomial in q and the normalization function $\Sigma(L)$, Eq. (6), is the polynomial of the order six in L, then

$$P(L, q) \rightarrow L^{-1} \prod \left(1, x_q = \frac{q}{L}\right) \approx p(x_q) \Delta x_q,$$

$$x_q \in \langle 1/L, 2/L, \dots, 3 \rangle, \quad \Delta x_q = \frac{1}{L}. \tag{9}$$

Even though the distribution $P(L, q)$ is exact for any L, for $L \gg 1$ we retain only terms of the first order in $1/L$ in the functional form of Π. In the limit $L \rightarrow \infty$, the discrete set $\langle 1/L, 2/L, \dots, 3 \rangle$ transforms in a dense subset $\langle 0, 3 \rangle \subset \Re$ and

$$p(x) = \lim_{L \to \infty} \lim_{x_q \to x} p(x_q) \tag{10}$$

becomes the exact probability density function of the continuous variable $0 \le x \le 3$. Due to the piecewise nature of $N(L, q)$, see Eq. (2), the function $p(x)$ has the following form

$$p(x) = \begin{cases} 4(1-x)x^2 + x^4 - \frac{1}{15}x^5; & 0 \le x \le 1, \\ -\frac{31}{5} + 21x - 22x^2 + 10x^3 - 2x^4 + \frac{2}{15}x^5; & 1 < x \le 2, \\ \frac{81}{5} - 27x + 18x^2 - 6x^3 + x^4 - \frac{1}{15}x^5; & 2 < x \le 3, \\ 0; & \text{elsewhere} \end{cases} \tag{11}$$

The function $p(x)$ is shown in Fig. 3. With the use of (11) we obtain the following exact formula for the moments $\mu_k = \int_\Re x^k p(x) dx$ of the random variable x

$$\mu_k = 24 \frac{3^{k+5} - (k+8)2^{k+5} + k^2 + 13k + 43}{(k+1)(k+2)(k+3)(k+4)(k+5)(k+6)}. \tag{12}$$

From the above equation we see that the mean value $\mu_1 = 1$ and that the moments diverge with $k \rightarrow \infty$.

When a SAW moves freely, i.e. without constraints imposed by a set of binding centers, the end-to-end distance distribution f is given by [17, 18]

$$f(x) = ax^{d-1}x^\sigma \exp(-bx^\delta), \tag{13}$$

where in three dimensions: $d = 3, \sigma = 0.33$ and $\delta = 2.43$. The normalization condition for the function (13) relates parameters a and b:

Fig. 3 The probability density function (pdf) of end-to-end distance x. Equation (11): continuous line, Eq. (15) with $d = 3, \sigma = 0.33, \delta = 2.43$ and $b^{1/\delta} = 1$: *dashed line*. The *dash-dotted line*: Eq. (15) with the same values of d, σ and δ but with $b^{1/\delta} = 1.04$

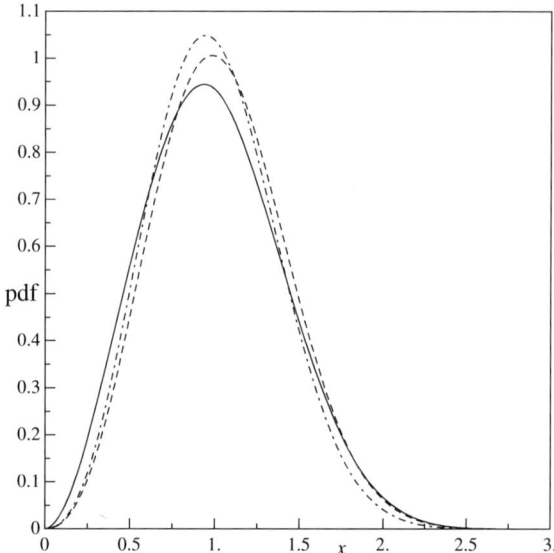

$$a = \delta \cdot \frac{b^{(d+\sigma)/\delta}}{\Gamma\left(\frac{d+\sigma}{\delta}\right)}. \tag{14}$$

Thus, the above relation yields the probability density function that varies depending on which of the value of the space-scale-factor $b^{1/\delta}$ is chosen, i.e.

$$f(x) = \delta \cdot \frac{b^{1/\delta}}{\Gamma\left(\frac{d+\sigma}{\delta}\right)} \cdot \left(b^{1/\delta}x\right)^{d+\sigma-1} \cdot \exp\left[-\left(b^{1/\delta}x\right)^{\delta}\right]. \tag{15}$$

For example, if we take $b^{1/\delta} = 1$ then the distribution (15) is represented by a dashed line in Fig. 3. We see that this curve is shifted slightly to the right compared to the solid line that represents the distribution (10). This shift vanishes when we equalize the mean value of x taken with respect to (15) with this one computed with the use of (11). According to (12) the first moment $\mu_1 = 1$. Since moments $m_k = \int_{\Re} x^k \cdot f(x) dx$ equal to

$$m_k = b^{-k/\delta} \frac{\Gamma\left(\frac{k+d+\sigma}{\delta}\right)}{\Gamma\left(\frac{d+\sigma}{\delta}\right)} \tag{16}$$

then from $\mu_1 = m_1 = 1$ we get the value of $b^{1/\delta} = 1.04$. The resulting probability density function is depicted in Fig. 3.

Even though our approach relies on geometry arguments alone, i.e. geometry of binding-center-lattice is the only constraint that reduces the number of accessible polymer conformations, we may get some insight into the end-to-end distance

Fig. 4 Comparision of
probability-density-functions
(*pdf*) of end-to-end distance x
for 2D substrates. Geometry-
based pdf, Eq. (17):
continuous line versus
distribution with
experimentally adjusted pdf
of DNA end-to-end distance,
Eq. (15) with $d = 1.55, \sigma =$
$0.44, \delta = 2.8$ and $b^{1/\delta} = 1$:
dashed line

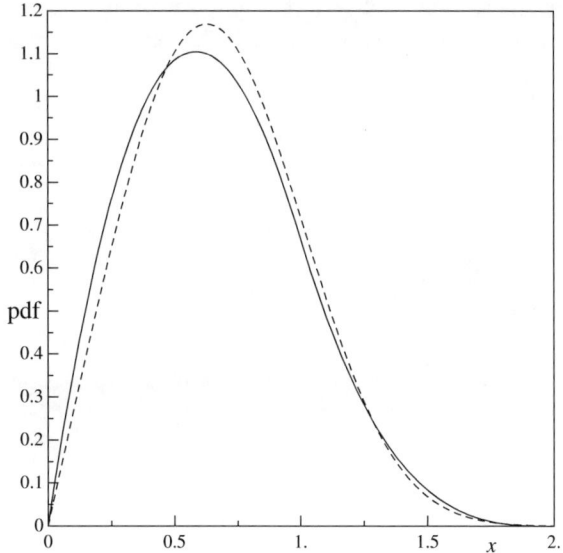

distribution of a real polymer. To see that consider distributions of such distance for
different contour length of DNA molecules adsorbed on a surface. As reported in
[18] these distributions were fitted using Eq. (13) with $\sigma = 0.44$ and experimentally
measured $\delta = 2.58 \pm 0.76$. Since in 3D a self-avoiding walk is a fractal object with
dimension $d_f = \delta/(\delta - 1)$ we compare that fit with the function [14]

$$g(x) = \begin{cases} 4(1-x)x + \frac{2}{3}x^3; & 0 \le x \le 1, \\ \frac{2}{3}(2-x)^3; & 1 < x \le 2, \\ 0; & \text{elsewhere.} \end{cases} \qquad (17)$$

which is an analog of (11) in two dimensions. Figure 4 shows a similarity between
$g(x)$ and the distribution based on measurements reported in [18]. From this figure
it is seen that the both curves are concave down for distances smaller than their
appropriate mean values. It is not the case for the distributions presented in Fig. 3.

3.2 Final Remark

The analytical approach employed in this work rests on the assumption that the
taxicab metric properly characterizes trajectories traced out by the functionalized
polymer-chain when it moves close to, or inside of a volume with periodically
distributed binding centers. Here, only geometrical aspects were considered, i.e. the
number of chain-like polymer conformations was scaled down by the space

available only. When a real polymer chain is confined in a reduced volume the number of its conformations is further lowered due to the interactions with the boundary and the interactions among its own segments. This fact is reflected in the shape of the end-to-end distance distribution we have obtained. Our distribution is slightly wider than distributions related to physically founded models of polymer. Nevertheless, our rough, purely geometrical approach enables us to construct distributions of end-to-end distance which may serve as envelope distributions.

References

1. R. Amin, S. Hwang, S.H. Park, Nanobiotechnology: an Interface Between nanotechnology and biotechnology. Nano: Brief Rep Rev, **6**(2), 101–111 (2011)
2. R.J. Kershner, L.D. Bozano, ChM Micheel, A.M. Hung, A.R. Fornof, J.N. Cha, ChT Rettner, M. Bersani, J. Frommer, P.W.K. Rothemund, G.M. Wallraff, Placement and orientation of individual DNA shapes on litographically patterned surfaces. Nat. Nanotechnol. **4**(9), 557–561 (2009)
3. Hh Lin, Y. Liu, S. Rinker, H. Yan, DNA tile based self-assembly: Building complex nanoarchitectures. Chem. Phys. Chem. **7**(8), 1641–1647 (2006)
4. S.M. Douglas, A.H. Marblestone, S. Teerapittayanon, A. Vazquez, G.M. Church, W.M. Shih, Rapid prototyping of 3D DNA-origami shapes with caDNAno. Nucl Acids Res **37**(15), 5001–5006 (2009)
5. P.W.K. Rothemund, Folding DNA to create nanoscale shapes and patterns. Nature **440**, 297–302 (2006)
6. S. Wang, H. Wang, J.J. Jiao, K.J. Chen, G.E. Owens, K. Kamei, J. Sun, D.J. Sherman, ChP Behrenbruch, H. Wu, H.R. Tseng, Three-dimensional nanostructured substrates toward efficient capture of circulating tumor cells. Angew. Chem. **121**, 9132–9135 (2009)
7. H.J. Yoon, T.H. Kim, Z. Zhang, E. Azizi, T.M. Pham, C. Paoletti, J. Lin, N. Ramnath, M.S. Wicha, D.F. Hayes, D.M. Simeone, S. Ngrath, Sensitive capture of circulating tumor cells by functionalized graphene oxide nanosheets. Nat. Nanotechnol. **8**(10), 735–741 (2013)
8. H. Otsuka, Nanofabrication of nonfouling surfaces for micropatterning of cell and microtissue. Molecules **15**(8), 5525–5546 (2010). www.mdpi.com/1420-3049/15/8/5525
9. I. Teraoka, *Polymer Solutions: An Introduction to Physical Properties* (Wiley, Brooklyn, 2002)
10. A.D. Sokal, monte carlo methods for the self-avoiding walk, in *Monte Carlo and Molecular Dynamics Simulations in Polymer Science*, ed. by K. Binder (Oxford University Press, New York, 1995)
11. G. Slade, Self-avoiding walks. Math. Intelligencer **16**(1), 29–35 (1994). www.math.ubc.ca/slade/intelligencer.pdf
12. N.W. Ashroft, N.D. Mermin, *Solid State Physics* (Harcourt Inc., Orlando, 1976)
13. F.F. Dragan, Estimating all pairs shortest paths in restricted graph families: A unified approach. J. Algorithms **57**(1), 1–21 (2005)
14. Z. Domański, N. Sczygiol, Distribution of the distance between receptors of ordered micropatterned substrates, eds. by H.K. Kim, S.I. Ao, B.B. Rieger. *IAENG Transactions on Engineering Technologies* (Springer, Dordrecht, 2011) (Special Edition of the World Congress on Engineering and Computer Science)
15. Z. Domański, Geometry-induced transport properties of two dimensional networks, ed. by M. Schmidt. Advances in Computer Science and Engineering (InTech, Rijeka, 2011)

16. Z. Domański, N. Sczygiol, Distribution of node-to-node distance in a cubic lattice of binding centers, lecture notes in engineering and computer science, in *Proceedings of The International MultiConference of Engineers and Computer Scientists 2014*, IMECS, 12–14 Mar 2014, Hong Kong, pp. 900–903
17. A.Y. Grosberg, A.R. Khoklov, *Statistical Physics of Marcomolecules* (American Institute of Physics, Woodbury, 1994)
18. F. Valle, M. Favre, P. De Los Rios, A. Rosa, G. Dietler, Scaling exponents and probability distributions of DNA end-to-end distance. Phys. Rev. Lett. **95**, 158105 (2005)

Using Modern Multi-/Many-core Architecture for the Engineering Simulations

Grzegorz Michalski and Norbert Sczygiol

Abstract The paper presents authorial method for computer simulation of the process of the casting solidification process using modern multi-/many-core architectures. The authors have focused above all on the GPU compatible with CUDA architecture. Presented in the article method enables division the process of the building a global matrix of coefficients in two independent phases. The first one is performed only once before the start of a computer engineering simulation, the second phase is performed at each time step of computer simulation. The reason for such division is decoupling of computations realised on the GPU from the spatial informations (e.g. physical coordinates of the nodes) contained in the mesh of finite elements. Application of presented method enables the computer scientific simulation in a variety of multicore architectures, such as multi-core CPUs, GPUs, APUs and Intel Phi. The use of computer engineering simulations multi-core processor allows significant acceleration computation, thereby also reduces costs. Presented in the work results have been obtained by using authorial software built by authors from scratch.

Keywords Casting · Computer simulation · CUDA · Enthalpy · Graphics processor · High performance computing · Parallel processing · Solidification processing

1 Introduction into Modern Processors Architecture

The progress of science has an impact on all areas of life, including the development of computer technology. The tasks appear before engineers from different fields of science are increasingly complex and require a considerable amount of time to solve

G. Michalski · N. Sczygiol (✉)
Institute of Computer and Information Sciences, Czestochowa University of Technology,
Dabrowskiego 69, 42201 Czestochowa, Poland
e-mail: norbert.sczygiol@icis.pcz.pl

G. Michalski
e-mail: grzegorz.michalski@icis.pcz.pl

© Springer Science+Business Media Dordrecht 2015
G.-C. Yang et al. (eds.), *Transactions on Engineering Technologies*,
DOI 10.1007/978-94-017-9588-3_5

them. In recent years the dynamic development of multicore processors has had a direct impact on the availability of advanced high performance solutions for engineers. A few years ago, computers with a high-end graphics card were intended primarily for computer game players and computer graphics designers. Nowadays, the situation has dramatically changed. Graphic processors (GPUs—Graphics Processing Unit) are increasingly being used in high performance computations. A single GPU has a theoretical computational power several times higher than the fastest general purpose processors available today. Figure 1 shows the increase in the theoretical computational power of graphic processors and general purpose processors in recent years. Computations using a GPU working as a computing accelerator do not require any additional equipment, such as specialized workstations, and can be made on an ordinary personal computer equipped with a graphic card that supports CUDA or OpenCL. As graphic processors, in accordance with their original purpose, have been designed to efficiently perform mathematical operations on two-dimensional matrices, they should be capable of performing numerical simulations. The application of modern multi- and many-core architectures, such as graphic processors, for computational purposes allows huge systems of equations, which may consist of many millions of variables, to be solved very fast. The algorithms used for solving such systems of equations are based primarily on standard arithmetic operations on matrices and vectors. Such algorithms permit efficient parallelization and implementation for many-cores architecture like graphic processors [1–3].

In this article, the authors present a computer simulation of the casting solidification process (enthalpy formulation), one that belongs to a group of unsteady processes. For processes of this type, a system of equations is built from scratch in each time step. The matrix of the coefficients of this system of equations and a vector

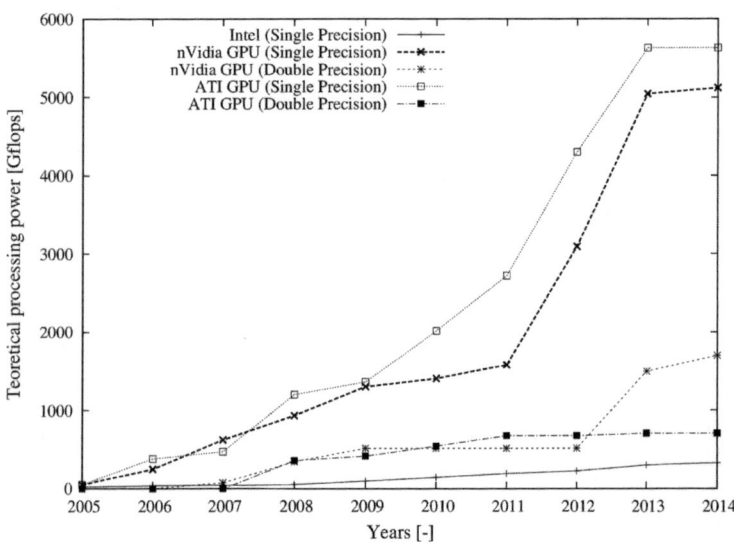

Fig. 1 An increase in theoretical computational power of the GPU and CPU in the recent years

of right sides are built on the basis of several factors: a description of the finite elements mesh, the boundary conditions, the material properties and the results obtained from previous time steps. These operations require a large amount of data in each time step of the simulation to be processed. Repeatedly sending huge data sets to the memory of the graphic card may create a real problem. The specific character of this data (mainly the finite element mesh) makes it difficult to efficiently parallelize the matrix building process. This difficulty derives from the lack of regularity of the data that are to be processed. This problem can be clearly seen in the case of GPUs and SIMT architecture (Single Instruction Multiple Thread), where the conditional execution of certain parts of the program code has a strong negative impact on the efficiency of computations. An additional problem is the need to synchronize the write operations performed during the construction of the system of equations [4–6].

2 Building the System of Equations on Graphic Processors

While performing simulations of unsteady processes the system of equations has to be built many times (for each time step). Owing to the specific nature of the graphic processors architecture, this can constitute a real problem and can significantly decrease the efficiency of computations. This relates to the slow data transfer from the global memory of the graphic devices (device memory) to the system memory (host memory), synchronize mechanism and conditional instructions.

As the data transfer from the system memory to the global memory of the graphic devices should be reduced to a minimum, a good solution is to transfer the description of the finite element mesh to the graphic device global memory and build the system of equations using the graphic processor [5–7]. However, it should be remembered that the finite element meshes that are used nowadays often consist of a few million nodes. Transferring such a large amount of data is a very time-consuming process and, moreover this data can reduce the limited resources of the device.

The parallelization operation of the matrix building process on the basis of the finite element mesh is very complicated. This process requires a lot of conditional instructions and synchronize blocks, which causes a significant decrease in the performance of computations made on the graphic processor (nVidia GT200 in this case).

The values of elements in i-row of the coefficient matrix depend on the finite elements, which include the node connected with the node with index i. A single node in the finite element mesh can belong to several finite elements. No direct method exists to identify these finite elements solely on the basis of the node index. These indices can only be read from the finite element mesh. Figure 2 shows the process of building the global coefficient matrix. Non-zero elements of the global coefficients matrix are determined as the sum of several values which depend on those finite elements which include the pair of nodes with indexes corresponding to the indexes of the row and column of those elements. Constructing the global matrix of coefficient in this way definitely makes the parallelization process more difficult.

Fig. 2 Process of building
the global coefficient matrix

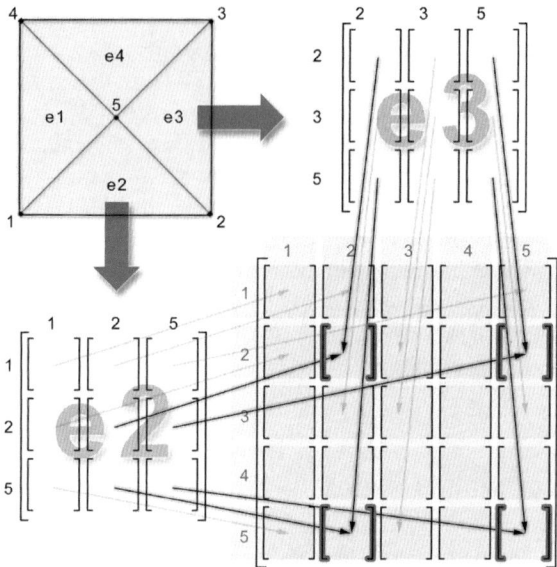

The approach presented in this paper uses a modified way of building the system of equations which divides this process into two separate phases. The first requires information from the finite element mesh. This phase is performed only once at the beginning of the computer simulation. An appropriate transformation of the matrix obtained from the Finite Element Method makes this part of the matrix independent from the nodal temperatures and enthalpy values, determined in subsequent time steps. The second phase is performed in each time step and is dependent on the nodal temperature values and enthalpy from previous steps.

2.1 New Two-Step Building of the System of Linear Equation Method

The system of equations resulting from the transformation of a differential equation, which describes the process of solidification, into the numerical model using the Finite Element Method can be finally written in matrix formulation (1) [8–10].

$$\mathbf{M}\mathbf{H}^{n+1} + \Delta t\mathbf{K}^n\mathbf{T}^{n+1} = \mathbf{M}\mathbf{H}^n + \Delta t\mathbf{b}^{n+1} \tag{1}$$

The elements of (1), which are built on the basis of the finite element mesh description are: matrix **M** (2) and conductivity matrix **K** (3) [11].

$$\mathbf{M}^e = \int_{\Omega^e} \mathbf{N}^T \mathbf{N} d\Omega \qquad (2)$$

$$\mathbf{K}^e(T) = \int_{\Omega^e} \lambda(T) \nabla^T \mathbf{N} \cdot \nabla \mathbf{N} d\Omega \qquad (3)$$

Matrix $\mathbf{M}$ is independent from the nodal temperature and enthalpy values in subsequent time steps. This matrix may be determined at the beginning of the simulation, before the first time step is executed. This matrix can be used in the same form in the subsequent time steps. Matrix $\mathbf{M}$ is a diagonal matrix, and it is preferable to store it in the form of a simple vector, as this helps to save priceless graphic device memory.

In contrast, conductivity matrix $\mathbf{K}$ is a temperature function. The element which causes this dependency is thermal conductivity coefficient λ. The value of the thermal conductivity coefficient is determined during the building of the matrix of coefficients for each finite element. If thermal conductivity coefficient λ is not dependent on the spatial coordinates, it can be taken out before the integral. After completing these steps, Eq. (3) assumes the form (4).

$$\mathbf{K}^e = \lambda(T) \int_{\Omega^e} \nabla^T \mathbf{N} \cdot \nabla \mathbf{N} d\Omega \qquad (4)$$

After integration, conductivity matrix $\mathbf{K}$ (for triangular elements) takes the form (5), where $\mathbf{K}^e$ is a local coefficient matrix for the finite element, A the surface area of the finite element, C_{ij} are coefficients which depend on the spatial coordinates of the nodes belonging to the finite element.

The value of thermal conductivity coefficient λ for the finite element is calculated as the average value of the thermal conductivity coefficients determined for the nodes belonging to that finite element. After building the local coefficient matrices, they are assembled into a global matrix of coefficients $\mathbf{K}$, which contains elements that are the sum of the products (their factors are different thermal conductivity coefficient λ). These are average values calculated for each finite element that includes the corresponding nodes. Such a situation makes it impossible to separate this part of the matrix, which is temperature-dependent, from that part which is built on the basis of information contained in the finite element mesh.

In order to find a solution to this problem the authors have developed an alternative approach to the building of the system of linear equations. This approach allows the two parts of conductivity matrix $\mathbf{K}$ to be separated. It involves the introduction into the local matrices, values of the thermal conductivity coefficient λ determined for the nodes and not for the finite elements as in the original approach. After this change is introduced into the Eq. (5) it takes the form (8).

$$\mathbf{K}^e = \frac{\lambda(T)}{4A} \begin{bmatrix} C_{21}^2 + C_{31}^2 & C_{21}C_{22} + C_{31}C_{32} & C_{21}C_{23} + C_{31}C_{33} \\ C_{21}C_{22} + C_{31}C_{32} & C_{22}^2 + C_{32}^2 & C_{22}C_{23} + C_{32}C_{33} \\ C_{21}C_{23} + C_{31}C_{33} & C_{22}C_{23} + C_{32}C_{33} & C_{23}^2 + C_{33}^2 \end{bmatrix} \quad (5)$$

Equations (6) and (7) presents: surface area and coefficients C_{ij} for a triangular finite element:

$$A = \frac{1}{2}(x_1(y_2 - y_3) + x_2(y_3 - y_1) + x_3(y_1 - y_2)) \quad (6)$$

$$\begin{aligned} C_{21} = \tfrac{y_2-y_3}{2A} \quad C_{22} = \tfrac{y_3-y_1}{2A} \quad C_{21} = \tfrac{y_1-y_2}{2A} \\ C_{31} = \tfrac{x_3-x_2}{2A} \quad C_{32} = \tfrac{x_1-x_3}{2A} \quad C_{33} = \tfrac{x_2-x_1}{2A} \end{aligned} \quad (7)$$

$$\mathbf{K}^e = \frac{1}{4A} \begin{bmatrix} \lambda_1(C_{21}^2 + C_{31}^2) & \lambda_1(C_{21}C_{22} + C_{31}C_{32}) & \lambda_1(C_{21}C_{23} + C_{31}C_{33}) \\ \lambda_2(C_{21}C_{22} + C_{31}C_{32}) & \lambda_2(C_{22}^2 + C_{32}^2) & \lambda_2(C_{22}C_{23} + C_{32}C_{33}) \\ \lambda_3(C_{21}C_{23} + C_{31}C_{33}) & \lambda_3(C_{22}C_{23} + C_{32}C_{33}) & \lambda_3(C_{23}^2 + C_{33}^2) \end{bmatrix}$$
$$(8)$$

This approach permits the removal of thermal conductivity coefficient λ before the parenthesis in each element of global matrix $\mathbf{K}$. After these steps the solidification equation in matrix formulation (1) takes the form described in Eq. (9).

$$\mathbf{MH}^{n+1} + \Delta t \lambda^n \mathbf{K}^* \mathbf{T}^{n+1} = \mathbf{MH}^n + \Delta t \mathbf{b}^{n+1} \quad (9)$$

where λ is the diagonal matrix of the thermal conductivity coefficient for each node of the finite element mesh, determined on the basis of nodal temperatures from the appropriate time-step, $\mathbf{K}^*$ is a matrix of coefficients built on the basis of the finite element mesh description. Matrix $\mathbf{K}^*$ is built only once at the beginning of the computer simulation.

This approach allows the process of building the global coefficient matrix to be divided into two phases. The first phase is independent from the nodal temperatures values, and thus simultaneously independent from the time steps. Phase one is performed only once before the first time step of the simulation is performed. In this stage, matrix $\mathbf{K}^*$ is built on the basis of information from the finite element mesh. Additionally, matrix $\mathbf{B}$ (calculated on the basis of the boundary conditions), diagonal matrix $\mathbf{M}$ and the vector (calculated on the basis of the boundary conditions) with values necessary to build the vector right sides in each time step are created in this stage.

After this step, the information stored in the mesh of finite elements (coordinates of the nodes, finite element descriptions, edges and areas) is no longer required for the process of computer simulation. The second phase of the matrix building process consists of determining conductivity coefficient λ for each node. As a result of this step diagonal matrix λ is created. This matrix is multiplied by matrix $\mathbf{K}^*$. As a result of the implementation of these operations the non-zero elements of matrix $\mathbf{K}^*$

are multiplied by λ values determined for the node whose index corresponds to the number of the row of the matrix. Since conductivity coefficient matrix λ is temperature dependent, this operation must be performed in each subsequent time-step.

This approach simplifies the process of building the global matrix of coefficients on graphic processors by allowing matrix $\mathbf{K}^*$ to be built on a general purpose processor.

3 Computer Experimental Numerical Verification

Simulations of the solidification process were carried out for the regions shown in Fig. 3. Computations were performed with the use of 6 finite element meshes

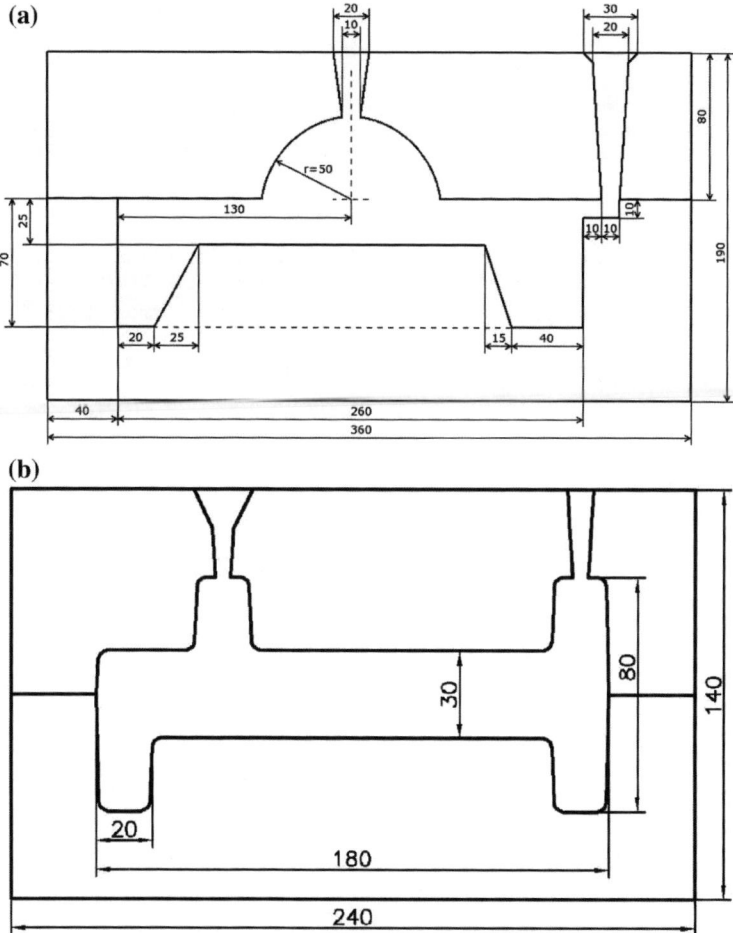

Fig. 3 Analyzed shapes

(for each one shape), whose sizes ranged from several thousand to several hundred thousand nodes.

Figure 4a, b shows time required to execute of 200 time steps simulation of solidification for each analyzed shape (in sequential version of authorial software). Software with the use of the two-step building of the system of linear equation

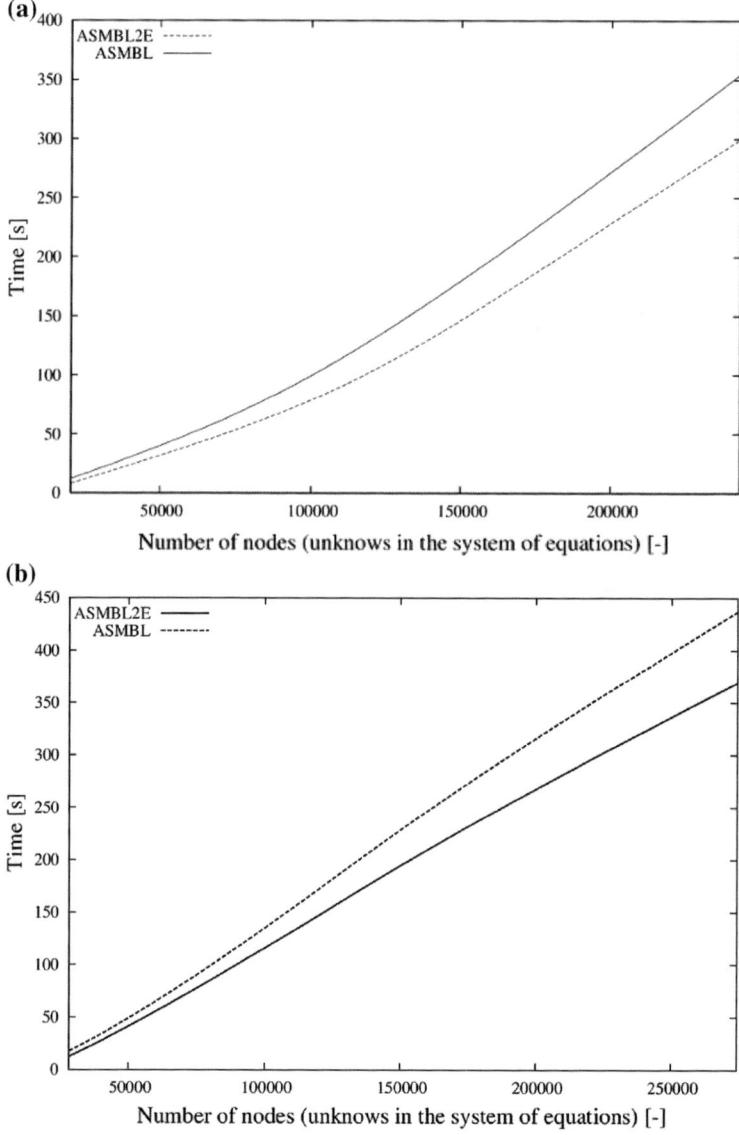

Fig. 4 Execution time of 200 time steps simulation of solidification for each shape (in the sequential version of authorial software)

(*ASMBL2E*), shown in Sect. 2.1, are performed in a shorter time than the unmodified version (*ASMBL*). This is due to the extract unchangeable coefficients from the process of the building a system of equations during subsequent time steps of simulation and determining them before the start of the simulation process.

Times of simulations performed with the GPUs were compared to the results obtained with the authorial CPU-based sequential software. This software uses the approach to building the system of equations which was presented in previous section. The authors used a Boost Math library for matrix and vector computations in the sequential implementation. It should also be noted that all matrices are stored as sparse matrices using Compressed Sparse Row format. Authorial software (CUDA version) was realized from scratch without use a outside libraries.

The computer simulation on the graphics processor was performed an nVidia GT200. The GPU-based casting solidification simulation software was run on the graphic devices:

1. nVidia TESLA C1060 (4 GiB GDDR),
2. GeForce GTX 260 (896 GiB GDDR).

Both the above are equipped with an nVidia GT 200 processor with 240 CUDA cores which has a peak performance of about 1 Tflops in single float operations. The authors used CUDA version 3.2 in the graphic processor implementations and Debian Operating System.

The casting material is aluminum alloy with the addition of copper and the mold is assumed to be made from steel. The material properties are listed in Table 1. The initial temperature of pouring was equal to 960 K and this value was the value of the initial condition temperature for the region of the cast in the performed computations. The initial temperature of the mold was set at 560 K. Newton's boundary

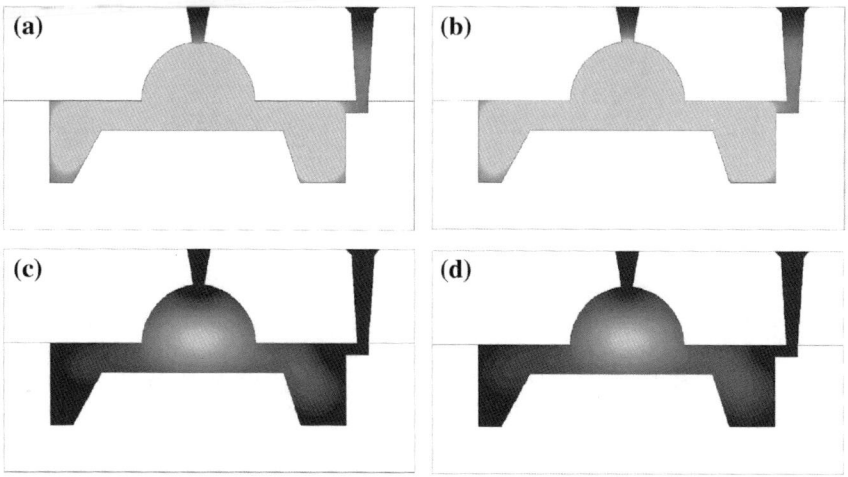

Fig. 5 Distribution of the solid phase (on shape (3a)). **a** after 200 time-steps (CPU). **b** after 200 time-steps (GPU). **c** after 6,000 time-steps (CPU). **d** after 6,000 time-steps (GPU)

Table 1 Physical properties of cast material (Al-2%Cu), and mold

Quantity	Unit symbol	Value
Thermal conductivity of solid phase	$\frac{W}{mK}$	262
Thermal conductivity of liquid phase	$\frac{W}{mK}$	104
Density of solid phase	$\frac{kg}{m^3}$	2,824
Density of liquid phase	$\frac{kg}{m^3}$	2,498
Specific heat of solid phase	$\frac{J}{kgK}$	1,077
Specific heat of liquid phase	$\frac{J}{kgK}$	1,275
Solidus temperature	K	853
Liquidus temperature	K	926
Melting temperature of pure metal	K	933
Latent heat of solidification	$\frac{J}{kgK}$	390,000
Thermal conductivity of mold	$\frac{W}{mK}$	40
Density of mold	$\frac{kg}{m^3}$	7,500
Specific heat of mold	$\frac{J}{kgK}$	620

condition is assumed on all the surfaces of the mold, assuming heat exchange with the environment with a coefficient equal to 100 $\frac{W}{m^2K}$. The ambient temperature has a value of 300 K in all boundary conditions. The heat exchange between the casting and the mold is obtained from a type IV boundary condition, which assumes the heat exchange through the insulation layer of the conductivity coefficient of the separating layer to be 800 $\frac{W}{m^2K}$.

Figure 6a, b shows that speed up depended on the size of the task (number of nodes in the finite element mesh) for each analyzed shapes. In small tasks the speed up is minor. This can be explained by the time taken to copy the data from the system memory to GPU memory and to copy the results into the system memory from the graphic device memory (first and last time-step). No differences were noted in the process of the computer simulation implemented on a general purpose processor and graphics processor. The simulation results (temperature and part of the solid phase) show that there is a slight difference between the results obtained with the GPU-based and CPU-based software. These differences are minor and do not exceed 0.1 % and have no effect on the simulation process. This is illustrated in Fig. 5 which shows a part of the solid phase after 200 and 6 000 time-steps of computer simulation of the casting solidification process realized on CPU (Fig. 5a, c) and GPU (Fig. 5b, d). Comparing these figures it can be seen that the simulation process in both cases is the same.

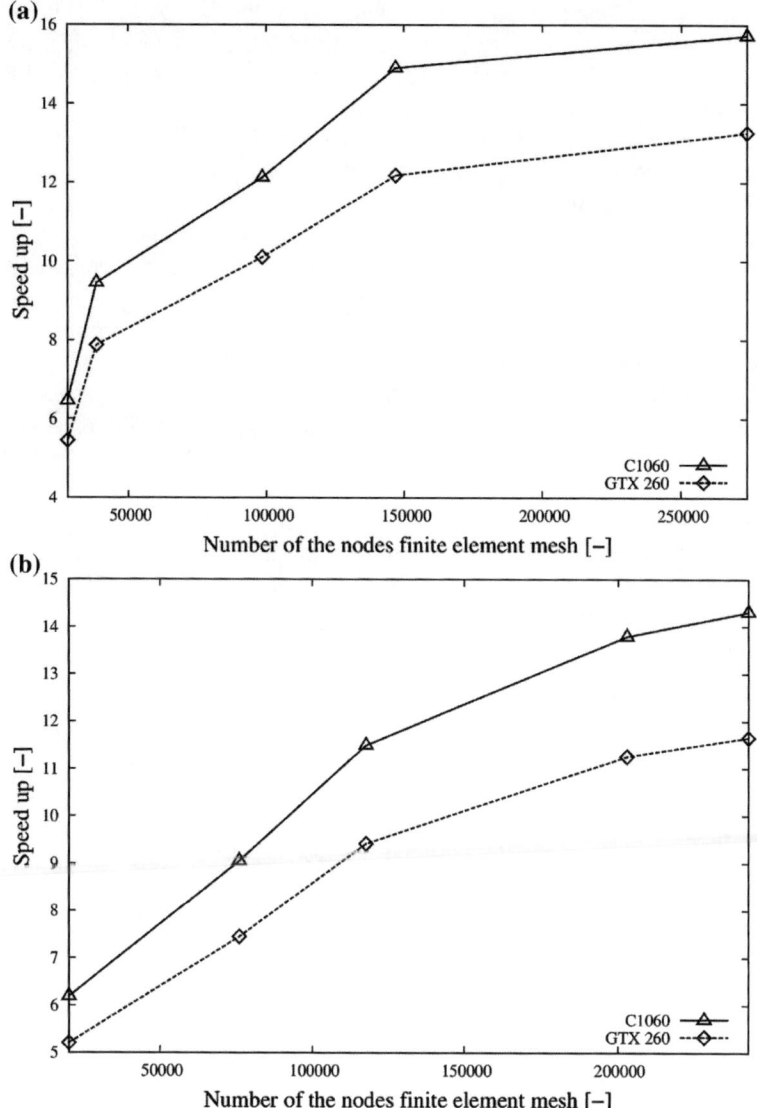

Fig. 6 Speed up obtained with the use of GPU-based simulation software for the two different analyzed shape

4 Remarks

In this article, the authors present a new method of parallelization for the computer simulation of the casting solidification process and its implementation on graphics processors compatible with CUDA architecture. The proposed method divides the process of building the system of equations into the two phases. This solution improves the efficiency with which the available system resources are used. The great advantage of the developed solution is that it is easily adapted to the different architectures of multicore processors.

The speed up observed during the computer simulation of the casting solidification process confirms that the use of graphic processors in engineering simulations brings significant benefits. It was also noted that with an increase in the number of unknowns in the system of equations, the time needed to solve such a system increases linearly. At the same time, as the size of the task increases so does the speed up observed in the computations with the graphic processor. Different way to speed-up computation in the scientific simulation is a proper using an eigenvalues and modifying the size of time step [12, 13], the speed-up obtained in this way are minor.

The modifications to the numerical model produce slight differences in the temperatures in the nodes. However, these are just minor differences that do not affect the simulation of the casting solidification process. These differences result from the different way in which the value of the coefficient of thermal conductivity sensor is determined for the node, and not for the finite elements. This leads to a speed up in computations performed in a sequential way (on the CPU).

Having regard to the above results it can be stated that using graphic processors in engineering simulations seems to be a viable solution as this approach can significantly reduce the time needed for research.

References

1. H.T. Meng, B.L. Nie, S. Wong, C. Macon, J.M. Jin, GPU accelerated finite-element computation for electromagnetic analysis. IEEE Antennas Propag. Mag. **56**, 39–62 (2014)
2. G. Michalski, N. Sczygiol, Using CUDA architecture for the computer simulation of the casting solidification process. Lect Notes Eng. Comp.: Proc. Int. MultiConf. Eng. Comput. Sci. **2**, 933–937 (2014)
3. R. Strzodka, M. Dogger, A. Kolb, Scientific computation for simulations on programmable graphics hardware. Simul. Model. Pract. Theory **13**, 667–680 (2005)
4. S. Che, M. Boyer, J. Meng, D. Tarjan, J.W. Sheaffer, K. Skadron, A performance study of general-purpose applications on graphics processors using CUDA. J. Parallel Distrb. Com. **68**, 1370–1380 (2008)
5. nVidia: CUDA C best practices guide. Design guide v. 6.0 (2014). http://docs.nvidia.com/cuda/#axzz398vseVcQ Cited 01 Aug 2014
6. P. Pospíchal, J. Schwarz, J. Jaroš, Parallel genetic algorithm solving 0/1 knapsack problem running on the GPU, in *16th International Conference on Soft Computing MENDEL 2010.* pp. 64–70
7. Ch. Lee, X. Wei, J.W. Kysar, J. Hone, Measurement of the elastic properties and intrinsic strength of monolayer graphene. Science **321**, 385–388 (2008)
8. N. Sczygiol, *Numerical modelling of thermo-mechanical phenomena in a solidifying casting and mold* (Wyd PCz, Czestochowa, 2000). (in Polish)
9. N. Sczygiol, G. Szwarc, Application of enthalpy formulations for numerical simulation of castings solidification. Comput. Assist. Mech. Eng. Sci. **8**, 99–120 (2001)

10. O. Zienkiewicz, The finite element method, vol I, the Basis, 5th ed. Oxford: Butterworth-Heinemann
11. N. Sczygiol, G. Szwarc, R. Wyrzykowski, Numerical modelling of equiaxed structure formation during solidification of a two-component alloy, in *Second European Conference on Computational Mechanics*, Kraków, pp. 820–821
12. E. Gawronska, N. Sczygiol, Application of mixed time partitioning methods to raise the efficiency of solidification modeling, in *12th International Symposium on Symbolic and Numeric Algorithms for Scientific Computing (SYNASC)*, doi:http://dx.doi.org/10.1109/SYNASC.2010.24
13. E. Gawronska, N. Sczygiol, Relationship between eigenvalues and size of time step in computer simulation of thermomechanics phenomena. Lect. Notes Eng. Comp.: Proc. Int. MultiConf. Eng. Comput. Sci. **2**, 881–885 (2014)

A Viscosity Approximation Method for the Split Feasibility Problems

Jitsupa Deepho and Poom Kumam

Abstract In this paper, we discuss the strong convergence of the viscosity approximation method for solving the split feasibility problem in Hilbert spaces. Consider also the iteration process $\{x_n\}$, where $x_0 \in C$ is arbitrary and $x_{n+1} = (1 - \alpha_n)P_C(I - \xi A^*(I - P_Q)A)x_n + \alpha_n f(x_n), n \geq 1$ where $\alpha_n \in (0, 1)$. The main result present in this paper improve and extend some recent result done by Xu [Iterative methods for the split feasibility problem in infinite-dimensional Hilbert space, Inverse Problem 26 (2010) 105018] and some others.

Keywords CQ algorithm · Metric projection · Split feasibility problem · Strong convergence theorem · Variational inequality problem · Viscosity approximation method

1 Introduction

Throughout this paper, we assume that H is a real Hilbert space with inner product, $\langle \cdot, \cdot \rangle$ and norm $\| \cdot \|$ and let C be a nonempty closed convex subset of H. A mapping $S : C \rightarrow C$ is called *nonexpansive* if $\|Sx - Sy\| \leq \|x - y\|, \forall x, y \in C$. We use $F(S)$ to denote the set of fixed points of S, that is, $F(S) = \{x \in C : Fs = s\}$. It is assumed throughout the paper that S is a nonexpansive mapping such that $F(S) \neq \emptyset$. Recall that a self-mapping $f : C \rightarrow C$ is a contraction on C if there exists a constant $\rho \in (0, 1)$ and $x, y \in C$ such that $\|fx - fy\| \leq \rho\|x - y\|$.

J. Deepho · P. Kumam (✉)
Department of Mathematics, Faculty of Science, King Mongkut's University of Technology
Thonburi, 126 Pracha Uthit Road, Bang Mod, Thung Khru, Bangkok 10140, Thailand
e-mail: poom.kum@kmutt.ac.th

J. Deepho
e-mail: jitsupa.deepho@mail.kmutt.ac.th

© Springer Science+Business Media Dordrecht 2015
G.-C. Yang et al. (eds.), *Transactions on Engineering Technologies*,
DOI 10.1007/978-94-017-9588-3_6

In this paper, we review the computation of fixed points of such general operators, T, by mean so-called *viscosity approximation method* which formally consists of the sequence $\{x_n\} \in C$ given by the iteration (see [1–5])

$$x_{n+1} = \alpha_n f(x_n) + (1 - \alpha_n)Tx_n \tag{1}$$

where $\alpha_n \in (0, 1)$ is a slowly vanishing sequence, i.e., $\lim_{n\to\infty} \alpha_n = 0$ and $\sum_n \alpha_n = \infty$.

The *split feasibility problem* (SFP) was introduced by Censor and Elving [6] for modeling inverse problems which arise from phase retrievals and in medical image reconstruction [7], and may well known iterative algorithms has been invented to solve it [7].

In this work, the SFP is formulated as finding a point $\hat{x}$ satisfying the following property:

$$\hat{x} \in C, \ A\hat{x} \in Q, \tag{2}$$

where C and Q are the nonempty closed convex subsets of the infinite-dimensional real Hilbert spaces H_1 and H_2, respectively, and $A \in B(H_1, H_2)$ (i.e., A is a bounded linear operator from H_1 to H_2).

The SFP was introduced by Censor and Elving [6] for modeling inverse problems which arise from phase retrievals and in medical image reconstruction [7], and may well known iterative algorithms has been invented to solve it [7].

We use Γ to denote the solution set of SFP :

$$\Gamma = \{\hat{x} \in C : A\hat{x} \in Q\}, \tag{3}$$

and assume that the SFP (2) is consistent [i.e., (2) has a solution] so that Γ is closed, convex and nonempty, it is not hard to see that $x \in C$ solves (2) if and only if it solves the following fixed point equation;

$$x = P_C(I - \gamma A^*(I - P_Q)A)x, \quad x \in C, \tag{4}$$

where P_C and P_Q are the metric projections onto C and Q, respectively, $\gamma > 0$ is any positive constant and A^* denotes the adjoint of A. Moreover, for sufficiently small $\gamma > 0$, the operator $P_C(I - \gamma A^*(I - P_Q)A)$ which defines the fixed point equation in (4) is nonexpansive.

To solve the SFP (2), Byrne [7] proposed his *CQ* algorithm (see also [8]) which generates a sequence $\{x_n\}$ by

$$x_{n+1} = P_C(I - \gamma A^*(I - P_Q)A)x_n, \quad n \geq 0, \tag{5}$$

where $\gamma \in (0, 2/\lambda)$ with λ being the spectral radius of the operator A^*A.

Very recently, Xu [9] has viewed the *CQ* algorithm for averaged mappings and applied Mann's algorithm to solving the SFP, and he also proved that an averaged *CQ* algorithm is weakly convergent to a solution of the SFP.

In 2014, Deepho and Kumam [10] suggest a hybrid extragradient method for finding a common element of the set of fixed point sets of an infinite family of nonexpansive mappings and the solution set of the SFP in real Hilbert spaces.

In this paper, we also regard the CQ algorithm as a fixed point algorithm for averaged mapping, and try to study the SFP by the following modified viscosity approximation method;

$$\begin{cases} x_0 \in C, \\ x_{n+1} = (1 - \alpha_n)P_C(I - \xi A^*(I - P_Q)A)x_n + \alpha_n f(x_n), n \geq 1, \end{cases} \tag{6}$$

where $\{\alpha_n\} \subset (0, 1)$. Furthermore, our result extends and improves the result of Xu [9] from weak to strong convergence theorems.

2 Preliminaries

Throughout the paper, we adopt the following notation: Let $\{x_n\}$ be a sequence and x be a point in a normed space X. We use $x_n \rightarrow x$ and $x_n \rightharpoonup x$ to denote strong and weak convergence to x of the sequence $\{x_n\}$, respectively.

Let H be a real Hilbert space with inner product $\langle \cdot, \cdot \rangle$ and norm $\| \cdot \|$, respectively, and let C be a nonempty closed convex subset of H. For every point $x \in H$, there exists a unique nearest point in C, denoted by $P_C x$, such that

$$\|x - P_C x\| \leq \|x - y\|, \quad \forall y \in C. \tag{7}$$

P_C is called the metric projection of H onto C. It is well known that P_C is a nonexpansive mapping of H onto C and satisfies

$$\langle x - y, P_C x - P_C y \rangle \geq \|P_C x - P_C y\|^2, \tag{8}$$

for every $x, y \in H$. Moreover, $P_C x$ is characterized by the following properties: $P_C x \in C$ and

$$\langle x - P_C x, y - P_C x \rangle \leq 0, \tag{9}$$

$$\|x - y\|^2 \geq \|x - P_C x\|^2 + \|y - P_C x\|^2 \tag{10}$$

for all $x \in H, y \in C$.

Some important properties of projections are gathered in the following proposition.

Proposition 1 *Given $x \in H$ and $z \in C$. Then $z = P_C x$ if and only if*

$$\langle x - z, y - z \rangle \leq 0, \quad \forall y \in C. \tag{11}$$

We also need other sorts of nonlinear operators which are introduced below. Let $T, A : H \to H$ be the nonlinear operators.

(i) T is nonexpansive if $\|Tx - Ty\| \leq \|x - y\|$, $\forall x, y \in H$.
(ii) T is firmly nonexpansive if $2T - I$ is nonexpansive. Equivalent, $T = (I + S)/2$, where $S : H \to H$ is nonexpansive. Alternatively, T is firmly nonexpansive if and only if

$$\langle x - y, Tx - Ty \rangle \geq \|Tx - Ty\|^2, \quad x, y \in H. \tag{12}$$

(iii) T is averaged if $T = (1 - \alpha)I + \alpha S$, where $\alpha \in (0, 1)$ and $S : H \to H$ is nonexpansive. In this case, we also say that T is α-averaged. A firmly nonexpansive mapping is $\frac{1}{2}$-averaged.

It is well known that both P_C and $I - P_C$ are firmly nonexpansive and $\frac{1}{2}$-ism.

Proposition 2 *[7, 11] We have the following assertions.*

1. *T is nonexpansive if and only if the complement $I - T$ is $\frac{1}{2}$-ism.*
2. *If T is v-ism and $\gamma > 0$, then γT is $\frac{v}{\gamma}$-ism.*
3. *T is averaged if and only if the complement $I - T$ is v-ism, for some $v > \frac{1}{2}$. Indeed, for $\alpha \in (0, 1), T$ is α-averaged if and only if $I - T$ is $\frac{1}{2\alpha}$-ism.*
4. *If T_1 is α_1-averaged and T_2 is α_2-averaged, where $\alpha_1, \alpha_2 \in (0, 1)$, then the composite $T_1 T_2$ is α-averaged, where $\alpha = \alpha_1 + \alpha_2 - \alpha_1 \alpha_2$.*
5. *If T_1 and T_2 are averaged and have a common fixed point, then $Fix(T_1 T_2) = Fix(T_1) \cap Fix(T_2)$.*

Lemma 3 *[12] Assume that $\{a_n\}$ is a sequence of nonnegative real numbers such that*

$$a_{n+1} \leq (1 - \gamma_n)a_n + \delta_n, \tag{13}$$

where $\{\gamma_n\}$ is a sequence in $(0, 1)$ and $\{\delta_n\}$ is a sequences such that

1. $\sum_{n=0}^{\infty} \gamma_n = \infty$;
2. $\limsup_{n \to \infty} \delta_n / \gamma_n \leq 0$ or $\sum_{n=0}^{\infty} |\delta_n| < \infty$.

Then, $\lim_{n \to \infty} a_n = 0$.

Lemma 4 [13] *Let C be a nonempty closed convex subset of a real Hilbert space H and T be nonexpansive mapping on C with Fix(T) $\neq \emptyset$. If $\{x_n\}$ is a sequence in C which converges weakly to x and if $\{(I - T)x_n\}$ converges strongly to y, then $y = (I - T)x$. In particular, if $y = 0$, then $x \in Fix(T)$.*

3 Main Results

Let C be a nonempty closed and convex subset of a real Hilbert space H. We define the sequence $\{x_n\}$ by

$$\begin{cases} x_0 \in C, \\ x_{n+1} = (1 - \alpha_n)P_C(I - \xi A^*(I - P_Q)A)x_n + \alpha_n f(x_n), n \geq 1, \end{cases} \qquad (14)$$

where $\{\alpha_n\} \subset (0, 1)$ satisfies
(C1) $\alpha_n \to 0$;
(C2) $\sum_{n=0}^{\infty} \alpha_n = \infty$;
(C3) $\sum_{n=0}^{\infty} |\alpha_{n+1} - \alpha_n| < \infty$.

Theorem 1 *Suppose that the SFP is consistent and $0 < \xi < \frac{2}{\|A\|^2}$. Let $f : C \to C$ be a contraction with constant $\rho \in (0, 1)$. Let $\{x_n\}$ be a sequence defined as in (14). If the following assumptions are satisfied (C1)–(C3) then $x_n \to \tilde{x}$, where $\tilde{x}$ is the unique solution of the variational inequality*

$$\langle (I - f)\tilde{x}, x - \tilde{x} \rangle \geq 0, \quad x \in \Gamma. \qquad (15)$$

Proof First we show that the sequence $\{x_n\}$ is bounded. For our convenience, we take $T := P_C(I - \xi A^*(I - P_Q)A)$. Then, for any $p \in \Gamma$, we have $Tp = p$. Now, we observe that

$$\begin{aligned} \|x_{n+1} - p\| &\leq (1 - \alpha_n)\|Tx_n - p\| + \alpha_n\|f(x_n) - p\| \\ &\leq (1 - \alpha_n)\|Tx_n - p\| + \alpha_n(\|f(x_n) - f(p)\| + \|f(p) - p\|) \qquad (16) \\ &\leq (1 - \alpha_n)\|Tx_n - p\| + \alpha_n(\rho\|x_n - p\| + \|f(p) - p\|). \end{aligned}$$

Now, we note that the condition $0 < \xi < \frac{2}{\|A\|^2}$ implies that the operator $P_C(I - \xi A^*(I - P_Q)A)$ is averaged. Since $I - P_Q$ is firmly nonexpansive mappings and so is $\frac{1}{2}$-average, which is 1-ism. Also observe that $A^*(I - P_Q)A$ is $\frac{1}{\|A\|^2}$-ism so that $\xi A^*(I - P_Q)A$ is $\frac{1}{\xi\|A\|^2}$-ism. Further, from the fact that $I - \xi A^*(I - P_Q)A$ is

$\frac{\xi\|A\|^2}{2}$-averaged and P_C is $\frac{1}{2}$-averaged, we may obtain that $P_C(I - \xi A^*(I - P_Q)A)$ is $\mathscr{U}$-averaged, where

$$\mathscr{U} = \frac{1}{2} + \frac{\xi\|A\|^2}{2} - \frac{1}{2} \cdot \frac{\xi\|A\|^2}{2} = \frac{2 + \xi\|A\|^2}{4}. \tag{17}$$

This implies that $T = \mathscr{U}I + (1 - \mathscr{U})S$, where $\mathscr{U} = \frac{2+\xi\|A\|^2}{4} \in (0,1)$ for some nonexpansive mappings S. Note that T is also nonexpansive mappings. Hence, we have

$$\|Tx_n - p\| = \|Tx_n - Tp\| \le \|x_n - p\|. \tag{18}$$

From the inequalities (16) and (18), we have

$$\begin{aligned}
\|x_{n+1} - p\| &\le (1 - \alpha_n)\|x_n - p\| + \alpha_n(\rho\|x_n - p\| + \|f(p) - p\|) \\
&= (1 - \alpha_n + \alpha_n\rho)\|x_n - p\| + \alpha_n\|f(p) - p\| \\
&\le (1 - (1 - \rho)\alpha_n)\|x_n - p\| + \alpha_n\|f(p) - p\| \\
&\le \max\{\|x_n - p\|, \frac{1}{1 - \rho}\|f(p) - p\|\}.
\end{aligned}$$

By induction

$$\|x_n - p\| \le \max\{\|x_0 - p\|, \frac{1}{1 - \rho}\|f(p) - p\|\}, n \ge 0, \tag{19}$$

and $\{x_n\}$ is bounded, so are $\{Tx_n\}$ and $\{f(x_n)\}$.

Next, we claim that

$$\|x_{n+1} - x_n\| \to 0. \tag{20}$$

Indeed we have (for some appropriate constant $M > 0$)

$$\begin{aligned}
\|x_{n+1} - x_n\| &= \|(1 - \alpha_n)Tx_n + \alpha_n f(x_n) - ((1 - \alpha_{n-1})Tx_{n-1} + \alpha_{n-1}f(x_{n-1}))\| \\
&= \|(1 - \alpha_n)(Tx_n - Tx_{n-1}) + (\alpha_n - \alpha_{n-1})(f(x_{n-1}) - Tx_{n-1}) \\
&\quad + \alpha_n(f(x_n) - f(x_{n-1}))\| \\
&\le (1 - \alpha_n)\|x_n - x_{n-1}\| + |\alpha_n - \alpha_{n-1}|M + \rho\alpha_n\|x_n - x_{n-1}\| \\
&= (1 - (1 - \rho)\alpha_n)\|x_n - x_{n-1}\| + |\alpha_n - \alpha_{n-1}|M.
\end{aligned}$$

By Lemma 3, we have $\|x_{n+1} - x_n\| \to 0$.

We now show that

$$\|x_n - Tx_n\| \to 0. \tag{21}$$

Indeed from (20) and (C1), we get

$$\|x_n - Tx_n\| \le \|x_n - x_{n+1}\| + \|x_{n+1} - Tx_n\| \tag{22}$$

$$= \|x_n - x_{n+1}\| + \alpha_n \|Tx_n - f(x_n)\| \to 0. \tag{23}$$

Next, we will show that

$$\limsup_{n\to\infty} \langle \tilde{x} - x_n, \tilde{x} - f(\tilde{x}) \rangle \le 0. \tag{24}$$

Indeed take a subsequence $\{x_{n_k}\}$ of $\{x_n\}$ such that

$$\limsup_{n\to\infty} \langle \tilde{x} - x_n, \tilde{x} - f(\tilde{x}) \rangle = \limsup_{k\to\infty} \langle \tilde{x} - x_{n_k}, \tilde{x} - f(\tilde{x}) \rangle. \tag{25}$$

We may assume that $\{x_{n_k}\} \rightharpoonup \bar{x}$. It follows from Lemma 4 and (21) that is $\bar{x} \in Fix(T) = \Gamma$. Hence from (15), we obtain

$$\limsup_{n\to\infty} \langle \tilde{x} - x_n, \tilde{x} - f(\tilde{x}) \rangle = \langle \tilde{x} - \bar{x}, \tilde{x} - f(\tilde{x}) \rangle \le 0. \tag{26}$$

Finally, we will show that $x_n \to \tilde{x}$. As a matter of fact, we get

$$\begin{aligned}
\|x_{n+1} - \tilde{x}\|^2 &= \|(1 - \alpha_n)(Tx_n - \tilde{x}) + \alpha_n(f(x_n) - \tilde{x})\|^2 \\
&= (1 - \alpha_n)^2 \|Tx_n - \tilde{x}\|^2 + \alpha_n^2 \|f(x_n) - \tilde{x}\|^2 + 2\alpha_n(1 - \alpha_n)\langle Tx_n - \tilde{x}, f(x_n) - \tilde{x} \rangle \\
&\le (1 - 2\alpha_n + \alpha_n^2)\|x_n - \tilde{x}\|^2 + \alpha_n^2 \|f(x_n) - \tilde{x}\|^2 + 2\alpha_n(1 - \alpha_n)\langle Tx_n - \tilde{x}, f(x_n) - f(\tilde{x}) \rangle \\
&\quad + 2\alpha_n(1 - \alpha_n)\langle Tx_n - \tilde{x}, f(\tilde{x}) - \tilde{x} \rangle \\
&\le (1 - 2\alpha_n + \alpha_n^2)\|x_n - \tilde{x}\|^2 + \alpha_n^2 \|f(x_n) - \tilde{x}\|^2 + 2\rho\alpha_n(1 - \alpha_n)\|x_n - \tilde{x}\|^2 \\
&\quad + 2\alpha_n(1 - \alpha_n)\langle Tx_n - \tilde{x}, f(x_n) - f(\tilde{x}) \rangle \\
&= (1 - 2\alpha_n + \alpha_n^2 + 2\rho\alpha_n(1 - \alpha_n))\|x_n - \tilde{x}\|^2 \\
&\quad + \alpha_n[2(1 - \alpha_n)\langle Tx_n - \tilde{x}, f(\tilde{x}) - \tilde{x} \rangle) + \alpha_n \|f(x_n) - f(\tilde{x})\|^2] \\
&= (1 - \gamma_n)\|x_n - \tilde{x}\|^2 + \gamma_n \delta_n,
\end{aligned}$$

where

$$\gamma_n = \alpha_n(2 - \alpha_n - 2\rho(1 - \alpha_n)),$$

$$\delta_n = \frac{2(1 - \alpha_n)\langle Tx_n - \tilde{x}, f(\tilde{x}) - \tilde{x} \rangle + \alpha_n \|f(x_n) - \tilde{x}\|^2}{2 - \alpha_n - 2\rho(1 - \alpha_n)}.$$

We can see that $\gamma_n \to 0$, $\sum_{n=1}^{\infty} \gamma_n = \infty$ and $\limsup_{n\to\infty} \delta_n \le 0$ by (24). By Lemma 3, we conclude that $x_n \to \tilde{x}$. $\qquad\square$

Letting $f(x_n) \equiv u$ of iterative scheme (14) in Theorem 1, then we obtain the followings corollary;

Corollary 5 [14] *For any $u, x_0 \in C$, we define the sequence $\{x_n\}$ by*

$$x_{n+1} = (1 - \alpha_n)P_C(I - \xi A^*(I - P_Q)A)x_n + \alpha_n u, n \geq 1, \tag{27}$$

where $\{\alpha_n\}$ is a sequence in $(0, 1)$. Suppose that the SFP is consistent and $0 < \xi < \frac{2}{\|A\|^2}$. Let $\{x_n\}$ be defined as in (27). If the following assumptions are satisfied (C1)–(C3)

(C1) $\alpha_n \to 0$;
(C2) $\sum_{n=0}^{\infty} \alpha_n = \infty$;
(C3) $\sum_{n=0}^{\infty} |\alpha_{n+1} - \alpha_n| < \infty$.

Then $\{x_n\}$ converges to a solution of the SFP (2).

Remark 6 Theorem 1 *and* Corollary 5 extend and improve the result of Xu [9] from weak to strong convergence theorems by using the viscosity approximation method.

Acknowledgment The first author was supported by the Thailand Research Fund through the Royal Golden Jubilee Ph.D. Program (Grant No. PHD/0033/2554) and the King Mongkut's University of Technology Thonburi. Moreover, this study was supported by the Higher Education Research Promotion and National Research University Project of Thailand, Office of the Higher Education Commission.

References

1. B. Halpern, Fixed points of nonexpanding maps. Bull. Amer. Math. Soc. **73**, 957–961 (1967)
2. P.L. Lions, Approximation de points fixes de contractions. C. R. Acad. Sci. Ser. A-B Paris **284**, 1357–1359 (1977)
3. A. Moudafi, Viscosity approximations methods for fixed point problems. J. Math. Anal. Appl. **241**, 46–55 (2000)
4. R. Wittman, Approximation of fixed points of nonexpansive mappings. Arch. Math. **58**, 486–491 (1992)
5. H.K. Xu, Viscosity approximations methods for nonexpansive mappings. J. Math. Anal. Appl. **298**, 279–291 (2004)
6. Y. Censor, T. Elving, A multiprojection algorithm using Bregman projections in product space. Numer. Algorithm **8**(2–4), 221–239 (1994)
7. C. Byrne, Iterative oblique projection onto convex subsets and the split feasibility problem. Inverse Prob. **18**(2), 441–453 (2002)
8. Q. Yang, The relaxed CQ algorithm solving the split feasibility problem. Inverse Prob. **20**(4), 1261–1266 (2004)
9. H.K. Xu, Iterative methods for the split feasibility problem in infinite-dimensional Hilbert spaces. Inverse Prob. 26 (2010) 105018 (p. 17)
10. J. Deepho, P. Kumam, The hybrid extragradient method for the split feasibility and fixed point problems, in *Proceedings of The International Multiconference of Engineers and Computer Scientists 2014*, vol I, IMECS, 12–14 Mar 2014, Hong Kong, pp. 558–563

11. H.K. Xu, Averaged mappings and the gardient-projection algorithm. J. Optim. Theory Appl. **150**(2), 360–378 (2011)
12. H.K. Xu, Viscosity approximation methods for nonexpansive mapping. J. Math. Anal. Appl. **298**(1), 279–291 (2004)
13. F.E. Browder, Fixed point theorems for noncompact mappings in Hilbert spaces. Proc. Natl. Acad. Sci. USA **53**(6), 1272–1276 (1965)
14. J. Deepho and P. Kumam, A modified Halpern's iterative scheme for solving split feasibility problems. Abstract Appl. Anal. 2012(876069), 8 (2012)

Analytic Method for Solving Heat and Heat-Like Equations with Classical and Non Local Boundary Conditions

Ahmed Cheniguel

Abstract In this paper, heat and heat-like equations with classical and non local boundary conditions are presented and a homotopy perturbation method (HPM) is utilized for solving the problems. The obtained results as compared with previous works are highly accurate. Also HPM provides continuous solutions in contrast to traditional methods, like finite difference method, which only provides discrete approximations. It is found that this method is a powerful mathematical tool and can be applied to a large class of linear and non linear problems in different fields of science and technology.

Keywords Diffusion equation · Exact solution · Heat-like equation · Homotopy perturbation method · Initial boundary value problems · Non local boundary conditions · Partial differential equations

1 Introduction

Recently, new analytical methods have gained the interest of researchers for finding approximate solutions to partial differential equations. This interest was driven by the needs from applications both in industry and sciences. Theory and numerical methods for solving initial boundary value problems were investigated by many researchers see for instance [1–9] and the reference therein. In the last decade, there has been a growing interest in the new analytical techniques for linear and non linear initial boundary value problems. The widely applied techniques are perturbation methods. He [10] has proposed a new perturbation technique coupled with the homotopy technique, which is called the homotopy perturbation method (HPM) for solving non linear problems. In contrast to the traditional perturbation methods,

A. Cheniguel (✉)
Department of Mathematics and Computer Science, Faculty of Sciences,
Kasdi Merbah University Ouargla, Ouargla, Algeria
e-mail: cheniguelahmed@yahoo.fr

© Springer Science+Business Media Dordrecht 2015
G.-C. Yang et al. (eds.), *Transactions on Engineering Technologies*,
DOI 10.1007/978-94-017-9588-3_7

a homotopy is constructed with an embedding parameter $p \in [0, 1]$, which is considered as a small parameter. Homotopy perturbation method has gained reputation as being a powerful tool for solving linear or non linear partial differential equations. He [11], applied HPM to solve initial boundary value problems which is governed by the non linear ordinary (partial) differential equations, the method has been shown to effectively, easily and accurately solve a large class of linear and non linear problems with components converging rapidly to exact solutions. Thus the main goal of this work is to apply the homotopy perturbation method (HPM) for solving heat and heat-like equations subject to different type of boundary conditions. The obtained results are more accurate than those obtained recently by Damrongsak et al. [12]. In this paper we consider a one-dimensional heat equation, one-dimensional and three-dimensional heat-like equations. The implementation of the method has shown reliable results in that few terms are needed to obtain either exact solution or to find an approximate solution of a reasonable degree of accuracy in real physical models. Numerical examples are presented to illustrate the efficiency of the homotopy perturbation method, the obtained results are all in good agreement with exact ones.

2 The Linear Heat Equation with Dirichlet and Neumann Conditions

2.1 Problem Definition

We consider the diffusion equation given by

$$\frac{\partial u}{\partial t} = \alpha \frac{\partial^2 u}{\partial x^2}, \quad 0 < x < l, \quad t > 0 \tag{1}$$

subject to the Initial condition:

$$u(x, 0) = u_0(x), \quad 0 < x < a \tag{2}$$

and the boundary conditions:

$$u(0, t) = g_0(t), \quad t > 0 \tag{3}$$

$$u(1, t) = g_1(t), \quad t > 0 \tag{4}$$

$$u_x(0, t) = g_2(t), \quad t > 0 \tag{5}$$

$$u_x(1, t) = g_3(t), \quad t > 0 \tag{6}$$

where the diffusion coefficient α is positive, $u(x,t)$ represents the the temperature at point (x,t) and $f(x,t)$, $g_0(t)$, $g_1(t)$, $g_2(t)$, $g_3(t)$ are sufficiently smooth known functions.

2.2 Analysis of Homotopy Perturbation Method

To illustrate the basic ideas, let X, *and* Y be two topological spaces. If f *and* g are continuous maps of the spaces X *into* Y, it is said that f is homotopic to g, if there is continuous map $F : X \times [0,1] \rightarrow Y$ such that $F(x,0) = f(x)$ *and* $F(x,1) = g(x)$ *for each xεX*, then the map is called homotopy between f *and* g. We consider the following nonlinear partial differential equation:

$$A(u) - f(r) = 0, \ r \in \Omega \qquad (7)$$

subject to the boundary conditions

$$B(u, \partial u / \partial \eta) = 0, \ r \in \Gamma \qquad (8)$$

where A is a general differential operator, f is a known analytic function, Γ is the boundary of Ω and $\partial / \partial \eta$ denotes directional derivative in outward normal direction to Ω. The operator A, generally divided into two *parts*, L and N, where L is linear while N is nonlinear. Using $A = L + N$, Eq. (7) can be rewritten as follows:

$$L(v) + N(v) - f(r) = 0 \qquad (9)$$

by the homotopy technique, we construct a homotopy defined as

$$H(v,p) : \Omega \times [0,1] \rightarrow R \qquad (10)$$

which satisfies:

$$H(v,p) = (1-p)(L(v) - L(u_0)) + p(A(v) - f(r)),$$
$$p \in [0,1], r \in \Omega. \qquad (11)$$

Or

$$H(v,p) = L(v) - L(u_0) + pL(u_0) + p(N(v) - f(r)) = 0,$$
$$p \in [0,1], \ r \in \Omega. \qquad (12)$$

Where p is an embedding parameter, u_0 is an initial approximation of Eq. (7), which satisfies the boundary conditions. It follows from Eq. (12) that:

$$H(v,0) = L(v) - L(u_0) = 0 \tag{13}$$

$$H(v,1) = A(v) - f(r) = 0 \tag{14}$$

The changing process of p from 0 to 1 monotonically is a trivial problem. $H(v,0) = L(v) - L(u_0) = 0$ is continuously transformed to the original problem

$$H(v,1) = A(v) - f(r) = 0 \tag{15}$$

In topology, this process is known as continuous deformation.

$L(v) - L(u_0)$ and $A(v) - f(r)$ are called homotopic. We use the embedding parameter p as a small parameter, and assume that the solution of Eq. (12) can be written as power series of p:

$$v = p^0 v_0 + p^1 v_1 + p^2 v_2 + p^3 v_3 + \cdots + p^n v_n + \cdots \tag{16}$$

Setting $p = 1$ we obtain the approximate solution of Eq. (7) as:

$$u = \lim_{p \to 1} v = v_0 + v_1 + v_2 + v_3 + \cdots + v_n + \cdots \tag{17}$$

The series of Eq. (17) is convergent for most of the cases. But the rate of the convergence depends on the linear operator $N(v)$. He [13] has suggested that:

1. The second derivative of $N(v)$ with respect to v should be small because the parameter may be relatively large i.e. $p = 1$.
2. The norm of $L^{-1}(\partial N / \partial v)$ must be smaller than one so that the series converges.

2.3 Solution Procedure

The solution is considered in the form below:

$$v = p^0 v_0 + p^1 v_0 + p^2 v_2 + \cdots \tag{18}$$

Setting $p = 1$, we obtain the approximate solution of Eq. (1) as follows:

$$u = \lim_{p \to 1} v = v_0 + v_1 + v_2 + \cdots \tag{19}$$

Substituting Eq. (18) into Eq. (12) and comparing the coefficient of like powers of p, we have

$$p^0 : (v_0)_t - (u_0)_t = 0, \quad v_0 = u_0 = u(x,0)$$
$$p^1 : (v_1)_t - (v_0)_{xx} - s(x,t) = 0$$
$$v_1 = \int_0^t ((v_0)_{xx} - s(x,t))dt, \quad v_1(x,0) = 0$$
$$p^2 : (v_2)_t - (v_1)_{xx} = 0 \Rightarrow v_2 = \int_0^t (v_1)_{xx}dt, \quad v_2(x,0) = 0 \tag{20}$$
$$p^3 : (v_3)_t - (v_2)_{xx} = 0 \Rightarrow v_3 = \int_0^t (v_2)_{xx}dt, \quad v_3(x,0) = 0$$
$$\vdots$$

Hence the approximate or exact solution of problem (1) is obtained as:

$$u(x,t) = v_0 + v_1 + v_2 + v_3 + \cdots \tag{21}$$

3 The One Dimensional Heat-Like Equation

3.1 Problem Definition

We consider the problem in two cases one-dimensional heat-like equation given by:

$$u_t = a(x) + b(x)u_{xx}, \quad 0 < x < 1, \tag{22}$$

subject to the initial condition

$$u(x,0) = x^2 \tag{23}$$

and the boundary conditions

$$u(0,t) = \int_0^1 u(x,t)dx + c_1 = c(t) \tag{24}$$

$$u(1,t) = \int_0^1 u(x,t)dt + c_2 = d(t) \tag{25}$$

3.2 Solution Procedure

Writing the approximate solution in the series form as the following:

$$v = p^0 v_0 + p^1 v_1 + p^2 v_2 + \cdots + p^n v_n + \cdots \tag{26}$$

Substituting Eq. (26) into Eq. (22) and equating the coefficients of the same powers of p we get the system of equations as follows:

$$
\begin{aligned}
&v_{0t} - u_{0t} = 0 \Rightarrow v_0 = u_0 \\
&v_{1t} - a(x) - b(x)v_{0xx} = 0, \quad v_1(x,0) = 0 \\
&v_1 = \int_0^t (a(x) - b(x)v_{0xx})dt \\
&v_{2t} - v_{1xx} = 0, \quad v_2(x,0) = 0 \\
&v_2 = \int_0^t v_{1xx}dt \\
&\vdots
\end{aligned}
\tag{27}
$$

and so on, we obtain the approximate solution in a series form as below:

$$u(x,t) = \sum_{i=0}^{\infty} v_i.$$

4 Three-Dimensional Heat-Like Equation

4.1 Problem Definition

Consider the three-dimensional heat-like equation as

$$u_t = p(x)q(y)r(z) + \alpha(x)u_{xx} + \beta(y)u_{yy} + \gamma(z)u_{zz}, \quad 0 < x, y, z < 1, \ 0 < t \le T \tag{28}$$

subject to the initial and boundary conditions

$$u(x,y,z,0) = f(x,y,z) \tag{29}$$

$$u(0, y, z, t) = \int_0^1 \int_0^1 \int_0^1 u(x, y, z, t)dxdydz + g_1 = k_1(t)$$

$$u(1, y, z, t) = \int_0^1 \int_0^1 \int_0^1 u(x, y, z, t)dxdydz + g_2 = k_2(t)$$

$$u(x, 0, z, t) = \int_0^1 \int_0^1 \int_0^1 u(x, y, z, t)dxdydz + g_3 = k_3(t)$$

$$u(x, 1, z, t) = \int_0^1 \int_0^1 \int_0^1 u(x, y, z, t)dxdydz + g_4 = k_4(t) \tag{30}$$

$$u(x, y, 0, t) = \int_0^1 \int_0^1 \int_0^1 u(x, y, z, t)dxdydz + g_5 = k_5(t)$$

$$u(x, y, 1, t) = \int_0^1 \int_0^1 \int_0^1 u(x, y, z, t)dxdydz + g_6 = k_6$$

4.2 Solution Procedure

We just consider three-dimensional equation which includes two other cases. Substituting Eq. (18) into Eq. (28) and equating the terms with identical powers of p, we have

$$p^0:(v_0)_t - (u_0)_t = 0 \Rightarrow v_0 = u(x, 0) \tag{31}$$

$$p^1:(v_1)_t - p(x)q(y)r(z) - ((\alpha(x)v_0)_{xx} + (\beta(y)v_0)_{yy} + (\gamma(z)v_0)_{zz}) = 0$$

$$v_1 = \int_0^t (p(x)q(y)r(z) + ((\alpha(x)v_0)_{xx} + (\beta(y)v_0)_{yy} + (\gamma(z)v_0)_{zz}))dt \tag{32}$$

$$p^2:(v_2)_t - (p(x)q(y)r(z) - (\alpha(x)v_1)_{xx} - (\beta(y)v_1)_{yy} - (\gamma(z)v_1)_{zz}) = 0$$

$$v_2 = \int_0^t (p(x)q(y)r(z) + ((\alpha(x)v_1)_{xx} + (\beta(y)v_1)_{yy} + (\gamma(z)v_1)_{zz}))dt$$

$$p^3:(v_3)_t - p(x)q(y)r(z) - (\alpha(x)v_2)_{xx} - (\beta(y)v_2)_{yy} - (\gamma(z)v_2)_{zz} = 0$$

$$v_3 = \int_0^t (p(x)q(y)r(z) + ((\alpha(x)v_2)_{xx} + (\beta(y)v_2)_{yy} + (\gamma(z)v_2)_{zz}))dt$$

.

.

So we can calculate the terms of $\sum_{k=0}^{\infty} v_k$, term by term and the series solution thus entirely determined. However, in many cases the exact solution in a closed form may be obtained. For numerical purposes, we can use the approximation

$$u(x, y, z, t) = \lim_{m \to \infty} \emptyset_m$$

where

$$\emptyset_m = \sum_{k=0}^{m-1} v_k \tag{33}$$

It is worth to mention that the errors are getting smaller with the growing number of terms in the sum (33).

5 Numerical Examples

5.1 Example 1: One Dimensional Homogeneous Heat Equation

We consider the one-dimensional diffusion equation:

$$\frac{\partial u}{\partial t} = \frac{\partial^2 u}{\partial x^2}, \quad 0 \le x \le 1, \ t > 0 \tag{34}$$

with the Initial condition:

$$u(x, 0) = \sin(\pi x). \tag{35}$$

and the boundary conditions

$$u(0, t) = 0, \quad u(1, t) = 0 \tag{36}$$

To solve (34) with initial condition (35), according to the homotopy perturbation technique, we construct the following homotopy:

$$H(v, p) = (1 - p)((v_0)_t - (u_0)_t) + p(v_t - v_{xx}) = 0 \tag{37}$$

Substituting of Eq. (16) into Eq. (37) and then equating the terms with like powers of p, we get the following

$$(v_0)_t - (u_0)_t = 0, \quad v_0 = u(x,0) = \sin(\pi x)$$
$$(v_1)_t - (v_0)_{xx} = 0, \quad v_1(x,0) = 0$$
$$(v_1)_t = -\pi^2 \sin(\pi x)$$
$$v_1 = \int_0^t (-\pi^2 \sin(\pi x)) dt = -\pi^2 \sin(\pi x) \times t$$
$$(v_2)_t - (v_1)_{xx} = 0, \quad v_2(x,0) = 0$$
$$v_2 = \int_0^t \pi^4 \sin(\pi x) \times t \, dt = \pi^4 \sin(\pi x) \times \frac{t^2}{2!} \qquad (38)$$
$$(v_3)_t - (v_2)_{xx} = 0, \quad v_3(x,0) = 0$$
$$v_3 = \int_0^t -\pi^6 \sin(\pi x) \times \frac{t^2}{2!} dt = -\pi^6 \sin(\pi x) \times \frac{t^3}{3!}$$
$$\vdots$$

and so on, we can calculate v_n as follows:

$$(v_n)_t - (v_{n-1})_t = 0, \quad v_n(x,0) = 0$$

$$v_n = \int_0^t (-1)^n \pi^{2n} \sin(\pi x) \times \frac{t^{n-1}}{(n-1)!} dt = (-1)^n \pi^{2n} \sin(\pi x) \times \frac{t^n}{n!}$$

Finally, we obtain the approximate solution as follows:

$$u = \lim_{p \to 1} v = v_0 + v_1 + v_2 + v_3 + \cdots + v_n + \cdots$$

And this leads to the following solution

$$u(x,t) = \sin(\pi x) e^{-\pi^2 t} \qquad (39)$$

Substituting Eq. (39) into Eq. (34), we conclude that the approximate solution coincides with the exact one.

5.2 Example 2

Consider the diffusion problem:

$$u_t = u_{xx}, \quad 0 \le x \le 1, \quad t > 0 \qquad (40)$$

subject to the Initial condition

$$u(x,0) = \cos(\pi x) \tag{41}$$

and the boundary conditions:

$$u_x(0,t) = 0, \quad u_x(1,t) = 0 \tag{42}$$

solving the Eq. (40) with the initial condition (41), yields:

$$(v_0)_t - (u_0)_t = 0, \quad v_0 = u_0 = \cos(\pi x) \tag{43}$$

$$(v_1)_t - (v_0)_{xx} = 0, \quad v_1(x,0) = 0$$
$$v_1 = -\pi^2 \cos(\pi x) \times t$$
$$(v_2)_t - (v_1)_{xx} = 0, \quad v_2(x,0) = 0$$
$$v_2 = -\pi^4 \cos(\pi x) \times t^2/2!$$

The next components v_k, $k \geq 3$ are calculated as the following:

$$(v_k)_t - (v_k)_{xx} = 0, \quad v_k(x,0) = 0$$
$$v_k = (-1)^k \pi^{2k} \cos(\pi x) \times t^k/k! \tag{44}$$

Combining all the terms in the above gives

$$u(x,t) = \cos(\pi x)(1 - \pi^2 t/1! + (\pi^2 t)^2/2! - (\pi^2 t)^3/3! + \cdots)$$

The series solution is:

$$u(x,t) = e^{-\pi^2 t} \cos(\pi x) \tag{45}$$

5.3 Example 3: One Dimensional Non Homogeneous Heat Equation

Consider the non homogeneous diffusion equation:

$$u_t = u_{xx} + (\pi^2 - 1)e^{-t} \times \cos(\pi x) + 4x - 2 \tag{46}$$

with the initial condition

$$u(x,0) = \cos(\pi x) + x^2 \tag{47}$$

and the boundary conditions

$$u(0,t) = e^{-t}, \quad u(1,t) = -e^t + 4t + 1 \tag{48}$$

according to HPM algorithm, we have

$$H(p,v) = (1-p)((v_0)_t - (u_0)_t) + p(v_t - v_{xx} - f) = 0 \tag{49}$$

where $f = (\pi^2 - 1)e^{-t} + 4x - 2$

by equating the terms with the identical powers of p, yields

$$(v_0)_t - (u_0)_t = 0, \quad v_0 = u_0 = \cos(\pi x) + x^2$$
$$(v_1)_t - (v_0)_{xx} - 4x + 2 - (\pi^2 - 1)e^{-t} = 0, \quad v_1(x,0) = 0$$
$$v_1 = 4xt + \cos(\pi x)(-\pi^2 t + (\pi^2 - 1)(1 - e^{-t}))$$
$$v_{2t} - v_{1xx} = 0, \quad v_{2t} = \cos(\pi x)(\pi^4 t - \pi^2(\pi^2 - 1)(1 - e^{-t}))$$
$$v_2 = \cos(\pi x)((\pi^4 - \pi^2)(1 - t/1! - e^{-t}) + (\pi^2 t)^2/2!))$$

continuing like-wise we get:

$$v_3 = \cos(\pi x)((\pi^6 - \pi^4)(1 - t/1! + t^2/2! - e^{-t}) - (\pi^2 t)^3/3!))$$
$$v_4 = \cos(\pi x)((\pi^8 - \pi^6)(1 - t/1! + t^2/2! - t^3/3! - e^{-t}) + (\pi^2 t)^4/4!)$$

and so on we then have

$$u_{5hpm} = x^2 + 4xt + \cos(\pi x)(\pi^8(1 - t/1! + t^2/2! - t^3/3! - e^{-t}) + e^{-t}) \tag{50}$$

From this result we deduce that the series solution converges to the exact one:

$$u(x,t) = x^2 + 4xt + \cos(\pi x)e^{-t}$$

5.4 Example 4

Once again, consider the non-homogeneous heat equation with non-homogeneous Neumann boundary conditions:

$$u_t = u_{xx} + (\pi^2/2)e^{(-\pi^2/2)t}\cos(\pi x) + x - 2, \quad 0 \le x, \le 1, t > 0 \tag{51}$$

$$u_x(0,t) = t, \quad u_x(1,t) = 2 + t$$

and the initial condition

$$u(x,0) = x^2 + \cos(\pi x) \tag{52}$$

The theoretical solution is:

$$u(x,t) = x^2 + xt + e^{(-\pi^2/2)t}\cos(\pi x)$$

now, applying the homotopy perturbation method we get:

$$p^0 : v_{0t} - u_{0t} = 0, \quad v_0 = u_0 = \cos(\pi x) + x^2$$

$$p^1 : v_{1t} - v_{0xx} - (\pi^2/2)e^{(-\pi^2/2)t}\cos(\pi x) - x + 2 = 0, \quad v_1(x,0) = 0$$

$$v_1 = xt + \cos(\pi x)(1 - \pi^2 t - e^{(-\pi^2/2)t})$$

$$p^2 : v_{2t} - v_{1xx} = 0, \quad v_2(x,0) = 0 \tag{53}$$

$$v_2 = \cos(\pi x)(2 - \pi^2 t + (\pi^2 t)^2/2! - 2e^{(-\pi^2/2)t})$$

$$p^3 : v_{3t} - v_{2xx} = 0, \quad v_3(x,0) = 0$$

$$v_3 = \cos(\pi x)(4 - 2\pi^2 t + (\pi^2 t)^2/2! - (\pi^2 t)^3/3! - 4e^{(-\pi^2/2)t})$$

$$p^4 : v_{4t} - v_{3t} = 0, \quad v_4(x,0) = 0$$

$$v_4 = \cos(\pi x)(8 - 4(\pi^2 t) + (\pi^2 t)^2 - (\pi^2 t)^3/3! + (\pi^2 t)^4/4! - 8e^{(-\pi^2/2)t})$$

Continuing in this way, we obtain

$$u(x,t) = \lim_{p \to 1} v = v_0 + v_1 + v_2 + v_3 + v_4 + \cdots$$

or

$$u(x,t) = x^2 + xt + \cos(\pi x)e^{(-\pi^2/2)t}$$
$$+ 15\left[\left\{1 - (\pi^2/2)t/1! + ((\pi^2/2)\,t)^2/2! - ((\pi^2/2)t)^3/3!\right.\right.$$
$$\left.\left. + ((\pi^2/2)t)^4/4! - . + \cdots)\right\} - e^{(-\pi^2/2)t}\right]$$

and this leads to the following solution

$$u(x,t) = x^2 + xt + \cos(\pi x)e^{(-\pi^2/2)t} \tag{54}$$

this solution coincides with the exact one.

5.5 Example 5: One Dimensional Non Homogeneous Heat-Like Equation

Consider the problem

$$u_t = x^5 + 1/20(x^2 u_{xx}), \quad 0<x<1, \ 0<t\le \tag{55}$$

subject to the initial condition

$$u(x,0) = 0 \tag{56}$$

and the boundary conditions

$$u(0,t) = \int_0^1 u(x,t)dx + g_1 = 1/6(e^t - 1), \quad g_1 = 0$$
$$\tag{57}$$
$$u(1,t) = \int_0^1 u(x,t)dx + g_2 = (1/6)e^t, \quad g_2 = 1/6$$

After substitution of Eq. (18) into Eq. (55) and identifying the terms of the same powers of p, we obtain the system of equations:

$$p^0: v_{0t} - u_{0t} = 0, \quad v_0 = u_0 = 0$$
$$p^1: v_{1t} - (1/20)x^2 v_{0xx} = 0, \quad v_1(x,0) = 0$$
$$v_1 = x^5 t$$
$$p^2: v_{2t} - (1/20)x^2 v_{1t} = 0, \quad v_2(x,0) = 0$$
$$v_2 = x^5 t^2/2!$$
$$p^3: v_{3t} - (1/20)x^2 v_{2t} = 0, \quad v_3(x,0) = 0$$
$$v_3 = x^5 t^3/3!$$
$$\vdots$$

$$p^n: v_{nt} - (1/20)x^2 v_{(n-1)t} = 0, \quad v_n(x,0) = 0$$
$$v_n = x^5 t^n/n!$$

Hence the series solution is given by:

$$u(x,t) = v_0 + v_1 + v_2 + v_3 + \cdots + v_n + \cdots$$

or

$$u(x,t) = x^5\left(1 + t/1! + t^2/2! + t^3/3! + \cdots + t^n/n! + \cdots\right) - x^5$$

and in a closed form:

$$u(x,t) = x^5(e^t - 1) \tag{58}$$

5.6 Example 6: Three Dimensional Non Homogeneous Heat-Like Equation

Let us consider the problem

$$u_t = x^5 y^5 z^5 + (1/60)(x^2 u_{xx} + y^2 u_{yy} + z^2 u_{zz}), \quad 0 < x, y, z < 1, 0 < t \leq T \tag{59}$$

with the following initial condition:

$$u(x,y,z,0) = 0 \tag{60}$$

and the boundary conditions

$$u(0,y,z,t) = \int_0^1\int_0^1\int_0^1 u(x,y,z,t)dxdydz + g_1 = 1/216(e^t - 1), \quad g_1 = 0$$

$$u(1,y,z,t) = \int_0^1\int_0^1\int_0^1 u(x,y,z,t)dxdydz + g_2 = 1/216(e^t - 1) + (1/2)t, \quad g_2 = (1/2)t$$

$$u(x,0,z,t) = \int_0^1\int_0^1\int_0^1 u(x,y,z,t)dxdydz + g_3 = 1/216(e^t), \quad g_3 = 1/216$$

$$u(x,1,z,t) = \int_0^1\int_0^1\int_0^1 u(x,y,z,t)dxdydz + g_4 = (1/216)(e^t + 3), \quad g_4 = (4/216)$$

$$\tag{61}$$

$$u(x,y,0,t) = \int_0^1\int_0^1\int_0^1 u(x,y,z,t)dxdydz + g_5 = (1/216)(e^t + 4), \quad g_5 = 5/216$$

$$u(x,y,1,t) = \int_0^1\int_0^1\int_0^1 u(x,y,z,t)dxdydz + g_6 = 1/216(e^t + 1), \quad g_6 = 1/108$$

According to Eqs. (18) and (59) the following terms are calculated successively:

$$p^0 : v_{0t} - u_{0t} = 0, \quad v_0 = u_0 = 0$$
$$p^1 : v_{1t} - x^5 y^5 z^5 - (1/60)(x^2 v_{0xx} + y^2 v_{0yy} + z^2 v_{0zz}) = 0, \quad v_1(x, y, z, 0) = 0$$
$$v_1 = x^5 y^5 z^5 (t/1!)$$
$$p^2 : v_{2t} - (1/60)(x^2 v_{1xx} + y^2 v_{1yy} + z^2 v_{1zz}) = 0, \quad v_2(x, y, z, 0) = 0$$
$$v_2 = x^5 y^5 z^5 (t^2/2!)$$

$$\vdots$$

$$p^n : v_{nt} - (1/60)(x^2 v_{(n-1)xx} + y^2 v_{(n-1)yy} + z^2 v_{(n-1)zz}) = 0$$
$$v_n = x^5 y^5 z^5 (t^n/n!)$$

$$(62)$$

Hence, the approximate solution is given by:

$$u(x, y, z, t) = v_0 + v_1 + v_2 + \cdots + v_n + \cdots$$

Or

$$u(x, y, z, t) = x^5 y^5 z^5 (1 + t/1! + (t^2/2!) + \cdots (t^n/n!) + \cdots) - x^5 y^5 z^5$$

The solution in the closed form is given as

$$u(x, y, z, t) = x^5 y^5 z^5 (e^t - 1)$$

This result is in good agreement with the exact one (Tables 1, 2 and 3).

Table 1 Example 1
$h_x = 0.1, \ h_t = 0.004,$
3-iterates

x_i	u_{ex}	u_{hpm}	$\left| u_{ex} - u_{hpm} \right|$
0.0	0.0	0.0	0.0
0.1	0.2971	0.2971	0.0
0.2	0.5650	0.5650	0.0
0.3	0.7777	0.7777	0.0
0.4	0.9142	0.9142	0.0
0.5	0.9613	0.9613	0.0
0.6	0.9142	0.9142	0.0
0.7	0.7777	0.7777	0.0
0.8	0.5650	0.5650	0.0
0.9	0.2971	0.2971	0.0
1.0	0.0	0.0	0.0

Table 2 Example 2
$h_x = 0.1$, $h_t = 0.004$,
3-Iterates

x_i	u_{ex}	u_{hpm}	$\left| u_{ex} - u_{hpm} \right|$
0.0	0.9613	0.9613	0.0
0.1	0.9142	0.9142	0.0
0.2	0.7777	0.7777	0.0
0.3	0.5650	0.5650	0.0
0.4	0.2971	0.2971	0.0
0.5	0.0	0.0	0.0
0.6	−0.2971	−0.2971	0.0
0.7	−0.5650	−0.5650	0.0
0.8	−0.7777	−0.7777	0.0
0.9	−0.9142	−0.9142	0.0
1.0	−0.9613	−0.9613	0.0

Table 3 Example 3
$h_x = 0.1$, $h_t = 0.004$,
2-Iterates

x_i	u_{ex}	u_{hpm}	$\left| u_{ex} - u_{hpm} \right|$
0.0	0.9960	0.9960	0.0
0.1	0.9589	0.9589	0.0
0.2	0.8490	0.8490	0.0
0.3	0.6802	0.6802	0.0
0.4	0.4742	0.4742	0.0
0.5	0.2580	0.2580	0.0
0.6	0.0618	0.0618	0.0
0.7	−0.0842	−0.0842	0.0
0.8	−0.1530	−0.1530	0.0
0.9	−0.1229	−0.1229	0.0
1.0	0.0200	0.0200	0.0

6 Conclusion

The main concern of this work has been to construct an approximate solution to heat and heat-like equations with different types of boundary conditions using homotopy perturbation method (HPM). Our approach differs from existing traditional methods like, finite differences, finite elements, spectral method, … etc., in that we find the solution in a closed form without, linearization, discretization, transformation or restrictive assumptions. The problems solved using (HPM) gave satisfactory results in comparison to those recently obtained other researchers (Figs. 1, 2 and 3).

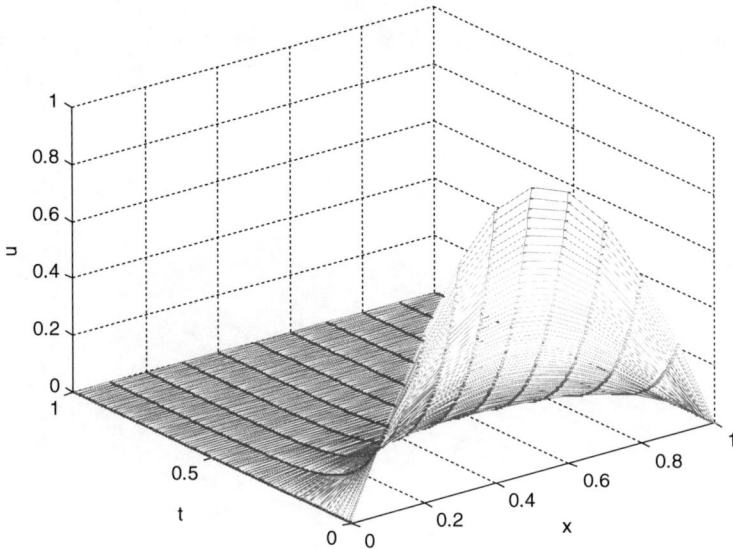

Fig. 1 *Example 1* Variation of the approximate solution for different values of x and t

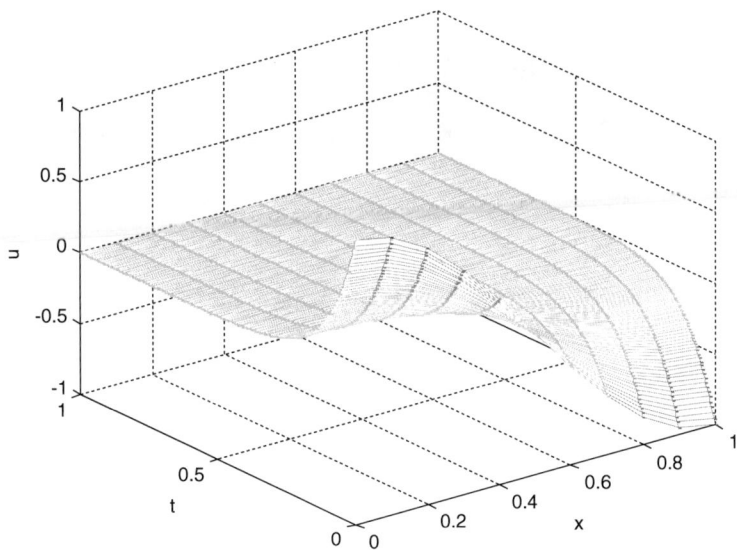

Fig. 2 *Example 2* Variation of the approximate solution for different values of x and t

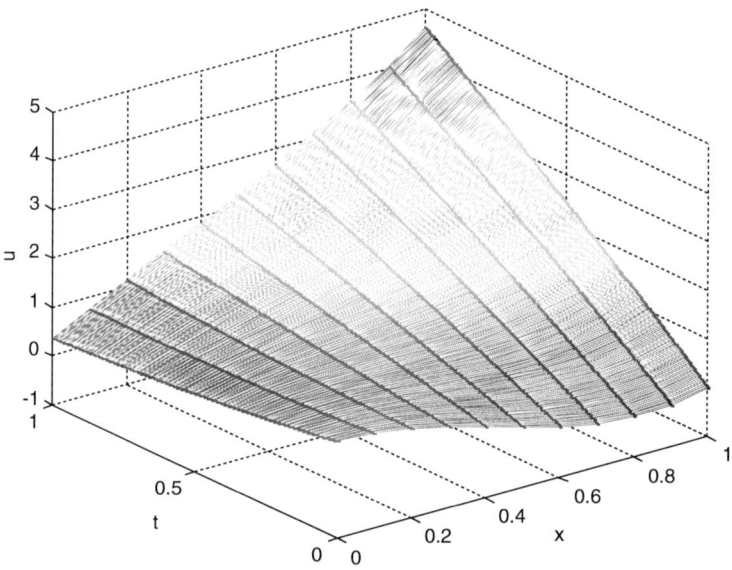

Fig. 3 *Example 3* Variation of the approximate solution for different values of x, y and z when $t = 0.004$

References

1. A. Cheniguel, Numerical method for the heat equation with Dirichlet and Neumann boundary conditions, in *Proceedings of the International Multi-Conference and Computer Scientists 2014*, vol I, 12–14 Mar 2014, (IMECS, Hong Kong, 2014), pp. 535–539
2. A. Cheniguel, Numerical method for solving wave equation with non local boundary conditions. Lect. Notes Eng. Comput. Sci. **2203**(1), 1190–1193 (2013)
3. A. Cheniguel, M. Reghioua, On the numerical solution of three-dimensional diffusion equation with an integral condition, in *Proceedings of the World Congress on Engineering and Computer Science 2013*, vol II, 21–23 Oct 2013 (WCECS, San Francisco, 2013), pp. 1017–1021
4. A. Cheniguel, Numerical method for solving heat equation with derivative boundary conditions. Lect. Notes Eng. Comput. Sci. **2194**(1), 983–985 (2011)
5. A. Cheniguel, A. Ayadi, Solving non homogeneous heat equation by the adomian decomposition method. Int. J. Numer. Methods Appl. **4**(2), 89–97 (2010)
6. A. Cheniguel, A. Ayadi, Numerical method for non local problem. Sci. Technol. **A-N30**, 15–18 (2009)
7. S. Momani, Analytical approximate solution for fractional heat-like and wave-like equations with variable coefficients using the decomposition method. Appl. Math. Comput. **165**(2), 459–472 (2005)
8. G. Ekolin, Finite difference methods for a non local boundary value problem for the heat equation. BIT **31**, 245–261 (1991)
9. G. Adomian, A review of the decomposition method in applied mathematics. J. Math. Anal. Appl. **135**, 501–544 (1988)
10. J.H. He, A coupling method of homotopy technique for non linear problems. Int. J. Non Linear Mech. **35**, 37–43 (2000)

11. J.H. He, Homotopy perturbation method for solving boundary value problems. Phys. Lett. A **350**, 87–88 (2006)
12. Damrongsak et al, Deferred correction technique to construct high-order schemes for the heat equation with Dirichlet and Neumann boundary conditions. Eng. Lett. **21**(2), 61–67 (2013)
13. J.H. He, Homotopy perturbation technique. Comput. Methods Appl. Mech. Eng. **178**(3/4), 257–262 (1999)

Iterative Algorithms for the Linear Matrix Equation $X + A^*XA = I$ and Some Properties

Sana'a A. Zarea, Salah M. El-Sayed and Amal A.S. Al-Marshdy

Abstract Two effective iterative methods are constructed to solve the linear matrix equation of the form $X + A^*XA = I$. Some properties of a positive definite solution of the linear matrix equation and the iterates generated by first Algorithm are discussed. Necessary and sufficient conditions for existence of a positive definite solution are derived for $\|A\| < 1$ and $\|A\| > 1$. Necessary and sufficient conditions for existence of a positive definite solution are derived for $\|A\| < 1$ and $\|A\| > 1$. Several numerical examples are given to show the efficiency of the presented iterative methods.

Keywords Algorithm · Fixed point iteration · Linear matrix equation · Numerical analysis · Positive definite solutions · Properties · Two sided iteration

1 Introduction

Considering the linear matrix equation

$$X + A^*XA = I, \tag{1}$$

with unknown matrix X, where $A \in C^{n \times n}$, I is the identity matrix of order n. Equation (1) could be viewed as a special case of the symmetric matrix equations

S.A. Zarea (✉)
Princess Nourah Bint Abdulrahman University, Riyad 11643, KSA
e-mail: sazarea@pnu.edu.sa

S.M. El-Sayed
Benha University, Benha 13518, Egypt
e-mail: ms4elsayed@fci.bu.edu.eg

A.A.S. Al-Marshdy
Hail University, Hail 1826, KSA
e-mail: lnd7y@hotmail.com

© Springer Science+Business Media Dordrecht 2015
G.-C. Yang et al. (eds.), *Transactions on Engineering Technologies*,
DOI 10.1007/978-94-017-9588-3_8

$$X \pm A_1^* X A_1 \pm \cdots \pm A_m^* X A_m = Q. \tag{2}$$

where Q is a positive definite matrix [1]. There are many linear matrix equations which were studied by some authors [1–16]. Two effective iterative methods for computing a positive definite solution of this equation are proposed. The first one is fixed point iteration method and the second one is two sided iteration method of the fixed point iteration method. These two iterative methods are used for computing a positive definite solution of nonlinear matrix equations, see [17–25].

This chapter aims to find the positive definite solution of the matrix Eq. (1) for all values of $\|A\| \neq 1$, for this purpose we investigated two iterative methods, the first one is based on fixed point iteration and the second is based on two sided iteration method, also to derive necessary and sufficient conditions for the existence of the solution of Eq. (1).

Section 2 describes some properties of positive definite solutions of the Eq. (1). Section 3, presents a first iterative method (Fixed point iteration method) for obtaining the solution of our problem. Also, it presents theorems for obtaining the necessary and sufficient conditions for the existence of a solution of matrix Eq. (1) and some properties of the iterates generated by a method are discussed. Section 4 represents the second iterative method (Two sided iteration method of the fixed point iteration method) for obtaining the solution of the problem and theorems for the sufficient conditions for the existence of a positive definite solution of (1). Numerical examples in Sect. 5 illustrate the effectiveness of these methods. Conclusion drawn from the results obtained in this chapter are in Sect. 6.

We'll use mathematical induction technique in the most proofs.

The notation $X > 0$ means that X is a positive definite Hermitian matrix and $A > B$ is used to indicate that $A - B > 0$. A^* denotes the complex conjugate transpose of A. Finally, throughout the chapter, $\|.\|$ will be the spectral norm for square matrices unless otherwise noted.

2 Some Properties of the Solutions

This section discusses some properties of positive definite solutions of the matrix Eq. (1).

2.1 Theorem [15]

If m and M are the smallest and the largest eigenvalues of a solution X of (1), respectively, and λ is an eigenvalue of A, then $\sqrt{\frac{1-M}{M}} \le |\lambda| \le \sqrt{\frac{1-m}{m}}$.

2.2 Theorem [15]

If (1) has a positive definite solution X, then $A^*A + (AA^*)^{-1} > I$.

3 The First Iteration Method (Fixed Point Iteration Method)

This section establishes the first iterative method which is suitable for obtaining a positive definite solution of (1) when $\|A\| < 1$.

3.1 Algorithm

Take $X_0 = \alpha I$. For $k = 0, 1, 2, \ldots$, compute

$$X_{k+1} = I - A^*X_kA. \tag{3}$$

Our theorems give necessary and sufficient conditions for the existence of a positive definite solution of (1).

3.1.1 Theorem [15]

Let the sequence $\{X_k\}$ be determined by the Algorithm 3.1 and

$$\|A\|^2 < 1 \tag{4}$$

If (1) has a positive definite solution, then $\{X_k\}$ converges to X, which is a solution of (1) for all numbers $\alpha > 1$. Moreover, if $X_k > 0$ for every k, then (1) has a positive definite solution.

Proof It is convenient to write proof as in [15]. Let (1) has a positive definite solution. From Algorithm 3.1, we have
$X_0 = \alpha I > I > X_1 = I - \alpha A^*A$, $X_0 = \alpha I > I > X_2 = I - A^*X_1A > I - \alpha A^*A = X_1$, i.e., $X_0 > X_2 > X_1$. To prove $X_s > X_1$ if $X_{s-1} < X_0$ for all s, we have $X_s = I - A^*X_{s-1}A > I - A^*X_0A = I - \alpha A^*A = X_1$. We will find the relation between X_2, X_3, X_4, X_5. Since $X_1 < X_2$, then $X_2 = I - A^*X_1A > I - A^*X_2A = X_3$ and $X_4 = I - A^*X_3A > I - A^*X_2A = X_3$, since $X_3 > X_1$, then $I - A^*X_1A = X_2$ and $X_5 = I - A^*X_4A > I - A^*X_2A = X_3$.
Also, since $X_4 > X_3$, then $X_5 = I - A^*X_4A < I - A^*X_3A = X_4$. Thus, we get $X_0 = \alpha I > X_2 > X_4 > X_5 > X_3 > X_1 = I - \alpha A^*A$. We will prove that $X_0 > X_{s+1} > X_s$,

if we have $X_0 > X_{s-1} > X_s$, thus $X_0 = \alpha I > I > I - A^* X_{s-1} A < I - A^* X_s A \Rightarrow$ $X_0 > X_{s+1} > X_s$. Also, we can prove that $X_0 > X_s > X_{s+1}$, if we have $X_0 > X_s > X_{s-1}$, $X_0 = \alpha I > I - A^* X_{s-1} A > I - A^* X_s A \Rightarrow X_0 > X_s > X_{s+1}$. Therefore, we have $X_0 = \alpha I > X_{2r} > X_{2r+2} > X_{2s+3} > X_{2s+1} > X_1 = I - \alpha A^* A$, for every positive integers r, s. Consequently, the subsequences $\{X_{2r}\}, \{X_{2s+1}\}$ are monotonic and bounded, and $\lim_{s \to \infty} X_{2s}, \lim_{s \to \infty} X_{2s+1}$ exist.

To prove these sequences have a common limit, we have

$$\|X_{2s} - X_{2s+1}\| = \|I - A^* X_{2s-1} A - I + A^* X_{2s} A\|$$
$$= \|A^* (X_{2s} - X_{2s-1}) A\| \le \|A\|^2 \|X_{2s} - X_{2s-1}\|.$$

Let $q = \|A\|^2 < 1$, and we get $\|X_{2s} - X_{2s+1}\| \le q \|X_{2s} - X_{2s-1}\| \le q^2$ $\|X_{2s-2} - X_{2s-1}\| \le \cdots \le q^{2s} \|X_0 - X_1\| \le q^{2s}(2\alpha - 1)$. Since $q < 1$ and $(2\alpha - 1) > 0$, $\|X_{2s} - X_{2s+1}\| \to 0$ as $s \to \infty$, that is, $\{X_{2r}\}$ and $\{X_{2s+1}\}$ have the same limit X and $X_{2s} > X > X_{2s+1}$, $s = 1, 2, \ldots$. Taking the limit of the sequence $\{X_k\}$ generated by Algorithm 3.1 leads to $X = I - A^* X A$, which is a solution of (1). Assuming that $X_k > 0$ for every k. We proved that the sequences have a common limit X. Since $X_{k+1} = I - A^* X_k A > 0$, taking the limits of both sides as k approaches to ∞, we get $X = I - A^* X A > 0$. Hence (1) has a positive definite solution. $\qquad\square$

3.1.2 Theorem [15]

Let X_k be the iterates in Algorithm 3.1. If $q = \|A\|^2 < 1$. Then $\|X_k - X\| < q^k (2\alpha - 1)$, for all real number $\alpha > 1$ where X is a positive definite solution of (1).

3.1.3 Corollary [15]

Suppose that (1) has a solution. If $q = \|A\|^2 < 1$, then $\{X_k\}$ converges to X with at least the linear convergence rate.

3.1.4 Theorem [15]

If (1) has a positive definite solution and after k iterative steps of Algorithm 3.1, we have $\|I - X_k^{-1} X_{k-1}\| < \varepsilon$, then $\|X_k + A^* X_k A - I\| < \alpha \varepsilon \|A\|^2$, where X_k is the iterates in Algorithm 3.1.

3.1.5 Theorem

If A is regular matrix, then $AX_k = X_kA$, $k = 0, 1, 2, \ldots$, where X_k is the iterates in Algorithm 3.1.

Proof From Algorithm 3.1, since $X_0 = \alpha I$, we have $AX_0 = X_0A$, using $AA^* = A^*A$, we get $\alpha AA^* = \alpha A^*A$, $A - \alpha AA^*A = A - \alpha A^*AA$, $A(I - \alpha A^*A) = (I - \alpha A^*A)A$, $AX_1 = X_1A$, to prove $AX_k = X_kA$ for all k.

Assume that $AX_s = X_sA$ is true when $k = s$, we'll prove that $AX_{s+1} = X_{s+1}A$ when $k = s + 1$, by multiplying $AX_s = X_sA$ from right and left sides by A^* and A, respectively and subtracting it from A, we get $A - A^*AX_sA = A - A^*X_sAA$ and $A(I - A^*X_sA) = (I - A^*X_sA)A$, that is, $AX_{s+1} = X_{s+1}A$, thus, $AX_k = X_kA$, $k = 0, 1, 2, \ldots$. $\qquad\qquad\square$

3.1.6 Theorem

If A is regular matrix, then $X_{k+1}X_k = X_kX_{k+1}$, $k = 0, 1, 2, \ldots$, where X_k is the iterates in Algorithm 3.1.

Proof From Algorithm 3.1, $X_0 = \alpha I$ and $X_1 = I - \alpha A^*A$, then $X_0X_1 = X_1X_0$, also

$$X_1X_2 = (I - \alpha A^*A)(I - A^*X_1A) = I - \alpha A^*A - A^*X_1A + \alpha A^*AA^*X_1A, \quad (5)$$

and

$$X_2X_1 = (I - A^*X_1A)(I - \alpha A^*A) = I - \alpha A^*A - A^*X_1A + \alpha A^*X_1AA^*A, \quad (6)$$

By comparing the right hand side of each of (5) and (6), to show that $X_1X_2 = X_2X_1$, we must have $\alpha A^*X_1AA^*A = \alpha A^*AA^*X_1A$, since A is regular we have $A^*AAA^* = AA^*A^*A$, by multiplying both sides by α and subtracting it from AA^*, we get

$$AA^* - \alpha A^*AAA^* = AA^* - \alpha AA^*A^*A,$$
$$(I - \alpha A^*A)AA^* = AA^*(I - \alpha A^*A),$$

$X_1AA^* = AA^*X_1$, multiplying from right and left sides by A^* and A, respectively, we get

$$A^*X_1AA^*A = A^*AA^*X_1A$$

$\alpha A^*X_1AA^*A = \alpha A^*AA^*X_1A$, thus $X_1X_2 = X_2X_1$.

Suppose $X_{s+1}X_s = X_sX_{s+1}$, $s = 0, 1, 2, \ldots$ is true for $k = s$. From Theorem 3.1.5, $AX_s = X_sA$,

$$X_s X_{s-1} = X_{s-1} X_s. \tag{7}$$

Multiplying (7) from right by A^*, we get

$$AX_s A^* = X_s AA^*$$
$$X_s AA^* = AA^* X_s \tag{8}$$

and multiplying (8) from left side by X_{s-1}, we get

$$X_{s-1} X_s AA^* = X_{s-1} AA^* X_s$$
$$X_{s-1} X_s AA^* = AA^* X_{s-1} X_s \tag{9}$$

From (7), we have $X_{s-1} X_s AA^* = AA^* X_s X_{s-1}$ and from (9), we have

$$X_{s-1} AA^* X_s = X_s AA^* X_{s-1}. \tag{10}$$

Multiplying (10) from right and left sides by A and A^*, respectively, we get $A^* X_{s-1} AA^* X_s A = A^* X_s AA^* X_{s-1} A$. Consequently, $(I - A^* X_s A)(I - A^* X_{s-1} A) = (I - A^* X_{s-1} A)(I - A^* X_s A)$, thus $X_{s+1} X_s = X_s X_{s+1}$, that is, $X_{k+1} X_k = X_k X_{k+1}$, $k = 0, 1, 2, \ldots$.   □

4 The Second Iteration Method (Two Sided Iteration Method of the Fixed Point Iteration Method)

This section establishes the second iterative method which is suitable for obtaining a positive definite solution of (1), see [15].

4.1 Algorithm [15]

Take $X_0 = \alpha I$, $Y_0 = \beta I$. For $k = 0, 1, 2, \ldots$, compute

$$X_{k+1} = I - A^* X_k A \text{ and } Y_{k+1} = I - A^* Y_k A. \tag{11}$$

Next theorems provide necessary and sufficient conditions for the existence of a solution of (1) when $\|A\| < 1$.

4.1.1 Theorem

If (1) has a positive definite solution, the sequences $\{X_k\}$ and $\{Y_k\}$ are determined by Algorithm 4.1 and

$$\|A\|^2 < 1,$$

then the two sequences $\{X_k\}$, $\{Y_k\}$ converge to the positive definite solution X for all real numbers α, β such that $\beta > \alpha > 0$. On the other hand, if X_k, $Y_k > 0$ for every k, $\|A\|^2 < 1$ and $\beta > \alpha > 0$, then (1) has a positive definite solution.

4.1.2 Theorem

For the Algorithm 4.1, if there exist a positive real numbers α and β such that $\beta > \alpha$ and $q = \|A\|^2 < 1$, then $\|X_k - X\| < q^k(2\alpha - 1)$, $\|Y_k - X\| < q^k(2\beta - 1)$ and $\|X_k - X\| < \|Y_k - X\| < q^k(2\beta - 1)$, where X is a positive definite solution of (1) and X_k, Y_k, $k = 0, 1, 2, \ldots$ are defined in (11).

4.1.3 Corollary

Suppose that (1) has a solution. If $q = \|A\|^2 < 1$, then $\{X_k\}$ and $\{Y_k\}$ converge to X with at least the linear convergence rate.

4.1.4 Theorem

If the (1) has a solution and after k-iterative steps of the Algorithm 4.1, we have $\|I - X_k^{-1} X_{k-1}\| < \varepsilon$ and $\|I - Y_k^{-1} Y_{k-1}\| < \varepsilon$. Then $\|X_k + A^* X_k A - I\| < \alpha \varepsilon \|A\|^2$ and $\|Y_k + A^* Y_k A - I\| < \beta \varepsilon \|A\|^2$.

Also, $\|X_k + A^* X_k A - I\| < \|Y_k + A^* Y_k A - I\| < \beta \varepsilon \|A\|^2$, where X_k, Y_k, $k = 0, 1, 2, \ldots$ are defined in (11) and $\varepsilon > 0$.

4.2 Algorithm

Take $X_0 = \alpha I$, $Y_0 = \beta I$. For $k = 0, 1, 2, \ldots$, compute

$$X_{k+1} = B^*(I - X_k)B \text{ and } Y_{k+1} = B^*(I - Y_k)B. \tag{12}$$

where $B = A^{-1}$, $B^* = A^{*-1}$.

Next theorems provide necessary and sufficient conditions for the existence of a solution of (1) when $\|A\| > 1$, see [15].

4.2.1 Theorem

If (1) has a positive definite solution, the sequences $\{X_k\}$ and $\{Y_k\}$ are determined by the Algorithm 4.2 and the inequalities

(i) $B^*B < \alpha I, \beta I$ and $B^*B < \beta I, 0 < \alpha < \beta$,

(ii) $q = \|B\|^2 < 1$,

are satisfied, then $\{X_k\}$, $\{Y_k\}$ converge to a positive definite solution X. Moreover, if $X_k > 0$ and $Y_k > 0$ for every k, $B^*B < \alpha I$, $B^*B < \beta I$ and $\alpha, \beta > 0$, then (1) has a positive definite solution.

4.2.2 Theorem

For the Algorithm 4.2, if there exist positive numbers α and β such that $0 < \alpha < \beta$ and the following two conditions are hold

(i) $B^*B < \alpha I$ and $B^*B < \beta I$,

(ii) $q = \|B\|^2 < 1$,

then $\|X_k - X\| < q^k(2\alpha - 1)$, $\|Y_k - X\| < q^k(2\beta - 1)$ and $\|X_k - X\| < \|Y_k - X\|$ $< q^k(2\beta - 1)$, where X is a positive definite solution of (1) and $X_k, Y_k, k = 0, 1, 2, \ldots$ is defined in Algorithm 4.2.

4.2.3 Corollary

Suppose that (1) has a solution. If $q = \|B\|^2 < 1$, then $\{X_k\}$ and $\{Y_k\}$ converge to X with at least the linear convergence rate.

4.2.4 Theorem

If (1) has a positive definite Solution and after k iterative steps of the Algorithm 4.2, we have $\|I - X_k^{-1}X_{k-1}\| < \varepsilon$ and $\|I - Y_k^{-1}Y_{k-1}\| < \varepsilon$, then

(i) $\|X_k + B^*X_kB - B^*B\| < \alpha\varepsilon\|B\|^2$ and $\|Y_k + B^*Y_kB - B^*B\| < \beta\varepsilon\|B\|^2$.

(ii) $\|X_k + B^*X_kB - B^*B\| < \|Y_k + B^*Y_kB - B^*B\| < \beta\varepsilon\|B\|^2$.

where $X_k, Y_k, k = 0, 1, 2, \ldots$ are the iterates generated by Algorithm 4.2 and $\varepsilon > 0$.

5 Numerical Experiments

In this section the numerical experiments are used to display the flexibility of the methods. The solutions are computed for some different matrices A with different sizes n. For the following examples, practical stopping criterion $\|X - X_k\| \le 10^{-9}$ and obtains the maximal solution $X = X_{500}$.

5.1 Numerical Experiments for the First Method (Algorithm 3.1)

In the following tables we denote
$q = \|A\|^2$, $\varepsilon_1(X) = \|X - X_k\|$, $\varepsilon_2(X) = \|X_k + A^* X_k A - I\|$, where X the solution which is obtained by the iterative method (Algorithm 3.1)

I. Example

Let $\alpha = 2.5$ and

$$A = \left(a_{ij}\right): \quad a_{ij} = \begin{cases} \frac{i}{2m} & i = j, \\ \frac{i+\sqrt{i}}{[(m+i)(m+j)]^2} & i \ne j. \end{cases}$$

for $m = 35, 70, 140$, $\|A\| = 0.485714, 0.492857, 0.496429$, respectively, see Tables 1, 2 and 3.

5.2 Numerical Experiments for the Second Method (Algorithm 4.1)

The following tables denotes $q = \|A\|^2$, $\varepsilon_1(X) = \|X - X_k\|$, $\varepsilon_2(X) = \|Y - Y_k\|$, $\varepsilon_3(X) = \|X_k - Y_k\|$, $\varepsilon_4(X) = \|X_k + A^* X_k A - I\|$, $\varepsilon_5(X) = \|Y_k + A^* Y_k A - I\|$, X and Y are the solutions which are obtained by the iterative method (Algorithm 4.1).

Table 1 Example I, $m = 35$, $\|A\| = 0.485714$, $q = 0.235918$

K	ε_1	q^k	$q^k(2\alpha - 1)$	ε_2
0	1.69089	1.00000	4.00000	2.0898
4	5.23795E−03	3.09775E−03	1.2391E−02	6.47368E−03
8	1.62259E−05	9.59608E−06	3.83843E−05	2.00539E−05
12	5.02638E−08	2.97263E−08	1.18905E−07	6.21219E−08
16	3.14255E−10	1.85104E−10	7.40417E−10	3.917E−10

Table 2 Example I, $m = 70$, $\|A\| = 0.492857$, $q = 0.242908$

k	ε_1	q^k	$q^k(2\alpha - 1)$	ε_2
0	1.69544	1.0	4.0	2.10727
4	5.90269E−03	3.48152E−03	1.39261E−02	7.3365E−03
8	2.05503E−05	1.2121E−05	4.84838E−05	2.55421E−05
12	7.15462E−08	4.21993E−08	1.68797E−07	8.89254E−08
16	2.49089E−10	1.46918E−10	5.8767E−10	3.09595E−10

Table 3 Example I, $m = 140$ $\|A\| = 0.496429$, $q = 0.246441$

K	ε_1	q^k	$q^k(2\alpha - 1)$	ε_2
0	1.69772	1.00000	4.00000	2.1161
4	6.26209E−03	3.68854E−03	1.47541E−02	7.80533E−03
8	2.30979E−05	1.36053E−05	5.44212E−05	2.87902E−05
12	8.51976E−08	5.01837E−08	2.00735E−07	1.06194E−07
16	1.55705E−10	9.20848E−11	3.68339E−10	1.92439E−10

II. Example

Let $\alpha = 6.5$, $\beta = 17$ and

$$A = \frac{1}{32}\begin{pmatrix} 0.2 & -0.1 & -0.5 & 0.1 \\ -0.1 & 0.6 & -0.5 & 0.7 \\ -0.5 & -0.5 & 0.1 & 0.8 \\ 0.1 & 0.7 & 0.8 & 0.5 \end{pmatrix}$$

$\|A\| = 0.0412375$, $q = 0.00170053 < 1$, see Table 4.

Table 4 Example II

k	ε_1	q^k	$q^k(2\alpha - 1)$	ε_4
0	2.9895	1.0	5.0	5.12086
3	1.67396E−03	4.7126E−02	0.23563E−01	9.8887E−01
6	9.58545E−07	2.22086E−03	1.11043E−02	9.88152E−01
9	5.48883E−10	1.0466E−04	5.23301E−04	9.88152E−01
k	ε_2	ε_3	$q^k(2\beta - 1)$	ε_5
0	3.9895	1.0	7.0	7.18129
3	2.24658E−03	5.72621E−04	3.29882E−01	9.89115E−01
6	1.28644E−06	3.27895E−07	1.5546E−02	9.88152E−01
9	7.36643E−10	1.8776E−10	7.32621E−04	9.88152E−01

Table 5 Example III

K	ε_1	q^k	$q^k(2\alpha - 1)$	ε_4
0	4.99918	1.0	9.0	5.02206
1	2.63485E−02	5.28161E−03	4.75344E−02	26672E−02
2	1.38995E−04	2.78954E−05	2.51058E−04	6.65329E−04
3	7.33231E−07	1.47332E−07	1.32599E−06	7.33359E−04
4	3.86797E−09	7.78151E−10	7.00336E−09	7.32986E−04
5	2.04045E−11	4.10989E−12	3.6989E−11	7.32988E−04
K	ε_2	ε_3	$q^k(2\beta - 1)$	ε_5
0	6.99918	2.0	13.0	7.03309
1	3.6899E−02	1.05505E−02	6.86609E−02	3.72743E−02
2	1.94651E−04	5.56563E−05	3.6264E−04	6.40508E−04
3	1.02683E−06	2.93601E−07	1.91532E−06	7.33508E−04
4	5.41679E−09	1.54881E−09	1.0116E−08	7.32985E−04
5	2.85749E−11	8.17037E−12	5.34285E−11	7.32988E−04

5.3 Numerical Experiments for the Second Method (Algorithm 4.2)

The following tables denotes $q = \|B\|^2$, $\varepsilon_1(X) = \|X - X_k\|$, $\varepsilon_2(X) = \|Y - Y_k\|$, $\varepsilon_3(X) = \|X_k - Y_k\|$, $\varepsilon_4(X) = \|X_k + B^*X_kB - B^*B\|$, $\varepsilon_5(X) = \|Y_k + B^*Y_kB - B^*B\|$, X and Y are the solutions obtained by the iterative method (Algorithm 4.2).

III. Example [15]

Let $\alpha = 5$, $\beta = 7$ and

$$A = \begin{pmatrix} -12.74 & 5.755 \\ -5.755 & -32.21 \end{pmatrix},$$

$\|A\| = 32.9351$, $\|B\| = 0.0726747$ and $q = 0.00528161 < 1$, see Table 5.

6 Conclusion

In this paper, the positive definite solution of the linear matrix equation $X + A^*XA = I$, which is a special case of the symmetric matrix Eq. (2) for $\|A\| \neq 1$ was obtained. Two effective iterative methods for computing a positive definite solution of this equation were proposed. The first one is fixed point iteration method when $\|A\| < 1$ and the second one is two sided iteration method of the fixed point iteration method when $\|A\| < 1$ and $\|A\| > 1$. By Algorithm 3.1, for initial matrix $X_0 = \alpha I$ and Algorithms 4.1 and 4.2, for initial matrices $X_0 = \alpha I$, $Y_0 = \beta I$ satisfying the hypothesis of theorems (3.1) in Sect. 3, (4.1.1) and (4.2.1) in Sect. 4, a positive definite solution X can be obtained in finite iteration, with at least the linear convergence rate. The given numerical examples show that the proposed iterative algorithms are efficient.

References

1. A. Ran, M. Beurings, The symmetric linear matrix equation. Electron. J. Linear Algebra **9**, 93–107 (2002)
2. Z. Bing, Y. Hu Lijuan, S. Dragana, The congruence class of the solutions of the some matrix equations. Comput. Math Appl. **57**, 540–549 (2009)
3. A. Bouhamidi, K. Jbilou, A note on the numerical approximate solutions for generalized Sylvester matrix equations with applications. Appl. Math. Comput. **206**, 687–694 (2008)
4. D. Hua, P. Lancaster, Linear matrix equation from an inverse problem of vibration theory. Linear Algebra Appl. **246**, 31–47 (1996)
5. L. Hung, The explicit solution and solvability of the linear matrix equations. Linear Algebra Appl. **311**, 195–199 (2000)
6. J. Jiang, H. Liu, Y. Yuan, Iterative solutions to some linear matrix equations. World Acad. Sci. Eng. Technol. **76**, 925–930 (2011)
7. G. Konghua, X. Hu, L. Zhang, A new iteration method for the matrix equation $AX = B^*$. Appl. Math. Comput. **178**, 1434–1441 (2007)
8. D. Mirko, On minimal solution of the matrix equation $AX - YB = 0$. Linear Algebra Appl. **325**, 81–99 (2001)
9. Z.-Y. Li, B. Zhou, Y. Wang, G.-R. Duan, Numerical solution to linear matrix equation by finite steps iteration. IET Control Theory Appl. **4**(7), 1245–1253 (2010)
10. X. Peng, Y. Hu, L. Zhang, An iteration method for symmetric solutions and the optimal approximation solution of the matrix equation $AXB = C^*$. Appl. Math. Comput. **160**, 763–777 (2005)
11. T. Stykel, Stability and inertia theorems for generalized Lyapunov equations. Linear Algebra Appl. **355**, 297–314 (2002)
12. M. Wang, X. Cheng, M. Wei, Iterative algorithms for solving the matrix equation $AXB + CX^TD = E^*$. Appl. Math. Comput. **187**, 622–629 (2007)
13. Y.O. Vorontsov, Numerical algorithm for solving the matrix equation $AX + X^* B = C$. Moscow Univ. Comput. Math. Cybern. **37**(1), 1–7 (2013)
14. H. Yong, Ranks of solutions of the linear matrix equation $AX + YB = C$. Comput. Math. Appl. **52**, 861–872 (2006)
15. S.A. Zarea, S.M. El-Sayed, A.A.S. Al-Marshdy, On positive definite solutions of the linear matrix equation $X + A^*A = I$, in *Proceedings of the International Multi Conference of Engineers and Computer Scientists 2014, IMECS 2014*, Hong Kong, 12–14 Mar 2014. Lecture Notes in Engineering and Computer Science, pp. 551–557
16. X. Zhang, The general common Hermitian nonnegative-definite solution to the matrix equations $AXA^* = BB^*$ and $CXC^* = DD^*$ with applications in statistics. J. Multivar. Anal. **93**, 257–266 (2005)
17. W. Anderson, T. Morley, G. Trapp, Positive solution to $X = A - BX^{-1}B^*$. Linear Algebra Appl. **134**, 53–62 (1990)
18. M.S. El-Sayed, An algorithm for computing the extremal positive definite solutions of a matrix equation $X + A^TX^{-1}A = I$. Int. J. Comput. Math. **80**, 1527–1534 (2003)
19. M.S. El-Sayed, On the positive definite solutions for a nonlinear matrix equation. Int. Math. J. **4**, 27–42 (2003)
20. J. Engwerda, On the existence of the positive definite solution of the matrix equation $X + A^TX^{-1}A = I$. Linear Algebra Appl. **194**, 91–108 (1993)

21. C. Guo, P. Lancaster, Iterative solution of two matrix equations. Math. Comput. **68**, 1589–1603 (1999)
22. B. Meini, Efficient computation of the extreme solutions of $X + A^*X^{-1}A = Q$ and $X + A^*X^{-1}A = Q$. Math. Comput. **239**, 1189–1204 (2002)
23. M.S. El-Sayed, Two sided iteration methods for computing positive definite solutions of a nonlinear matrix equation. J. Aust. Math. Soc., Ser B **44**, 145–152 (2003)
24. S.A. Zarea, M.S. El-Sayed, On the matrix equation $X + A^* X^{-a} A = I$. Int. J. Comput. Math. Numer. Simul. **1**(10), 89–97 (2008)
25. S.A. Zarea, M.S. El-Sayed, A.A. Al-Eidan, Iterative algorithms for the nonlinear matrix equation $X + A^*X^{-r}A = I$. Assiut Univ. J. Math. Comput. Sci. **38**(1), 11–24 (2009)

Multiobjective Fuzzy Random Linear Programming Problems Based on E-Model and V-Model

Hitoshi Yano and Kota Matsui

Abstract In this paper, an interactive decision making method for multiobjective fuzzy random linear programming problems based on an expectation model (E-model) and a variance minimization model (V-model) is proposed. In the proposed method, it is assumed that the decision maker intends to not only maximize the expected degrees of possibilities that the original objective functions attain the corresponding fuzzy goals, but also minimize the standard deviations for such possibilities, and such fuzzy goals are quantified by eliciting the corresponding membership functions. Using the fuzzy decision, both the expected degrees of possibilities and the membership functions of the standard deviations are integrated, and an EV-Pareto optimality concept is introduced. In the integrated membership space, a satisfactory solution is obtained from among an EV-Pareto optimal solution set through the interaction with the decision maker.

Keywords Expectations · Fuzzy decision · Fuzzy random variables · Interactive method · Multiobjective programming · Standard deviations

1 Introduction

In the real world decision making situations, we often have to make a decision under uncertainty. In order to deal with decision problems involving uncertainty, stochastic programming approaches [1–4] and fuzzy programming approaches [5–7] have been developed. Recently, mathematical programming problems with

H. Yano (✉)
Graduate School of Humanities and Social Sciences, Nagoya City University,
Nagoya 467-8501, Japan
e-mail: yano@hum.nagoya-cu.ac.jp

K. Matsui
Graduate School of Information Science, Nagoya University, Nagoya 4464-8601, Japan
e-mail: matsui@math.cm.is.nagoya-u.ac.jp

© Springer Science+Business Media Dordrecht 2015   113
G.-C. Yang et al. (eds.), *Transactions on Engineering Technologies*,
DOI 10.1007/978-94-017-9588-3_9

fuzzy random variables [8] have been formulated [9–11], whose concept includes both probabilistic uncertainty and fuzzy one simultaneously. Extensions to multi-objective fuzzy random linear programming problems (MOFRLP) have been done and interactive methods to obtain a satisfactory solution for the decision maker have been proposed [10, 12, 13]. In their methods, it is required in advance for the decision maker to specify permissible possibility levels in a probability maximization model or permissible probability levels in a fractile optimization model [14]. However, it seems to be very difficult for the decision maker to specify such permissible levels appropriately. From such a point of view, a fuzzy approach to MOFRLP has been proposed [15], in which the decision maker specifies not the values of permissible levels but the membership functions for the fuzzy goals of permissible levels. In the proposed method, it is assumed that the decision maker adopts the fuzzy decision [6, 16] to integrate the membership functions. As a natural extension of such methods, interactive fuzzy decision making methods for MOFRLP to obtain a satisfactory solution from among an extended Pareto optimal solution set have been proposed [17–20]

On the other hand, some decision maker may prefer to adopt an expectation model (E-model) or a variance model (V-model) rather than a probability maximization model or a fractile optimization model to deal with MOFRLP, because the expectation value or the variance is a well-known statistical quantity. From such a point of view, Katagiri et al. [21, 22] proposed interactive decision making methods for MOFRLP to obtain a satisfactory solution of the decision maker using E-model and V-model [23]. However, when adopting E-model, the effects for the variance of the random variable coefficients of fuzzy random variables are ignored. Similarly, when adopting V-model, although the effects for the variance of the random variable coefficients of fuzzy random variables are considered in the formulation processes of MOFRLP, the decision maker must specify in advance a permissible expectation level for each objective function of MOFRLP subjectively. In general, the minimization of a permissible expectation level in a minimization problem conflicts with the minimization of the corresponding variance. Therefore, it seems to be difficult for the decision maker to specify appropriately a permissible expectation level for each objective function of MOFRLP.

In this paper, it is assumed that the decision maker intends to not only maximize the expected degrees of possibilities [16] that the original objective functions involving fuzzy random variable coefficients attain the corresponding fuzzy goals, but also minimize the standard deviations for such possibilities in MOFRLP [21, 22, 24]. In order to deal with such decision making situations in MOFRLP, we introduce an EV-Pareto optimal solution concept, in which both the expected degrees of possibilities and the corresponding standard deviations for such possibilities are integrated through the fuzzy decision [6, 16]. To obtain an EV-Pareto optimal solution, the minmax problem is formulated. An interactive algorithm is proposed to obtain a satisfactory solution from among an EV-Pareto optimal solution set [24]. The proposed method is applied to the numerical example [24], and the interactive processes under the hypothetical decision maker are demonstrated. Moreover, the proposed method is compared with the V-model based method.

2 Multiobjective Fuzzy Random Linear Programming Problems

In this section, we focus on multiobjective programming problems involving fuzzy random variable coefficients in objective functions called multiobjective fuzzy random linear programming problem (MOFRLP).

[MOFRLP]

$$\min_{x \in X} \widetilde{\overline{C}}x = (\widetilde{\overline{c}}_1 x, \ldots, \widetilde{\overline{c}}_k x) \tag{1}$$

where $x = (x_1, \ldots, x_n)^T$ is an n dimensional decision variable column vector. X is a linear constraint set with respect to x. $\widetilde{\overline{c}}_i = (\widetilde{\overline{c}}_{i1}, \ldots, \widetilde{\overline{c}}_{in}), i = 1, \ldots, k$ are coefficient vectors of objective function $\widetilde{\overline{c}}_i x$, whose elements are fuzzy random variables (The symbols "-" and "~" mean randomness and fuzziness respectively).

In this paper, we assume that under the occurrence of each scenario $\ell_i \in \{1, \ldots, L_i\}$, $\tilde{c}_{ij\ell_i}$ is a realization of a fuzzy random variable $\widetilde{\overline{c}}_{ij}$ which is a fuzzy number whose membership function is defined as follows [10].

$$\mu_{\tilde{c}_{ij\ell_i}}(t) = \begin{cases} \max\left\{1 - \frac{d_{ij\ell_i} - t}{\alpha_{ij}}, 0\right\}, & t \le d_{ij\ell_i} \\ \max\left\{1 - \frac{t - d_{ij\ell_i}}{\beta_{ij}}, 0\right\}, & t > d_{ij\ell_i} \end{cases} \tag{2}$$

where the parameters $\alpha_{ij} > 0, \beta_{ij} > 0$ are constants and $d_{ij\ell_i}$ varies depending on which a scenario ℓ_i occurs. Moreover, we assume that a scenario ℓ_i occurs with a probability $p_{i\ell_i}$, where $\sum_{\ell_i=1}^{L_i} p_{i\ell_i} = 1$ for $i = 1, \ldots, k$.

By Zadeh's extension principle, the realization $\tilde{c}_{i\ell_i} x$ becomes a fuzzy number which characterized by the following membership function.

$$\mu_{\tilde{c}_{i\ell_i} x}(y) = \begin{cases} \max\left\{1 - \frac{d_{i\ell_i} x - y}{\alpha_i x}, 0\right\}, & y \le d_{i\ell_i} x \\ \max\left\{1 - \frac{y - d_{i\ell_i} x}{\beta_i x}, 0\right\}, & y > d_{i\ell_i} x \end{cases} \tag{3}$$

where $d_{i\ell_i} = (d_{i1\ell_i}, \ldots, d_{in\ell_i})$, $\alpha_i = (\alpha_{i1}, \ldots, \alpha_{in}) \ge 0$, $\beta_i = (\beta_{i1}, \ldots, \beta_{in}) \ge 0$.

Considering the imprecise nature of the decision maker's judgment, it is natural to assume that the decision maker has a fuzzy goal for each objective function in MOFRLP. In this paper, it is assumed that such a fuzzy goal $\widetilde{G}_i$ can be quantified by eliciting the corresponding membership function defined as follows.

$$\mu_{\widetilde{G}_i}(y_i) = \begin{cases} 1 & y_i < z_i^1 \\ \frac{y_i - z_i^0}{z_i^1 - z_i^0} & z_i^1 \le y_i \le z_i^0 \\ 0 & y_i > z_i^0 \end{cases} \tag{4}$$

where z_i^0 represents the minimum value of an unacceptable level of the objective function, and z_i^1 represents the maximum value of a sufficiently satisfactory level of the objective function. By using a concept of possibility measure [16], the degree of possibility that the objective function value $\widetilde{c}_i x$ satisfies the fuzzy goal $\widetilde{G}_i$ is expressed as follows [13].

$$\Pi_{\widetilde{c}_{ix}}(\widetilde{G}_i) \overset{\text{def}}{=} \sup_y \min \{\mu_{\widetilde{c}_{ix}}(y), \mu_{\widetilde{G}_i}(y)\} \tag{5}$$

It should be noted here that if a scenario ℓ_i occurs with probability $p_{i\ell_i}$ then the value of possibility measure can be represent as

$$\Pi_{\widetilde{c}_{i\ell_i}x}(\widetilde{G}_i) \overset{\text{def}}{=} \sup_y \min \{\mu_{\widetilde{c}_{i\ell_i}x}(y), \mu_{\widetilde{G}_i}(y)\}. \tag{6}$$

Using the above possibility measure, MOFRLP can be transformed into the following multiobjective stochastic programming problem (MOSP).

[MOSP]

$$\max_{x \in X} (\Pi_{\widetilde{c}_1 x}(\widetilde{G}_1), \dots, \Pi_{\widetilde{c}_k x}(\widetilde{G}_k)) \tag{7}$$

3 E-Model and V-Model for MOFRLP

Katagiri et al. [21, 22] formulated MOFRLP as the multiobjective programming problems through an expectation model (E-model) and a variance model (V-model) respectively. At First, we explain E-model for MOFRLP formulated as follows.

[MOP-E1]

$$\max_{x \in X} (E[\Pi_{\widetilde{c}_1 x}(\widetilde{G}_1)], \dots, E[\Pi_{\widetilde{c}_k x}(\widetilde{G}_k)]) \tag{8}$$

where $E[\cdot]$ denotes the expectation operator. In order to deal with MOP-E1, we introduce an E-Pareto optimal solution concept.

Definition 1 $x^* \in X$ is said to be an E-Pareto optimal solution to MOP-E1, if and only if there does not exist another $x \in X$ such that $E[\Pi_{\widetilde{c}_i x}(\widetilde{G}_i)] \geq E[\Pi_{\widetilde{c}_i x^*}(\widetilde{G}_i)]$, $i = 1, \dots, k$ with strict inequality holding for at least one i.

It should be noted here that (6) can be represented as follows [10].

$$\Pi_{\widetilde{c}_{i\ell_i}x}(\widetilde{G}_i) = \frac{\sum_{j=1}^n (\alpha_{ij} - d_{ij\ell_i})x_j + z_i^0}{\sum_{j=1}^n \alpha_{ij}x_j - z_i^1 + z_i^0} \tag{9}$$

Since the probability that a scenario ℓ_i occurs is $p_{i\ell_i}$, $E[\Pi_{\tilde{c}_ix}(\tilde{G}_i)]$ can be computed as follows.

$$E[\Pi_{\tilde{c}_ix}(\tilde{G}_i)] = \frac{\sum_{j=1}^n (\alpha_{ij} - \sum_{\ell_i=1}^{L_i} p_{i\ell_i} d_{ij\ell_i})x_j + z_i^0}{\sum_{j=1}^n \alpha_{ij}x_j - z_i^1 + z_i^0} \tag{10}$$

$$\overset{\text{def}}{=} Z_i^E(x)$$

Then, MOP-E1 can be transformed into MOP-E2.
[MOP-E2]

$$\max_{x \in X}(Z_1^E(x), \ldots, Z_k^E(x)) \tag{11}$$

Next, consider V-model for MOFRLP. The multiobjective programming problem based on V-model can be formulated as follows.
[MOP-V1]

$$\min_{x \in X}(V[\Pi_{\tilde{c}_1x}(\tilde{G}_1)], \ldots, V[\Pi_{\tilde{c}_kx}(\tilde{G}_k)]) \tag{12}$$

subject to

$$E[\Pi_{\tilde{c}_ix}(\tilde{G}_i)] \geq \xi_i, \quad i = 1, \cdots, k \tag{13}$$

where $V[\cdot]$ denotes the variance operator, and ξ_i represents a permissible expectation level for $E[\Pi_{\tilde{c}_ix}(\tilde{G}_i)]$. Now, we denote the feasible set of MOP-V1 as

$$X(\xi) \overset{\text{def}}{=} \{x \in X | E[\Pi_{\tilde{c}_ix}(\tilde{G}_i)] \geq \xi_i, i = 1, \ldots, k\}. \tag{14}$$

Similar to E-model, in order to deal with MOP-V1, a V-Pareto optimal solution concept is defined.

Definition 2 $x^* \in X(\xi)$ is said to be a V-Pareto optimal solution to MOP-V1, if and only if there does not exist another $x \in X(\xi)$ such that $V[\Pi_{\tilde{c}_ix}(\tilde{G}_i)] \leq V[\Pi_{\tilde{c}_ix^*}(\tilde{G}_i)]$, $i = 1, \ldots, k$ with strict inequality holding for at least one i.

It should be noted here that $V[\Pi_{\tilde{c}_ix}(\tilde{G}_i)]$ can be represented as follows [10].

$$V[\Pi_{\tilde{c}_ix}(\tilde{G}_i)] = \frac{1}{(\sum_{j=1}^n \alpha_{ij}x_j - z_i^1 + z_i^0)^2} x^T V_i x$$

$$\overset{\text{def}}{=} Z_i^V(x) \tag{15}$$

where V_i is the variance-covariance matrix of $\bar{d}_i$ expressed by

$$V_i = \begin{pmatrix} v_{11}^i & v_{12}^i & \cdots & v_{1n}^i \\ v_{21}^i & v_{22}^i & \cdots & v_{2n}^i \\ \vdots & \vdots & \ddots & \vdots \\ v_{n1}^i & v_{n2}^i & \cdots & v_{nn}^i \end{pmatrix}, \quad i = 1 \ldots, k, \tag{16}$$

and

$$v_{jj}^i = V[\bar{d}_{jj}] = \sum_{\ell_i=1}^{L_i} p_{i\ell_i} d_{ij\ell_i}^2 - \left(\sum_{\ell_i=1}^{L_i} p_{i\ell_i} d_{ij\ell_i} \right)^2, \quad j = 1, \ldots, n, \tag{17}$$

$$\begin{aligned} v_{jr}^i = \mathrm{Cov}[\bar{d}_{ij}, \bar{d}_{ir}] &= E[\bar{d}_{ij} \cdot \bar{d}_{ir}] - E[\bar{d}_{ij}]E[\bar{d}_{ir}] \\ &= \sum_{\ell_i=1}^{L_i} p_{i\ell_i} d_{ij\ell_i} d_{ir\ell_i} - \sum_{\ell_i=1}^{L_i} p_{i\ell_i} d_{ij\ell_i} \sum_{\ell_i=1}^{L_i} p_{i\ell_i} d_{ir\ell_i}, \end{aligned} \tag{18}$$

$$j, r = 1, \ldots, n, j \neq r$$

Furthermore, the inequalities (13) can be expressed by the following forms.

$$\sum_{j=1}^{n} \left(\sum_{\ell_i=1}^{L_i} p_{i\ell_i} d_{ij\ell_i} - (1 - \xi_i)\alpha_{ij} \right) x_j \leq z_i^0 - \xi_i(z_i^0 - z_i^1), \quad i = 1, \ldots, k \tag{19}$$

Then, MOP-V1 can be transformed into MOP-V2.
[MOP-V2]

$$\min_{x \in X}(Z_1^V(x), \ldots, Z_k^V(x)) \tag{20}$$

subject to

$$\sum_{j=1}^{n} \left(\sum_{\ell_i=1}^{L_i} p_{i\ell_i} d_{ij\ell_i} - (1 - \xi_i)\alpha_{ij} \right) x_j \leq z_i^0 - \xi_i(z_i^0 - z_i^1),$$

$$i = 1, \ldots, k$$

From the fact that $\sum_{j=1}^{n} \alpha_{ij} x_j - z_i^1 + z_i^0 > 0$, $x^T V_i x > 0$, due to the positive-semidefinite property of V_i, MOP-V2 can be equivalently transformed to MOP-V3.
[MOP-V3]

$$\min_{x \in X}(Z_1^{SD}(x), \ldots, Z_k^{SD}(x)) \tag{21}$$

subject to

$$\sum_{j=1}^{n}\left(\sum_{\ell_i=1}^{L_i} p_{i\ell_i} d_{ij\ell_i} - (1-\xi_i)\alpha_{ij}\right)x_j \leq z_i^0 - \xi_i(z_i^0 - z_i^1),$$

$$i = 1,\ldots,k,$$

where

$$Z_i^{SD}(x) \stackrel{\text{def}}{=} \frac{\sqrt{x^T V_i x}}{\sum_{j=1}^{n}\alpha_{ij}x_j - z_i^1 + z_i^0}$$

It should be noted here that $Z_i^E(x)$ and $Z_i^{SD}(x)$ are the statistical values for the same random function $\Pi_{\widetilde{c}_i x}(\widetilde{G}_i)$. When solving MOFRLP, it is natural for the decision maker to consider both $Z_i^E(x)$ and $Z_i^{SD}(x)$ for each objective function $\Pi_{\widetilde{c}_i x}(\widetilde{G}_i)$ of MOSP simultaneously, rather than considering either of them. From such a point of view, in the following sections, we propose the hybrid model for MOFRLP, in which E-model and V-model are incorporated simultaneously, and define an EV-Pareto optimality concept.

4 EV-Model for MOFRLP

In this section, we consider the following hybrid model for MOFRLP, where both E-model and V-model are considered simultaneously.

[MOP-EV1]

$$\max_{x\in X}\left(Z_1^E(x),\ldots,Z_k^E(x), -Z_1^{SD}(x),\ldots,-Z_k^{SD}(x)\right) \qquad (22)$$

In MOP-EV1, $Z_i^E(x)$ and $Z_i^{SD}(x)$ mean the expected value and the standard deviation of the objective function $\Pi_{\widetilde{c}_i x}(\widetilde{G}_i)$ in MOSP. It should be noted here that $Z_i^E(x)$ can be interpreted as an expected value of the satisfactory degree for $\Pi_{\widetilde{c}_i x}(\widetilde{G}_i)$, but $Z_i^{SD}(x)$ does not mean the satisfactory degree itself. In MOP-EV1, we assume that the decision maker has fuzzy goals for $Z_i^{SD}(x), i = 1,\ldots,k$, and the corresponding linear membership functions are defined as $\mu_i^{SD}(Z_i^{SD}(x)), i = 1,\ldots,k$.

$$\mu_i^{SD}(s_i) = \begin{cases} 1 & s_i < q_i^1 \\ \frac{s_i - q_i^0}{q_i^1 - q_i^0} & q_i^1 \leq s_i \leq q_i^0 \\ 0 & s_i > q_i^0 \end{cases} \qquad (23)$$

where q_i^0 represents the minimum value of an unacceptable level of $Z_i^{SD}(x)$, and q_i^1 represents the maximum value of a sufficiently satisfactory level of $Z_i^{SD}(x)$.

From the point of view that both $Z_i^E(x)$ and $\mu_i^{SD}(Z_i^{SD}(x))$ mean the satisfactory degrees for $\Pi_{\tilde{c}_{ix}}(\tilde{G}_i)$, we introduce the integrated membership function in which the both satisfactory levels $Z_i^E(x)$ and $\mu_i^{SD}(Z_i^{SD}(x))$ are incorporated simultaneously through the fuzzy decision [6, 16].

$$\mu_{D_i}(x) \stackrel{def}{=} \min\{Z_i^E(x), \mu_i^{SD}(Z_i^{SD}(x))\} \tag{24}$$

Then, MOP-EV1 can be transformed into the following multiobjective programming problem.

[MOP-EV2]

$$\max_{x \in X}\left(\mu_{D_1}(x), \ldots, \mu_{D_k}(x)\right) \tag{25}$$

$\mu_{D_i}(x)$ can be interpreted as an overall satisfactory degree for the fuzzy goal $\tilde{G}_i$. For MOP-EV2, we introduce an EV-Pareto optimal solution concept defined as follows.

Definition 3 $x^* \in X$ is an EV-Pareto optimal solution to MOP-EV2, if and only if there does not exist another $x \in X$ such that $\mu_{D_i}(x) \geq \mu_{D_i}(x^*)$, $i = 1, \ldots, k$ with strict inequality holding for at least one i.

In order to generate a candidate of a satisfactory solution from among an EV-Pareto optimal solution set, the decision maker is asked to specify the reference membership values [6]. For the reference membership values $\hat{\mu} = (\hat{\mu}_1, \ldots, \hat{\mu}_k)$, the corresponding EV-Pareto optimal solution is obtained by solving the following minmax problem.

[MINMAX ($\hat{\mu}$)]

$$\min_{x \in X, \lambda \in \Lambda} \lambda \tag{26}$$

subject to

$$\hat{\mu}_i - Z_i^E(x) \leq \lambda, \quad i = 1, \ldots, k \tag{27}$$

$$\hat{\mu}_i - \mu_i^{SD}(Z_i^{SD}(x)) \leq \lambda, \quad i = 1, \ldots, k \tag{28}$$

where

$$\Lambda \stackrel{def}{=} \left[\max_{i=1,\ldots,k} \hat{\mu}_i - 1, \max_{i=1,\ldots,k} \hat{\mu}_i\right] = [\lambda_{\min}, \lambda_{\max}] \tag{29}$$

The relationships between the optimal solution (x^*, λ^*) of MINMAX ($\hat{\mu}$) and EV-Pareto optimal solutions of MOP-EV2 can be characterized by the following theorems.

Theorem 1 *If $x^* \in X$, $\lambda^* \in \Lambda$ is an unique optimal solution of MINMAX $(\hat{\mu})$ then x^* is an EV-Pareto optimal solution of MOP-EV2.*

Proof Let us assume that $x^* \in X$ is not an EV-Pareto optimal solution of MOP-EV2. Then, there exists $x \in X$ such that $\mu_{D_i}(x) \geq \mu_{D_i}(x^*), i = 1, \ldots, k$, with strict inequality holding for at least one i. This implies that

$$
\begin{aligned}
\mu_{D_i}(x) \geq \mu_{D_i}(x^*) &\Leftrightarrow \hat{\mu}_i - \min\{Z_i^E(x), \mu_i^{SD}(Z_i^{SD}(x))\} \\
&\leq \hat{\mu}_i - \min\{Z_i^E(x^*), \mu_i^{SD}(Z_i^{SD}(x^*))\} \\
&\Leftrightarrow \max\{\hat{\mu}_i - Z_i^E(x), \hat{\mu}_i - \mu_i^{SD}(Z_i^{SD}(x))\} \\
&\leq \max\{\hat{\mu}_i - Z_i^E(x^*), \hat{\mu}_i - \mu_i^{SD}(Z_i^{SD}(x^*))\} \leq \lambda^*, \quad i = 1, \ldots, k.
\end{aligned}
$$

This contradicts the assumption that $x^* \in X$, $\lambda^* \in \Lambda$ is an unique optimal solution of MINMAX $(\hat{\mu})$. □

Theorem 2 *If $x^* \in X$ is an EV-Pareto optimal solution of MOP-EV2, then there exists reference membership values $\hat{\mu} = (\hat{\mu}_1, \ldots, \hat{\mu}_k)$ such that $x^* \in X$, $\lambda^* = \hat{\mu}_i - \mu_{D_i}(x^*), i = 1, \ldots, k$ is an optimal solution of MINMAX($\hat{\mu}$).*

Proof Let us assume that $x^* \in X$, $\lambda^* = \hat{\mu}_i - \mu_{D_i}(x^*) = \max\{\hat{\mu}_i - Z_i^E(x^*), \hat{\mu}_i - \mu_i^{SD}(Z_i^{SD}(x^*))\}, i = 1, \ldots, k$, is not an optimal solution of MINMAX $(\hat{\mu})$. Then, there exists $x \in X$ and $\lambda < \lambda^*$ such that

$$
\begin{cases}
\hat{\mu}_i - Z_i^E(x) \leq \lambda < \lambda^* \\
\hat{\mu}_i - \mu_i^{SD}(Z_i^{SD}(x)) \leq \lambda < \lambda^*
\end{cases}
$$
$$
\begin{aligned}
&\Leftrightarrow \hat{\mu}_i - \mu_{D_i}(x) \leq \lambda < \lambda^* \\
&\Leftrightarrow \hat{\mu}_i - \mu_{D_i}(x) < \hat{\mu}_i - \mu_{D_i}(x^*) \\
&\Leftrightarrow \mu_{D_i}(x) > \mu_{D_i}(x^*)
\end{aligned}
$$

for all $i = 1, \ldots, k$. This contradicts the fact that $x^* \in X$ is an EV-Pareto optimal solution of MOP-EV2. □

5 An Interactive Algorithm

In Theorem 1, if the optimal solution (x^*, λ^*) of MINMAX $(\hat{\mu})$ is not unique, the EV-Pareto optimality can not be guaranteed. In order to guarantee the EV-Pareto optimality for (x^*, λ^*), we formulate the EV-Pareto optimality test problem. Before formulating such a test problem, without loss of generality, we assume that the following inequalities hold at the optimal solution $x^* \in X, \lambda^* \in \Lambda$.

$$
Z_i^E(x^*) \leq \mu_i^{SD}(Z_i^{SD}(x^*)), \quad i \in I_1 \tag{30}
$$

$$Z_i^E(x^*) > \mu_i^{SD}(Z_i^{SD}(x^*)), \quad i \in I_2 \tag{31}$$

$$I_1 \cup I_2 = \{1, \cdots, k\}, \quad I_1 \cap I_2 = \phi \tag{32}$$

Under the above conditions, we formulate the following EV-Pareto optimality test problem.

[EV-Pareto optimality test problem]

$$\max_{x \in X, \varepsilon_i \geq 0, i=1,\ldots,k} \sum_{i=1}^{k} \epsilon_i$$

subject to

$$Z_i^E(x) \geq Z_i^E(x^*) + \epsilon_i, i \in I_1$$

$$\mu_i^{SD}(Z_i^{SD}(x)) \geq Z_i^E(x^*) + \epsilon_i, i \in I_1$$

$$Z_i^E(x) \geq \mu_i^{SD}(Z_i^{SD}(x^*)) + \epsilon_i, i \in I_2$$

$$\mu_i^{SD}(Z_i^{SD}(x)) \geq \mu_i^{SD}(Z_i^{SD}(x^*)) + \epsilon_i, \in I_2$$

The following theorem shows the relationships between the optimal solution of EV-Pareto optimality test problem and the EV-Pareto optimal solution for MOP-EV2.

Theorem 3 *Let $\breve{x} \in X$, $\breve{\epsilon}_i \geq 0$, $i = 1, \ldots, k$ be an optimal solution of the EV-Pareto optimality test problem for (x^*, λ^*). If $\sum_{i=1}^{k} \breve{\epsilon}_i = 0$, then $x^* \in X$ is an EV-Pareto optimal solution.*

Proof Assume that $\breve{\epsilon}_i = 0$, $i = 1, \ldots, k$. If $x^* \in X$ is not an EV-Pareto optimal solution, there exists some $x \in X$ such that $\mu_{D_i}(x) \geq \mu_{D_i}(x^*)$, $i = 1, \ldots, k$, with strict inequality holding for at least one i. From the inequalities (30) and (31), this is equivalent to the following relations.

$$\min\{Z_i^E(x), \mu_i^{SD}(Z_i^{SD}(x))\} \geq \min\{Z_i^E(x^*), \mu_i^{SD}(Z_i^{SD}(x^*))\}$$
$$= \begin{cases} Z_i^E(x^*), & i \in I_1 \\ \mu_i^{SD}(Z_i^{SD}(x^*)), & i \in I_2 \end{cases}$$

As a result, the following inequalities holds.

$$\begin{cases} Z_i^E(x) \geq Z_i^E(x^*), & i \in I_1 \\ \mu_i^{SD}(Z_i^{SD}(x)) \geq Z_i^E(x^*), & i \in I_1 \\ Z_i^E(x) \geq \mu_i^{SD}(Z_i^{SD}(x^*)), & i \in I_2 \\ \mu_i^{SD}(Z_i^{SD}(x)) \geq \mu_i^{SD}(Z_i^{SD}(x^*)), & i \in I_2 \end{cases}$$

with strict inequality holding for at least one $i \in I_1 \cup I_2$. Hence, there must exist at least one i such that $\breve{\epsilon}_i > 0$. This contradicts the assumption that $\breve{\epsilon}_i = 0$, $i = 1, \ldots, k$. $\quad \square$

Now, following the above discussions, we can construct the interactive algorithm in order to derive a satisfactory solution from among an EV-Pareto optimal solution set.

[An interactive algorithm]

Step 1 The decision maker sets the membership function $\mu_{\tilde{G}_i}(y)$, $i = 1, \ldots, k$ for the fuzzy goals of the objective functions in MOFRLP.

Step 2 The decision maker sets the membership function $\mu_i^{SD}(Z_i^{SD}(x))$, $i = 1, \ldots, k$.

Step 3 Set the initial reference membership values as $\hat{\mu}_i = 1$, $i = 1, \ldots, k$.

Step 4 Solve MINMAX($\hat{\mu}$), and obtain the optimal solution $x^* \in X, \lambda^* \in \Lambda$. In order to guarantee EV-Pareto optimality, solve the EV-Pareto optimality test problem for $x^* \in X$.

Step 5 If the decision maker is satisfied with the current value of the EV-Pareto optimal solution $x^* \in X$, then stop. Otherwise, the decision maker updates his/her reference membership values $\hat{\mu}_i$, $i = 1, \ldots, k$ and return to Step 4.

6 Numerical Example

In order to demonstrate the proposed method and the interactive processes, we consider the three-objective linear programming problem with fuzzy random variable coefficients which is defined as a numerical example in [24]. Let us assume that the hypothetical decision maker sets the membership functions $\mu_{\tilde{G}_i}(\cdot)$, $\mu_i^{SD}(\cdot)$, $i = 1, 2, 3$ as follows.

$$\mu_{\tilde{G}_i}(y) = \frac{y - z_i^0}{z_i^1 - z_i^0}, \quad z_i^1 \leq y \leq z_i^0, \quad i = 1, 2, 3$$

$$\mu_i^{SD}(s) = \frac{s - q_i^0}{q_i^1 - q_i^0}, \quad q_i^1 \leq s \leq q_i^0, \quad i = 1, 2, 3$$

where

$$\left(z_1^0, z_2^0, z_3^0 \right) = (-7.5, -0.937, 33.5),$$

$$\left(z_1^1, z_2^1, z_3^1 \right) = (-23.818, -16.25, 9.375),$$

$$\left(q_1^0, q_2^0, q_3^0 \right) = (0.6, 0.45, 0.4),$$

$$\left(q_1^1, q_2^1, q_3^1 \right) = (0.25, 0.12, 0.15).$$

Table 1 Interactive
processes

Iteration	1	2	3
$\hat{\mu}_1$	1	1	0.9
$\hat{\mu}_2$	1	0.8	0.8
$\hat{\mu}_3$	1	0.8	0.8
x_1^*	3.666	1.843	3.167
x_2^*	2.991	4.563	3.758
x_3^*	0	0.732	0.052
$Z_1^E(x^*)$	0.593	0.729	0.668
$Z_2^E(x^*)$	0.593	0.530	0.568
$Z_3^E(x^*)$	0.593	0.529	0.568
$\mu_1^{SD}(Z_1^{SD}(x^*))$	0.593	0.729	0.668
$\mu_2^{SD}(Z_2^{SD}(x^*))$	0.593	0.529	0.568
$\mu_3^{SD}(Z_3^{SD}(x^*))$	0.593	0.529	0.568
$\mu_{D_1}(x^*)$	0.593	0.729	0.668
$\mu_{D_2}(x^*)$	0.593	0.529	0.568
$\mu_{D_3}(x^*)$	0.593	0.529	0.568

The interactive processes under the hypothetical decision maker are summarized in Table 1.

Let us compare the proposal method based on EV-model with V-model proposed by Katagiri et al. [22]. According to V-model (MOP-V3), we set the parameters ξ_i heuristically as $(\xi_1, \xi_2, \xi_3) = (0.8, 0.7, 0.7)$. The results are summarized in Table 2. It is shown that the proper balance between the membership functions $\mu_{D_i}(x), i = 1, 2, 3$ is attained in the proposed method.

Table 2 Comparison
between EV-model and V-
model

Model	EV-model	V-model
$\hat{\mu}_1$	1	1
$\hat{\mu}_2$	1	1
$\hat{\mu}_3$	1	1
$Z_1^E(x^*)$	0.593	0.8
$Z_2^E(x^*)$	0.593	0.7
$Z_3^E(x^*)$	0.593	0.7
$\mu_1^{SD}(Z_1^{SD}(x^*))$	0.593	0.390
$\mu_2^{SD}(Z_2^{SD}(x^*))$	0.593	0.567
$\mu_3^{SD}(Z_3^{SD}(x^*))$	0.593	0.766
$\mu_{D_1}(x^*)$	0.593	0.390
$\mu_{D_2}(x^*)$	0.593	0.567
$\mu_{D_3}(x^*)$	0.593	0.7

7 Conclusion

In this paper, under the assumption that the decision maker intends to not only maximize the expected degrees of possibilities that the original objective functions attain the corresponding fuzzy goals, but also minimize the corresponding standard deviations for such possibilities, an interactive decision making method for MOFRLP is proposed. In the proposed method, a satisfactory solution is obtained from among an EV-Pareto optimal solution set through the interaction with the decision maker.

References

1. J.R. Birge, F. Louveaux, *Introduction to Stochastic Programming* (Springer, London, 1997)
2. A. Charnes, W.W. Cooper, Chance constrained programming. Manage. Sci. **6**, 73–79 (1959)
3. G.B. Danzig, Linear programming under uncertainty. Manage. Sci. **1**, 197–206 (1955)
4. P. Kall, J. Mayer, *Stochastic Linear Programming Models, Theory, and Computation* (Springer, New York, 2005)
5. V.J. Lai, C.L. Hwang, *Fuzzy Mathematical Programming* (Springer, Berlin, 1992)
6. M. Sakawa, *Fuzzy Sets and Interactive Multiobjective Optimization* (Plenum Press, New York, 1993)
7. H.-J. Zimmermann, *Fuzzy Sets, Decision-Making and Expert Systems* (Kluwer Academic Publishers, Boston, 1987)
8. H. Kwakernaak, Fuzzy random variable-1. Inf. Sci. **15**, 1–29 (1978)
9. M.K. Luhandjula, M.M. Gupta, On fuzzy stochastic optimization. Fuzzy Sets Syst. **81**, 47–55 (1996)
10. M. Sakawa, I. Nishizaki, H. Katagiri, *Fuzzy Stochastic Multiobjective Programming* (Springer, Berlin, 2011)
11. G.-Y. Wang, Z. Qiao, Linear programming with fuzzy random variable coefficients. Fuzzy Sets Syst. **57**, 295–311 (1993)
12. H. Katagiri, M. Sakawa, Interactive multiobjective fuzzy random linear programming through the level set-based probability model. Inf. Sci. **181**, 1641–1650 (2011)
13. H. Katagiri, M. Sakawa, K. Kato, I. Nishizaki, Interactive multiobjective fuzzy random linear programming: maximization of possibility and probability. Eur. J. Oper. Res. **188**, 530–539 (2008)
14. M. Sakawa, H. Yano, and I. Nishizaki, *Linear and Multiobjective Programming with Fuzzy Stochastic Extensions* (Springer, Berlin, 2013)
15. H. Yano, K. Matsui, Fuzzy approaches for multiobjective fuzzy random linear programming problems through a probability maximization model, in *Lecture Notes in Engineering and Computer Science: Proceedings of The International Multiconference of Engineers and Computer Scientists 2011*, pp. 1349–1354 (2011)
16. D. Dubois, H. Prade, *Fuzzy Sets and Systems* (Academic Press, New York, 1980)
17. H. Yano, Interactive decision making for fuzzy random multiobjective linear programming problems with variance-covariance matrices through probability maximization, in *Proceedings of The 6th International Conference on Soft Computing and Intelligent Systems and 13th International Symposium on Advanced Intelligent Systems*, pp. 965–970 (2012)
18. H. Yano, Fuzzy decision making for fuzzy random multiobjective linear programming problems with variance covariance matrices. Inf. Sci. **272**, 111–125 (2014)

19. H. Yano, M. Sakawa, Interactive multiobjective fuzzy random linear programming through fractile criteria. Adv. Fuzzy Syst. **2012**, 1–9 (2012)
20. H. Yano, M. Sakawa, Interactive fuzzy programming for multiobjective fuzzy random linear programming problems through possibility-based probability maximization. Oper. Res. Int. J. **2013**(07), 1–19 (2013)
21. H. Katagiri, M. Sakawa, H. Ishii, Multiobjective fuzzy random linear programming using E-model and possibility measure, in *Joint 9-th IFSA World Congress and 20-th NAFIPS International Conference*, Vancouver, pp. 2295–2300 (2001)
22. H. Katagiri, M. Sakawa, S. Osaki, *An interactive satisficing method through the variance minimization model for fuzzy random linear programming problems.* Multi-Objective Programming and Goal-Programming: Theory and Applications (Advances in Soft Computing) (Springer, Berlin, 2003), pp. 171–176
23. D.J. White, *Optimality and Efficiency* (Wiley, New York, 1982)
24. H. Yano, K. Matsui, M. Furuhashi, Interactive Decision Making for Multiobjective Fuzzy Random Linear Programming Problems Using Expectations and Coefficients of Variation, in *Lecture Notes in Engineering and Computer Science: Proceedings of The International MultiConference of Engineers and Computer Scientists 2014*, IMECS, 12–14 Mar 2014, Hong Kong, pp. 1251–1256 (2014)

Fixed Point Theorem and Stability for (α, ψ, ξ)-Generalized Contractive Multivalued Mappings

Supak Phiangsungnoen, Nopparat Wairojjana and Poom Kumam

Abstract In this paper, we introduce and prove a fixed point theorem for (α, ψ, ξ)-generalized contractive multivalued mappings on collections of non-empty closed subsets. We also prove the ξ-generalized Ulam-Hyers stability results for fixed point inclusion. Finally, we provide illustrative example to support our main result.

Keywords Admissible mapping · (α, ψ, ξ)-generalized contractive · Fixed point · Generalized Ulam-Hyer stability · Hausdorff metric · Multivalued mappings

1 Introduction

Fixed point theory has many applications in functional analysis. The contractive conditions on underlying functions play an important role for finding solutions of metric fixed point problems. Over the years, there have been generalized contractive in single valued and multivalued mappings by several mathematicians (see [1–5] and references therein). In 2012, Samet et al. [6] first introduced the concept of admissible mapping for single valued mapping. Afterwards, Asl et al. [7] extended the concept of admissibility for single valued mappings to multivalued mappings. Very,

S. Phiangsungnoen · P. Kumam (✉)
Department of Mathematics, Faculty of Science, King Mongkut's University of Technology
Thonburi (KMUTT), 126 Pracha Uthit Road, Bang Mod, Thung Khru, Bangkok 10140,
Thailand
e-mail: poom.kum@kmutt.ac.th

S. Phiangsungnoen
e-mail: supuk_piang@hotmail.com

N. Wairojjana
Faculty of Science and Technology, Valaya Alongkorn Rajabhat University under the Royal
Patronage, Number 1, Moo 20, Phaholyothin Road, Klong Neung Subdistrict,
Klong Luang District, Pathumthani Province 13180, Thailand
e-mail: Noparatw@windowslive.com

© Springer Science+Business Media Dordrecht 2015   127
G.-C. Yang et al. (eds.), *Transactions on Engineering Technologies*,
DOI 10.1007/978-94-017-9588-3_10

recently, Ali et al. [8] introduce the notion of (α, ψ, ξ)-contractive multivalued mappings to generalize and extend the notion of α-ψ-contractive mappings to closed valued mappings.

On the other hand, the stability problem of functional equations originated with a question of Ulam [9] concerning the stability of group homomorphisms. Afterward, Hyers [10] gave a first affirmative partial answer to the question of Ulam for Banach spaces, this type of stability is called Ulam-Hyers stability. Several authors consider Ulam-Hyers stability results in fixed point theory and remarkable result on the stability of certain classes of functional equations via fixed point approach (see [11–16] and references therein). Very recently, Phiangsungnoen and Kumam [17] proved the existence of fixed point for generalized multivalued almost contractions and prove the generalized Ulam-Hyers stability of fixed point problems for multi-valued operators.

In this paper, we introduce an (α, ψ, ξ)-generalized contractive multivalued mappings and we prove the existence of fixed point on collections of non-empty closed subsets. We also prove the generalized Ulam-Hyers stability results for fixed point inclusion. We provide illustrative example to support our main result.

2 Preliminaries

Throughout in this paper the letters $\mathbf{N}$ and $\mathbf{R}$ will denote the set of positive integer numbers and the set of real numbers, respectively. Let X be a non empty set and (X, d) be a metric space, $\mathscr{C}\mathscr{L}(X)$ denote the class of all nonempty closed subsets of X. For every $A, B \in \mathscr{C}\mathscr{L}(X)$, let

$$H(A, B) = \begin{cases} \max\{\sup_{x \in A} d(x, B), \sup_{y \in B} d(y, A)\}, & \text{if the maximum exists;} \\ \infty, & \text{otherwise.} \end{cases}$$

Such a map H is called the generalized Hausdorff metric induced by the metric d. Let Ψ be a set of nondecreasing functions, $\psi : [0, \infty) \to [0, \infty)$ such that $\sum_{n=1}^{\infty} \psi^n(t) < \infty$ for each $t > 0$, where ψ^n is the nth iterate of ψ. It is known that for each $\psi \in \Psi$, we have $\psi(t) < t$ for all $t > 0$ and $\psi(0) = 0$ for $t = 0$.

Let Ξ be a family of function $\xi : [0, \infty) \to [0, \infty)$ satisfying the following conditions:

1. ξ is continuous;
2. ξ is nondecreasing on $[0, \infty)$;
3. $\xi(0) = 0$ and $\xi(t) > 0$ for all $t \in (0, \infty)$;
4. ξ is subadditive.

Lemma 1 [8] *Let (X, d) be a metric space and let $\xi \in \Xi$. Then $(X, \xi \circ d)$ is a metric space.*

Lemma 2 [8] *Let (X, d) be a metric space and let $\xi \in \Xi$ and let $B \in \mathscr{CL}(X)$. Assume that there exits $x \in X$ such that $\xi(d(x, B)) > 0$. Then there exists $y \in B$ such that*

$$\xi(d(x, y)) < q\xi(d(x, B)),$$

where $q > 1$.

On the other hand, Asl et al. [7] introduced the concept of α_*-admissible mapping for multivalued mappings which motivated the concept of Samet et al. [6].

Definition 3 ([7]) Let X be a nonempty set, $S : X \to 2^X$, where 2^X is a collection of nonempty subsets of X and $\alpha : X \times X \to [0, \infty)$. We say that S is α_*-admissible mapping if

$$for \; x, y \in X, \quad \alpha(x, y) \geq 1 \Rightarrow \alpha_*(Sx, Sy) \geq 1,$$

where

$$\alpha_*(Sx, Sy) := \inf\{\alpha(a, b) : a \in Sx \text{ and } b \in Sy\}.$$

3 Fixed Point Results

In this section, we introduce a generalized (α, ψ, ξ)-contractive for multivalued mappings and prove the existence of fixed point theorem in generalized Hausdorff metric spaces.

Definition 4 Let (X, d) be a metric space. A mapping $S : X \to \mathscr{CL}(X)$ is called a generalized (α, ψ, ξ)-contractive if there exist three functions $\psi \in \Psi$, $\xi \in \Xi$ and $\alpha : X \times X \to [0, \infty)$ such that

$$\alpha(x, y) \geq 1 \Rightarrow \xi(H(Sx, Sy)) \leq \psi(\xi(M(x, y))) \\ + L\min\{d(x, Sx), d(y, Sy), d(x, Sy), d(y, Sx)\}. \tag{1}$$

for all $x, y \in X, t > 0$, where $L \geq 0$ and

$$M(x, y) = \max\{d(x, y), d(x, Sx), d(y, Sy), \frac{d(x, Sy) + d(y, Sx)}{2}\}.$$

In case when $\psi \in \Psi$ is strictly increasing, the (α, ψ, ξ)-contractive mapping is called a strictly (α, ψ, ξ)-contractive mapping.

Theorem 5 *Let (X, d) be a complete metric space. The mapping $S : X \to \mathscr{CL}(X)$ is a generalized (α, ψ, ξ)-contractive mapping satisfying the following assertions:*

(a) *S is α_*-admissible mapping;*
(b) *there exists $x_0 \in X$ and $x_1 \in Tx_0$ such that $\alpha(x_0, x_1) \geq 1$;*
(c) *either S is continuous or for any sequence $\{x_n\}$ in X with $\alpha(x_n, x_{n+1}) \geq 1$ for all $n \in \mathbf{N} \cup \{0\}$ and $x_n \to x$ as $n \to \infty$, we have $\alpha(x_n, x) \geq 1$ for all $n \in \mathbf{N} \cup \{0\}$.*

Then S has a fixed point.

Proof For arbitrary $x_0 \in X$ and $x_1 \in Sx_0$, we have

$$\alpha(x_0, x_1) \geq 1.$$

□

If $Sx_0 = Sx_1$, then $x_1 \in Sx_1$ is a fixed point of S. Assume that $Sx_0 \neq Sx_1$ and $x_1 \notin Sx_1$. By condition (1), we have

$$0 < \xi(H(Sx_0, Sx_1))$$
$$\leq \psi(\xi(M(x_0, x_1))) + L \min\{d(x_0, Sx_0), d(x_1, Sx_1), d(x_0, Sx_1), d(x_1, Sx_0)\}.$$

Since

$$M(x_0, x_1) = \max\{d(x_0, x_1), d(x_0, Sx_0), d(x_1, Sx_1), \frac{d(x_0, Sx_1) + d(x_1, Sx_0)}{2}\}$$
$$\leq \max\{d(x_0, x_1), d(x_1, Sx_1), \frac{d(x_0, x_1) + d(x_1, Sx_1)}{2}\}$$
$$= \max\{d(x_0, x_1), d(x_1, Sx_1)\}.$$

So, we have

$$0 < \xi(H(Sx_0, Sx_1))$$
$$\leq \psi(\xi(\max\{d(x_0, x_1), d(x_1, Sx_1)\}))$$
$$\quad + L \min\{d(x_0, x_1), d(x_1, Sx_1), d(x_0, Sx_1), d(x_1, x_1)\}$$
$$\leq \psi(\xi(\max\{d(x_0, x_1), d(x_1, Sx_1)\})).$$

Assume that $\max\{d(x_0, x_1), d(x_1, Sx_1)\} = d(x_1, Sx_1)$. Since $\psi \in \Psi$, we get

$$0 \leq \xi(d(Sx_0, Sx_1))$$
$$\leq \psi(H(Sx_0, Sx_1))$$
$$\leq \psi(\xi(\max\{d(x_0, x_1), d(x_1, Sx_1)\}))$$
$$\leq \psi(\xi(d(x_1, Sx_1)))$$
$$< \xi(d(x_1, Sx_1))$$

which is a contradiction. Thus, $\max\{d(x_0, x_1), d(x_1, Sx_1)\} = d(x_0, x_1)$. So, we have

$$0 < \xi(d(x_1, Sx_1)) \leq \xi(H(Sx_0, Sx_1)) \leq \psi(\xi(d(x_0, x_1))). \tag{2}$$

By Lemma 2, for $p > 1$ and Sx_1 is a nonempty closed subset of X, there exists $x_2 \in Sx_1$ such that

$$0 < \xi(d(x_1, x_2)) < p\xi(d(x_1, Sx_1)). \tag{3}$$

Obtain (3) in (2), we get

$$0 < \xi(d(x_1, x_2)) < p\psi(\xi(d(x_0, x_1))). \tag{4}$$

Using ψ in (4), we have

$$0 < \psi(\xi(d(x_1, x_2))) < \psi(p\psi(\xi(d(x_0, x_1)))). \tag{5}$$

Let $p_1 = \frac{\psi(p\psi(\xi(d(x_0, x_1))))}{\psi(\xi(d(x_1, x_2)))}$, then $p_1 > 1$. Since, S is α_*-admissible mapping, we have

$$\alpha(Sx_0, Sx_1) = \alpha(x_1, x_2) \geq 1.$$

For $x_2 \in Sx_1$, again by (1) and $x_2 \notin Sx_2$, we get

$$0 < \xi(H(Sx_1, Sx_2))$$
$$\leq \psi(\xi(M(x_1, x_2))) + L\min\{d(x_1, Sx_1), d(x_2, Sx_2), d(x_1, Sx_2), d(x_2, Sx_1)\}.$$

$$M(x_1, x_2) = \max\{d(x_1, x_2), d(x_1, Sx_1), d(x_2, Sx_2), \frac{d(x_1, Sx_2) + d(x_2, Sx_1)}{2}\}$$
$$\leq \max\{d(x_1, x_2), d(x_2, Sx_2), \frac{d(x_1, x_2) + d(x_2, Sx_2)}{2}\}$$
$$= \max\{d(x_1, x_2), d(x_2, Sx_2)\}.$$

So, we have

$$0 < \xi(H(Sx_1, Sx_2))$$
$$\leq \psi(\xi(\max\{d(x_1, x_2), d(x_2, Sx_2)\}))$$
$$\quad + L\min\{d(x_1, x_2), d(x_2, Sx_2), d(x_1, Sx_2), d(x_2, x_2)\}$$
$$\leq \psi(\xi(\max\{d(x_1, x_2), d(x_2, Sx_2)\})).$$

Assume that $\max\{d(x_1, x_2), d(x_2, Sx_2)\} = d(x_2, Sx_2)$. Since $\psi \in \Psi$, we get

$$0 < \xi(d(x_2, Sx_2)) \leq \xi(H(Sx_1, Sx_2))$$
$$\leq \psi(\xi(\max\{d(x_1, x_2), d(x_2, Sx_2)\}))$$
$$\leq \psi(\xi(d(x_2, Sx_2)))$$
$$< \xi(d(x_2, Sx_2))$$

which is a contradiction. Then, $\max\{d(x_1, x_2), d(x_2, Sx_2)\} = d(x_1, x_2)$. So, we have

$$0 < \xi(d(x_2, Sx_2)) \leq \xi(H(Sx_1, Sx_2)) \leq \psi(\xi(d(x_1, x_2))). \tag{6}$$

By Lemma 2, for $p_1 > 1$ and Sx_2 is a nonempty closed subset of X, there exists $x_3 \in Sx_2$ such that

$$0 < \xi(d(x_2, x_3)) < p_1 \xi(d(x_2, Sx_2)). \tag{7}$$

By (7) and (6), we get

$$0 < \xi(d(x_2, x_3)) < p_1 \psi(\xi(d(x_1, x_2))) < \psi(p\psi(\xi(d(x_0, x_1)))). \tag{8}$$

Using ψ in (9), we have

$$0 < \psi(\xi(d(x_2, x_3))) < \psi(p_1 \psi(\xi(d(x_1, x_2)))) < \psi^2(p\psi(\xi(d(x_0, x_1)))). \tag{9}$$

Let $p_2 = \frac{\psi^2(p\psi(\xi(d(x_0, x_1))))}{\psi(\xi(d(x_2, x_3)))}$, then $p_2 > 1$. Continuing this process, we can define a sequence $\{x_n\}$ in X by $x_{n+1} \in S_n$ such that $\alpha(x_n, x_{n+1}) \geq 1$ for all $n \geq 0$. By condition (1), we have

$$
\begin{aligned}
0 < \xi(H(Sx_{n-1}, Sx_n)) \\
\leq \psi(\xi(M(x_{n-1}, x_n))) + L\min\{d(x_{n-1}, Sx_{n-1}), d(x_n, Sx_n), \\
d(x_{n-1}, Sx_n), d(x_n, Sx_{n-1})\}.
\end{aligned}
$$

On the other hand,

$$
\begin{aligned}
M(x_{n-1}, x_n) &= \max\{d(x_{n-1}, x_n), d(x_{n-1}, Sx_{n-1}), d(x_n, Sx_n), \frac{d(x_{n-1}, Sx_n) + d(x_n, Sx_{n-1})}{2}\} \\
&\leq \max\{d(x_{n-1}, x_n), d(x_n, Sx_n), \frac{d(x_{n-1}, x_n) + d(x_n, Sx_n)}{2}\} \\
&= \max\{d(x_{n-1}, x_n), d(x_n, Sx_n)\}.
\end{aligned}
$$

So, we have

$$
\begin{aligned}
0 < \xi(H(Sx_{n-1}, Sx_n)) \\
\leq \psi(\xi(M(x_{n-1}, x_n))) + L\min\{d(x_{n-1}, Sx_{n-1}), d(x_n, Sx_n), \\
d(x_{n-1}, Sx_n), d(x_n, Sx_{n-1}) \\
\leq \psi(\xi(\max\{d(x_{n-1}, x_n), d(x_n, Sx_n)\})) + L\min\{d(x_{n-1}, Sx_{n-1}), \\
d(x_n, Sx_n), d(x_{n-1}, Sx_n), d(x_n, Sx_{n-1})
\end{aligned}
$$

Assume that $\max\{d(x_{n-1}, x_n), d(x_n, Sx_n)\} = d(x_n, Sx_n)$. Since $\psi \in \Psi$, we get

$$0 < \xi(d(x_n, Sx_n))$$
$$\leq \xi(H(Sx_{n-1}, Sx_n))$$
$$\leq \psi(\xi(\max\{d(x_{n-1}, x_n), d(x_n, Sx_n)\}))$$
$$\leq \psi(\xi(d(x_n, Sx_n))))$$
$$< \xi(d(x_n, Sx_n))$$

which is a contradiction. Hence, for all $n \in \mathbf{N}$, we have

$$0 < \xi(d(x_n, Sx_n)) \leq \xi(H(Sx_{n-1}, Sx_n)) \leq \psi(\xi(d(x_{n-1}, x_n))). \tag{10}$$

By Lemma 2, for $p_{n-1} > 1$ and Sx_n is a nonempty closed subset of X, there exists $x_{n+1} \in Sx_n$ such that

$$0 < \xi(d(x_n, x_{n+1})) < p_{n-1}\xi(d(x_n, Sx_n)). \tag{11}$$

By (10) and (11), we get

$$0 < \xi(d(x_n, x_{n+1})) < p_{n-1}\psi(\xi(d(x_{n-1}, x_n))) < \psi^{n-1}(p\psi(\xi(d(x_0, x_1)))). \tag{12}$$

Using ψ in (12), we have

$$0 < \psi(\xi(d(x_n, x_{n+1}))) < \psi^n(p\psi(\xi(d(x_0, x_1)))). \tag{13}$$

For $\varepsilon > 0$, there exists $N \in \mathbf{N}$ such that

$$\sum_{n \geq N} \psi^n(p\psi(\xi d(x_n, x_{n+1}))) < \infty$$

for all $n \geq N$. Let $m, n \in \mathbf{N}$ with $m > n \geq N$ and using the triangular inequality, we get

$$\xi(d(x_n, x_m)) \leq \sum_{k=n}^{m-1} \xi(d(x_k, x_{k+1})) \leq \sum_{n \geq N} \psi^n(p\psi(\xi d(x_n, x_{n+1}))).$$

This implies that, $\lim_{m,n\to\infty} \xi(d(x_n, x_m)) = 0$ and $\lim_{m,n\to\infty} d(x_n, x_m) = 0$. Hence $\{x_n\}$ is a Cauchy sequence. Since X is complete, there exists $v \in X$ such that $x_n \to v$ as $n \to \infty$. Suppose that S is continuous, we have

$$d(v, Sv) = \lim_{n\to\infty} d(x_{n+1}, Sv) \leq \lim_{n\to\infty} H(Sx_n, Sv) = 0.$$

Thus v is a fixed point of S. On the other hand, if $\alpha(x_n, x_{n+1}) \geq 1$ for all $n \in \mathbf{N} \cup \{0\}$ and $x_n \to v$ as $n \to \infty$, we have $\alpha(x_n, v) \geq 1$ for all $n \in \mathbf{N} \cup \{0\}$. Since ξ is subadditive and by condition (1), we have

$$
\begin{aligned}
\xi(d(v, Sv)) &\leq \xi(d(v, x_{n+1}) + d(x_{n+1}, Sv)) \\
&\leq \xi(d(v, x_{n+1})) + \xi(H(Sx_n, Sv)) \\
&\leq \xi(d(v, x_{n+1})) + [\psi(\xi(M(x_n, v))) + L\min\{d(x_n, Sx_n), \\
&\quad d(v, Sv), d(x_n, Sv), d(v, Sx_n)\}].
\end{aligned} \tag{14}
$$

Since

$$
M(x_n, v) = \max\{d(x_n, v), d(x_n, Sx_n), d(v, Sv), \frac{d(x_n, Sv) + d(v, Sx_n)}{2}\}. \tag{15}
$$

Letting $n \to \infty$ in (15), we get $\lim_{n \to \infty} M(x_n, v) = d(v, Sv)$. Now, letting $n \to \infty$ in (14), we have $\xi(d(v, Sv)) \leq \psi(\xi(d(v, Sv)))$. If $\xi(d(v, Sv)) > 0$, since $\psi \in \Psi$, we get

$$
\xi(d(v, Sv)) \leq \psi(\xi(d(v, Sv))) < \xi(d(v, Sv)).
$$

This is not possible. Thus, $\psi(\xi(d(v, Sv))) = 0$, that means, $\xi(d(v, Sv)) = 0$, which implies that $d(v, Sv) = 0$. Therefore, v is a fixed point of S. This completes the proof.

Now, we give an illusive example to support our main result.

Example 6 Let $X = \mathbf{R}$ and define $d(x, y) = |x - y|$. Define the mapping $S : X \to \mathscr{CL}(X)$ by

$$
Sx = \begin{cases} \{0\}, & x < 1 \\ \frac{x+3}{4}, & 1 \leq x \leq 5 \\ x^2, & x > 5. \end{cases}
$$

Define $\alpha : X \times X \to [0, \infty)$ *by*

$$
\alpha(x, y) = \begin{cases} 2, & 1 \leq x, y \leq 5, \\ 0, & otherwise. \end{cases}
$$

One can easy to check that S is an α_*-admissible mapping. Setting $\psi \in \Psi$, $\psi(t) = \frac{1}{2}t$ and $\xi(t) = \sqrt{t}$. By the assumption, we get

$$
H(Sx, Sy) = |\frac{x+3}{4} - \frac{y+3}{4}| = \frac{1}{4}|x - y| = \frac{1}{4}d(x, y).
$$

Moreover, we get

$$M(x,y) = \max\{d(x,y), d(x, Sx), d(y, Sy), \frac{d(x, Sy) + d(y, Sx)}{2}\} = d(x,y).$$

Thus, we get

$$\begin{aligned} \xi(H(Sx, Sy)) &= \frac{1}{2}\sqrt{d(x,y)} = \psi(\xi(d(x,y))) = \frac{1}{2}\sqrt{d(x,y)} \\ &\leq \psi(\xi(d(x,y))) + L\min\{d(x, Sx), d(y, Sy), \\ &\quad d(x, Sy), d(y, Sx)\}. \end{aligned}$$

for all $x, y \in X, t > 0$, where $L > 0$.

Therefore all conditions in Theorem 5 hold, thus there exists $1 \in X$ is a fixed point of T. This completes the proof.

Remark 7 From Example 6, if S is not an α_*-admissible mapping. Setting $x = 0$, $y = 6$, $\xi(t) = \sqrt{t}$ and $\psi(t) = \frac{1}{2}t$, we have

$$M(x,y) = \max\{d(x,y), d(x, Sx), d(y, Sy), \frac{d(x, Sy) + d(y, Sx)}{2}\} = 30$$

and $\min\{d(x, Sx), d(y, Sy), d(x, Sy), d(y, Sx)\} = 0$. Thus, we have

$$\xi(H(Sx, Sy)) = \sqrt{30} > \psi(\xi(d(x,y))) = \frac{1}{2}\sqrt{6}.$$

Also, if $\xi(t) = t$ and $\psi(t) = kt$, then

$$\xi(H(Sx, Sy)) = H(Sx, Sy) = 30 > \psi(\xi(d(x,y))) = kd(x,y) = 6k.$$

for all $0 < k < 1$. Therefore, Nadler's fixed point theorem is not applicable.

If $L = 0$ in Theorem 5, we have the following result of Ali et al. [8].

Corollary 8 *Let (X, d) be a complete metric space. The mapping $S : X \to \mathscr{CL}(X)$ i following conditions hold:*

(a) S is α_*-admissible mapping;
(b) For $x, y \in X$ there exist three functions $\psi \in \Psi$, $\xi \in \Xi$ and $\alpha : X \times X \to [0, \infty)$ such that

$$\alpha(x, y) \geq 1 \Rightarrow \xi(H(Sx, Sy)) \leq \psi(\xi(M(x, y)));$$

(c) there exists $x_0 \in X$ and $x_1 \in Tx_0$ such that $\alpha(x_0, x_1) \geq 1$;
(d) either S is continuous or for any sequence $\{x_n\}$ in X with $\alpha(x_n, x_{n+1}) \geq 1$ for all $n \in \mathbf{N} \cup \{0\}$ and $x_n \to x$ as $n \to \infty$, we have $\alpha(x_n, x) \geq 1$ for all $n \in \mathbf{N} \cup \{0\}$.

Then S has a fixed point.

By taking $\xi(t) = t$ for each $t > \geq 0$ in Corollary 8, we have the following result of Asl et al. [7].

Corollary 9 *Let* (X, d) *be a complete metric space. The mapping* $S : X \to \mathscr{CL}(X)$ *i following conditions hold:*

(a) S *is* α_*-*admissible mapping;*
(b) *For* $x, y \in X$ *there exist two functions* $\psi \in \Psi$ *and* $\alpha : X \times X \to [0, \infty)$ *such that*

$$\alpha(x, y) \geq 1 \Rightarrow H(Sx, Sy) \leq \psi(M(x, y));$$

(c) *there exists* $x_0 \in X$ *and* $x_1 \in Tx_0$ *such that* $\alpha(x_0, x_1) \geq 1;$
(d) *either* S *is continuous or for any sequence* $\{x_n\}$ *in* X *with* $\alpha(x_n, x_{n+1}) \geq 1$ *for all* $n \in \mathbf{N} \cup \{0\}$ *and* $x_n \to x$ *as* $n \to \infty$, *we have* $\alpha(x_n, x) \geq 1$ *for all* $n \in \mathbf{N} \cup \{0\}$.

Then S has a fixed point.

4 The Ulam-Hyers Stability of Fixed Point Inclusion

We start this section by presenting the Ulam-Hyers stability concepts for the fixed point problem associated to a multivalued operator. Now, we recall the following (generalized) functional is used in this section.

The gap functional $D : X \times X \to [0, \infty]$

$$D(A, B) = \begin{cases} \inf\{d(a, b)|a \in A,\ b\}, & A \neq \emptyset \neq B, \\ 0, & A = \emptyset = B, \\ +\infty, & otherwise. \end{cases}$$

Definition 10 Let (X, d) be metric space and $S : X \to \mathscr{CL}(X)$ be an operator. By definition, the fixed point inclusion

$$x \in Sx \tag{16}$$

for all $x \in X$ and for each $\varepsilon > 0$ real number the following inequality

$$D(y, Sy) \leq \varepsilon, \tag{17}$$

is said to be ξ-generalized Ulam-Hyers stable. If there exists an increasing operator $\varphi : [0, \infty) \to [0, \infty)$, continuous at 0, $\varphi(0) = 0$ and $\xi \in \Xi$ such that for each $\varepsilon > 0$

real number and each ε-solution $y^* \in X$ an solution of the inequality (16) there exists a solution $x^* \in X$ of the fixed point inclusion (16) such that

$$d(y^*, x^*) \leq \varphi(\xi(\varepsilon)). \tag{18}$$

If $\xi(t) = t$ for each $t \geq 0$ then the fixed point inclusion (16) is said to be generalized Ulam-Hyers stable. If there exists $c > 0$ such that $\varphi(t) := ct$, for each $t \in [0, \infty)$, then the fixed point inclusion (16) is said to be Ulam-Hyers stable.

Now, we prove the ξ-generalized Ulam-Hyers stability for fixed point problems which Theorem 5 hold.

Theorem 11 *Let (X, d) be a complete metric space. Suppose that all the hypotheses of* Theorem 5 *hold and also that the function $\varphi : [0, \infty) \rightarrow [0, \infty)$ defined by $\varphi(t) := \xi(t) - \psi(\xi(t))$ is strictly increasing and onto. If $\alpha(v, v^*) \geq 1$ for all $v^* \in X$ which is an ε-solution, then, the fixed point inclusion (16) is ξ-generalized Ulam-Hyers stable.*

Proof By Theorem 5, we have $v \in Sv$, that is, $v \in X$ is a solution of the fixed point inclusion (16). Let $\varepsilon > 0$ and $v^* \in Sv^*$ is a solution of the inequality (17), that is

$$D(v^*, Sv^*) \leq \varepsilon.$$

Since $v, v^* \in X$ are ε-solution, we have

$$\alpha(v, v^*) \geq 1.$$

Now, we obtain

$$
\begin{aligned}
\xi(d(v, v^*)) = \xi(D(Sv, v^*)) \\
\leq \xi(D(Sv, Sv^*) + D(Sv^*, v^*)) \\
\leq \xi(D(Sv, Sv^*)) + \xi(D(Sv^*, v^*)) \\
\leq \xi(H(Sv, Sv^*)) + \xi(\varepsilon) \\
\leq \psi(\xi(M(v, v^*) + L\min\{d(v, Sv), d(v^*, Sv^*), d(v, Sv^*), d(v^*, Sv)\})) + \xi(\varepsilon) \\
\leq \psi(\xi(M(v, v^*))) + \xi(\varepsilon).
\end{aligned}
$$

Since

$$M(v, v^*) = \max\{d(v, v^*), d(v, Sv), d(v^*, Sv^*), \frac{d(v, Sv^*) + d(v^*, Sv)}{2}\} = d(v, v^*).$$

So, we have

$$\xi(d(v,v^*)) \leq \psi(\xi(d(v,v^*))) + \xi(\varepsilon).$$

It follows that

$$\xi(d(v,v^*)) - \psi(\xi(d(v,v^*))) \leq \xi(\varepsilon).$$

Since $\varphi(t) := \xi(t) - \psi(\xi(t))$, we have

$$\varphi(d(v,v^*)) := \xi(d(v,v^*)) - \psi(\xi(d(v,v^*))).$$

It implies that

$$d(v,v^*) \leq \varphi^{-1}(\xi(\varepsilon)).$$

Notice that $\varphi^{-1} : [0,\infty) \to [0,\infty)$ exists, is increasing, continuous at 0 and $\varphi^{-1}(0) = 0$. Therefore, the fixed point inclusion (16) is ξ-generalized Ulam-Hyers stable. This completes the proof. □

Corollary 12 *Let (X,d) be a complete metric space. Suppose that all the hypotheses of Corollary 8 hold and also that the function $\varphi : [0,\infty) \to [0,\infty)$ defined by $\varphi(t) := \xi(t) - \psi(\xi(t))$ is strictly increasing and onto. If $\alpha(v,v^*) \geq 1$ for all $v^* \in X$ which is an ε-solution, then, the fixed point inclusion (16) is ξ-generalized Ulam-Hyers stable.*

Acknowledgments The authors were supported by the Higher Education Research Promotion and National Research University Project of Thailand, Office of the Higher Education Commission (NRU-CSEC No.55000613).

References

1. J. von Neumann, Zur theorie der gesellschaftsspiele. Math. Ann. **100**(1), 295–320 (1928)
2. S.B. Nadler Jr, Multivalued contraction mapping. Pac. J. Math. **30**(2), 475–488 (1969)
3. R. Kannan, Some results on fixed points. Bull. Calcutta Math. Soc. **60**, 71–76 (1968)
4. I.A. Rus, *Generalized Contractions and Applications* (Cluj University Press, Cluj-Napoca, 2001)
5. V. Berinde, On the approximation of fixed points of weak contractive operators. Fixed Point Theory **4**(2), 131–142 (2003)
6. B. Samet, C. Vetro, P. Vetro, Fixed point theorems for α-ψ-contractive type mappings. Nonlinear Anal. **75**, 2154–2165 (2012)
7. J.H. Asl, S. Rezapour, N. Shahzad, On fixed points of α-ψ-contractive multifunctions. Fixed Point Theory Appl. **2012**, 212 (2012)
8. M.U. Ali, T. Kamran, E. Karapınar , (α, ψ, ξ)-contractive multivalued mappings. Fixed Point Theory Appl. **2014**, 7 (2014)
9. S.M. Ulam, *Problems in Modern Mathematics* (Wiley, New York, 1964)
10. D.H. Hyers, On the stability of the linear functional equation. Proc. Natl. Acad. Sci. U.S.A. **27** (4), 222–224 (1941)

11. M.F. Bota-Boriceanu, A. Petruşel, Ulam-Hyers stability for operatorial equations, Analel Univ. Al. I. Cuza, Iaşi, **57**, 65–74 (2011)
12. L. Cădariu, L. Găvruţa, P. Găvruţa, Fixed points and generalized Hyers-Ulam stability. Abstract Appl. Anal. **2012** (712743), 10 (2012)
13. I.A. Rus, The theory of a metrical fixed point theorem: theoretical and applicative relevances. Fixed Point Theory **9**(2), 541–559 (2008)
14. I.A. Rus, Remarks on Ulam stability of the operatorial equations. Fixed Point Theory **10**(2), 305–320 (2009)
15. W. Sintunavarat, Generalized Ulam-Hyers stability, well-posedness and limit shadowing of fixed point problems for α-β-contraction mapping in metric spaces. Sci. World J. **2014** (569174), 7 (2014)
16. RH. Haghi, M. Postolache, Sh. Rezapour, On T-stability of the Picard iteration for generalized ψ-contraction mappings. Abstr. Appl. Anal. **2012**(658971), 7 (2012)
17. S. Phiangsungnoen, P. Kumam, Ulam-Hyers Stability results for fixed point problems via generalized multivalued almost contraction, in *Proceedings of the International Multiconference of Engineers and Computer Scientists 2014*, pp. 1222–1225 (2014)

A System Development for Laboratory Assignment Problem with Rotations: A Mixed Integer Programming Approach

Takeshi Koide

Abstract Our department offers a course called pre-semi to junior students. The course plays a role as a preliminary seminar and the students are assigned to three different laboratories to experience research activities. An assignment of students to three laboratories has been conducted manually by a department faculty member in turns considering both student's preference for laboratories and some conditions for laboratories. The author has constructed a spreadsheet-based system to execute the assignment task efficiently. The assignment task is modeled mathematically as a mixed integer programming and its optimal solution is derived by the execution of external optimization software. The system has been modified repeatedly in response to opinions from department members. This paper reports the developed system and mathematical models. Numerical results are also demonstrated to show the efficiency of the system and how to seek a suitable assignment satisfied with requests from department members.

Keywords Laboratory assignment · Mixed integer programming · Operations research · Optimization · Spreadsheet · System development

1 Introduction

In our university department, students are assigned to a laboratory to launch their research in the second semester of the junior year. In the first semester of the same academic year, a preliminary seminar, called pre-semi, is provided for junior students where the students are temporally assigned to three different laboratories and experience introductory research activities. The course plays a role to assist students to determine their preferable laboratories.

T. Koide (✉)
Department of Intellegence and Informatics, Konan University, Okamoto 8-9-1,
Higashinada-ku, Kobe 658-8501, Japan
e-mail: koide@konan-u.ac.jp

The assignment in the pre-semi course is conducted by a faculty selected by rotation among department members. The assignment is essentially the task to determine which three slots each student is assigned to. The slot is defined as a pair between of a laboratory and a cycle, and a semester is divided into three cycles. The faculty member in charge of the assignment makes an effort to construct a better assignment with taking student's preference and some other conditions into account. The assignment task is troublesome for many faculties and it takes several hours to complete it. A system hence has been required to execute the assignment in a short time and conveniently.

The laboratory assignment is related to the matching problem where the aim is to find desirable pairs between two groups of agents. In 2012, Gale and Shapley [1] received the Nobel Prize in economics for their algorithm to derive a stable matching for stable marriage problem and its extended problems have been widely researched [2]. The matching algorithm assigns students to laboratories efficiently if laboratories have their preferences for students. In our target problem, however, students have their preferences for laboratories but laboratories do not have the preferences. Furthermore, our problem has to consider the rotation of students' assignment, which disables us to apply the matching algorithm to the target problem.

The timetabling problem is the most popular problem among researches associated with optimal assignments in universities [3]. Course scheduling and classroom assignment are mainly considered in typical timetabling problems. Each timetabling problem has individual situations and a specialized algorithm is usually proposed to deal with the problem. Various approaches are adopted for the timetable problems such as Tabu search [4], genetic algorithm [4, 5], evolutional computation [6], ant colony algorithm [7], artificial bee colony algorithm [8], mixed integer programming [9–12], and heuristic approach [13]. The target problem in this paper, the laboratory assignment problem, can be regarded as a variation of the timetabling problem.

This paper reports the system development for the pre-semi assignment from a theoretical viewpoint as well as a practical viewpoint. Mixed integer programming specified for the assignment is adopted because of its easiness to understand even for faculty members unfamiliar to computer algorithms. Spreadsheet software is adopted as an interface of the system due to its familiarity. The system is gradually updated through several acceptance tests with using practical data.

Section 2 introduces the practical task for the pre-semi assignment. The outline of the developed system is explained in Sect. 3. Some preprocessing task before a main optimization is achieved, mentioned in Sect. 4. The mathematical models adopted in this research are shown in Sect. 5. Section 6 reveals results in numerical experiments and finally Sect. 7 concludes the paper.

2 Pre-semi Assignment Task

This section explains the whole task of the pre-semi assignment in our department. The task is executed once a year by a department faculty by rotation. A faculty rarely conducts the task for a long time and a framework for the task is established to facilitate the assignment for any faculties.

First, all faculty members prepare the syllabuses of pre-semi which mainly describe the content of research activity and the acceptable capacity for students. Then, prospective students for the pre-semi course are supposed to submit questionnaires in which students' preferences for laboratories are written in the form of rankings. The submitted students' preferences by a due date are compiled in a summarized table.

A faculty in charge of the assignment achieves the task in two stages. In the first stage, students are assigned to laboratories without considering cycles. If a laboratory is selected as top three choices by more than three times students as many as the capacity of the laboratory, the corresponding faculty is supposed to select three times students as many as the capacity manually.

Table 1 represents an example of a summarized table for students' preference. In this example, only top three preferences are focused. Laboratory A is selected by 71 students within top two choices while its acceptable capacity is 60, which is three times of 20. The faculty for Laboratory A selects 17 students to be assigned among 28 students in the second choice. Consequently, 12 students who selected Laboratory A as the third choice are not assigned. As for Laboratory B, totally 39 students who selected within top two choices are assigned and 10 students in the third choice are not assigned. In the case of Laboratory C, all 18 students within top two choices are assigned and the corresponding faculty selects 6 students to be assigned among 16 students in the third choice. All 5 students who selected Laboratory D within their top three choices are assigned.

After finishing the assignment according to students' top three choices, the faculty in charge of the assignment task determines the rest of assignment considering the following constraints.

(c1) A student must be assigned to three different laboratories.
(c2) The number of assigned students to a laboratory must not exceed three times as many as the capacity of the laboratory.
(c3) Reverse assignments are prohibited.

Table 1 Summarized table for students' preferences

Lab.	Capacity	1st	2nd	3rd
A	20	43	28	12
B	13	18	21	10
C	8	9	9	16
D	8	1	2	2

Table 2 Reverse assignment

Student	1st	2nd	3rd	4th	5th
#01	Lab. A	Lab. B	Lab. C	Lab. D	Lab. E
#02	Lab. B	Lab. C	Lab. A	Lab. E	Lab. D
#03	Lab. A	Lab. C	Lab. F	Lab. E	Lab. D

The reverse assignment is defined as the assignment where student A is not assigned to laboratory X but student B whose preference for laboratory X is lower than that of student A is assigned to laboratory X. It is no problem that student A is assigned to three laboratories for which his/her preferences are higher than the preference for laboratory X. Table 2 shows an example of reverse assignment where the table represents students' preferences and satisfied choices are expressed by the underlined. In the example, a reverse assignment occurs for the assignment of Students #01 and #02 to Laboratory B; the first choice of Student #02 is denied and the second choice of Student #01 is accepted. With respect to Laboratory D, the fourth choice of Student #01 is rejected and the fifth choice of Student #03 is accepted. The assignment is not referred to as a reverse assignment because Student #01 is assigned to his/her more preferred three laboratories than Laboratory D.

Among feasible assignments which satisfy constraints (c1) through (c3), better assignments are sought considering the next goal:

(g1)   Student should be assigned to laboratories with their higher preferences.

In this process, the worst assigned rank is desperately considered. When all students are assigned to laboratories with their at worst kth choices and at least one student is assigned to his/her kth preferred laboratory, the worst assigned rank is defined as k. The faculty in charge attempts to assign students to improve the worst assigned rank.

In the second stage, students are assigned to slots based on the results in the first stage. The following constraints are taken into account in this stage:

(c4)   A student must be assigned to a laboratory in each cycle.
(c5)   The number of assigned students to a laboratory must not exceed the capacity of the laboratory in any cycles.

At the end of this stage, students' timetables have been fixed.

After creating a feasible assignment, the faculty in charge tries to improve the assignment considering the next goals:

(g2)   The number of closed slots should be raised.
(g3)   The number of assigned students to open slots in a laboratory should be equalized.

The slots without assigned students are referred to as closed slots. On the contrary, the open slot is defined as the slot to which at least one student is assigned. The goals (g2) and (g3) are generated from the standpoint of time efficiency and workload balance, respectively. It is complicated for many faculties to improve the goals and the resulting assignment constructed in the first stage is sometimes modified to improve the goals.

The laboratory assignment task has been basically accomplished through the above two stages. Some students sometimes submit their questionnaires after due date. The students are additionally assigned to unoccupied slots in the resulting assignment considering their preferences. In the additional assignment, the constraint (c3) is relaxed for the delayed students. The resulting assignment is confirmed by all faculty members in a department meeting.

3 System Outline

It is important that a system is easily comprehensible for system users to utilize the system continuously. The developed system in this study, hence, adopts spreadsheet software as system interface, which is familiar to department faculty members. The system procedures are written in VBA, Visual Basic for Application, and the optimization is executed by CPLEX, optimization software [14]. The target laboratory assignment task is formulated as a mathematical model so that users simply understand the factors considered in the optimization as well as how to revise resulting solutions. Meta-heuristic approaches are judged inappropriate in this study because the approaches are complicated to comprehend their schemes for other members to operate the developed system.

Figure 1 illustrates a configuration diagram of the developed system. First, some required fundamental constants for the laboratory assignment task are inputted into a designated spreadsheet, such as the number of students, the number of laboratories, the number of cycles, and the number of student's choices of laboratories in a questionnaire. After the constants, laboratory information and student information are inputted. The weights in the objective function in a mathematical model are also required for optimization.

Before the optimization, some preprocessing is executed. As mentioned in the previous section, the top three choices of students are determined to be satisfied or not to be satisfied based on decisions of some faculties. Some other assignments of students to laboratories are also determined in the preprocessing, mentioned in detail in the next section.

Then, clicking a button executes the optimal laboratory assignment. The system makes some inputted data suitable for CPLEX, writes them down to a data file, and calls CPLEX. After-mentioned mathematical models for the assignment are written in model files and CPLEX loads the data and a model. After executing an optimization, a derived optimal solution is outputted to the data file. Finally, the system loads the resulting solution, creates some summarized tables for the result, and represents the derived assignment and the tables to the system operator.

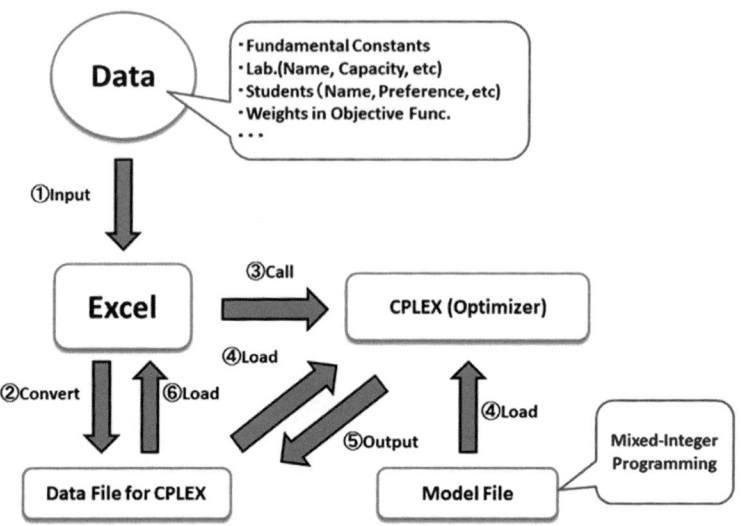

Fig. 1 Configuration diagram of developed system

4 Preprocessing

Some preprocessing procedure determines a part of assignments and it reduces computational time of optimization for the whole assignment.

As mentioned in Sect. 2, when a laboratory is selected by not more than three times students as many as the capacity of the laboratory within top three preferences, the students are determined to be assigned to the laboratory. The developed system provides a button to execute such trivial assignments. When a laboratory is selected by more than three times students as many as its capacity within top three preferences, as mentioned in Sect. 2, a corresponding faculty has to decide which students are to be assigned. The faculty member in charge of the assignment task inputs the decisions to the system.

The preferences of students within top three choices are determined to be satisfied or not to be by the above-mentioned procedures. When a student is determined to be assigned to his/her most preferable three laboratories, it means that his/her remaining lower choices are redundant and obviously determined not to be assigned. Such determinations are also executed by the system in the preprocessing.

5 Mathematical Models

This section introduces the mathematical models adopted in this research. The developed system has been revised repeatedly in order to deal with requirements from department faculty members.

5.1 Model M1

After preprocessing described in the previous section, the system solves an optimization problem to derive a final assignment, that is, the system assigns students to three different slots. The problem is formulated as Model M1 [15]. Model M1 considers the constraints (c1), (c3), (c4), and (c5) while the constraints (c2) are included in (c5). The objective function in Model M1 consists of the three goals described in the Sect. 2.

In order to indicate Model M1, let us define design variables, constants, and sets in advance.

Design variables:
x_{ijk} 1 if student i is assigned to laboratory j in the kth cycle, 0 otherwise
y_{jk} 1 if a student is assigned to laboratory j in the kth cycle, 0 otherwise
z_j positive minimum among the numbers of assigned students to laboratory j in each cycle
s_{il} 1 if student i is assigned to his/her lth preferred laboratory, 0 otherwise
$\bar{s}_{il}$ 1 if student i is assigned to $|K|$ laboratories within his/her top l choices, 0 otherwise

Constants:
a_{ij} satisfaction level of student i occurred by the assignment to laboratory j (positive)
b_l satisfaction level of a student for the assignment to his/her lth preferred laboratory (positive)
c_j capacity of laboratory j
f_{ij} 1/0 if student i is determined to/not to be assigned to laboratory j in the preprocessing procedure, -1 otherwise
r_{il} laboratory that student i selects as the lth preferred laboratory
w_h weight in the objective function ($h = 1, ..., 3$)

Sets:
I index set for students: $\{1, 2, ..., |I|\}$
J index set for laboratories: $\{1, 2, ..., |J|\}$
K index set for cycles: $\{1, 2, ..., |K|\}$ ($|K| = 3$ by default)
L index set for choices in the student's preference: $\{1, 2, ..., |L|\}$
Ω set of all pairs between student i and laboratory j
Ω_0 subset of Ω with $f_{ij} = 0$
Ω_1 subset of Ω with $f_{ij} = 1$

By utilizing the above definitions, Model M1 is shown as follows:

Model M1

Maximize

$$w_1 f_1 + w_2 f_2 - w_3 f_3 \tag{1}$$

Subject to

$$f_1 = \sum_{i \in I} \sum_{j \in J} \sum_{k \in K} a_{ij} x_{ijk} \tag{2}$$

$$f_2 = \sum_{j \in J} z_j \tag{3}$$

$$f_3 = \sum_{j \in J} \sum_{k \in K} y_{jk} \tag{4}$$

$$\sum_{k \in K} x_{ijk} = 1, \quad i \in I, \, j \in J, \, (i,j) \in \Omega_1 \tag{5}$$

$$\sum_{k \in K} x_{ijk} = 0, \quad i \in I, \, j \in J, \, (i,j) \in \Omega_0 \tag{6}$$

$$\sum_{k \in K} x_{ijk} \leq 1, \quad i \in I, \, j \in J, \, (i,j) \in \Omega - \Omega_0 - \Omega_1 \tag{7}$$

$$\sum_{j \in J} x_{ijk} = 1, \quad i \in I, \, k \in K \tag{8}$$

$$\sum_{i \in I} x_{ijk} \leq c_j y_{jk}, \quad j \in J, \, k \in K \tag{9}$$

$$z_j - c_j \left(1 - y_{jk}\right) \leq \sum_{i \in I} x_{ijk}, \quad j \in J, \, k \in K \tag{10}$$

$$s_{il} = \sum_{k \in K} x_{i r_{il} k}, \quad i \in I, \, l \in L \tag{11}$$

$$\bar{s}_{il} = 0, \quad i \in I, \, l = 1, 2, \ldots, |K| - 1 \tag{12}$$

$$|K| \bar{s}_{il} \leq \sum_{l'=1}^{l} s_{il'}, \quad i \in I, \, l = |K|, \ldots, L \tag{13}$$

$$\sum_{k \in K} x_{ijk} \leq \sum_{k \in K} x_{i'jk} + \bar{s}_{i'l'}, \quad i, i' \in I, \, j = r_{il} = r_{i'l'}, l, l' \in L, \, l > l' \tag{14}$$

$$x_{ijk} \in \{0, 1\}, \quad i \in I, \, j \in J, k \in K \tag{15}$$

$$y_{jk} \in \{0, 1\}, \quad j \in J, \, k \in K \tag{16}$$

$$z_j \geq 0, \quad j \in J \tag{17}$$

$$s_{il}, \bar{s}_{il} \in \{0, 1\}, \quad i \in I, \, l \in L \tag{18}$$

The objective function (1) in Model M1 contains three criteria f_1, f_2, and f_3 defined by Eqs. (2), (3), and (4), respectively.

The first criterion f_1 measures the sum of students' satisfaction level based on the goal (g1). In fact, students do not submit their satisfaction levels for each laboratory in the questionnaires. The system operator, hence, has to determine appropriate values for the constant b_l. It is noticeable that the values of b_l are assumed to be in common for all students and it naturally holds that $b_{l_1} \geq b_{l_2}$ for $l_1 < l_2$. The value of a_{ijk} is set by the following equation:

$$a_{ij} = \begin{cases} b_l & r_{il} = j, \quad \exists l \in L \\ 0 & r_{il} \neq j, \quad \forall l \in L \end{cases} \quad i \in I \tag{19}$$

Since students select $|L|$ laboratories in a questionnaire, the satisfaction levels are given to their top $|L|$ laboratories, otherwise the satisfaction level is set to zero.

The second criterion f_2 in the objective function (1) is settled for the goal (g3), that is, equalization of the number of assigned students in each laboratory. The design variable z_j implies the minimum among the number of assigned students in each slot for laboratory j except for closed slots. When no students are assigned to all of three slots for laboratory j, the value of z_j is assumed to be equal to its capacity c_j as a special case. Since the design variable z_j should be increased in order to increase the value of f_2, the equalization among three cycles in laboratory j is encouraged. The criterion f_2 in Eq. (3) is given as the sum of z_j for all laboratories but actually it can be in different form such as the product of z_j or the minimum of z_j.

The third criterion f_3 defined in Eq. (4) indicates the number of open slots, which should be reduced represented as the goal (g2).

Equations (5) and (6) show the results of the preprocessing task, that is, some students have been determined to be or not to be assigned to certain laboratories. It is noticeable that the assignment of students to slots has not determined in the preprocessing task and the determination is executed by the optimization for Model M1. Inequality (7) prohibits the assignment of a student to a certain laboratory in multiple cycles. Equation (8) guarantees that a student is surely assigned to a laboratory in any cycle. Inequalities (9) and (10) define the design variables y_{jk} and z_j, respectively. Since y_{jk} is a 0–1 variable, it must have 1 if a student is assigned to laboratory j in cycle k. When a student is assigned to laboratory j in cycle k, z_j can be increased up to the number of assigned students to laboratory j in cycle k. When no student is assigned, z_j can be increased up to c_j, the capacity of laboratory j. Consequently, z_j can be increased up to the positive minimum among the number of assigned student to laboratory j in each cycle. Equation (11) defines the design variable s_{il}. The design variable $\bar{s}_{il}$ is necessary to judge the reverse assignment.

As mentioned in Sect. 2, if a student has been assigned to $|K|$ laboratories within his/her top l choices, his/her choices below lth choice are excluded from consideration of reverse assignment. Equation (12) obviously holds due to the definition of $\bar{s}_{il}$. Inequality (13) leads that $\bar{s}_{il}$ can have 1 if student i is assigned to $|K|$ laboratories within his/her top l choices, otherwise $\bar{s}_{il}$ is equal to 0. Inequality (14) indicates the prohibition of the reverse assignment. Consider the situation that student i and i' respectively select laboratory j in lth and l'th choices where $l > l'$. Student i can be assigned to laboratory j if student i' is also assigned to laboratory j, when the first term in the right side is 1, or student i' has been assigned to $|K|$ laboratories within his/her top l' choices, when $\bar{s}_{i'l'}$ can have 1.

At the beginning, the model does not involve the prohibition of the reverse assignment explicitly and appropriate settings for the values of weights w_h in the objective function as well as the satisfaction level b_l interfere the reverse assignment [16]. The approach was basically successful to derive optimal solution in a short time but the incorporation of strict prohibition of reverse assignment into mathematical models was requested from some faculties.

5.2 Revised Model: Model M2

The derived assignment by Model 1 depends on the values of some parameters such as weights w_h in the objective function. It is hard for a system operator to settle adequate values for desirable results. The optimization process, therefore, have been divided into three sub-processes in order to reduce the workforce for the assignment task as well as waiting time for the execution of optimization.

In the first process, the system optimizes the worst assigned rank, mentioned in Sect. 2, with satisfying the feasibility of assignment of students to slots. Model M2-1 is the mathematical model for the first stage and it needs an additional design variable:

t_l 1 if a student is assigned to his/her lth preferred laboratory, 0 otherwise

The objective function in Model M2-1 is as follows:

$$\text{Minimize} \quad \sum_{l \in L} (b_1 - b_l)t_l \tag{20}$$

The constraints in Model M2-1 are all of the constraints in Model M1 except the conditions (9) and (10).

Model M2-1 derives an optimal assignment just for the goal (g1) and it ignores the goals (g2) and (g3). Let R_w be the worst assigned rank among the derived assignment. In other word, Model M2-1 guarantees the capability of assignment where all students are assigned to three laboratories within their top R_w choices at worst. The value of R_w is one of the most noticeable measurements to estimate resulting assignment in the practical assignment task.

In the second process, an additional optimization is executed to deal with the goals (g2) and (g3). The mathematical model, named Model M2-2, in this process is almost exactly the same as Model M1 and it is constructed by modifying the set Ω_0 as follows:

$$\Omega_0 = \{(i,j)\,|\,f_{ij} = 0\} \cup \{(i, r_{il})\,|\,l > R_w\}, \tag{21}$$

The redefinition (21) indicates that the assignment is executed only using students' top R_w choices. The restriction reduces the size of search space and it contributes to the reduction of computational time for the optimization.

In the final third process, the system deals with some specialized situations. For instance, a faculty sometimes wants to open more slots for any reason. Such a requirement can be satisfied by adding the following constraint:

$\bar{y}_j$ required number of the open slots in laboratory j

$$\sum_{k \in K} y_{jk} \geq \bar{y}_j, \quad j \in J \tag{22}$$

The modification by the new constraint is basically acceptable unless the goal (g1), the satisfactory level of students, is deteriorated.

Some students sometimes submit their questionnaires after due date. Such students are assigned to unoccupied slots through additional execution of optimization where the resulting assignments obtained by Model M2-2 are assumed to be fixed and the sets Ω_0 and Ω_1 are modified in accordance with the obtained assignment. Moreover, constraints (11) through (14) are excluded from Model M2-2 since the prohibition of reverse assignment is relaxed for the delayed student. The condition with respect to R_w is also relaxed for the delayed students.

6 Numerical Experiments

Numerical experiments were conducted using practical past data for two academic years in order to estimate the performance of the developed system. The fundamental constants were set as shown Table 3. The experiments were conducted using a PC with Intel Core i5-2400 (3.1 GHz) CPU, 4 GB memory, Windows 7 (64 bit). The versions of Excel and CPLEX are 2010 and 12.2, respectively.

Table 3 Values of fundamental constants in utilized data sets

Data set	Year	$|I|$	$|J|$	$|K|$	$|L|$	b_l
A	2011	141	19	3	11	2^{12-l}
B	2012	185	18	3	11	2^{12-l}

First, Model M2-1 derived R_w, worst assigned rank, for two data sets. The values of R_w were 8 and 6 in data sets A and B, respectively. Then, Model M2-2 was executed with weights $(w_1, w_2, w_3) = (1, 0, 0)$ to derive the assignment optimal just for students' satisfaction f_1.

The influence of the weight w_2 on f_2, the equalization of number of assigned students, was investigated on the resulting assignment and computational time. The value of w_2 was set to several values while $w_1 = 1$ and $w_3 = 0$. Figure 2 illustrates the results in numerical experiments. The closed slot rate is referred to as the rate of closed slots against all slots. The hatched area represents that the value of f_1 is deteriorated compared with that with $w_2 = 0$. For instance, Fig. 2a shows that the setting $w_2 = 5$ improves the closed slot rate with preserving the students' satisfaction f_1 but the setting $w_2 = 10$ deteriorates f_1 while it furthermore improves the closed slot rate slightly. The results explain that the increment of the value of w_2 not only improves the equalization of the number of assigned students but also increases the number of closed slots. The computational time for optimization tends to increase with respect to w_2 but the trend does not always continue as shown $w_2 = 20$ in Fig. 2a and $w_2 = 50$ in Fig. 2b. The irregular fluctuation occurs due to the discontinuity of discrete optimization. Figure 3 shows the influence of the weight w_3 on f_3, the

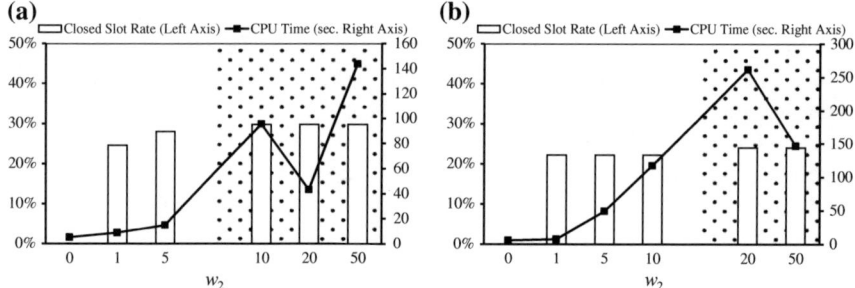

Fig. 2 Closed slot rate and computational time for various values of w_2. **a** Data set A, **b** data set B

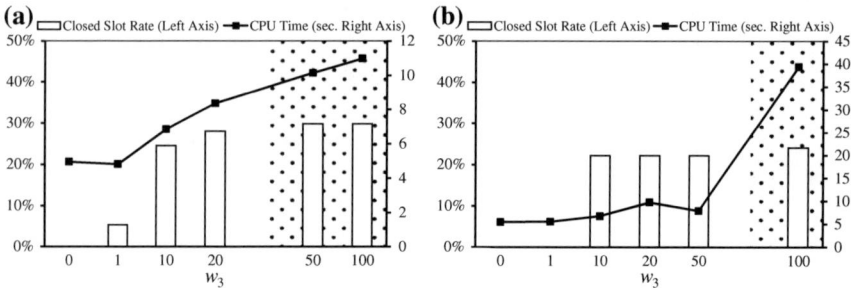

Fig. 3 Closed slot rate and computational time for various values of w_3. **a** Data set A, **b** data set B

Table 4 Number of assigned students in data set A

(w_1, w_2, w_3)		(1, 0, 0)			(1, 0, 20)			(1, 5, 0)		
Lab.	Capacity	C1	C2	C3	C1	C2	C3	C1	C2	C3
A	8	3	3	2	8	0	0	0	0	8
B	8	8	8	8	–	–	–	–	–	–
C	20	3	4	1	0	9	0	–	–	–
D	12	12	12	12	–	–	–	–	–	–
E	20	3	5	6	0	0	14	–	–	–
F	6	6	1	3	0	5	5	5	5	0
G	20	20	20	20	–	–	–	–	–	–
H	13	12	13	13	–	–	–	–	–	–
I	20	1	2	4	0	7	0	7	0	0
J	10	10	10	10	–	–	–	–	–	–
K	6	1	4	2	3	0	4	3	4	0
L	15	15	15	15	–	–	–	–	–	–
M	12	5	2	6	0	0	12	0	12	0
N	10	5	7	9	4	9	8	7	7	7
O	12	12	9	12	11	10	12	11	11	11
P	8	4	6	3	6	7	0	0	7	6
Q	8	8	8	8	–	–	–	–	–	–
R	20	7	4	3	14	0	0	–	–	–
S	10	6	8	4	10	8	0	9	0	9

number of closed slots, where $w_2 = 0$. The tendency on w_3 is similar to that on w_2 with respect to the closed slot rate as well as computational time.

Table 4 shows the transition of number of assigned students according to the change of weight values where Ck denotes Cycle k for $k = 1, 2, 3$. The symbol '–' means that the corresponding value is the same as in the left setting. The setting (1, 0, 20) increases the number of closed slot and the setting (1, 5, 0) additionally executes the equalization. Since system users prefer more simple operations, the result suggests that the weight w_2 should be emphasized rather than w_3.

7 Conclusions

This paper has introduced a development of a spreadsheet-based system for the pre-semi assignment task. The task is formulated as a mixed integer programming problem and an optimal solution is computed by optimizer software. The system helps a department faculty to accomplish the task in a short time. The proposed mathematical model can be applied to the personnel assignment task with rotations or a variation of the staff rostering problem.

The system almost reaches completion and should be treated a few improvements on usability especially on the search of desirable results by changing the weights in objective function. Since the system derives the preferable assignment in substantially shorter time than ever before, some silent requests to the assignment have been gradually revealed. The system could be upgraded repeatedly for its continuous usage.

Acknowledgments This work was partially supported by JSPS KAKENHI Grant number 25285131 and by MEXT, Japan.

References

1. D. Gale, L. Shapley, College admission and the stability of marriage. Am. Math. Mon. **69**, 9–15 (1962)
2. A.E. Roth, M.A.O. Sotomayor, *Two-Sided Matching: A Study in Game-Theoretic Modeling and Analysis* (Cambridge University Press, Cambridge, 1990)
3. International Timetabling Competition (2011), http://www.utwente.nl/ctit/hstt/itc2011/welcome/
4. S. Abdullah, H. Turabieh, On the use of multi neighbourhood structures within a Tabu-based memetic approach to university timetabling problems. Inf. Sci. **191**, 146–168 (2012)
5. L.E. Agustín-Blas, S. Salcedo-Sanz, E.G. Ortiz-García, A. Portilla-Figueras, Á.M. Pérez-Bellido, A hybrid grouping genetic algorithm for assigning students to preferred laboratory groups. Expert Syst. Appl. **36**(3), 7234–7241 (2009)
6. G.N. Beligiannis, C.N. Moschopoulos, G.P. Kaperonis, S.D. Likothanassis, Applying evolutionary computation to the school timetabling problem: the Greek case. Comput. Oper. Res. **35**(4), 1265–1280 (2008)
7. T. Thepphakorn, P. Pongcharoen, C. Hicks, An ant colony based timetabling tool. Int. J. Prod. Econ. **149**, 131–144 (2014)
8. C.W. Fong, H. Asmuni, B. McCollum, P. McMullan, S. Omatu, A new hybrid imperialist swarm-based optimization algorithm for university timetabling problems. Inf. Sci. **283**, 1–21 (2014)
9. J.A. Ferland, S. Roy, Timetabling problem for university as assignment of activities to resources. Comput. Oper. Res. **12**(2), 207–218 (1985)
10. S. Daskalaki, T. Birbas, E. Housos, An integer programming formulation for a case study in university timetabling. Eur. J. Oper. Res. **153**(1), 117–135 (2004)
11. S. Daskalaki, T. Birbas, Efficient solutions for a university timetabling problem through integer programming. Eur. J. Oper. Res. **160**(1), 106–120 (2005)
12. M. Dimopoulou, P. Miliotis, Implementation of a university course and examination timetabling system. Eur. J. Oper. Res. **130**(1), 202–213 (2001)
13. J.M. Mulvey, A classroom/time assignment model. Eur. J. Oper. Res. **9**(1), 64–70 (1982)
14. IBM CPLEX Optimizer, http://www-01.ibm.com/software/commerce/optimization/cplex-optimizer/
15. T. Koide, Improvement on spreadsheet-based system for seminar assignment problem with rotations, in *Proceedings of the International Multiconference of Engineering and Computer Scientists 2014, IMECS 2014*, Hong Kong, 12–14 March 2014. Lecture Notes in Engineering and Computer Science, pp. 1183–1185
16. T. Koide, A spreadsheet optimization system for seminar assignment problem with rotation, in *Proceedings of the 14th Asia Pacific Industrial Engineering and Management System, APIEMS 2013*, Cebu, Philippines, 3–6 Dec 2013, 7 pages

Scheduling the Finnish Major Ice Hockey League Using the PEAST Algorithm

Kimmo Nurmi, Jari Kyngäs, Dries Goossens and Nico Kyngäs

Abstract Good schedules have many benefits for the league, such as higher incomes, lower costs and more interesting and fairer seasons. Generating a schedule for a professional sports league is an extremely demanding task and requires computational intelligence to generate an acceptable schedule. There are a multitude of stakeholders with varying requests (and often requests vary significantly year on year). This paper presents the format played in the Finnish major ice hockey league in the 2013–2014 season. The paper describes the PEAST algorithm which have been used to schedule the league since the 2008–2009 season. We report our computational results especially for the 2013–2014 season.

Keywords Local search · Metaheuristics · PEAST algorithm · Real-world scheduling · Round robin tournament · Sports scheduling

1 Introduction

In the past decades professional sports leagues have become big businesses; at the same time the quality of the schedules have become increasingly important. This is not surprising, since the schedule directly impacts the revenue of all involved parties.

K. Nurmi (✉) · J. Kyngäs · N. Kyngäs
Satakunta University of Applied Sciences, Tiedepuisto 3, 28600 Pori, Finland
e-mail: cimmo.nurmi@samk.fi

J. Kyngäs
e-mail: jari.kyngas@samk.fi

N. Kyngäs
e-mail: nico.kyngas@samk.fi

D. Goossens
Faculty of Economics and Business Administration, Ghent University,
Sint-Pietersnieuwstraat 25, 9000 Ghent, Belgium
e-mail: Dries.Goossens@UGent.be

© Springer Science+Business Media Dordrecht 2015
G.-C. Yang et al. (eds.), *Transactions on Engineering Technologies*,
DOI 10.1007/978-94-017-9588-3_12

For instance, the number of spectators in the stadiums, and the traveling costs for the teams are influenced by the schedule. TV networks that pay for broadcasting rights want the most attractive games to be scheduled at commercially interesting times in return. Furthermore, a good schedule can make a tournament more interesting for the media and the fans, and fairer for the teams. Nurmi et al. [1] report a growing number of cases where academic researchers have been able to close a scheduling contract with a professional sports league owner. An excellent overview of sports scheduling can be found in [2] and an annotated bibliography in [3].

In a sports tournament, n teams play against each other over a period of time according to a given timetable. The teams belong to a *league*, which organizes games between the teams. Each *game* consists of an ordered pair of *teams*, denoted (i, j) or i-j, where team i plays *at home*—that is, uses its own *venue* (stadium) for a game—and team j plays *away*. Games are scheduled in *rounds*, which are played on given days. A *schedule* consists of games assigned to rounds. A schedule is *compact* if it uses the minimum number of rounds required to schedule all the games; otherwise it is *relaxed*. If a team plays two home or two away games in two consecutive rounds, it is said to have a *break*. In general, for reasons of fairness, breaks are to be avoided. However, a team can prefer to have two or more con-secutive away games if its stadium is located far from the opponent's venues, and the venues of these opponents are close to each other. A series of consecutive away games is called an *away tour*.

In a *round robin tournament* each team plays against each other team a fixed number of times. Most sports leagues play a double round robin tournament (*2RR*), where the teams meet twice (once at home, once away), but quadruple round robin tournaments (*4RR*) are also quite common. A *mirrored* double round robin tour-nament (*M2RR*) is a tournament where every team plays against every other team once in the first $n - 1$ rounds, followed by the same games with reversed venues in the last $n - 1$ rounds.

Table 1 shows an example of a compact double round robin tournament with six teams. The schedule has no breaks for team 1, three-in-a-row home games for team 6 and a four-game away tour for team 4.

Sports scheduling involves three main problems. First, the problem of finding a schedule with the *minimum number of breaks* is the easiest one. de Werra [4] has presented an efficient algorithm to compute a minimum break schedule for a 1RR.

Table 1 A double round robin tournament with six teams

R1	R2	R3	R4	R5
1–6	3–1	1–5	2–1	1–4
2–5	5–4	2–4	5–3	3–2
4–3	6–2	3–6	6–4	6–5
R6	R7	R8	R9	R10
4–2	1–2	3–4	1–3	2–3
5–1	3–5	5–2	2–6	4–1
6–3	4–6	6–1	4–5	5–6

If n is even, it is always possible to construct a schedule with $n - 2$ breaks. For an M2RR, it is always possible to construct a schedule with exactly $3n - 6$ breaks.

Second, the problem of finding a schedule that *minimizes the travel distances* is called the Traveling Tournament Problem (TTP) [5]. In TTP the teams do not return home after each away game but instead travel from one away game to the next. However, excessively long away trips as well as home stands should be avoided. The TTP is recently shown to be strongly NP-complete [6].

Third, most professional sports leagues introduce many additional requirements in addition to minimizing breaks and travel distances. We call the problem of finding a schedule which *satisfies given constraints* [1] the Constrained Sports Scheduling Problem (CSSP). The goal is to find a feasible solution that is the most acceptable for the sports league owner—that is, a solution that has no hard constraint violations and that minimizes the weighted sum of the soft constraint violations.

Scheduling the Finnish major ice hockey league is an example of a CSSP. It is very important to minimize the number of breaks. The fans do not like long periods without home games, consecutive home games reduce gate receipts and long sequences of home or away games might influence the team's current position in the tournament. It is also very important to minimize the travel distances. Some of the teams do not return home after each away game but instead travel from one away game to the next. There are also around a dozen more other criteria that must be optimized.

Section 2 presents the format played in the Finnish major ice hockey league in the 2013–2014 season. The section also introduces the requirements, requests and other constraints that the format implies. In Sect. 3 we describe the PEAST algorithm which has been used since the 2008–2009 season to schedule the league. Section 4 reports some statistical findings and our computational results especially for the 2013–2014 season.

We are not aware of any sports scheduling papers dealing with such a broad class of constraints that arise in the Finnish major ice hockey league. We believe that the model and the solution method help sports scheduling researchers to evaluate, compare and exchange their equivalent ideas.

2 The Finnish Major Ice Hockey League Format and the Constraint Model

Ice hockey is the biggest sport in Finland, both in terms of revenue and the number of spectators. The spectator average per game for the current season (2012–2013) is about 5,200. In the Saturday rounds 1 % of the Finnish population (age 15–70) attended the games in the ice hockey arenas.

The Finnish major ice hockey league has 14 teams (see Table 2). Seven of the teams in the league are located in big cities (over 100,000 citizens) and the rest in

Table 2 The fourteen teams in the Finnish major ice hockey league and their number of titles

#1	Jokerit	6	#8	Ilves	16
#2	HIFK	7	#9	HPK	1
#3	Blues	0	#10	JYP	2
#4	TPS	11	#11	Pelicans	0
#5	Ässät	3[a]	#12	SaiPa	0
#6	Lukko	1	#13	KalPa	0
#7	Tappara	15	#14	Kärpät	6[a]

[a] Ässät won the 2012–2013 season and Kärpät the 2013–2014 season

smaller cities. One team is quite a long way up north, two are located in the east and the rest in the south (see Fig. 1).

The format played in the league since the 2012–2013 season is somewhat eccentric. The competition starts with a regular season in September and ends with the playoffs from mid-March to mid-April. The league fixes the dates on which the games can be played. The last team of the regular season plays best out of seven elimination games against the best team of the Finnish 1st division ice hockey league. The six best teams of the regular season proceed directly to quarter-finals. Teams placing between 7th and 10th play preliminary playoffs best out of three. The two winners take the last two quarter-final slots. Teams are paired up for each playoff round according to the regular season standings, so that the highest-ranking team plays against the lowest-ranking, and so on. The playoffs are played best out of seven. The winner of the playoffs receives the Canada Bowl, the championship trophy of the League.

Fig. 1 The fourteen teams on the map of Finland

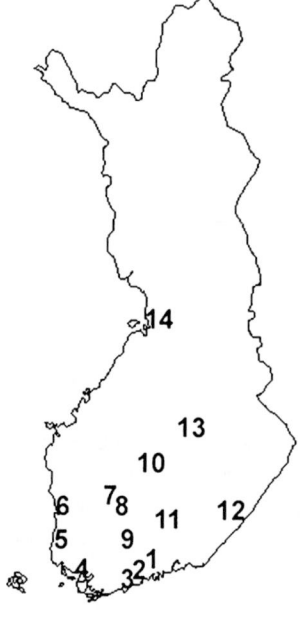

The basis of the regular season is a quadruple round robin tournament resulting in 52 games for each team. In addition, the teams are divided into two groups of seven teams in order to get a few more games to play. The teams in the groups are selected based on fairness, i.e. the strengths of the teams are most likely to be equal. These teams play a single round robin tournament resulting in 6 games. The home teams of the games are decided so that in two consecutive seasons each team has exactly one home game against every other team in the group.

Finally, the so-called "January leveling" adds two extra games for each team. In January, in the middle of the season, the last team on the current standings selects an opponent against which it plays once at home and once away on two consecutive days on Friday and on Saturday. The opponent selects the day for its home game. Then, the second last team (or the third last if the second last was selected by the last team) selects its opponent from the rest of teams and so on. The teams can choose to select their opponents either by maximizing the winning possibilities or by maximizing the ticket sales.

The quadruple round robin, six group games and two extra games per team total to 60 games for each team and 420 games overall. The format includes several other interesting features than those mentioned earlier. First, the standard game days are Tuesday, Friday and Saturday. The schedule should maximize the number of games on Fridays and on Saturdays in order to maximize the revenue. For the same reason consecutive home games are not allowed on Fridays and Saturdays. Furthermore, due to the travel distances between some venues, certain combinations of a Friday home team playing a Saturday away game against the given team are not allowed. Table 3 shows the forbidden pairs. The games that cannot be scheduled on Fridays due to these restrictions, are played on Thursdays.

Second, every team should play in the two rounds before and the two rounds after New Year's Eve and adhere to the traveling rules given in the Table 3. Third, the schedule should include a weekend when seven pairs of teams play against each

Table 3 The forbidden combinations of a Friday home team playing a Saturday away game against the given team		
	Ässät	KalPa, Kärpät, SaiPa
	Blues	KalPa
	HIFK	KalPa
	HPK	KalPa
	Ilves	Kärpät
	Jokerit	KalPa
	JYP	Kärpät, TPS
	KalPa	Ässät, Blues, HIFK, HPK, Jokerit, Lukko
	Kärpät	Ässät, Ilves, Lukko, SaiPa, Tappara, TPS
	Lukko	KalPa, Kärpät, SaiPa
	Pelicans	
	SaiPa	Ässät, Kärpät, Lukko, TPS
	Tappara	Kärpät
	TPS	KalPa, Kärpät, SaiPa

Table 4 The local rivals

Blues, HIFK, Jokerit
Ässät, Lukko, TPS
HPK, Ilves, Pelicans, Tappara
JYP, Kärpät, KalPa, SaiPa

Table 5 The number of games that should be played on Fridays and Saturdays

Jokerit	HIFK	4
Jokerit	Blues	3
HIFK	Blues	4
Ilves	Tappara	5
Ässät	Lukko	5
JYP	KalPa	4
Kärpät	KalPa	4
SaiPa	Pelicans	5
HPK	Pelicans	4

other on consecutive rounds on Friday and on Saturday. These match-up games are called "back-to-back games". Fourth, local rivals (see Table 4) should play as many games as possible against each other in the first two rounds.

Fifth, the number of Friday and Saturday games between some local rivals (see Table 5) should be maximized. Sixth, the Ässät, HPK, JYP, KalPa, Kärpät, Lukko, Pelicans, SaiPa and TPS teams should play at least one Friday or Saturday home game against the Jokerit and HIFK teams. These two teams guarantee the best revenue for the home team.

Seventh, the traveling distances between some of the venues require some teams to make away tours. That is, they should play away either on Tuesday and on Thursday or on Thursday and on Saturday. The teams that make away tours, their possible opponents and the minimum number of tours required are given in Table 6.

Eighth, the Tappara and Ilves teams cannot play at home on the same day because they share a venue. Also the Jokerit and HIFK teams cannot play at home on the same day because they share the same (businessmen) spectators. Ninth, some of the teams cannot play at home in certain days because their venues are in use for some other event. A total of 69 home game restrictions existed in the 2012–2013 season. Finally, in the last two rounds each team should play exactly one home game.

Next, we present the constraint model of the Finnish major ice hockey league scheduling problem for the 2013–2014 season. Nurmi et al. [1] present a collection

Table 6 The teams that make away tours, their possible opponents and the minimum number of tours required

Kärpät	HPK + Ilves, TPS + Ässät, TPS + Ilves, Pelicans + Tappara, SaiPa + HIFK	6
KalPa	HPK + Blues, Ässät + Lukko, HIFK + Pelicans	4
SaiPa	Ässät + Lukko	2
Ässät	SaiPa + KalPa	2
Lukko	SaiPa + KalPa	2

of typical constraints that are representative of many scheduling scenarios in sports scheduling. In the following problem description, we refer to this constraint classification. The detailed problem file will be found on the sports scheduling web site [7]. The objective is to find a feasible solution that is the most acceptable for the sports league owner. That is, a solution that has no hard constraint violations and that minimizes the weighted sum of the soft constraint violations.

The hard constraints are the following:

C01.   There are at most 90 rounds available for the tournament
C04.   Team t cannot play at home in round r (43 cases)
C07.   The *Tappara* and *Ilves* teams and the *Jokerit* and *HIFK* teams cannot play at home in the same round
C08.   A team cannot play at home on two consecutive calendar days
C12.   A break cannot occur in the *second* and *last* round
C41.   The schedule should include *one* weekend where *seven* pairs of teams play against each other on consecutive rounds on *Friday* and on *Saturday*

The soft constraints are the following:

C09.   Team t wants to play at least m_1 away tours (see Table 6)
C13.   Teams cannot have more than *two* consecutive home games
C14.   Teams cannot have more than *two* consecutive away games
C15.   The total number of breaks must not be larger than 140
C19.   There must be at least *five* rounds between two games with the same opponents
C22.   Two teams cannot play against each other in series of HHAA, AAHH, HAAAH or AHHHA
C23.   Team t wishes to play at least m_1 and at most m_2 home games on *weekday*$_1$, m_3–m_4 on *weekday*$_2$ and so on (see [7])
C26.   The difference between the number of played home and away games for each team must not be larger than *two* in any stage of the tournament
C27.   The difference in the number of played home games between the teams must not be larger than *two* in any stage of the tournament
C37.   A team t_1 cannot play away against team t_2 if it played at home against team t_2 on the previous round, and the two rounds are on consecutive calendar days (see Table 3). Note that this constraint is also used for the two rounds before and two rounds after New Year's Eve
C38.   Teams in the *first two* groups should play *two* games against each other and teams in the *last two* groups should play *four* games against each other between rounds *one* and *two* (see Table 4)
C39.   At least m games between teams t_1 and t_2 should be played on *Fridays* and *Saturdays* (see Table 5) (10)
C40.   The schedule should include at least 200 games played either on *Friday* or on *Saturday* (10)

Some of the given soft constraints are actually goals. These goals are presented as exact numbers in the constraint model:

- minimize the number of breaks (C15)
- the defined local rivals should play as many games as possible in the first two rounds (C38)
- the number of Friday and Saturday games between some local rivals should be maximized (C39)
- the schedule should maximize the number of games on Fridays and on Saturdays (C40).

3 The PEAST Algorithm

This section describes the PEAST algorithm which is used to schedule the league. It has been used to solve several real-world scheduling problems (see e.g. [8–11]) and it is in industrial use.

The PEAST algorithm is a population-based local search method. The heart of the algorithm is the local search operator called greedy hill-climbing mutation (GHCM). The GHCM operator is used to explore promising areas in the search space to find local optimum solutions. Another important feature of the algorithm is the use of shuffling operators. They assist in escaping from local optima in a systematic way. Furthermore, simulated annealing and tabu search are used to avoid staying stuck in promising search areas too long. We next discuss these and other important characteristics briefly. For the detailed discussion we refer to [12]. The pseudo-code of the algorithm is given in Fig. 2.

The GHCM operator is based on similar ideas to the Lin-Kernighan procedures [13] and ejection chains [14]. The basic hill-climbing step is extended to generate a sequence of moves in one step, leading from one solution candidate to another. The GHCM operator moves an object, o_1, from its old position, p_1, to a new position, p_2, and then moves another object, o_2, from position p_2 to a new position, p_3, and so on, ending up with a sequence of moves.

Picture the positions as cells as shown in Fig. 3. The initial object is selected by tournament selection with $k = 7$. In the (deterministic) tournament selection we randomly pick k objects and then we choose the best one. The cell that receives the object is selected by considering all the possible cells and selecting the one that causes the least increase in the objective function when only considering the relocation cost. Then, another object from that cell is selected by considering all the objects in that cell and picking the one for which the removal causes the biggest decrease in the objective function when only considering the removal cost. Next, a new cell for that object is selected, and so on. The sequence of moves stops if the last move causes an increase in the objective function value and if the value is larger than that of the previous non-improving move, or if the maximum number of moves

Set the iteration limit t, cloning interval c, shuffling interval s, ADAGEN
update interval a and the population size n
Generate a random initial population of schedules S_i for $1 <= i <= n$
Set $best\_sol = null$, $round = 1$
WHILE $round \leq t$
   $index = 1$
   WHILE $index++ <= n$
      Apply GHCM to schedule S_{index} to get a new schedule
      IF $Cost(S_{index}) < Cost(best\_sol)$ THEN Set $best\_sol = S_{index}$
   END REPEAT
   Update simulated annealing framework
   IF $round \equiv 0$ (mod a) THEN **Update the ADAGEN framework**
   IF $round \equiv 0$ (mod s) THEN **Apply shuffling operators**
   IF $round \equiv 0$ (mod c) THEN **Replace the worst schedule with the**
                                         best one
   Set $round = round + 1$
END WHILE
Output $best\_sol$

Fig. 2 The pseudo-code of the PEAST algorithm

Fig. 3 A sequence of moves
in the GHCM heuristic

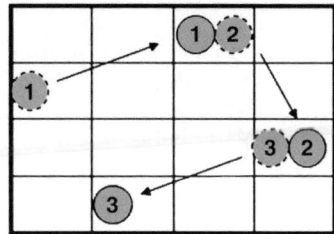

is reached. Then, a new sequence of moves is started. The maximum number of
moves in the sequence is 10.

It may sound surprising that the best way to select the new cell for the object is
to consider all possible cells and select the best one. Moreover, the best way to
select a new object from that cell is again to consider all the objects in that period.
Very often a not so greedy strategy ends up with better results.

We improve the GHCM operator by introducing a tabu list which prevents
reverse order moves in the same sequence of moves. i.e. if we move an object
o from position p_1 to position p_2, we do not allow the object o to be moved back to
position p_1 before a new sequence of moves begins.

We use a simulated annealing refinement to decide whether or not to commit to a
sequence of moves in the GHCM operator. This refinement is different from the
standard simulated annealing. It is used in a three-fold manner. Firstly, when

choosing an object to be moved from a cell, a random object is chosen with probability $\exp(-1/T_k)$ instead of choosing the least fit object. Secondly, when choosing the cell where to move the object, a random cell is chosen with probability $\exp(-1/T_k)$ instead of choosing the fittest cell. Lastly, when the sequence of moves is cut short (i.e. a worsening move is made, and it worsens the solution more than the previous worsening move did), the whole sequence will still be committed with probability $\exp(-\text{costDiff}/T_k)$ instead of rolling back to the best position (i.e. the position at which the objective function value is the lowest) of the sequence. The cooling scheme T_k can be found in [12].

For most PEAST applications we introduce a number of shuffling operators—simple heuristics used to perturb a solution into a potentially worse solution in order to escape from local optima—that are called upon according to some rule. The idea of shuffling is the same as in hyperheuristics [15] but the other way around. A hyperheuristic is a mechanism that chooses a heuristic from a set of simple heuristics, applies it to the current solution to get a better solution, then chooses another heuristic and applies it, and continues this iterative cycle until the termination criterion is satisfied. We introduce a number of simple heuristics that are used to worsen the solution instead of improving it.

In sports league scheduling we use five shuffling operations:

1. Select a random game and move it to a random round, and do this k_1 times.
2. Swap two random games, and do this k_2 times.
3. Select a random round and move k_3 random games from that round to random rounds.
4. Swap all the games in two random rounds.
5. Select a random game A-B and swap it with the game B-A, and do this k_4 times.

The best results have been obtained using the values $k_1 = 3$, $k_2 = 2$, $k_3 = 3$ and $k_4 = 2$.

No crossover operators are applied to the population of schedules. Every c iterations the least fit individual is replaced with a clone of the fittest individual. This operation is completely irrespective of the globally fittest schedule (*best_sol* in Fig. 1) found. The PEAST algorithm uses ADAGEN, the adaptive genetic penalty method introduced in [16]. A traditional penalty method assigns positive weights (penalties) to the soft constraints and sums the violation scores to the hard constraint values to get a single value to be optimized. The ADAGEN method assigns dynamic weights to the hard constraints based on the constant weights assigned to the soft constraints. The soft constraints are assigned fixed weights according to their significance. The hard constraint weights are updated every kth generation using the method given in [11].

Random initial solutions work best in all the real-world cases where we have applied PEAST algorithm (see e.g. [12–15]). We have found no evidence that a sophisticated initial solution improves results. On the contrary, random initial solutions seem to yield superior or at least as good results.

4 Computational Results

We believe that scheduling the Finnish major ice hockey league is one of the most difficult sports scheduling problems because it combines break minimization and traveling distance restrictions with dozens of constraints that must be satisfied. We have used the PEAST algorithm and its predecessors to schedule the league since the 2008–2009 season. This section reports our computational results for the 2013–2014 season. We start with some interesting statistical findings from the earlier seasons.

In addition to scheduling the league we also contribute to the process of improving the league format. For example the "January leveling" and the "back-to-back games" have been introduced to the format based on our ideas. Table 7 shows the increase in the number of spectators in the last five seasons.

The standard game days used to be Tuesday, Thursday and Saturday. From the 2011–2012 season the league decided to change Thursdays to Fridays to get more spectators. Friday games have had about 10 % more spectators.

However, playing at home both on Fridays and on Saturdays is not allowed. Due to this, the games that cannot be scheduled on Fridays are played on Thursdays. This on the other hand means that some teams play two consecutive games and some teams have a rest day before the Saturday game. In the last ten seasons the probability for a home team to defeat an away team that has had a rest day is 10 % smaller. Likewise, the probability for an away team to win a home team that has had a rest day is even 85 % smaller. Minimizing the number of breaks is very important because it is likely that having two consecutive home games on Thursday and on Saturday decreases the number of spectators. In the last ten seasons the number of spectators on Thursday has decreased by 3.5 % and on Saturday by 1.9 %.

Some teams desire away tours because of the traveling distances between their venue and some of the opponents' venues (see Table 6). For example, the Kärpät team wants to make at least six away tours. In the last ten seasons the probability for the team to win its second away game is 30 % smaller than to win any away game.

The process of scheduling the league takes about 2 months. First, we discuss the possible improvements to the format with the league's competition manager. Then, the format is accepted by the team CEOs. Next, all the restrictions, requirements and requests by the teams are gathered. Finally, the importance (penalty value) of

Table 7 The number of spectators in the league in the last eight seasons		
	2005–2006	1,958,843
	2006–2007	1,943,312
	2007–2008	1,964,626
	2008–2009	1,997,114
	2009–2010	2,015,080
	2010–2011	2,036,915
	2011–2012	2,145,462
	2012–2013	2,189,350

the constraints is decided. We ran the algorithm for 1 week and choose the best solution. The algorithm was run on an Intel Xeon X5690 3.47 GHz with 24 GB of RAM running Windows 7 Professional.

Recall that the objective is to find a feasible solution that is the most acceptable for the sports league owner. That is, a solution that has no hard constraint violations and that minimizes the weighted sum of the soft constraint violations. Table 8 shows the constraints used for the 2013–2014 season, whether they were decided to be hard or soft constraints, the importance (penalty value) of the soft constraints and how the constraint violations are calculated. The constraints were first introduced in [17].

Table 8 also shows the best solution found. The solution has no hard constraint violations and the penalty value for the soft constraint violations is 12. The schedule has 89 rounds (C01), 139 breaks (C15) and 220 games played either on Friday or on Saturday (C40). This was by far the most difficult schedule to generate compared to the earlier seasons. The league accepted the schedule and it will be used in the 2013–2014 season.

It should be noted that for the 2000–2009 seasons the average number of 3-breaks at home (C13) was 14 and the average number of cases when there were

Table 8 The best solution found (accepted by the league)

C01	Hard	There are at most 90 rounds available for the tournament	0
C04	Hard	A team cannot play at home in the given round (43 cases)	0
C07	Hard	Two pairs of teams cannot play at home in the same round (2 cases)	0
C08	Hard	A team cannot play at home on two consecutive calendar days	0
C12	Hard	A break cannot occur in the second and last rounds	0
C41	Hard	"Back-to-back games"	0
C09	Soft	Number of away tours not scheduled	0
C13	Soft	One violation for each case when a team has more than two consecutive home games	0
C14	Soft	One violation for each case when a team has more than two consecutive away games	0
C15	Soft	One violation for each break more than 140	0
C19	Soft	One violation for each round less than five	1
C22	Soft	One violation for each case when two teams meet in series of HHAA, AAHH, HAAAH or AHHHA	4
C23	Soft	One violation for each home game less or more than the requested number on given weekday	2
C26	Soft	One violation for each case when the difference is more than two	1
C27	Soft	One violation for each case when the difference is more than two	0
C37	Soft	One violation for each forbidden combination	0
C37	Soft	One violation for each forbidden combination (the compact rounds around New Year's Eve)	0
C38	Soft	One violation for each game less than 12	0
C39	Soft	One violation for each game not played on Fridays or on Saturdays	0
C40	Soft	One violation for each game less than 200	0

less than five rounds between two games with the same opponents (C19) was 10. Furthermore, in these seasons the maximum difference between the number of played home and away games (C26) was 5 and the difference in the number of played home games between the teams was 4.

5 Conclusion and Future Work

We presented the format played in the Finnish major ice hockey league in the 2013–2014 season. The format is very complicated requiring computational intelligence to generate an acceptable schedule. We presented computational results that show that our PEAST algorithm generated a good-quality schedule for the 2013–2014 season. The league owner accepted the schedule.

Our future work concentrates on solving the Australian Football League. One of the most interesting features of the League is that it includes only a single round robin with 18 teams. In addition, each team has 5 extra matches. The traveling is a big issue for the teams. The traveling distance is from 12,000 km up to 70,000 km per season per team. We will compare the schedules generated by the PEAST algorithm to the schedule currently running.

References

1. K. Nurmi, D. Goossens, T. Bartsch, F. Bonomo, D. Briskorn, G. Duran, J. Kyngäs, J. Marenco, C.C. Ribeiro, F.C.R. Spieksma, S. Urrutia, R. Wolf-Yadlin, A Framework for Scheduling Professional Sports Leagues, in *IAENG Transactions on Engineering Technologies*, vol. 5, ed. by S.I. Ao (Springer, USA, 2010)
2. P. Rasmussen, M. Trick, Round robin scheduling—a survey. Eur. J. Oper. Res. **188**, 617–636 (2008)
3. G. Kendall, S. Knust, C.C. Ribeiro, S. Urrutia, Scheduling in sports: an annotated bibliography. Comput. Oper. Res. **37**, 1–19 (2010)
4. D. de Werra, Scheduling in Sports, in *Studies on Graphs and Discrete Programming*, ed. by P. Hansen (Elsevier, Amsterdam, 1981), pp. 381–395
5. K. Easton, G. Nemhauser, M. Trick, The traveling tournament problem: description and benchmarks, in *Proceedings of the 7th International Conference on Principles and Practice of Constraint Programming*, Paphos, pp. 580–584 (2001)
6. C. Thielen, S. Westphal, Complexity of the traveling tournament problem. Theoret. Comput. Sci. **412**(4–5), 345–351 (2011)
7. K. Nurmi et al., *Sports Scheduling Problem* [Online]. Available http://www.samk.fi/ssp. Last update 7 Mar 2013
8. N. Kyngäs, K. Nurmi, J. Kyngäs, Solving the person-based multitask shift generation problem with breaks, in *Proceedings of the 5th International Conference on Modeling, Simulation and Applied Optimization (ICMSAO)*, Hammamet, Tunis (2013)
9. N. Kyngäs, K. Nurmi, E.I. Ásgeirsson, J. Kyngäs, Using the PEAST algorithm to Roster nurses in an intensive-care unit in a Finnish hospital, in *Proceedings of the 9th Conference on the Practice and Theory of Automated Timetabling*, Son, Norway (2012)

10. N. Kyngäs, K. Nurmi, J. Kyngäs, Optimizing large-scale staff rostering instances, in *Proceedings of the International MultiConference of Engineers and Computer Scientists 2012, IMECS 2012*, Hong Kong, 14–16 Mar 2012. Lecture Notes in Engineering and Computer Science, pp. 1524–1531

11. K. Nurmi, J. Kyngäs, A conversion scheme for turning a curriculum-based timetabling problem into a school timetabling problem, in *Proceedings of the 7th Conference on the Practice and Theory of Automated Timetabling (PATAT)*, Montreal, Canada (2008)

12. N. Kyngäs, K. Nurmi, J. Kyngäs, Crucial components of the PEAST algorithm in solving real-world scheduling problems, in *Proceedings of the 2nd International Conference on Software and Computer Applications*, Paris, France (2013)

13. S. Lin, B.W. Kernighan, An effective heuristic for the traveling salesman problem. Oper. Res. **21**, 498–516 (1973)

14. F. Glover, New ejection chain and alternating path methods for traveling salesman problems, in *Computer Science and Operations Research: New Developments in their Interfaces*, ed. by R. Sharda, O. Balci, S.A. Zenios (Elsevier, Amsterdam, 1992), pp. 449–509

15. P. Cowling, G. Kendall, E. Soubeiga, A hyperheuristic approach to scheduling a sales summit, in *Proceedings of the 3rd International Conference on the Practice and Theory of Automated Timetabling (PATAT)*, pp. 176–190 (2000)

16. K. Nurmi, Genetic Algorithms for Timetabling and Traveling Salesman Problems, Ph.D. Dissertation, Department of Applied Mathematics, University of Turku, Finland, 1998. Available http://www.bit.spt.fi/cimmo.nurmi/

17. K. Nurmi, J. Kyngäs, D. Goossens, N. Kyngäs, Scheduling a professional sports league using the PEAST algorithm, in *Proceedings of the International MultiConference of Engineers and Computer Scientists 2014, IMECS 2014*, Hong Kong, 12–14 Mar 2014. Lecture Notes in Engineering and Computer Science, pp. 1176–1182

Biobjective Sightseeing Route Planning with Uncertainty Dependent on Tourist's Tiredness Responding Various Conditions

Takashi Hasuike, Hideki Katagiri, Hiroe Tsubaki and Hiroshi Tsuda

Abstract This paper proposes a biobjective route planning problem for sightseeing with fuzzy random traveling times and satisfaction of sightseeing activities under general sightseeing constraints and various conditions. In general, traveling times among sightseeing sites and satisfactions of activities depend on weather and climate conditions. Furthermore, the satisfactions also depend on the tourist's tiredness. Therefore, not only fuzzy random variables for traveling times and satisfactions but also the tiredness-dependency is introduced. In addition, the tourist will like to do a route planning without drastically changing from the optimal route to each condition. A route planning problem is proposed to obtain the favorable common route supplying target satisfactions under various conditions. As a basic case of fuzzy numbers, trapezoidal fuzzy numbers and the order relation are introduced. From the order relation for fuzziness and the transformation for the biobjective, the proposed model is transformed into an extended model of network optimization problems.

T. Hasuike (✉)
Graduate School of Information Science and Technology, Osaka University,
2-1 Yamadaoka Suita, Osaka 565-0871, Japan
e-mail: thasuike@ist.osaka-u.ac.jp

H. Katagiri
Graduate School of Engineering, Hiroshima University, 1-4-1 Kagamiyama,
Higashi-Hiroshima, Hiroshima 739-8527, Japan
e-mail: katagiri-h@hiroshima-u.ac.jp

H. Tsubaki
Department of Data Science, The Institute of Statistical Mathematics,
10-3 Midorimachi, Tachikawa, Tokyo 190-8562, Japan
e-mail: tsubaki@ism.ac.jp

H. Tsuda
Faculty of Science and Engineering, Doshisha University,
1-3 Tatara Miyakodani, Kyotanabe, Kyoto 610-0321, Japan
e-mail: htsuda@mail.doshisha.ac.jp

© Springer Science+Business Media Dordrecht 2015
G.-C. Yang et al. (eds.), *Transactions on Engineering Technologies*,
DOI 10.1007/978-94-017-9588-3_13

169

Keywords Biobjective programming · Fuzzy random variable · Mathematical modeling · Network optimization · Sightseeing route planning · Tiredness-dependency

1 Introduction

Information and communication technologies (ICT) including internet technologies are rapidly developed in recent years, and tourists can collect a lot of information and plan personal sightseeing routes by themselves. Therefore, it is important to develop a decision support system for sightseeing route planning considering personal satisfactions under various conditions. Furthermore, traveling and sightseeing times are also important factors, and hence, it is necessary to do a suitable management of these times for the effective utilization of sightseeing activities. Thus, tourists should construct their appropriate routes in advance, considering the above-mentioned factors, transportation networks, personal context, properties of activities, etc.

Previous researches of tour planning problems are broadly divided into some groups; for instance, solving the mathematical programming problems, dynamically planning an optimal itinerary which is related to designing intelligent tour planning systems based on the personalized tour recommender, etc. There are various tour planning problems such as tourist trip design problem [13], tour planning problem in a multimodal and time-scheduled urban public transport network [18], and time-dependent tour planning methodology to design a time-limited tour based on the maximum total priority value [1].

However, previous sightseeing route planning problems do not include several important factors such as uncertainty of required traveling times and satisfaction values to sightseeing sites. For instance, from traffic data, tourists may estimate a traveling time between two sightseeing sites. However, the actual traveling times are often different from the predictions in some degree, and hence, it is generally difficult to set traveling times as constant values, and it is important to consider traveling times as uncertain variables. Furthermore, weather and climate conditions are also important factors to change traveling times and satisfactions to sightseeing sites. For instance, a zoo is a weather-sensitive sightseeing site due to outdoors. On the other hand, an aquarium is a climate or weather-insensitive sightseeing site due to indoors. Therefore, a satisfaction of zoo more drastically changes than that of aqua museum dependent on weather conditions, and tourists' satisfactions to sightseeing sites may be different according to each weather condition under uncertainty. As another important factor, we need to consider the tourist's tiredness during sightseeing. If the tourist does activities to stand and walk all the way with the much tired, the tourist may not enjoy very well even if the sightseeing site is much popular. In order to overcome the drawbacks in previous researches and to construct the more general framework of tour planning, a mathematical model of route planning with uncertain traveling times and tiredness-dependent satisfactions of activities need to be developed. In this paper, we consider a sightseeing route planning problem with uncertain traveling times and fuzzy random and tiredness-dependent satisfactions from a mathematical point of view.

In addition, tourists often want not to change their sightseeing routes even if weather conditions are changed. If these routes are utterly different each weather condition, tourists are confused how they select an appropriate plan in these routes, and this approach will not become a better decision support system for sightseeing. Therefore, it is important for the tourist to have the same satisfactory route plan to change adaptively according to current weather, climate and traffic conditions. Recently, we [7] have proposed a route planning system in terms of mathematical programming. It provides not only the optimal tour route of usual condition but also flexible routes under the other conditions which are the same or similar route of usual condition. On the other hand, it is also important to arrive at the final destination as soon as possible satisfying the target satisfactions, because the tourist can have free times to stay favorable site longer or to visit other sightseeing sites. Most recently, we [8] have constructed a route planning problems to obtain a favorable common route plan under several conditions. This previous model [8] has considered the only one objective function minimizing the maximum value of the total sightseeing and traveling time. However, maximizing the total satisfaction value is also one of most important factors in sightseeing route planning. Therefore, in this paper, we propose the above-mentioned biobjective sightseeing route planning problem.

This paper is organized as follows. In Sect. 2, we introduce fuzzy random variables for uncertain traveling times and satisfactions. As a specific case of fuzzy numbers, we introduce a trapezoidal fuzzy number and define the order relation between two trapezoidal fuzzy numbers, called Yager's ranking method. In Sect. 3, we explain some assumptions of our proposed model in this paper, and formulate a proposed route planning problem for sightseeing in mathematical programming. Since this formulated problem includes fuzzy inequalities, we transform the previous problem into an extended model of network optimization problems using the order relation in Sect. 2. Finally, in Sect. 4, we conclude this paper.

2 Fuzzy Random Variable

We first introduce a fuzzy set theory before defining a fuzzy random variable. The fuzzy set theory was proposed by Zadeh [17] as a means of representing and manipulating data that was not precise, but rather fuzzy, and it was specifically designed to mathematically represent uncertainty and vagueness. Therefore, it allows decision making with estimated values under incomplete or uncertain information [2]. The mathematical definition of fuzzy set is given as follows.

Definition 1 Let $\hat{X}$ be a nonempty set. A fuzzy set Φ in $\hat{X}$ is characterized by its membership function $\mu_\Phi : \hat{X} \to [0, 1]$ and μ_Φ is interpreted as the degree of membership of element x in fuzzy set Φ for each $x \in \hat{X}$.

Consider the degree to which the statement "x is in Φ" is true. This definition means that the value 0 is used to represent complete non-membership, the value 1 is used to represent complete membership, and values in between are used to represent intermediate degrees of membership.

Using the definition of fuzzy sets as well as the probability theory, a fuzzy random variable was first defined by Kwakernaak [9], and Puri-Ralescu [12] established the mathematical basis. In this paper, consider the case where the realization of random variable is a fuzzy set. Accordingly, fuzzy random variables are defined as follows:

Definition 2 Let (Ω, B, P) be a probability space, $F(\Re)$ be the set of fuzzy numbers with compact supports, and $\hat{X}$ be a measurable mapping $\Omega \to F(\Re)$. Then, $\hat{X}$ is a fuzzy random variable if and only if given $\omega \in \Omega$, $\hat{X}_h(\omega)$ is a random interval for any $h \in [0, 1]$ where $\hat{X}_h(\omega)$ is a h-level set of the fuzzy set $\hat{X}(\omega)$.

This definition corresponds to a special case of fuzzy random variables given by Kwakernaak and Puri-Ralesu. Though it is a simple definition, it would be useful for various applications.

As an assumption to simplify the following discussion, we consider that all fuzzy numbers are represented as trapezoidal fuzzy numbers $\tilde{a} = (a_L, a_U, \alpha, \beta)$ in this paper, whose membership functions are defined as follows:

$$
\mu_{\tilde{a}}(\omega) = \begin{cases} \dfrac{\omega - (a_L - \alpha)}{\alpha} & (a_L - \alpha \le \omega \le a_L) \\ 1 & (a_L \le \omega \le a_U) \\ \dfrac{(a_U + \beta) - \omega}{\beta} & (a_U \le \omega \le a_U + \beta) \\ 0 & \text{otherwise} \end{cases}
$$

where a_L, a_U, α, β are constant values set by the decision maker. The trapezoidal fuzzy number is viewed as a special fuzzy number. A convenient approach for solving fuzzy mathematical programming problems is to use the ranking between two fuzzy numbers, and hence, several rankling functions have been proposed to compare two fuzzy numbers until now. In this paper, we introduce Yager's ranking method [16]. This is one of linear ranking functions, and the calculation process is simple. Therefore, many studies of fuzzy mathematical programming problems adopt the Yager's ranking method. Yager's ranking function for trapezoidal fuzzy numbers is defined as follows:

$$
Y(\tilde{a}) = \frac{1}{2} \int_0^1 \left(\inf[\tilde{a}]_h + \sup[\tilde{a}]_h \right) dh \tag{1}
$$

$$
= \frac{1}{2} \left(a_L + a_U + \frac{\beta - \alpha}{2} \right)
$$

where $[\tilde{a}]_h = \{\omega | \mu_{\tilde{a}}(\omega) \geq h\}$ means the h-cut of fuzzy number. Using the Yager's ranking method, we define orders between two trapezoidal fuzzy numbers $\tilde{a}$ and $\tilde{b}$ as follows:

$$\tilde{a} \geq \tilde{b} \text{ if and only if } Y(\tilde{a}) \geq Y(\tilde{b})$$
$$\tilde{a} = \tilde{b} \text{ if and only if } Y(\tilde{a}) = Y(\tilde{b}) \tag{2}$$
$$\tilde{a} \leq \tilde{b} \text{ if and only if } Y(\tilde{a}) \leq Y(\tilde{b})$$

3 Formulation of Our Proposed Route Planning Problem for Sightseeing

Let $G = (V, E)$ be a connected graph to represent a sightseeing area including m sightseeing sites, one departure place 0, and one final destination M, i.e., node set $V = (0, 1, \ldots, m, M)$. c_{ij} denotes a positive deterministic number associated with arc (i, j) corresponding to costs necessary to traverse (i, j) from i to j and activity of sightseeing site j. In this paper, we assume the total of weather and climate conditions is K. As a fuzzy random variable, $\tilde{\bar{a}}_{ij}$ denotes the sum of satisfaction values to landscapes from sightseeing sites i to j and the activity sightseeing site j. Fuzzy random variable $\tilde{\bar{a}}_{ij}$ is defined as the following trapezoidal fuzzy number under each condition:

$$\tilde{\bar{a}}_{ij} = \begin{cases} \tilde{a}_{ij}^1 = \left(\underline{a}_{ij}^1, \bar{a}_{ij}^1, \alpha_{ij}^1, \beta_{ij}^1 \right) \\ \quad \vdots \\ \tilde{a}_{ij}^k = \left(\underline{a}_{ij}^k, \bar{a}_{ij}^k, \alpha_{ij}^k, \beta_{ij}^k \right) \quad , \quad (k = 1, 2, \ldots, K) \\ \quad \vdots \\ \tilde{a}_{ij}^K = \left(\underline{a}_{ij}^K, \bar{a}_{ij}^K, \alpha_{ij}^K, \beta_{ij}^K \right) \end{cases} \tag{3}$$

$$\Pr\left\{ \tilde{\bar{a}}_{ij} = \tilde{a}_{ij}^k \right\} = \gamma_k, \quad \sum_{k=1}^{K} \gamma_k = 1$$

where γ_k denotes a positive occurrence probability corresponding to kth condition.

In this paper, we assume that each satisfaction value of sightseeing site is dependent on tourist's tiredness. If the tourist visits the most favorable sightseeing site with the much tired, she or he may not enjoy this site very well. The sightseeing tiredness depends on the current total sightseeing time and the contents of sightseeing site such as stand and walk activity. Thus, it is important to consider the tourist's tiredness factor in the tour route planning problem. However, it is difficult to introduce this tiredness factor in a static network model based on the actual traffic network straightforwardly, because the static network cannot deal with time-dependent parameters directly. In order to overcome this weakness, we introduce the following network based on a tree structure as shown in Fig. 1.

This figure shows a partial given network with 5 sightseeing sites, particularly focused on route 0-1-5-2-M. For instance, the tourist can go to sightseeing sites and final destination M from the departure place 0, and hence, we set directed arcs from departure place 0 to each sightseeing site. Then, in the case that the tourist goes to sightseeing site 1, the next sightseeing sites are selected from sightseeing sites 2 to 5 except for sightseeing site 1, and hence, directed arcs from sightseeing site 1 to the other sightseeing sites. In addition, value 5 on directed arc from departure place 0 to sightseeing site 1 means the value of tiredness the tourist goes this route. By counting values of tiredness until the present location on the sightseeing route, tiredness-dependent satisfactions of sightseeing sites are straightforwardly obtained. In this paper, fuzzy random and tiredness-dependent satisfactions are defined as $\tilde{a}_{ij}^k(p) = \left(\underline{a}_{ij}^k(p), \bar{a}_{ij}^k(p), \alpha_{ij}^k(p), \beta_{ij}^k(p) \right)$ where p is a feasible path in the given tree structure-based network such as Fig. 1.

As another fuzzy random variable, $\tilde{t}_{ij}$ denotes the sum of necessary times to travel from sightseeing sites i to j and to do the activity for sightseeing site j, i.e., fuzzy random variable $\tilde{t}_{ij}$ is formulated as follows:

$$
\tilde{t}_{ij} = \begin{cases} \tilde{t}_{ij}^1 = \left(\underline{t}_{ij}^1, \bar{t}_{ij}^1, \varsigma_{ij}^1, \eta_{ij}^1 \right) \\ \quad \vdots \\ \tilde{t}_{ij}^k = \left(\underline{t}_{ij}^k, \bar{t}_{ij}^k, \varsigma_{ij}^k, \eta_{ij}^k \right) \quad , \quad (k = 1, 2, \ldots, K) \\ \quad \vdots \\ \tilde{t}_{ij}^K = \left(\underline{t}_{ij}^K, \bar{t}_{ij}^K, \varsigma_{ij}^K, \eta_{ij}^K \right) \end{cases} \tag{4}
$$

Fig. 1 Partial given traffic network with 5 sightseeing sites and tourist's tiredness under route 0-1-5-2-M

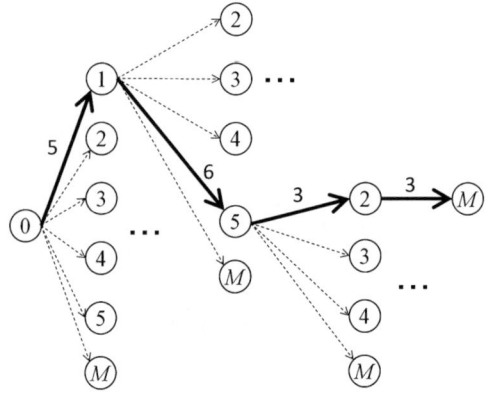

$$\Pr\left\{\tilde{\bar{t}}_{ij} = \tilde{t}_{ij}^k\right\} = \gamma_k, \quad \sum_{k=1}^{K} \gamma_k = 1$$

The other parameters, which are common to all conditions, denote as follows:

b minimum visiting points initially decided by the tourist

c_{ij} cost traveling from sightseeing sites i to j. c_{ii} means sightseeing cost at sightseeing site i

C total budget available for the tour

x_{ij} 0-1 decision variables to favorable common route satisfying the following condition;

$$x_{ij} = \begin{cases} 1, & \text{if the tourist travels from places } i \text{ to } j \\ 0, & \text{otherwise} \end{cases}$$

In the case of sightseeing site i, $x_{ii} = 1$ means that the tourists does sightseeing at sightseeing site i.

The main object of our propose model is both to maximize the minimum value of total satisfaction value and to minimize the maximum value of total traveling and sightseeing time under kth condition. Based on the above-mentioned situations and occurrence probabilities γ_k corresponding to kth condition, we formulate a biobjective sightseeing route planning problem as follows:

$$\text{Maximize} \min_{k} \left\{ w(\gamma_k) \sum_{(i,j) \in E} \tilde{a}_{ij}^k(p) x_{ij} \right\}$$

$$\text{Minimize} \max_{k} \left\{ w(\gamma_k) \sum_{(i,j) \in E} \tilde{t}_{ij}^k x_{ij} \right\}$$

subject to $x \in X$

$$X = \left\{ x \left| \begin{array}{l} \sum_{(i,j) \in E} c_{ij} x_{ij} \leq C \\[2mm] \sum_{(i,j) \in E} x_{ij} \geq b, \quad \sum_{(i,j) \in E} x_{ij} \leq 1, \\[4mm] \sum_{\{j|(i,j) \in E\}} x_{ij} - \sum_{\{j|(j,i) \in E\}} x_{ji} = \begin{cases} 1 & \text{if} \quad i = 0 \\ 0 & \text{if} \quad i \neq 0, M \\ & \quad\quad (i = 1, 2, \ldots, M) \\ -1 & \text{if} \quad i = M \end{cases} \\[6mm] x_{ij} \in \{0, 1\}, \forall (i,j) \in E, \, (k = 1, 2, \ldots, K) \end{array} \right. \right\} \tag{5}$$

where $w(\gamma_k)$ is the fixed weight value to probability γ_k. In feasible set X of our proposed model, the first constraint is the budget constraint, and the second constraint means that tourists visit more than b which is the target visiting number of sightseeing site. Third constraint means that the tourist can visit each sightseeing place only one time. The other constraints are derived from the general network flow for given network G. Tourists solve the proposed model, and obtain the favorable common route under several weather and climate conditions. The obtained route can be flexibly changed responding each condition, because the objective function is derived from the minimization of the total traveling time satisfying large satisfaction value derived from the first objective function in problem (5), and the tourist can have leeway to stay favorable place longer or to visit other sightseeing places. Therefore, it will be easy for tourists to do the more favorable sightseeing planning than the other existing tour planning problems.

In proposed problem (5), objective functions are equivalently transformed into the following form by introducing decision variable θ_s and θ_t:

$$
\text{Maximize } \min_k \left\{ w(\gamma_k) \sum_{(i,j)\in E} \tilde{a}_{ij}^k(p) x_{ij} \right\} \Leftrightarrow \begin{cases} \text{Maximize} & \theta_s \\ \text{subject to} & w(\gamma_k) \sum_{(i,j)\in E} \tilde{a}_{ij}^k(p) x_{ij} \geq \theta_s \end{cases}
$$

$$
\text{Minimize } \max_k \left\{ w(\gamma_k) \sum_{(i,j)\in E} \tilde{t}_{ij}^k x_{ij} \right\} \Leftrightarrow \begin{cases} \text{Maximize} & \theta_t \\ \text{subject to} & w(\gamma_k) \sum_{(i,j)\in E} \tilde{t}_{ij}^k x_{ij} \leq \theta_t \end{cases}
\tag{6}
$$

Furthermore, $w(\gamma_k) \sum_{(i,j)\in E} \tilde{a}_{ij}^k(p) x_{ij} \geq \theta_s$, and $w(\gamma_k) \sum_{(i,j)\in E} \tilde{t}_{ij}^k x_{ij} \leq \theta_t$ are transformed into appropriate inequalities in terms of ordering relation (2) to trapezoidal fuzzy numbers derived from the Yager's ranking method, and problem (5) is also transformed into the following problem:

Maximize θ_s

Minimize θ_t

subject to $x \in X'$

$$
X' = \left\{ x \middle| \begin{array}{l} \displaystyle\sum_{(i,j)\in E} \left(\underline{a}_{ij}^k(p) + \bar{a}_{ij}^k(p) + \frac{\beta_{ij}(p)-\alpha_{ij}(p)}{2} \right) \geq \frac{1}{w(\gamma_k)} \theta_s, \\[2ex] \displaystyle\sum_{(i,j)\in E} \left(\underline{t}_{ij}^k(p) + \bar{t}_{ij}^k(p) + \frac{\eta_{ij}(p)-\xi_{ij}(p)}{2} \right) \leq \frac{1}{w(\gamma_k)} \theta_t, \\[2ex] \displaystyle\sum_{(i,j)\in E} c_{ij} x_{ij} \leq C, \\[2ex] \displaystyle\sum_{(i,j)\in E} x_{ij} \geq b, \quad \sum_{(i,j)\in E} x_{ij} \leq 1, \\[2ex] \displaystyle\sum_{\{j|(i,j)\in E\}} x_{ij} - \sum_{\{j|(j,i)\in E\}} x_{ji} = \begin{cases} 1 & \text{if } i = 0 \\ 0 & \text{if } i \neq 0, M \\ & (i = 1, 2, \ldots, M) \\ -1 & \text{if } i = M \end{cases} \\[4ex] x_{ij} \in \{0,1\}, \forall (i,j) \in E, (k = 1, 2, \ldots, K) \end{array} \right\}
\tag{7}
$$

Each feasible set X' consists of a constrained network flow, and this problem is one extended model of network optimization problems. Satisfaction values $\underline{a}_{ij}^k(p), \bar{a}_{ij}^k(p), \alpha_{ij}^k(p), \beta_{ij}^k(p)$ are also obtained as constant values using given tree structure-based network such as Fig. 1 initially. However, this problem is a biobjective network optimization problem, and hence, it is hard to solve this problem directly since a complete optimal solution that simultaneously optimizes biobjective functions does not always exist. Various approaches to obtain a compromise solution for the multiobjective programming problem have been proposed. In this paper, we adopt a principle of compromise called Minkowski's Lp-metric, which is a standard approach with various applications. In problem (7), θ_s means the target total satisfaction for sightseeing, and θ_t also means the target total sightseeing and traveling time. We set the shortest distance from positive ideal values G_s^* and T^*, and the farthest distance from minimum target values 0 and T for Minkowski's Lp-metric. Minkowski's Lp-metric is defined by the following form using Lp-norm:

$$Z_p(x) = \left\{ w_s \left\{ \frac{G_s^* - \theta_s}{G_s^*} \right\}^p + w_t \left\{ \frac{\theta_s - T^*}{T - T^*} \right\}^p \right\}^{\frac{1}{p}} \tag{8}$$

where w_s and w_t are the relative importance of each objective function θ_s and θ_t satisfying $w_s + w_t = 1$, and p is the parameter of norm functions.

To simplify the discussion, we set $p = 1$ in this paper. The case $p = 1$ is operationally and practically important, which provides better credibility than others and emphasizes the sum of individual distances in the utility concept. Main problem (7) is transformed into the following single-objective programming problem:

$$\text{Minimize } w_s \left\{ \frac{G_s^* - \theta_s}{G_s^*} \right\} + w_t \left\{ \frac{\theta_s - T^*}{T - T^*} \right\} \tag{9}$$
$$\text{subject to } x \in X'$$

Since this problem is a single-objective network programming problem, it can be solved by network optimization method strictly or heuristic solution algorithms efficiently; branch-and-cut algorithm (Fischetti et al. [4]), 2-opt or n-opt algorithm [3], LP relaxation [11], Genetic algorithm [1], Neural network [15], Tabu-search [6], [14], Ant Colony Optimization [10], etc.

4 Conclusion

In this paper, we have proposed a biobjective sightseeing route planning model with fuzzy random traveling times among sightseeing sites and tiredness-dependent satisfactions. From general concepts that tourists like to do their satisfactory common route planning without changing the tour route even if the weather and climate condition changes and to have free times, we have formulated a mathematical model whose objective function are both maximizing the minimum value of the weighted total satisfaction values and minimizing the maximum value of the weighted total sightseeing traveling time under possible conditions. Furthermore, we have transformed the initial proposed problem into a constrained network optimization problem using the Yager's ranking method for trapezoidal fuzzy numbers in fuzzy random parameters and the Minkowski's Lp-metric for the biobjective.

Our proposed model includes various existing sightseeing route planning problems under uncertainty, but there are still some drawbacks. For instance, in terms of computation algorithm, the proposed problem is strongly NP-hard [5] due to mixed integer linear programming problem. Furthermore, with respect to the tiredness degree, we need to construct a tree structure with all sightseeing sites and conditions. Therefore, the proposed model with many sightseeing sites and conditions is more complex, and obviously, it is computationally difficult to solve large-scale tour planning problems. Therefore, we are now developing efficient heuristic solution algorithms based on soft computing approaches.

References

1. R.A. Abbaspour, F. Samadzadegan, Time-dependent personal tour planning and scheduling in metropolises. Expert Syst. Appl. **38**, 12439–12452 (2011)
2. C. Carlsson, R. Fuller, *Fuzzy Reasoning in Decision Making and Optimization* (Springer, Berlin, 2002)
3. I.M. Chao, B.L. Golden, E.A. Wasil, A fast and effective heuristic for the orienteering problem. Eur. J. Oper. Res. **88**(3), 475–489 (1996)
4. M. Fischetti, J.S. Gonzalez, P. Toth, Solving the orienteering problem through branch-and-cut. INFORMS J. Comput. **10**(2), 133–148 (1998)
5. M. Fischetti, J.J. Salazar-González, P. Toth, The Generalized Traveling Salesman and Orienteering Problems, in *The Traveling Salesman Problem and its Variations*, ed. by G. Gutin, A.P. Punnen (Kluwer Academic Publisher, Dordrecht, 2002), pp. 609–662
6. M. Gendreau, G. Laporte, F. Semet, A tabu search heuristic for the undirected selective traveling salesman problem. Eur. J. Oper. Res. **106**(2–3), 539–545 (1998)
7. T. Hasuike, H. Katagiri, H. Tsubaki, H. Tsuda, Flexible route planning for sightseeing with fuzzy random and fatigue-dependent satisfactions. J. Ref: J. Adv. Comput. Intell. Intell. Inf. **18**(2), 190–196 (2014)
8. T. Hasuike, H. Katagiri, H. Tsubaki, H. Tsuda, Sightseeing route planning responding various conditions with fuzzy random satisfactions dependent on tourist's tiredness, in *Proceedings of the International MultiConference of Engineers and Computer Scientists 2014, IMECS 2014*, Hong Kong, 12–14 Mar 2014. Lecture Notes in Engineering and Computer Science, pp. 1232–1236
9. H. Kwakernaak, Fuzzy random variable-I. Inf. Sci. **15**, 1–29 (1978)
10. L. Ke, C. Archetti, Z. Feng, Ants can solve the team orienteering problem. Comput. Ind. Eng. **54**(3), 648–665 (2008)

11. J.L. Kennington, C.D. Nicholson, The uncapacitated time-space fixed-charge network flow problem; an empirical investigation of procedures for arc capacity assignment. INFORMS J. Comput. **22**, 326–337 (2009)
12. M.L. Puri, D.A. Ralescu, Fuzzy random variables. J. Math. Anal. Appl. **114**, 409–422 (1986)
13. W. Souffriau, P. Vansteenwegen, J. Vertommen, G.V. Berghe, D.V. Oudheusden, A personalized tourist trip design algorithm for mobile tourist guides. Appl. Artif. Intell. **22**(10), 964–985 (2008)
14. H. Tang, E. Miller-Hooks, A tabu search heuristic for the team orienteering problem. Comput. Oper. Res. **32**, 1379–1407 (2005)
15. Q. Wang, X. Sun, B.L. Golden, J. Jia, Using artificial neural networks to solve the orienteering problem. Ann. Oper. Res. **61**, 111–120 (1995)
16. R.R. Yager, A procedure for ordering fuzzy subsets of the unit interval. Inf. Sci. **24**, 143–161 (1981)
17. L.A. Zadeh, Fuzzy sets. Inf. Control **8**, 338–353 (1965)
18. K.G. Zografos, K.N. Androutsopoulos, Algorithms for itinerary planning in multimodal transportation networks. IEEE Trans. Intell. Transp. Syst. **9**(1), 175–184 (2008)

Computing the Lower and Upper Bound Prices for Multi-asset Bermudan Options via Parallel Monte Carlo Simulations

Nan Zhang, Ka Lok Man and Tomas Krilavičius

Abstract We present our work on computing the lower and upper bound prices for multi-asset Bermudan options. For the lower bound price we follow the Longstaff-Schwartz least-square Monte Carlo method. For the upper bound price we follow the Andersen-Broadie duality-based nested simulation procedure. For case studies we computed the prices of Bermudan max-call options and Bermudan interest rate swaptions. The pricing procedures are parallelized through POSIX multi-threading. Times required by the procedures on ×86 multi-core processors are much shortened than those reported in previous work.

Keywords Interest rate Bermudan swaption · LIBOR market model · Multi-asset Bermudan options · Monte Carlo simulation · Multi-threaded programming · Parallel computing

1 Introduction

Prices of multi-asset Bermudan options depend on a collection of underlying state variables. Such state variables can be stock prices, interest rates, etc. To price such options, because of the high-dimensionality, the Black-Scholes formulas [3] can not

N. Zhang · K.L. Man (✉)
Department of Computer Science and Software Engineering,
Xi'an Jiaotong-Liverpool University, Suzhou, China
e-mail: ka.man@xjtlu.edu.cn

N. Zhang
e-mail: nan.zhang@xjtlu.edu.cn

T. Krilavičius
Informatics Faculty, Vytautas Magnus University, Kaunas, Lithuania
e-mail: t.krilavicius@bpti.lt

T. Krilavičius
Baltic Institute of Advanced Technology, Vilnius, Lithuania

© Springer Science+Business Media Dordrecht 2015
G.-C. Yang et al. (eds.), *Transactions on Engineering Technologies*,
DOI 10.1007/978-94-017-9588-3_14

be applied. Neither can be the various finite-difference schemes [8, 15], or the lattice-based methods [7, 12]. The pricing of such options is not only challenging, but also computationally demanding.

Regression-based methods [6, 13, 16] have been proposed to be applied with Monte Carlo simulations to compute lower bounds for the true prices of multi-asset Bermudan options. Because the exercise strategies generated by the regressions are sub-optimal, the prices computed by these algorithms are generally lower than the true price. The Longstaff-Schwartz algorithm [13] is such an example where it applies linear least-squares regression at every exercise time spot on observed continuation values on all in-the-money paths. The algorithm assumes a linear relationship between true continuation values and functions of the prices of the underlying assets. By regressing on the observed continuation values the algorithm approximates the linear relationship through finding the coefficients of the linear relationship.

To bind the prices at upper level Andersen and Broadie [2] have proposed a duality-based algorithm to work with any lower bound estimator. The Andersen-Broadie algorithm presents the upper bound as the sum of a lower bound and a penalty term. The penalty term is a non-negative quantity representing a compensation for the potential incorrect exercise decisions made by following a sub-optimal exercise strategy.

In this paper, we present our formulations and implementations of the Longstaff-Schwartz algorithm and the Andersen-Broadie algorithm in pricing Bermudan max-call stock options and Bermudan interest rate swaptions. The price of the former depends on the values of a basket of stocks, and that of the latter depends on a large number of forward rates. To accelerate the pricing procedures we parallelised the Monte Carlo simulations via POSIX multi-threading. The code is executable on shared-memory ×86 multi-core processors. Highly-optimised functions from Intel's Math Kernel Library (MKL) [9, 10] were used in the implementation for linear algebra operations and random number generation. Experiments on entry-level and main-stream multi-core machines showed that the computational times were shortened significantly compared to the results reported in previous work [5]. A preliminary conference presentation appeared as [17].

Organisation of the rest of this paper
Our formulations of the Longstaff-Schwartz algorithm and the Andersen-Broadie algorithm are presented in Sects. 2 and 3, respectively. The pricing of the multi-asset Bermudan max-call options is presented in Sect. 4. The pricing of the Bermudan interest rate swaptions is presented in Sect. 5. The parallelisation in the Monte Carlo simulations is explained in Sect. 6. Source-level optimisations in implementing the extended multi-factor LIBOR market model for pricing the swaptions is discussed in Sect. 7. Conclusions are drawn in Sect. 8.

2 Formulation of the Longstaff-Schwartz Lower Bound Algorithm

Assume a Bermudan option has d exercise opportunities within time period 0 to T. These exercise times are denoted by $0 < t_1 < t_2 < \cdots < t_d = T$. The Bermudan option's value depends on a basket of n assets whose state $\mathbf{S}_i = (S_i^1, S_i^2, \ldots, S_i^n)^1$ at time t_i, $i \in [1, d]$, follows a vector-valued Markov process on $\mathbb{R}^n$ with initial value $\mathbf{S}_0$. Let B_t denote the time t value of one unit of cash invested in a risk-free money market account at time 0. Let h_i denote the payoff from exercising the option at the stopping time t_i, where $t_i \in \mathscr{F} = \{t_1, t_2, \ldots, t_d\}$. The price Q_0 of the Bermudan option is, therefore,

$$Q_0 = \sup_{t_i \in \mathscr{F}} \mathbb{E}_0(h_i / B_i), \tag{1}$$

where $\mathbb{E}_i$ denotes the expectation conditional on the information available until time t_i. The Longstaff-Schwartz algorithm computes not Q_0 but L_0, where

$$L_0 = \mathbb{E}_0(h_i / B_i) \le Q_0. \tag{2}$$

The quantity L_0 is a low-biased price of the option obtained by following some specific exercise policy rather than the optimal one.

To compute L_0 using Monte Carlo simulations we have to estimate the continuation values at each exercise time, so that on a path j at time t_i, $i \in [1, d]$, if the immediate payoff exceeds the continuation value the option is assumed to be exercised. The computation starts from time $t_d = T$, at which it is assumed that the option is exercised for all in-the-money paths. The payoffs h_d on all simulated paths are then used as the observed continuation values for the estimation of the continuation value at time t_{d-1}. The algorithm assumes a linear relationship between the true continuation value C_{d-1} at time t_{d-1} and the state variable $\mathbf{S}_{d-1}$ such that

$$\begin{aligned} C_{d-1} = {} & e_{d-1}^1 f_1(\mathbf{S}_{d-1}) + e_{d-1}^2 f_2(\mathbf{S}_{d-1}) \\ & + \cdots + e_{d-1}^m f_m(\mathbf{S}_{d-1}). \end{aligned} \tag{3}$$

The m functions $f_1, f_2, \ldots, f_m$ are the pre-defined basis functions, and $e_{d-1}^1, e_{d-1}^2, \ldots, e_{d-1}^m$ are the coefficients whose values are estimated by the following least-squares regression.

Assume totally N_R simulation paths are launched and at time t_{d-1} there are p in-the-money paths. We use $1, 2, \ldots, p^2$ to index these in-the-money paths. We form a p-by-m (p rows and m columns) design matrix A_{d-1} whose jth row, $j \in [1, p]$, is

---

[1] We use the subscript i for t_i. So $\mathbf{S}_i$ is actually $\mathbf{S}_{t_i}$.

[2] These are not their indexes in the whole N_R simulated paths.

$$\mathbf{A}_{d-1}^{j} = (f_1(\mathbf{S}_{d-1}^{j}), f_2(\mathbf{S}_{d-1}^{j}), \ldots, f_m(\mathbf{S}_{d-1}^{j})), \tag{4}$$

where $\mathbf{S}_{d-1}^{j}$ is the n-dimensional vector-valued state variable at time t_{d-1} on the jth in-the-money path. Then we form the p-dimensional vector $\mathbf{V}_{d-1}$ as

$$\mathbf{V}_{d-1} = \left(B_{d-1}\frac{h_d^1}{B_d}, B_{d-1}\frac{h_{d1}^2}{B_d}, \ldots, B_{d-1}\frac{h_d^p}{B_d} \right), \tag{5}$$

where $B_{d-1}h_{d-1}^{j}/B_d$ is the observed discounted continuation value on the jth in-the-money path at time t_{d-1}. Now we can set up a linear system and solve the m-dimensional vector $\mathbf{E}_{d-1} = (e_{d-1}^1, e_{d-1}^2, \ldots, e_{d-1}^m)$ for the exercise time t_{d-1} by minimising the merit function

$$\chi^2 = \sum_{i=1}^{p} (V_{d-1}^i - \mathbf{A}_{d-1}^i \mathbf{E}_{d-1}^{\mathrm{T}})^2, \tag{6}$$

where V_{d-1}^i is the ith component of vector $\mathbf{V}_{d-1}$ and $\mathbf{A}_{d-1}^i$ is the ith row of the matrix A_{d-1}. The merit function is a measure for the aggregated difference between the observed continuation values and the estimated continuation values.

The linear system can be solved by using either QR or LQ factorisation, depending on whether it is over-determined or under-determined. Alternatively, the more robust singular value decomposition can be used at a higher computational cost. Once $\mathbf{E}_{d-1}$ has been computed the p-dimensional vector $\hat{\mathbf{C}}_{d-1}^{\mathrm{T}} = A_{d-1}\mathbf{E}_{d-1}^{\mathrm{T}}$ can be derived, which contains the estimated continuation values for the p in-the-money paths. This process proceeds backwards from time t_{d-1} to t_1 and outputs the d-by-m coefficient matrix E, whose ith row, $i \in [1, d-1]$, stores the coefficients for the exercise time t_i. The dth row of E is filled by zeroes.

After the estimated exercise policy coefficient matrix E is obtained, another N_L simulation paths are launched. On a jth path at the first exercise time t_i, $i \in [1, d]$, when $h_i^j > (f_1(\mathbf{S}_i^j), f_2(\mathbf{S}_i^j), \ldots, f_m(\mathbf{S}_i^j))\mathbf{E}_i^{\mathrm{T}}$, the option is assumed to be exercised, and the time 0 value of the resulted cash flow is h_i^j/B_i. The low-biased option price L_0 is the mean of such cash flows on all the N_L simulated paths.

3 Formulation of the Andersen-Broadie Upper Bound Algorithm

The upper bound U_0 of the option is defined in [2] as $U_0 = L_0 + \Delta_0$, where Δ_0 is a penalty term defined as $\Delta_0 = \mathbb{E}_0(\max_{t_i \in \mathscr{F}}(h_i/B_i - \pi_i))$. The process π_i at time t_{i+1}, $i \in [1, d-1]$ is a martingale defined in [5] by

$$\pi_{i+1} = \pi_i + L_{i+1}/B_{i+1} - Q_i/B_i. \tag{7}$$

At time 0 and time t_1 it is defined as $\pi_0 = L_0$ and $\pi_1 = L_1/B_1$. The process L_{i+1}/B_{i+1} is the discounted lower bound price at time t_{i+1} and is computed by $L_{i+1}/B_{i+1} = \mathbb{E}_{i+1}(h_{\tau_{i+1}}/B_{\tau_{i+1}})$, where τ_{i+1} is the first exercise time instance starting from time t_{i+1} at which exercise is indicated. The algorithm in [5] with sub-optimality checking suggests computing π_i only at those exercise times when exercise is suggested.

To estimate Δ_0 we simulate N_H paths, each of which has $d + 1$ time steps, corresponding to the n-dimensional vector state variables $S_0, S_1, \ldots, S_d$. On a jth simulated path, starting from time t_1 we ignore all exercise times at which continuation is suggested by the exercise policy. For a time t_i, $i \in [1, d]$, if exercise is suggested, we follow one of the two procedures listed below.

1. If time t_i is the first exercise time on the jth path at which exercise is suggested, we set $\pi_i^j = h_i^j/B_i$—the discounted option's payoff, which is also the discounted option value in this case. We then launch N_S simulation trials to estimate $Q_i/B_i = \mathbb{E}_i(h_{\tau_i}/B_{\tau_i})$. This is the discounted option's continuation value from time t_i if continuation is enforced. As above, stopping time τ_i is the first exercise time instance starting from time t_i at which exercise is suggested. The penalty term Δ_i^j is set to 0.

2. If, otherwise, time t_i is not the first exercise time on the jth path at which exercise is suggested, we set $\pi_i^j = \pi_l^j + h_i^j/B_i - Q_l/B_l$, where t_l is the previous exercise time on the jth path at which exercise is suggested. By the time when π_i^j is computed the values of π_l^j and Q_l/B_l are already available. We then launch N_S simulation trials to estimate $Q_i/B_i = \mathbb{E}_i(h_{\tau_i}/B_{\tau_i})$ for later use. The penalty term Δ_i^j in this case is set to $\Delta_i^j = h_i^j/B_i - \pi_i^j$.

The above procedure is deduced from Proposition 4.1 in [5]. Applying Eq. (15) in [5] to a Bermudan option, at exercise time t_i where exercise is suggested process π_i^j is defined as

$$\pi_i^j = \pi_{i-1}^j + h_i^j/B_i - Q_{i-1}/B_{i-1} \tag{8}$$

If time t_i is the first exercise time on the jth path at which exercise is suggested, according to Proposition 4.1 (1) in [5], $\pi_{i-1}^j = h_{i-1}^j/B_{i-1}$. At an exercise time where continuation is suggested such as time t_{i-1}, the option's payoff is equal to its continuation value, and, so, we have $\pi_{i-1}^j = Q_{i-1}/B_{i-1}$. Substitute Q_{i-1}/B_{i-1} for π_{i-1}^j in Eq. (8) we get $\pi_i^j = h_i^j/B_i$. The penalty term $\Delta_i = h_i^j/B_i - \pi_i^j = 0$. If, however, time t_i is an exercise time, but not the first one, at which exercise is suggested, according to Proposition 4.1 (2), $\pi_{i-1}^j = \pi_l^j - Q_l/B_l + h_{i-1}^j/B_{i-1}$, where time t_l is the previous exercise time on the path at which exercise is suggested. Because at time t_{i-1} the option's payoff equals to its continuation value, we have

$\pi_{i-1}^j = \pi_l^j - Q_l/B_l + Q_{i-1}/B_{i-1}$. Substitute this expression for π_{i-1}^j in Eq. (8) we get $\pi_i^j = \pi_l^j + h_i^j/B_i - Q_l/B_l$.

The penalty term Δ^j for the jth path is then set to $\Delta^j = \max(\Delta_1^j, \Delta_2^j, \ldots, \Delta_d^j)$. The penalty terms for the exercise times at which continuation is suggested are set to zeroes. After the penalty terms have been computed for all the N_H paths, the upper bound increment Δ_0 is set to the mean of all the penalty terms. The upper bound U_0 is therefore $U_0 = L_0 + \Delta_0$.

Now, if $\hat{L}_0$ is the lower bound estimation obtained by N_L simulation trials with a sample standard deviation $\hat{s}_L$ and $\hat{\Delta}_0$ is the estimation for the increment using N_H trials with a sample standard deviation $\hat{s}_\Delta$, as in [2], a $100(1 - \alpha)$ %-probability confidence interval for the price Q_0 of the Bermudan option can be computed as $[\hat{L}_0 - z_{1-\alpha/2}\hat{s}_L/\sqrt{N_L},\ \hat{L}_0 + \hat{\Delta}_0 + z_{1-\alpha/2}\sqrt{\hat{s}_L^2/N_L + \hat{s}_\Delta^2/N_H}]$ with z_x denoting the xth percentile of a standard Gaussian distribution.

4 Pricing the Max-Call Stock Options

The multi-asset symmetric Bermudan max-call options we priced are the same ones chosen from [2, 5]. Each of these Bermudan options depends on a basket of n uncorrelated equity assets. The risk-neutral dynamics of their prices all follow geometric Brownian motion processes. For a jth, $j \in [1, n]$, asset in the basket, its price dynamic at time t is modelled by

$$dS_t^j = (r - \delta)S_t^j dt + \sigma S_t^j dW_t^j, \tag{9}$$

where r is the annual risk-free interest rate, δ is the dividend yield of the equity asset, σ is the annual volatility and W_t^j is a standard Brownian motion process. The time t value of one unit of cash invested at time 0 is $B_t = \exp(rt)$. The d exercise times for the options are equally spaced at times $t_i = iT/d$, $i = 1, 2, \ldots, d$. On the simulated paths, from a time t_i to time t_{i+1}, $i \in [0, d - 1]$, the price of a jth asset is updated by

$$\begin{aligned} S_{i+1}^j = S_i^j \exp((r - \delta - \sigma^2/2)(t_{i+1} - t_i) \\ + \sigma\varepsilon\sqrt{t_{i+1} - t_i}), \end{aligned} \tag{10}$$

where ε is a $\mathcal{N}(0, 1)$ random number. The options' strike prices are K, and so the payoff of such an option at time t_i, $i \in [1, d]$, is

$$h_i(\mathbf{S}_i) = (\max(\mathbf{S}_i^1, \mathbf{S}_i^2, \ldots, \mathbf{S}_i^n) - K)^+. \tag{11}$$

For comparison purposes we used the same values for the market and simulation parameters as in [2, 5]. We set $K = 100$, $r = 0.05$, $\delta = 0.1$, $T = 3$, $\sigma = 0.2$, $d = 9$,

N_R = 200,000, N_L = 2,000,000, N_H = 1,500, N_S = 1,000 and $\mathbf{S}_0 = (S, S, \ldots, S)$. For the five-asset Bermudan options we used the same 18 basis functions found in Appendix A.2 [5]. For the three-asset Bermudan options we used the first 10 functions and the first 8 for the two-asset Bermudan options.

The sequential and parallel implementations were programed in C/C++. The native POSIX thread library (NPTL) 2.12.1 was used for the threading. The tests were made on Intel multi-core systems running Ubuntu Linux 10.10 (64 bit). The binary executables were compiled by the Intel compiler icpc 12.0 for Linux.

We first present in Table 1 the results from three groups of tests using the Bermudan max call options when $n = 2, 3, 5$. For comparison we include data from [2, 5]. These groups of tests were made on a 2.4 GHz dual-core Intel P8600 processor. We show the results on this entry-level multi-core processor to demonstrate that our sequential and parallel implementations are practical on common personal computers.

From the data reported in Table 1 we can see that the $\hat{L}_0$s and $\hat{U}_0$s produced by the parallel program are very close to the data reported in [2, 5], yet the runtimes were much shorter. We also ran tests using the seven five-asset Bermudan options on a 3.4 GHz (turbo boost to 3.8 GHz) quad-core (8 threads with hyperthreading) Intel Core i7 2600 processor. It took the sequential program 27.19 s to compute the $\hat{L}_0$s and $\hat{U}_0$s for the seven options and the parallel program 11.66 s using eight threads.

5 Pricing the Bermudan Interest Rate Swaptions

We apply the extended multi-factor LIBOR market model [1] to model the dynamics of the forward rates underlying the swaptions. This model is an extension to the LIBOR market model frameworks discussed in [4, 11, 14]. We define an increasing maturity structure $0 = t_0 < t_1 < t_2, \ldots, < t_K$. The discounting process $B(t_i, t_j)$ is the time t_i price of a zero-coupon bond paying off \$1 at time t_j for $i \leq j$ and $i, j \in \{0, 1, 2, \ldots, K\}$. The LIBOR market model in general does not put any restriction on the increasing maturity structure, but our implementation assumed that any two successive time spots in the structure span an equidistant accrual period, often 3 or 6 months in practice. While the function $B(,)$ can be defined on any time spots not necessarily coinciding with the dates in the maturity structure, we define it on dates in the maturity structure to simplify the problem and, thus, to serve the purpose of our implementation. The discrete forward rate $F_{t_j}(t_i)$ for any t_i and t_j when $i \leq j$ and $i, j \in \{0, 1, 2, \ldots, K - 1\}$ that applies to period between t_j and t_{j+1} observed at time t_i is defined as

$$F_{t_j}(t_i) = F_j(i) = \frac{1}{\delta_j} \left(\frac{B(t_i, t_j)}{B(t_i, t_{j+1})} - 1 \right),$$

$$delta_j = t_{j+1} - t_j$$

(12)

Table 1 Computational and timing results on Intel dual-core 2.4 GHz P8600

S_0	$\hat{L}_0$	T_L	$\hat{U}_0$	T_U	T_P	T_S	T'_S	95 % CI	95 % CI'
$n = 2, m = 8$									
90	8.048	1.80	8.051	0.48	2.28	2.59		[8.043, 8.068]	[8.053, 8.082]
100	13.886	1.71	13.891	0.52	2.23	2.88		[13.881, 13.912]	[13.892, 13.934]
110	21.341	1.82	21.345	0.59	2.41	3.14		[21.335, 21.370]	[21.316, 21.359]
Total runtime					6.91	8.61			
$n = 3, m = 10$									
90	11.254	2.06	11.257	0.70	2.75	4.00		[11.249, 11.277]	[11.265, 11.308]
100	18.653	2.45	18.660	0.76	3.21	4.50		[18.647, 18.684]	[18.661, 18.728]
110	27.546	2.66	27.556	0.83	3.49	4.82		[27.539, 27.584]	[27.512, 27.663]
Total runtime					9.44	13.31			
$n = 5, m = 18$									
70	3.899	2.74	3.901	1.02	3.76	5.63	46.2	[3.895, 3.912]	[3.880, 3.913]
80	9.024	3.30	9.029	1.07	4.37	6.52	54	[9.019, 9.046]	[8.984, 9.033]
90	16.612	4.25	16.618	1.17	5.42	7.88	85.2	[16.607, 16.641]	[16.599, 16.686]
100	26.125	5.10	26.143	1.23	6.33	9.03	121.8	[26.119, 26.170]	[26.093, 26.194]
110	36.746	5.30	36.771	1.28	6.58	9.33	151.2	[36.740, 36.802]	[36.681, 36.819]
120	47.890	5.31	47.909	1.33	6.63	9.44	168.6	[47.883, 47.943]	[47.816, 48.033]
130	59.352	5.33	59.379	1.33	6.66	9.47	198	[59.345, 59.415]	[59.199, 59.437]
Total runtime					39.75	57.29			

Notes The estimations $\hat{L}_0$, $\hat{U}_0$ and the 95 % CI were computed by our parallel program. T_L is the parallel runtime for computing the $\hat{L}_0$ and T_U is the parallel runtime for computing the $\hat{U}_0$. T_P is the total parallel runtime, $T_P = T_L + T_U$. T_S is the runtime of our sequential program in computing $\hat{L}_0$ and $\hat{U}_0$. T'_S is the run-time reported in [5] (in minutes), 95 % CI' for $n = 2, 3$ is from [2], and for $n = 5$ is from [5]. No timing results were found in from [2], and, therefore, data for T'_S when $n = 2, 3$ is not included in the table

With this definition for forward rates, the definition for $B(t_i, t_j)$ can be written as

$$B(t_i, t_j) = B(i,j) = \prod_{k=i}^{j-1}\left(\frac{1}{1+\delta_k F_k(i)}\right),$$

$$i \le j, \quad i,j \in \{0,1,2,\ldots,K\}$$

(13)

Note that in Eq. (13) the production is performed up to time t_{j-1}, because the forward rate $F_{j-1}(i)$ applies to the period from time t_{j-1} to time t_j. In the extended LIBOR market model [1], knowing the initial forward rates $F_j(0)$ for all $j \in \{0,1,2,\ldots,K-1\}$ forward rates observed at future times can be approximated by

$$\hat{F}_j(i+1) = \hat{F}_j(i) \exp\left(\frac{\varphi(\hat{F}_j(i))}{\hat{F}_j(i)}\lambda_j^{\mathrm{T}}(i)\right.$$

$$\left.\left[\left(\hat{\mathbf{u}}_j(i) - \frac{1}{2}\frac{\varphi(\hat{F}_j(i))}{\hat{F}_j(i)}\lambda_j(i)\right)\Delta_i + \varepsilon_i\sqrt{\Delta_i}\right]\right)$$

(14)

Equation (14) is obtained by applying Euler scheme to the dynamics of the forward rate in continuous time. For the equation to hold we have the obvious condition $i+1 \le j \le K-1$. Function $\varphi(.)$ in the equation is the skew function, $\lambda_j(i)$ is the m-dimensional volatility vector, and $\hat{\mathbf{u}}_j(i)$ is the m-dimensional drift vector defined as

$$\hat{\mathbf{u}}_j(i) = \sum_{k=i+1}^{j}\lambda_k(i)\frac{\delta_k\varphi(\hat{F}_k(i))}{1+\delta_k\hat{F}_k(i)}$$

(15)

The ε_i in Eq. (14) is a m-dimensional vector of independent standard Gaussian variables and $\Delta_i = t_{i+1} - t_i$. Using Eq. (14) with Monte Carlo simulation paths of forward rates can be generated. The equation guarantees the generated rates are positive. It should be pointed out that (1) in computing the drift term $\hat{\mathbf{u}}_j(i)$ the summation starts from time t_{i+1}, thus excluding the term $\lambda_i(i)\delta_i\varphi(\hat{F}_i(i))/(1+\delta_i\hat{F}_i(i))$, and (2) the same m-dimensional vector ε_i applies to the generation of all forward rates $F_j(i+1)$ for $j = i+1, i+2, \ldots, K-1$ from $F_j(i)$.

A Bermudan swaption gives the holder the right of exercising to enter a swap agreement in which the holder pays fixed cash flows $\theta\delta_{j-1}$ at time t_j for $j = s+1, s+2, \ldots, e$, in exchange for LIBOR on a \$1 notional, assuming t_s being the first exercise date and t_e being the last payment date. Payments for periods between t_j and t_{j+1} are exchanged at time t_{j+1} (paid in arrears). A Bermudan swaption is characterised by three dates: the lockout date t_s, the last exercise date t_x and the final swap maturity t_e. Our implementation assumed that all these three dates coincide with dates in the maturity structure and $t_s \le t_x = t_{e-1} = t_{K-1}$, that is, the last exercise date t_x is the second last date in the maturity structure and the swap matures at time t_K. A Bermudan swaption characterised by t_s, t_x and t_e (assuming $t_e = t_{x+1}$), can be exercised once at any time between t_s and t_x. The first payments are exchanged at time t_{s+1}, and the last at time $t_e = t_{x+1}$.

For such a Bermudan swaption exercised at t_j for $j \in \{s, s+1, \ldots, x = e-1\}$ we define a strictly positive process $Z(t_i, t_j)$ for $i \leq j$ as

$$Z(t_i, t_j) = Z(i,j) = \sum_{k=j}^{x} \delta_k B(i, k+1) \tag{16}$$

The par-rate $R(t_i, t_j)$ observed at time t_i if the swaption is exercised at time t_j assuming the swap maturing at time $t_e = t_{x+1}$ is

$$R(t_i, t_j) = R(i,j) = \frac{B(i,j) - B(i, x+1)}{Z(i,j)} \tag{17}$$

This rate is also known as the swap rate at time t_i that makes the payoff of the swap exercised at time t_j and maturing at time $t_e = t_{x+1}$ equal to zero, thus, being fair to both the parties. With the above definitions the payoff $h(t_j)$ of the swaption exercised at time t_j for the fixed payer is

$$h(t_j) = h(j) = Z(j,j)(R(j,j) - \theta)^+$$
$$= Z(j,j)\left(\frac{1 - B(j, x+1)}{Z(j,j)} - \theta\right)^+ \tag{18}$$

where θ is the fixed coupon rate. The time 0 value of this payoff $h(j)$ received at exercise time t_j is $B(0,j)h(j)$.

We priced the same Bermudan interest rate swaptions as in [2]. The tests were made on a quad-core 3.4 GHz Intel Core i7-2600 processor and two quad-core 2.0 GHz Intel Xeon E5405 processors. Both the systems ran Ubuntu Linux 10.10 64-bit version. The binary executables were compiled by Intel's C/C++ compiler icpc 12.0 for Linux with –O3 optimisation. Parallelisation was achieved through POSIX multi-threading. In all the tests, forward rates were generated in an extended two-factor LIBOR market model according to Eq. (14). The parameters were set to the same values as in [2]: $\varphi(x) = x$, $\delta = 0.25$, $F_j(0) = 0.1$ for $j \in \{t_0, t_1, \ldots, t_{x-1}\}$, and $\lambda_j(i) = (0.15, 0.15 - \sqrt{0.009(t_j - t_i)})$ for $i \leq j$. For the simulations we had $N_R = 5{,}000$, $N_L = 50{,}000$, $N_H = 750$ and $N_S = 300$. The testing results on the Intel Xeon E5405s are reported in Table 2. The runtimes were measured in seconds as wall clock times. Data from Table 4 in [2] are included for comparison. In all the tests we assumed $t_x = t_{e-1}$.

From the speedup (S_P) data it can be seen that the parallel program demonstrated significant accelerations against the sequential program. Comparing the data in the last two columns we can see that the figures are quite close, although not identical.

Table 2 Lower and upper bounds of Bermudan swaptions and timing results on two quad-core 2.0 GHz Intel Xeon E5405s

t_s	t_e	θ	$\hat{L}_0$	T_L	$\hat{\Delta}_0$	T_U	T_P	S_P	95 % CI	95 % CI'
0.25	1.25	0.08	183.6	0.12	0.022	0.23	0.35	2.14	[183.6, 183.6]	[183.9, 184.1]
0.25	1.25	0.10	42.3	0.05	0.031	0.12	0.17	2.71	[42.3, 42.3]	[43.1, 43.6]
0.25	1.25	0.12	5.2	0.05	0.010	0.11	0.16	2.0	[5.2, 5.2]	[5.5, 5.7]
1.00	3.00	0.08	341.5	0.13	0.094	0.54	0.67	4.13	[341.5, 341.6]	[339.2, 340.6]
1.00	3.00	0.10	126.1	0.09	0.214	0.35	0.44	4.07	[126.1, 126.3]	[125.1, 127.2]
1.00	3.00	0.12	36.8	0.14	0.217	0.27	0.41	2.76	[36.8, 37.0]	[36.4, 37.6]
1.00	6.00	0.08	751.0	0.33	0.966	3.24	3.57	6.64	[751.0, 752.0]	[749.0, 755.2]
1.00	6.00	0.10	315.9	0.36	3.152	1.66	2.02	5.41	[315.8, 319.0]	[315.6, 323.5]
1.00	6.00	0.12	130.8	0.35	1.957	0.89	1.23	4.45	[130.8, 132.8]	[126.5, 131.6]
1.00	11.00	0.08	1236.5	0.95	11.310	19.02	19.98	7.14	[1236.5, 1247.9]	[1245.1, 1269.0]
1.00	11.00	0.10	613.3	1.11	17.619	9.62	10.73	6.29	[613.3, 631.0]	[618.4, 645.0]
1.00	11.00	0.12	334.2	1.05	13.022	4.20	5.26	6.70	[334.2, 347.3]	[324.7, 345.0]
3.00	6.00	0.08	458.0	0.29	0.218	1.23	1.52	5.17	[458.0, 458.3]	[443.6, 446.6]
3.00	6.00	0.10	234.4	0.34	0.403	0.86	1.20	4.53	[234.4, 234.8]	[225.5, 229.5]
3.00	6.00	0.12	110.8	0.32	0.776	0.65	0.97	3.90	[110.8, 111.6]	[105.9, 109.0]

Notes The estimations $\hat{L}_0$ and $\hat{\Delta}_0$ and the confidence intervals are reported in basis points. T_L is the parallel runtime in computing $\hat{L}_0$, T_U is the parallel runtime in computing $\hat{\Delta}_0$, and $T_P = T_L + T_U$ is the total parallel runtime. S_P is the speedup of the parallel implementation in computing $\hat{L}_0$ and $\hat{\Delta}_0$ against the sequential implementation running under the same settings. For comparison we include CI' [2]

While exercise strategies were generated by an optimisation procedure in [2], we generated them using the regression-based method. In computing $\hat{L}_0$ the work in [2] used antithetic sampling, but we did not use any variance reduction technique.

On the quad-core 3.4 GHz Intel Core i7-2600 the same tests were made. For the 11-year contract with 0.08 fixed coupon rate the parallel program finished the whole computation in 11.63 s and demonstrated 5.2 times speedup against the sequential program.

6 Parallelisation of the Monte Carlo Simulations

In a parallel computer system having c processors, we denote the processors by $p_1, p_2, \ldots, p_c$. In our program, the parallelisation was achieved through POSIX multi-threading working on shared-memory multi-processor systems. However, the parallel algorithm is not confined by such platforms. The construction of the N_R simulation paths for estimating the coefficient matrix E needs ndN_R standardised normally distributed random numbers ($\mathcal{N}(0,1)$). With c processors, this task is equally divided such that each processor generates ndN_R/c random numbers and then constructs N_R/c paths. For each processor p_i, $i \in [1,c]$, it starts generating random numbers from the $(i-1)ndN_R/c$ position in the stream, so that there is no overlap between the processors.

The computing of the coefficient vectors $\mathbf{E}_{d-1}, \mathbf{E}_{d-2}, \ldots, \mathbf{E}_1$ needs $d-1$ iterations. In each of the iterations, cross-sectional information needs to be collected, which makes this phase not as easy to be parallelised as the others. However, the MKL function LAPACKE_dgels and other functions for linear algebra we used support automatic multi-threading, and as a result of that this phase was not explicitly threaded.

After obtaining the matrix E, the lower bound L_0 is computed over N_L paths where maximumly ndN_L random numbers may be needed. Each of the c processors completes an equal fraction of this task.

The estimation of Δ_0 is also equally divided among the c processors. Like the computation of L_0, the exact number of random numbers to be needed is unknown beforehand. However, in this case, the maximum number $nd^2N_HN_S$ is so large that, often, they cannot all be generated and stored as 8-byte double-precision floats. For this reason, we only generated ndN_HN_S random numbers with each processor generating an equal-lengthed segment without overlapping. For any processor p_i, $i \in [1,c]$, in the simulation if the numbers in its segment are used up, it will roll back to the beginning of the segment.

When the program starts, the main thread is bound onto processor p_1. When using the c processors for explicit threading, the main thread spawns $c-1$ child threads and binds each of them onto one of the processors $p_2, p_3, \ldots, p_c$. After generating the child threads, the main thread is given an equal fraction of the task to complete. During this course we disable the multi-threading feature of the MKL functions to avoid resource contention [9]. This can be done by using the MKL function mkl_set_num_threads to set the number of threads used by MKL to

1. However, when estimating the coefficient vectors we enable the multi-threading feature by setting the number to c. Nevertheless, the MKL functions may dynamically adjust the number of threads they actually use during the computation.

7 Optimisations in Implementing the Extended Multi-factor LIBOR Market Model

A path of forward rates conceptually is a two-dimensional structure. To save storage space in memory we map it onto an one-dimensional array. When generating a forward rate using Eq. (14), in computing the drift function $\hat{u}$, repetitive evaluation of common sub-expressions should be avoided. For example, in generating $\hat{F}_3(3)$ drift function $\hat{u}_3(2)$ needs to be evaluated according to Eq. (15). Let $D_j(i)$ denote the term inside the summation in Eq. (15), that is, $D_j(i) = \lambda_j(i)(\delta_j \varphi(\hat{F}_j(i)))/(1 + \delta_j \hat{F}_j(i))$. Using this notation we have $\hat{u}_3(2) = D_3(3)$. Now, when $\hat{F}_4(3)$ is computed from $\hat{F}_4(2)$, drift function $\hat{u}_4(2) = D_3(3) + D_4(3)$ needs to be evaluated. However, by this time $D_3(3)$ has already been computed in the generation of $\hat{F}_3(3)$, and, so, its value should not be evaluated again. In our implementation, whenever $\hat{F}_{i+1}(i+1)$ is computed we have a buffer initialised for drift function $\hat{u}_{i+1}(i) = D_{i+1}(i+1)$. Next, when generating $\hat{F}_j(i+1)$ for $t_j \in \{t_{i+2}, t_{i+3}, \ldots, t_x = t_{e-1}\}$ only $D_j(i+1)$ is computed and its value is added to the accumulated value in the buffer to form the value for drift function $\hat{u}_j(i)$. This optimisation significantly reduced execution times of the path generations, especially for the nested simulations in computing the upper bound penalty term Δ_0. The routine in our implementation for computing the payoff $h(j)$ defined in Eq. (18) not only returns the payoff but also the swap rate $R(j, j)$ to save the rate being computed separately.

In computing the penalty term Δ_0 the inner simulation generates forward rates based on rates in existing path. For example, if on one of the N_H paths at exercise time t_i exercise is suggested, the inner simulation will be launched, which will then generate N_S paths to estimate the discounted continuation value. These N_S paths all originate from the time instance t_i on the original path, and so they all have the same forward rates observed between time t_0 and time t_i along the original path.

8 Conclusions

We have presented our formulations for the Longstaff-Schwartz algorithm and the Andersen-Broadie algorithm. Used together these two algorithms compute lower and upper bounds for the true price of a multi-asset Bermudan option, as well as a confidence interval to enclose the true price. We have implemented the algorithms to compute the lower and upper bound prices for groups of multi-asset Bermudan stock options and interest rate swaptions. The implementation parallelised the Monte Carlo simulations through POSIX multi-threading. Intel's high-performance MKL functions were used in generating the random numbers, the QR/LQ factorisations and the vector and matrix calculations. The implementation was tested on three x86 multi-core machines. Significant speedups were observed compared to the timing results reported in the earlier work. In coding the extended multi-factor LIBOR market model source-level optimisations, such as elimination of common sub-expressions, were applied to the programs to speedup the simulations. Conceptually two-dimensional forward rate paths are mapped onto one-dimensional arrays in memory to save storage space.

Acknowledgments Research is partially funded by ESFA (VP1-3.2-ŠMM-01-K-02-002).

References

1. L. Andersen, J. Andreasen, Volatility skews and extensions of the Libor market model. Appl. Math. Finance **7**, 1–32 (2000)
2. L. Andersen, M. Broadie, Primal-dual simulation algorithm for pricing multidimensional American options. Manage. Sci. **50**(9), 1222–1234 (2004)
3. F. Black, M. Scholes, The pricing of options and corporate liabilities. J. Polit. Econ. **81**(3), 637–659 (1973)
4. A. Brace, D. Gatarek, M. Musiela, The market model of interest rate dynamics. Math. Finance **7**(2), 127–155 (1997)
5. M. Broadie, M. Cao, Improved lower and upper bound algorithms for pricing American options by simulation. Quant. Finance **8**(8), 845–861 (2008)
6. J.F. Carriere, Valuation of the early-exercise price for options using simulations and nonparametric regression. Insur.: Math. Econ. **19**(1), 19–30 (1996)
7. J.C. Cox, S.A. Ross, M. Rubinstein, Option pricing: a simplified approach. J. Finance Econ. **7** (3), 229–263 (1979)
8. J. Crank, *The Mathematics of Diffusion*, 2nd edn. (Oxford University Press, Oxford, 1980)
9. Intel Corporation. *Intel Math Kernel Library for Linux OS: User's Guide*, (2011). Document Number: 314774-018US
10. Intel Corporation. *Intel Math Kernel Library Reference Manual*, (2011). Document Number: 630813-044US
11. F. Jamshidian, LIBOR and swap market models and measures. Finance Stochast. **1**(4), 293–330 (1997)
12. D. Leisen, M. Reimer, Binomial models for option valuation-examining and improving convergence. Appl. Math. Finance **3**, 319–346 (1996)
13. F.A. Longstaff, E.S. Schwartz, Valuing American options by simulation: a simple least-squares approach. Rev. Financ. Stud. **14**(1), 113–147 (2001)
14. K.R. Miltersen, K. Sandmann, D. Sondermann, Closed form solutions for term structure derivatives with log-normal interest rates. J. Finance **52**, 409–430 (1997)
15. K.W. Morton, D.F. Mayers, *Numerical Solution of Partial Differential Equations: An Introduction*, 2nd edn. (Cambridge University Press, Cambridge, 2005)
16. J.N. Tsitsiklis, B. Van Roy, Optimal stopping of Markov processes: Hilbert space theory, approximation algorithms, and an application to pricing high-dimensional financial derivatives. IEEE Trans. Autom. Control **44**(10), 1840–1851 (1999)
17. N. Zhang, K.L. Man, Accelerating financial code through parallelisation and source-level optimisation, in *Proceedings of the International MultiConference of Engineers and Computer Scientists 2014, IMECS 2014*, Hong Kong, 12–14 Mar 2014. Lecture Notes in Engineering and Computer Science, pp. 805–806

PVT Insensitive I_{REF} Generation

Suhas Vishwasrao Shinde

Abstract In this paper, supply, process and temperature compensated, low voltage current reference for CMOS integrated circuits is presented. To minimize production cost, it uses no BJTs, external components or trimming procedures. This circuit is designed in Intel-22nm process and evaluated by computer simulations. The circuit behaviour is supported by theoretical expressions and is in agreement with simulation results. A comparison with most current references in the literature shows considerable tolerance improvement. Simulation results show PVT tolerance of ±10 % and 1σ standard deviation of 5 µA at mean of 82 µA. An autonomous, all MOS, low voltage and process technology independent features make it suitable for advanced sub micron processes like 14, 10 nm and beyond.

Keywords BGR · CTAT · PTAT · Sub threshold · Strong inversion · Weak inversion

1 Introduction

An ideal current reference circuit should be autonomous and should be used locally per analog block from modularity point of view. This avoids long reference current distribution network and associated issues of IR drop, noise coupling and current mirroring mismatch. A classical way of current reference generation is by a resistor as current defining element [1]. This requires precise bandgap reference voltage and off-chip precision resistor which consumes I/O pin. It also involves a feedback loop hence may break into oscillations if not meeting sufficient phase margin criteria. Although this current reference is very accurate it increases the system cost. Several integrated current reference circuits have been reported in the past which avoids explicit voltage reference and external components. However these current references either compensate for temperature or for process and not for both. In this

S.V. Shinde (✉)
Intel Mobile Communications, GmbH, Am Campeon 10-12, 85579 Neubiberg, Germany
e-mail: suhas.v.shinde@intel.com

© Springer Science+Business Media Dordrecht 2015
G.-C. Yang et al. (eds.), *Transactions on Engineering Technologies*,
DOI 10.1007/978-94-017-9588-3_15

195

chapter, MOS reference current circuit has proposed which has very low sensitivity for supply, process and temperature variations [2]. Further it enables modular design approach for large systems because it does not require bandgap reference (BGR) circuit, regulator or external components. This architecture exploits the sub threshold region properties of MOSFET to generate reference current. This sub threshold region operation allows use of low supply voltage (~ 1 V). Low voltage operation is gaining importance for advanced sub-micron processes due to devices shrinking down and becoming more fragile with reduced electrical stress limit.

The proposed scheme of reference current generation is realized by summation of process and supply compensated proportional to absolute temperature (PTAT) and complementary to absolute temperature (CTAT) currents. Precise reference current is generated by adjusting negative temperature coefficient of CTAT current such that it ideally cancels positive temperature coefficient of PTAT current. This architecture is feed forward system and hence no stability issue arises. All MOS implementation, low voltage operation and process technology independent operation makes it suitable for advanced sub-micron processes like 14, 10 nm and onwards.

This paper is organized into 6 sections. Section 2 describes theory and circuit of supply and process immune PTAT current generation. Section 3 explains theory and circuit of supply and process immune CTAT current generation. Section 4 develops new reference current generation circuit and Sect. 5 reports simulation results from Intel-22nm process, testing theory and validating proposed reference current circuit. Finally Sect. 6 gives concluding remark.

2 PTAT Current Generation

In this section, a standard voltage reference circuit [3] is described as supply and process insensitive PTAT current generating circuit. A PTAT current circuit is as shown in Fig. 1a. Here devices MP9 and MP10 are operated in strong inversion whereas MN7 and MN8 are operated in weak inversion.

Hence,

$$I_{P9} = \frac{\beta_{p9}}{2}(V_{GS9} - V_{TP})^2$$

$$I_{P10} = \frac{\beta_{p10}}{2}(V_{GS10} - V_{TP})^2$$

and

$$I_{N7} = I_0 \frac{W_{N7}}{L_{N7}} e^{(V_{GS7} - V_{TN})/\eta V_T}$$

$$I_{N8} = I_0 \frac{W_{N8}}{L_{N8}} e^{(V_{GS8} - V_{TN})/\eta V_T}$$

Fig. 1 **a** PTAT circuit and
[3] **b** start-up operating points

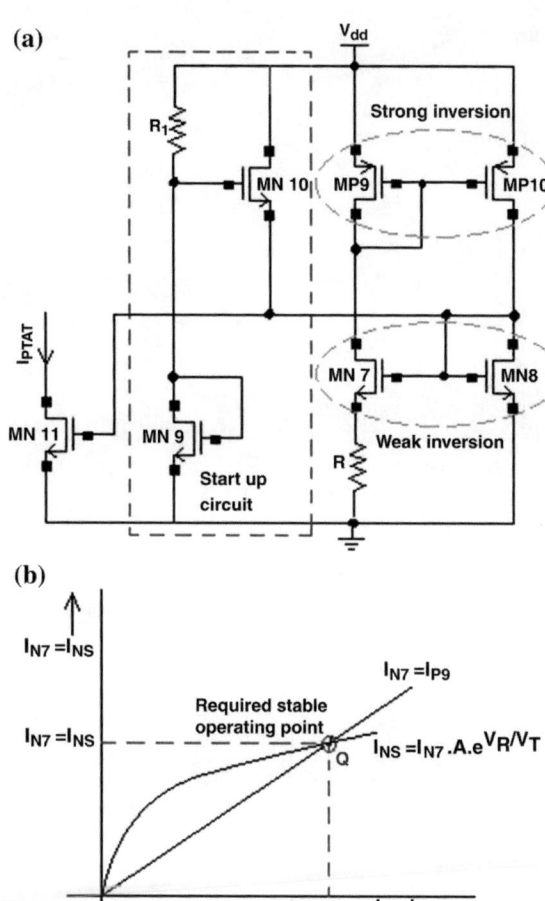

Designing,

$$\beta_{N8} = A\,\beta_{N7} \Rightarrow$$
$$\frac{W_{N8}}{L_{N8}} = A\,\frac{W_{N7}}{L_{N7}}$$

Therefore,

$$I_{N8} = I_0 A\,\frac{W_{N7}}{L_{N7}}\,e^{(V_{GS8} - V_{TN})/\eta V_T}$$
$$I_{N8} = I_0 A\,\frac{W_{N7}}{L_{N7}}\,e^{(V_{GS7} + V_R - V_{TN})/\eta V_T}$$

since

$$V_{GS8} = V_{GS7} + V_R$$

Therefore,

$$I_{N8} = I_0 A \frac{W_{N7}}{L_{N7}} e^{V_R/\eta V_T} \cdot e^{(V_{GS7} - V_{TN})/\eta V_T}$$

$$I_{N8} = I_0 \underbrace{\frac{W_{N7}}{L_{N7}} e^{(V_{GS7} - V_{TN})/\eta V_T}}_{\Downarrow \atop I_{N7}} \cdot A e^{V_R/\eta V_T}$$

$$\therefore I_{N8} = I_{N7} A e^{V_R/\eta V_T}$$

Designing for

$$I_{P10} = B I_{P9} \Rightarrow$$
$$I_{N8} = B I_{N7}$$

Therefore,

$$V_{REF} = \eta V_T \ln\left(\frac{S_{N7}}{S_{N8}} \cdot \frac{S_{P10}}{S_{P9}}\right)$$

where $A = S_{N8}/S_{N7}$, $B = S_{P10}/S_{P9}$ and

$$I_{PTAT} = V_{REF}/R$$

Therefore

$$I_{PTAT} = \frac{\eta V_T}{R} \ln\left(\frac{S_{N7}}{S_{N8}} \cdot \frac{S_{P10}}{S_{P9}}\right)$$
$$I_{PTAT} = \frac{\eta}{R} \frac{KT}{q} \ln\left(\frac{S_{N7}}{S_{N8}} \cdot \frac{S_{P10}}{S_{P9}}\right) \tag{1}$$

Since $V_T = KT/q$
where the slope,

$$m = \frac{\eta}{R} \frac{K}{q} \ln\left(\frac{S_{N7}}{S_{N8}} \cdot \frac{S_{P10}}{S_{P9}}\right) \tag{2}$$

where K = Boltzmann constant
q = Electronic charge
$\eta \sim 1$

The slope or temperature coefficient of CTAT current can be adjusted through design parameter SN7, SN8, SP10, SP9 and resistor R.

The circuit in Fig. 1a has two stable operating points; First, where no current flows in any branch of circuit and PTAT voltage is zero and second, where IN7 is equal to IN8. The circuit is brought up to this 2nd stable operating point by start up circuit comprised of MN9, MN10 and resistor R1.

3 CTAT Current Generation

This section presents supply and process insensitive CTAT current generation based on subtraction of two currents [4]. It uses natural variation of two currents to cancel out variation in the difference current. The current subtraction is such that their difference is non-zero but their process coefficient is the same. Hence their variation across process gets cancelled. Although process coefficient is made zero (theoretically!), variation with respect to another independent variable temperature exist. The observed temperature coefficient of current is negative giving rise to CTAT current. The underlying theory and model of process compensation of current for circuit in Fig. 2 is as follows.

$$I_{P8} \approx \beta Z_8 (V_{GS8} - V_{TP})^2 \text{ and}$$

$$I_{P7} \approx \beta Z_7 (V_{GS7} - V_{TP})^2$$

where reference current,

$$I_{REF} = I_{P8} - I_{P7} (I_{P8} \neq I_{P7})$$

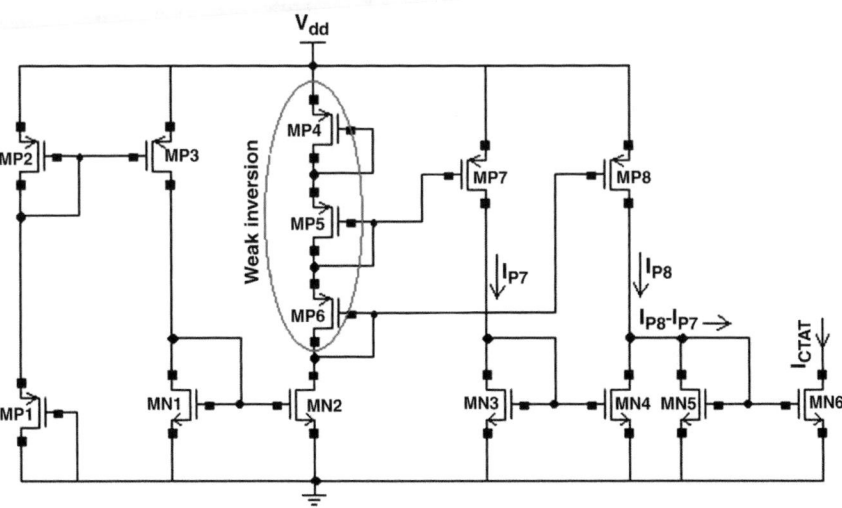

Fig. 2 CTAT current circuit [4]

For reference current variation with respect to process to be zero,

$$\frac{\partial I_{REF}}{\partial P} = 0 \Rightarrow \frac{\partial I_{P8}}{\partial P} = \frac{\partial I_{P7}}{\partial P}$$

Where
$$\frac{\partial I_{P8}}{\partial P} = z_8 (V_{GS8} - V_{TP})^2 \frac{\partial \beta}{\partial P} - 2\beta z_8 (V_{GS8} - V_{TP}) \frac{\partial V_{TP}}{\partial P} \text{ and}$$

$$\frac{\partial I_{P7}}{\partial P} = z_7 (V_{GS7} - V_{TP})^2 \frac{\partial \beta}{\partial P} - 2\beta z_7 (V_{GS7} - V_{TP}) \frac{\partial V_{TP}}{\partial P}$$

Therefore tuning z_7, z_8 such that following terms equate to cancel process variation in I_{REF}.

$$z_8 (V_{GS8} - V_{TP})^2 \frac{\partial \beta}{\partial P} = -2\beta z_7 (V_{GS7} - V_{TP}) \frac{\partial V_{TP}}{\partial P} \text{ and}$$

$$z_7 (V_{GS7} - V_{TP})^2 \frac{\partial \beta}{\partial P} = -2\beta z_8 (V_{GS8} - V_{TP}) \frac{\partial V_{TP}}{\partial P}$$

Therefore

$$\frac{z_8 (V_{GS8} - V_{TP})^2}{z_7 (V_{GS7} - V_{TP})^2} = \frac{z_7 (V_{GS7} - V_{TP})}{z_8 (V_{GS8} - V_{TP})}$$

$$\Rightarrow (\frac{z_8}{z_7})^2 = \frac{(V_{GS7} - V_{TP})^3}{(V_{GS8} - V_{TP})^3}$$

$$\Rightarrow \frac{z_8}{z_7} = \left(\frac{V_{GS7} - V_{TP}}{V_{GS8} - V_{TP}} \right)^{3/2}$$

$$\Rightarrow \frac{z_8}{z_7} = \left(\frac{aV_{TP} - V_{TP}}{bV_{TP} - V_{TP}} \right)^{3/2}$$

where $V_{GS7} = aV_{TP}$ and $V_{GS8} = bV_{TP}$

$$\Rightarrow \frac{z_8}{z_7} = \left(\frac{a - 1}{b - 1} \right)^{3/2}$$

For b = 3 and a = 2, z8/z7 = 0.3535 $\Rightarrow$ z7 = (2.84)z8

Thus choosing MP7 size 2.84 times MP8 creates zero process coefficient (ZPC). The process insensitive difference current is immune against supply variation as well because subtraction operation of IP7 and IP8 cancels out their variation partially. Although process and supply variation is cancelled the variation due to other independent variable called temperature still exist which exhibits CTAT behavior.

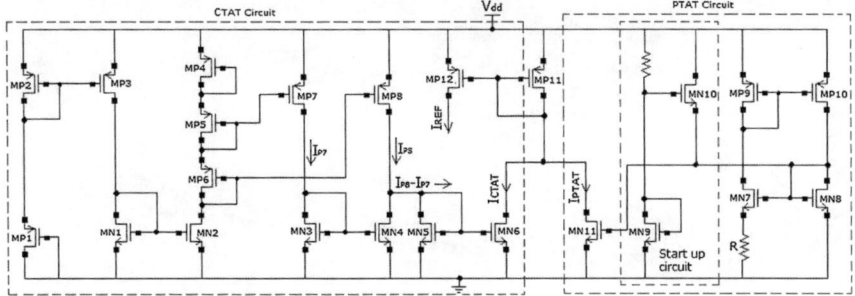

Fig. 3 Proposed reference current generation circuit

4 IREF Current Generation

The basic principle of the reference current generation is illustrated in Fig. 4 where supply and process insensitive PTAT and CTAT currents are added. Figure 3 shows complete circuit schematic of current reference, developed by combining PTAT and CTAT circuits from Sects. 2 and 3 respectively. It's all MOSFET integrated circuit design with 1 V supply. The negative temperature coefficient of CTAT current can be adjusted to match the positive temperature coefficient of PTAT current to nil temperature variation. From Eq. (2) design parameters SN7, SN8, SP10, SP9 and R can be tuned to adjust negative temperature coefficient of CTAT current. The design equations are solved and circuit is optimized considering extremes of supply, process and temperature corners for minimum reference current variation.

The low voltage operation of this circuit is an advantage over prior art architectures which require 2.5 V [5] and 3.5 V [6] supply voltages. Some reported current reference circuits [7] involve feedback loops hence critical from stability point of view. This circuit has no stability issues being feed-forward system.

Fig. 4 Basic principle of reference current generation

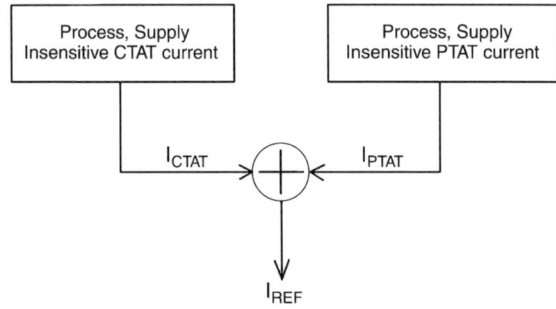

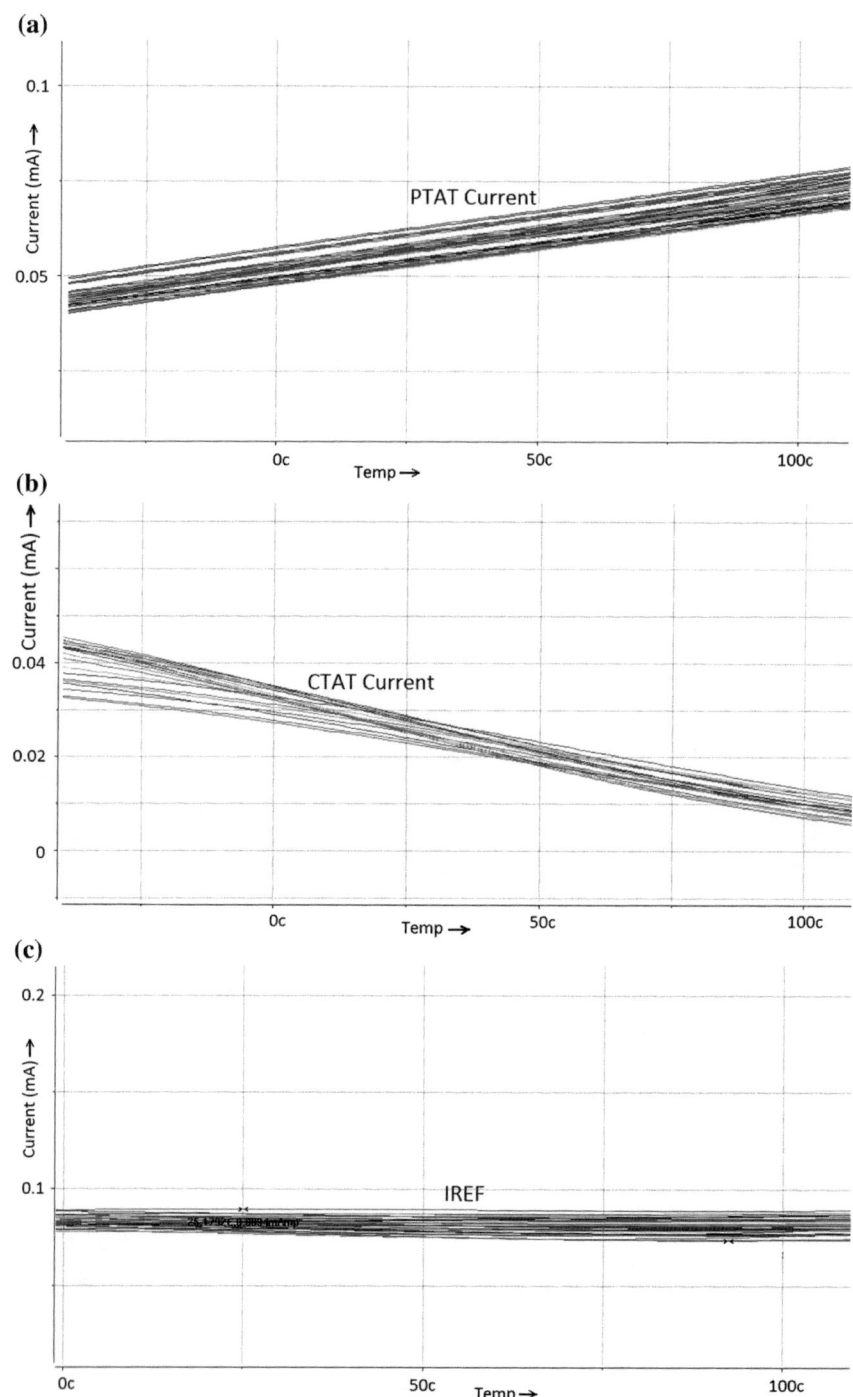

Fig. 5 **a** PTAT current **b** CTAT current and **c** IREF current

5 Simulation Results

This current reference is designed in Intel-22nm process and simulation results are reported to test mathematical model and validate the design. Figure 5a shows I-T characteristic curve of PTAT current generated by voltage reference circuit depicted in Fig. 1a. Figure 5b shows I-T characteristic curve of CTAT current generated by circuit architecture in Fig. 2. The addition of these two currents results into reference current characteristics as plotted in Fig. 5c. The extreme PVT variation observed is ±10 %. This tolerance is over temperature range of 0–110 °C, supply variation ±5 % and on-die calibrated resistor variation of ±5 %. The PVT performance comparison of this current reference with prior published work in [5, 7, 8] shows better results due to the fact that they do not address all variations together. The ±10 % tolerance is mainly attributed to non-linearity of PTAT and CTAT currents with respect to temperature, partial compensation of CTAT current with respect to supply variation and small but finite variation of calibrated resistor. A word of caution to designers about SPICE simulations! Precaution has to be taken with respect to SPICE simulations since sub threshold region models may not be accurate for their process. The incorrect modelling of weak inversion region tends to predict good results and hence silicon correlation becomes necessary. Finally Fig. 6 shows statistical simulation results for reference current under typical conditions, supply = 1 V, process = typical, and temperature = 50 °C. Observed 1σ standard deviation for 1,000 runs is 5 μA at mean value of 82 μA. This statistical spread, σ/μ = 0.06 is significantly lesser than reported results in [6].

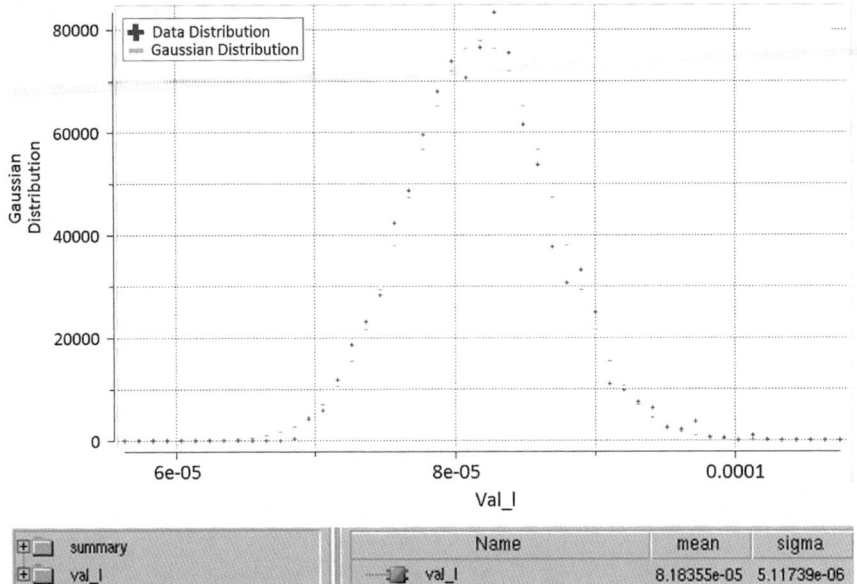

Fig. 6 Monte Carlo results showing normal Pdf with μ = 82 μA and 1σ SD = 5 μA

6 Conclusion and Future Work

This paper demonstrates an alternative approach of reference current generation with necessary mathematical expressions. It verifies approach through circuit simulations of current reference designed in Intel 22nm process. The proposed architecture provides immunity against supply, process and temperature variations. An autonomous nature makes it suitable for modular design style of large systems. This circuit is easy to use for biasing applications in analog circuits. An autonomous, low voltage, CMOS compatibility and process insensitive features make it suitable for advanced sub-micron processes.

Acknowledgments The author wishes to thank Intel Microelectronics for providing necessary access to tools and Intel 22nm process technology.

References

1. E. Vittoz, The design of high performance analog circuits on digital CMOS chips. IEEE J. Solid State Circuits **20**(3), 657–665 (1985)
2. S.V. Shinde, PVT insensitive reference current generation, in *Proceedings of the International MultiConference of Engineers and Computer Scientists 2014, IMECS 2014*, Hong Kong, 12–14 Mar 2014. Lecture Notes in Engineering and Computer Science, pp. 690–694
3. E. Vittoz, J. Fellrath, CMOS analog integrated circuits based on weak inversion operation. IEEE J. Solid-State Circuits **12**, 224–231 (1977)
4. S. Tang, S. Narendra, V. De, Temperature and process invariant MOS-based reference current generation circuits for sub-1 V operation, in *Proceedings of ISLPED'03*, pp. 199–204
5. Franco Fiori, Paolo Stefano Crovetti, A new compact temperature-compensated CMOS current reference. IEEE Trans. Circuits Syst. **52**(11), 724–728 (2005)
6. W.M. Sansen, F. Opt Eynde, M. Steyaert, A CMOS temperature-compensated current reference. IEEE J. Solid-State Circuits **23**(3), 821–824 (1988)
7. Z. Zhe, Z. Feng, H. Shengzhuan, All-CMOS temperature compensated current reference. J. Semicond. **31**(6), 065016 (2010)
8. H.J. Oguey, D. Aebischer, CMOS current reference without resistance. IEEE J. Solid-State Circuits **32**(7), 1132–1135 (1997)

Optimum Life Test Plans of Electrical Insulation for Thermal Stress Under the Arrhenius Law

Hideo Hirose and Naoki Tabuchi

Abstract We search for the optimum life test plans of electrical insulation for thermal stress assuming that the Arrhenius law holds between the thermal stress and the lifetime, and that the logarithmic lifetime follows some consistent probability distributions at a constant stress. The optimization target is to find the optimum number of test specimens at each test stress level, and we consider the case of the number of stress level is three. The criterion for optimality is measured by the root mean squared error for the lifetime in use condition. To take into account the reality, we used the parameter values in a real experimental case. Two situations are considered: one is the fundamental situation, the other is the weakest situation. Comparing the optimum results in the fundamental situation with those using the conventional test method where test specimens are equally allocated to each test stress level, we have found that the confidence interval for the predicted value in the optimum case becomes around 80–85 % of that in the conventional test. However, there is only a small difference between the optimum test result and the conventional test result if linearity of the Arrhenius plot is required. It would be useful to know the semi-optimum test plan in which the efficiency is close to that in the optimum one and the test condition is simple. In that sense, we have found that we may regard the conventional test plan as one of the semi-optimum test plans. In the weakest situation where the weakest specimen determines the system failure, similar properties are observed except the Arrhenius curve shift.

Keywords Arrhenius law · Electrical insulation · Generalized logistic distribution · Generalized Pareto distribution · Life test plans · Normal distribution · Thermal stress

H. Hirose (✉) · N. Tabuchi
Kyushu Institute of Technology, Kawazu 680-4, Iizuka, Fukuoka 820-8502, Japan
e-mail: hirose@ces.kyutech.ac.jp
URL: http://hirose.ces.kyutech.ac.jp

N. Tabuchi
e-mail: tabuchi@ume98.ces.kyutech.ac.jp

© Springer Science+Business Media Dordrecht 2015
G.-C. Yang et al. (eds.), *Transactions on Engineering Technologies*,
DOI 10.1007/978-94-017-9588-3_16

1 Introduction

It is well known that the Arrhenius law is dominant as the aging model due to the thermal stress in electrical insulation. Many researchers, such as Montsinger, Dakin, Simoni, Montanari, and Nelson referred to those modeling [1–9]. However, there are not so many references describing the probability distribution model for the thermal deterioration due to the thermal stress. Recently, the mathematical deterioration models due to the thermal stress are newly proposed [10–13], where three probability distribution models are combined with the Arrhenius law. In the mathematical models, we assume that the Arrhenius law holds between the thermal stress and the lifetime, and that the logarithmic lifetime follows some probability distributions at a constant stress. We considered the Pareto distribution, the generalized logistic distribution, and the normal distribution for such probability distribution models.

When the probability distribution models for lifetime are established, we can find the reliability of the target in use condition in a life model, such as the confidence intervals of the estimates for life model parameters. In addition, we may pursue the optimum accelerated life test design so that we can use less test time and less cost, i.e., the efficient test plan. In this paper, we search for the optimum number of test specimens at each test stress level in the accelerated lifetime test when the Arrhenius law and some probability distribution model is appropriately assumed. The criterion for optimality is measured by the root mean squared error for the lifetime in use condition. Such optimum test plans for thermal stress are investigated by Nelson, Meeker, and others [14–18]. In [8–17], optimum plan for two stress levels are discussed; in [15], three stress levels are incorporated but the number of test specimens at one level is fixed. We consider the case of the number of stress level is three. This is a new challenge.

However, if the optimum test obtained is too complex to perform, practitioners would be reluctant to use the method. Therefore, in this paper, we also aim at finding the semi-optimum test plan in which the efficiency is close to that in the optimum one and the test condition is simple. To do that, we compare the optimum results with those using the conventional test method where test specimens are equally allocated to each test stress level.

To take into account the reality, we first obtained the estimates of parameters as a typical model using a real experimental case. Then, using the parameters just obtained, we investigated the efficiency by using the simulation study for the three probability distribution models.

In actual cases, the failure in a system is observed when the weakest part in the system fails. Considering such a situation, we have mimicked the weakest situation in the simulation study.

2 Mathematical Model

2.1 Arrhenius Law, Thermal Stress

We assume that under a constant thermal stress of T [K] there is a relationship between the thermal stress T and the chemical reaction rate k such that

$$k = A \exp\left(-\frac{E}{RT}\right),\qquad(1)$$

where, E, R, and A are the activation energy, Boltzmann constant, and a constant. Then, the time to failure, t, can be given by

$$t = \frac{C}{k} = A' \exp\left(\frac{E}{RT}\right).\qquad(2)$$

Transforming this to the logarithmic formula, we obtain the logarithmic lifetime $y = \log t$, such that

$$y = \log t = \frac{E}{R}\frac{1}{T} - \log B.\qquad(3)$$

Then, we have a linear relationship between $y = \log t$ and $1/T$. This is called the well-known Arrhenius Law.

2.2 Probability Distribution Model for Deterioration

We assume some probability distribution models for thermal deterioration. These are, (1) the generalized Pareto distribution model, (2) generalized logistic distribution model, and (3) the normal distribution model as shown in [10, 11]. We usually observe the degradation phenomenon by the logarithmic time scale, we may transform y such that $y = \log t$ where t is time to failure.

2.2.1 Generalized Logistic Distribution Model

We assume that the logarithmic time to failure, $y = \log t$, follows the generalized logistic distribution function $F(y)$ such as

$$F(y) = \frac{1}{\{1 + \exp(-z)\}^{\beta}},\qquad(4)$$

where $z = (y - \mu)/\sigma$, $(\sigma > 0)$.

2.2.2 Generalized Pareto Distribution Model

We assume that the logarithmic time to failure, $y = \log t$, follows the generalized Pareto distribution function $F(y)$ such as

$$F(y) = 1 - \frac{1}{(1 + \xi z)^{1/\xi}}, \quad (1 + \xi z > 0), \tag{5}$$

where $z = (y - \mu)/\sigma$, $(\sigma > 0)$.

2.2.3 Normal Distribution Model

We assume that the logarithmic time to failure, $y = \log t$, follows the normal distribution function $F(y)$ such as

$$F(y) = \int_{-\infty}^{y} \frac{1}{\sqrt{2\pi}\sigma} \exp\left\{ -\frac{(s - \mu)^2}{2\sigma^2} \right\} ds. \tag{6}$$

2.3 Arrhenius Combined Mathematical Models

The Arrhenius law is combined with one of the probability distribution models mentioned above. These are (1) Arrhenius Pareto model, (2) Arrhenius logistic model, and (3) Arrhenius normal model.

3 Parameter Estimation Method

When we assume that underlying probability distribution for $y = \log t$ follows the generalized Pareto distribution, the generalized logistic distribution, or the normal distribution, we use one of the two methods to find the unknown parameters to be estimated; one is the maximum likelihood estimation method (MLE), and the other is the method of least squares (LS).

3.1 Maximum Likelihood Estimation Method

Assuming that $T_i(i = 1, \ldots, m)$ are the thermal stress levels, c_i are the censoring times under stress T_i, $t_{i,j}$ are the time to failure under stress T_i, and n_i are the number of specimens under stress T_i. We define $r_{i,j}$ such that $r_{i,j} = 0$ if the specimen failed, and $r_{i,j} = 1$ if the specimen did not fail until c_i.

Then, the maximum likelihood function L is given by

$$L \propto \prod_{i=1}^{m} \prod_{j=1}^{n_i} \left\{ g(t_{i,j})^{r_{i,j}} \cdot (1 - G(c_i))^{1-r_{i,j}} \right\}, \tag{7}$$

where, $g(x)$ and $G(x)$ are the density function and the cumulative probability distribution function, respectively, for the normal, the generalized logistic, or the generalized Pareto distribution models. We can obtain the unknown parameters when L becomes the maximum. In the probability distribution models, we assume that the Arrhenius model is incorporated.

3.2 Method of Least Squares

When censoring is not planned, the method of (non-linear) least squares is a useful estimation tool to obtain the unknown parameters. The optimization method we used here is the downhill simplex method [19]. We find the parameters so that RSS shown below becomes the minimum.

$$RSS = \sum_{i=1}^{m} \sum_{j=1}^{n_i} \left(\hat{G}(x_{i,j}) - G(x_{i,j}) \right)^2, \tag{8}$$

where $\hat{G}(x_{i,j})$ is the estimated value for $G(x_{i,j})$ in the normal distribution, the generalized logistic distribution, or the generalized Pareto distribution.

4 An Experimental Case

For some insulation material, we have an experimental test case. Test thermal stresses in testing are 250°, 270°, 290°. The thermal stresses in use are assumed to be 150°, 180°, 200°. Figure 1 shows the times to failure in experiments. The number of test specimens is 25 to each test stress level. Table 1 shows the estimates for the parameters in the mathematical models using this real test data case.

We do not know whether this test design was efficient or not. Thus, we next pursue the optimum test plan by mimicking this situation; that is, we use the similar test condition to one mentioned above, but a little bit different from it. The exception is the number of test specimens to each stress level. In conventional test cases, we use the same number of test specimens to each level. Let us check if the conventional test is efficient or not in the next section.

Fig. 1 Arrhenius plot of an
experimental case

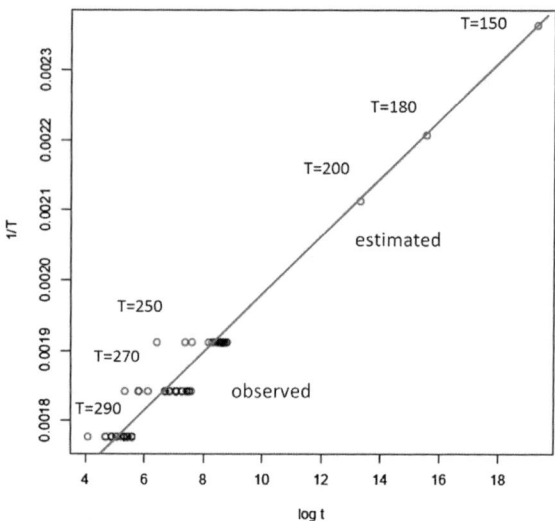

Table 1 Maximum
likelihood estimates for
parameters

Model	$\hat{E}$	$\log \hat{B}$	$\hat{\sigma}$	$\hat{\xi}$	$\hat{\beta}$
Pareto	2.12	40.6	4.45	−1.53	–
Logistic	2.06	37.0	0.126	–	0.306
Normal	2.06	37.3	0.502	–	–

5 Optimum Test Analysis

5.1 Test Condition for the Fundamental Situation

First, by mimicking the real case mentioned above, we set the test stress levels
T such that $(T_H, T_M, T_L) = (290, 270, 250)$, the stress levels in use T_U are 150, 180,
and 200, and the parameters such as $\hat{E}$ and $\log \hat{B}$ are the same to those in Table 2.
However, the total number of specimens N is set to be 30 which is smaller than that
in the experimental test case because we assume less test specimens. The number of
specimens at each levels (T_H, T_M, T_L) are denoted by (n_H, n_M, n_L), and we consider
all the combinatorial cases where the number of specimens (n_H, n_M, n_L) consists of
integer combinations with exceptions that two of $\{n_H, n_M, n_L\}$ are zero. That is, we
consider $493 = (31 \times 30)/2 - 3$ cases.

To evaluate the test efficiency, we use the root mean square error (*RMSE*) for T_U
by using the Monte Carlo simulations with 10,000 trials to each case. For comparison, we choose the case of $(n_H, n_M, n_L) = (10, 10, 10)$ as a standard, which is
commonly used in the conventional cases.

Table 2 Ratio of *RMSE* to that of the conventional case under the fundamental situation

Model	Case	$T_U = 150$	$T_U = 180$	$T_U = 200$
Pareto	(n_H, n_M, n_L) = optimum	0.858	0.835	0.816
	(n_H, n_M, n_L) = (10, 10, 10)	1	1	1
	(n_H, n_M, n_L) = (15, 10, 5)	1.21	1.23	1.25
	(n_H, n_M, n_L) = (5, 10, 15)	1.01	0.984	0.962
Logistic	(n_H, n_M, n_L) = optimum	0.839	0.848	0.860
	(n_H, n_M, n_L) = (10, 10, 10)	1	1	1
	(n_H, n_M, n_L) = (15, 10, 5)	1.13	1.15	1.16
	(n_H, n_M, n_L) = (5, 10, 15)	1.09	1.06	1.04
Normal	(n_H, n_M, n_L) = optimum	0.811	0.806	0.800
	(n_H, n_M, n_L) = (10, 10, 10)	1	1	1
	(n_H, n_M, n_L) = (15, 10, 5)	1.15	1.17	1.19
	(n_H, n_M, n_L) = (5, 10, 15)	1.04	1.02	1.00

5.2 Efficiency Analysis for the Fundamental Situation

Figure 2 shows the *RMSE* of $y(=\log t)$ for $T_U = 150, 180, 200$ for the combinatorial cases $\{n_H, n_M, n_L\}$ mentioned above in the generalized Pareto distribution, the generalized logistic distribution, and the normal distribution, respectively. In the figures, horizontal axis means the ratio p, the number of specimens to the total at T_H, and vertical axis means the ratio q, the number of specimens to the total at T_M. The figure indicates that the optimum cases are observed when $q = 0$ and $p < 0.5$ which means that $n_H < n_L$. In the figure, the points for $(n_H, n_M, n_L) = (10, 10, 10), (5, 10, 15), (15, 10, 5)$ are also shown as indices for simple comparisons.

To compare the efficiency of the optimum cases to the conventional cases, we computed the ratio of the *RMSE* of the optimum case to that of the conventional case. Table 2 shows the ratio of the optimum *RMSE* value of $y = \log t$ for T_U to the *RMSE* value in the conventional case. The table reveals us that the *RMSE* value of $y = \log t$ in the optimum case is around 80–85 % of *RMSE* value in the conventional test.

5.3 Test Condition for the Weakest Situation

In actual cases, the failure in a system is observed when the weakest part in the system fails. Considering such a situation, we have mimicked the weakest situation in the simulation study. The thermal stress condition and the stress level patterns are assumed to be exactly the same as above, and the parameters such as $\hat{E}$ and $\log \hat{B}$ are the same to those in Table 1. However, we assume that one test specimen consists of a set of elements, and the number of elements is assumed to 5, 10, 100, and 1,000. Failure of one of the elements will cause the system failure. The total

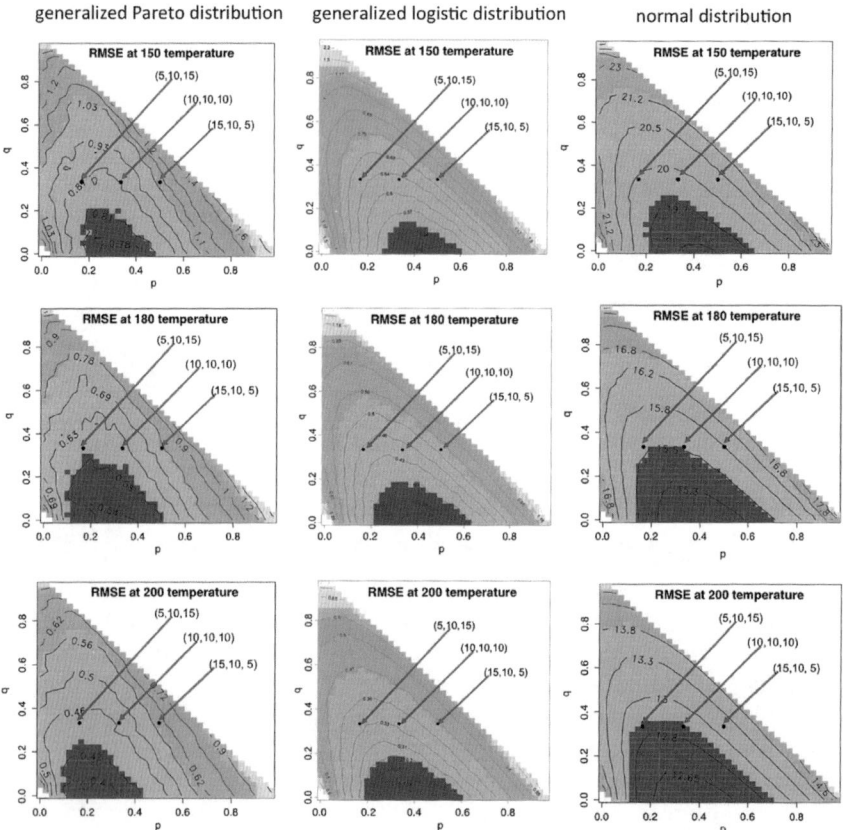

Fig. 2 *RMSE* of $y(= \log t)$ for $T_U = 150, 180, 200$ for the combinatorial cases $\{n_H, n_M, n_L\}$ when the underlying distribution is assumed to be the generalized Pareto distribution, the generalized logistic distribution, or the normal distribution under the fundamental situation

number of specimens N is set to be 6, 9, 15, 30. The number of specimens at each levels (T_H, T_M, T_L) are denoted by (n_H, n_M, n_L), and we consider all the combinatorial cases where the number of specimens (n_H, n_M, n_L) consists of integer combinations with exceptions that two of $\{n_H, n_M, n_L\}$ are zero. The data in this simulation were generated from the generalized logistic distribution.

5.4 Efficiency Analysis for the Weakest Situation

First, we show the differences of the probability distribution function and the density function between the original one sample failure and the system failure caused one of the elements in the system. Figure 3 shows the probability

Fig. 3 The probability
distribution function and the
density function for the
system failure, where n is the
number of elements in the
system

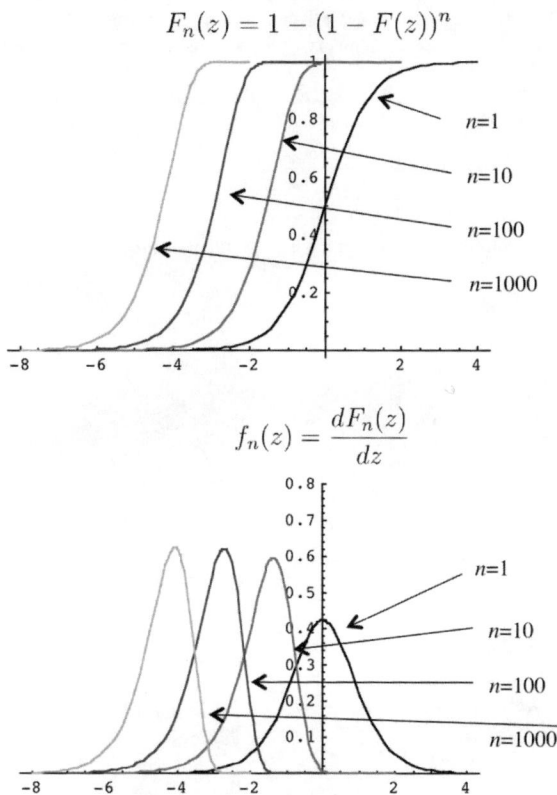

distribution function and the density function for the system failure, where n is the number of elements in the system. We assumed a case of the generalized logistic distribution. For illustration, the mean value is 0 and the standard deviation is 1. Contrary to our intuition that the weakest value (the failure value) becomes smaller to the original one and that the variance of the failure shrinks, the variance seems to remain the similar value. Actually, it becomes larger from 0.21 to 0.28 as n tends to large values in our model shown above.

However, this tendency does not remain for the other underlying probability distributions. Table 3 shows the standard errors for the system failure under the various probability distributions, where n is the number of elements in the system. Here, we have used the parameter values shown in Table 1. Contrary to the generalized logistic distribution case, the standard errors shrink in the cases of the normal distribution.

Figure 4 shows the *RMSE* of $y(= \log t)$ for $T_U = 150, 180, 200$ for the combinatorial cases $\{n_H, n_M, n_L\}$ mentioned above in the generalized logistic distribution. The number of elements in one specimen is 100. The figure indicates that the optimum cases are observed when $q = 0$ and $p < 0.5$ which means that $n_H < n_L$.

Table 3 Standard errors for the system failure under the various probability distributions, where n is the number of elements in the system

Model	$n = 1$	$n = 10$	$n = 100$
Logistic	0.212	0.261	0.276
Normal	0.243	0.0866	0.0462

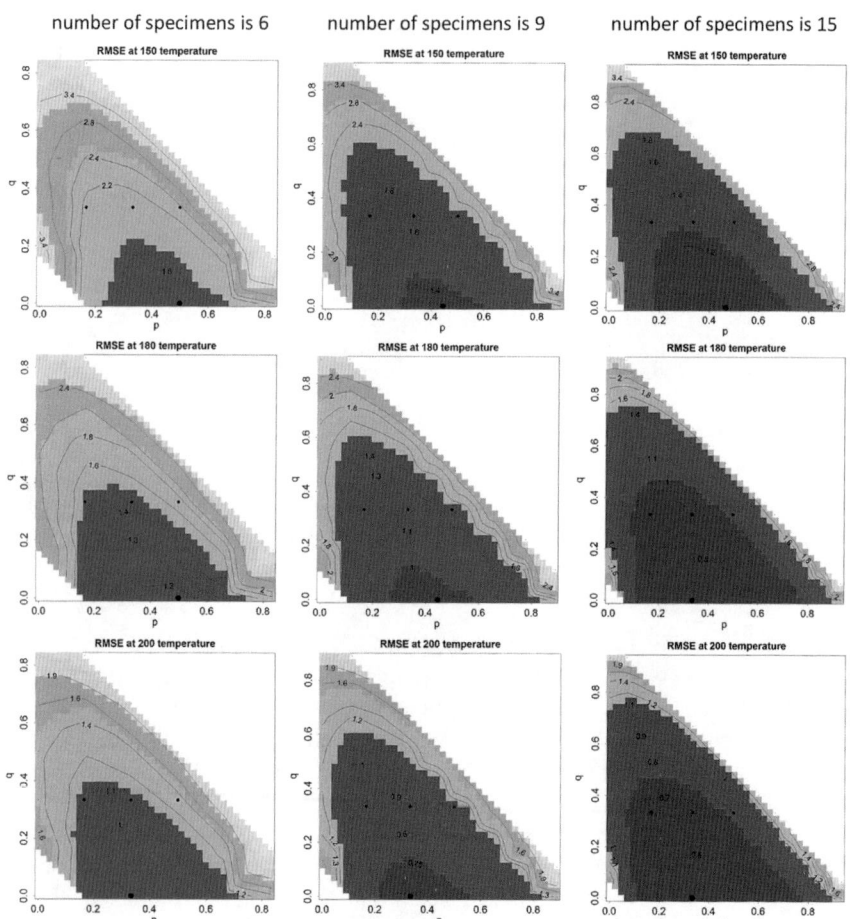

Fig. 4 *RMSE* of $y(= \log t)$ for $T_U = 150, 180, 200$ for the combinatorial cases $\{n_H, n_M, n_L\}$ when the underlying distribution is assumed to be the generalized logistic distribution under the weakest situation

To compare the efficiency of the optimum cases to the conventional cases, we computed the ratio of the *RMSE* of the optimum case to that of the conventional case when we consider this weakest situation. The number of elements in one

Table 4 Ratio of *RMSE* to that of the conventional case under the weakest situation

T_U	Number of specimens	Optimum allocation	RMSE ratio
200	30	$(n_H, n_M, n_L) = (11, 0, 19)$	0.7898
	15	$(n_H, n_M, n_L) = (5, 0, 10)$	0.8036
	9	$(n_H, n_M, n_L) = (3, 0, 6)$	0.8020
	6	$(n_H, n_M, n_L) = (2, 0, 4)$	0.7957
180	30	$(n_H, n_M, n_L) = (11, 0, 19)$	0.7999
	15	$(n_H, n_M, n_L) = (5, 0, 10)$	0.8184
	9	$(n_H, n_M, n_L) = (4, 0, 5)$	0.8100
	6	$(n_H, n_M, n_L) = (3, 0, 3)$	0.8088
150	30	$(n_H, n_M, n_L) = (13, 0, 17)$	0.8081
	15	$(n_H, n_M, n_L) = (7, 0, 8)$	0.8245
	9	$(n_H, n_M, n_L) = (4, 0, 5)$	0.8120
	6	$(n_H, n_M, n_L) = (3, 0, 3)$	0.8068

specimen is 100. Table 4 shows the ratio of the optimum *RMSE* value of $y = \log t$ for T_U to the *RMSE* value in the conventional case. The table reveals us that the *RMSE* value of $y = \log t$ in the optimum case is around 80 % of *RMSE* value in the conventional test.

6 Discussion

6.1 Do We Need More than Two Different Thermal Stresses?

As indicated in Figs. 2 and 4, the optimum cases do not require the test specimens at T_M. It is obvious that only two different x values are sufficient to make the linear regression model to be consistent. The simplest and the most efficient allocation of the specimen is to locate the specimens at T_H and T_L. The removal of specimens at T_M means the stability increase of the straight line of the Arrhenius plot. When we are interested in predicting the lifetime at T_U in use at lower temperature, it is naturally imagined that we need more specimens to T_L level than to T_H level. This tendency is indicated in Figs. 2 and 4. However, we cannot assume the linearity without absolute many evidences. It would be convenient that we can check if the linearity holds using the hypothesis testing. Then, the allocation of test specimens at T_M ($n_M \neq 0$) makes sense. In that sense, the conventional method that each number of specimens at each stress level is equivalently allocated is a good choice. Because we do not lose the efficiency much and the linearity can be assessed. For linearity check, we need the hypothesis testing.

6.2 Optimum Test and Semi-optimum Test

We have pursued the optimum test plan for the thermal stress accelerated test using the real experimental case. The practitioners use the conventional test such that the allocation of the test specimens to each stress level is equivalent. The optimum test using T_L and T_H becomes efficient regarding the standard deviation of the predicted (logarithmic) lifetime is around 80–85 %. This means that the total specimens in the optimum test can be saved about 30–35 % comparing to the conventional case. More concretely, the conventional test using 30 specimens is equivalent to the optimum test using 20 specimens. When the number of test specimens at T_L and T_H are 12 and 8, the optimality condition is attained. We may allocate, in addition, 10 test specimens to T_M, which will make the linearity check. Therefore, we can call this (conventional method) the semi-optimum test.

6.3 How to Assess the Prediction Accuracy of Lifetime?

When we measure the unbiased s.d. in logarithmic scale $y = \log t$, we can transform this to scale t on some assumption. When y is small,

$$\log y \approx y - 1.$$

Using this, the unbiased variance in logarithmic scale y can be transformed to

$$
\begin{aligned}
S_l &= \frac{1}{n-2} \sum_i (\log t_i - \log \tilde{t})^2 \\
&= \frac{1}{n-2} \sum_i (\log(t_i/\tilde{t}))^2 \\
&\approx \frac{1}{n-2} \sum_i (t_i/\tilde{t} - 1)^2 \\
&= \frac{1}{(n-2)\tilde{t}^2} \sum_i (t_i - \tilde{t})^2 \\
&= \frac{1}{\tilde{t}^2} S,
\end{aligned}
\tag{9}
$$

where, S measures the unbiased variance in scale t. We have used the ratio for the s.d. in logarithmic scale $y = \log t$ here.

6.4 Can We Expect Smaller Variance Under the Weakest Situation than that Under the Fundamental Situation?

The Arrhenius plot (in mean value or median value) under the weakest situation shows the curve shift to the left (smaller lifetime side). Intuitively, we think we can expect the smaller variance for the predicted value at the temperature of use. However, the prediction error changes its value from that under the fundamental situation. It depends on the underlying probability distribution; in the case of the generalized logistic distribution, it becomes larger; in the cases of the normal distribution and the generalized Pareto distribution, they become smaller.

7 Conclusions

Assuming that the Arrhenius law holds between the thermal stress and the lifetime, and that the logarithmic lifetime follows some consistent probability distributions at a constant stress. Under such life models, we have investigated the optimum life test plan. The optimization target is to find the optimum number of test specimens at each test stress level, and we consider the case of the number of stress level is three. The criterion for optimality is measured by the root mean squared error for the lifetime in use condition. To take into account the reality, we used the parameter values in a real experimental case. Two situations are considered: one is the fundamental situation, the other is the weakest situation. Comparing the optimum results with those using the conventional test method where test specimens are equally allocated to each test stress level, we have found that the *RMSE* value of in the optimum case is about 80–85 % of *RMSE* value in the conventional test. However, there is only a small difference between the optimum test result and the conventional test result if linearity of the Arrhenius plot is required. It would be useful to know the semi-optimum test plan in which the efficiency is close to that in the optimum one and the test condition is simple. In that sense, we have found that we may regard the conventional test plan as one of the semi-optimum test plans. In the weakest situation where the weakest specimen determines the system failure, similar properties are observed except the Arrhenius curve shift.

Acknowledgments The authors would like to express the deepest appreciation to Dr. Okamoto, Central Research Institute of Electric Power Industry, Dr. Sakumura, Chuo University, and Mr. Kiyosue, Kyushu Institute of Technology, for their cooperation to this work.

References

1. P. Cygan, J.R. Laghari, Models for insulation aging under electrical and thermal multistress. IEEE Trans. Dielectr. Electr. Insul. **25**, 923–934 (1990)
2. T.W. Dakin, Electrical insulation deterioration treated as a chemical rate phenomenon. AIEE Trans. **67**, 113–122 (1948)
3. G.C. Montanari, G. Mazzanti, L. Simoni, Progress in electrothermal life modeling of electrical insulation during the last decades. IEEE Trans. Dielectr. Electr. Insul. **9**, 730–745 (2002)
4. G.C. Montanari, F.J. Lebok, Thermal degradation of electrical insulating materials and the thermokinetic background: experimental data. IEEE Trans. Electr. Insul. **25**, 1037–1045 (1990)
5. G.C. Montanari, L. Simoni, Aging phenomenology and modeling. IEEE Trans. Dielectr. Electr. Insul. **28**, 755–776 (1993)
6. V.M. Montsinger, Loading transformers by temperature. AIEE Trans. **67**, 113–122 (1944)
7. W. Nelson, Analysis of accelerated life test data-part I: the Arrhenius model and graphical methods. IEEE Trans. Electr. Insul. **6**, 165–181 (1971)
8. W. Nelson, Analysis of accelerated life test data-part II: numerical methods and test planning. IEEE Trans. Electr. Insul. **7**, 36–55 (1972)
9. L. Simoni, A general approach to the endurance of electrical insulation under temperature and voltage. IEEE Trans. Electr. Insul. **16**, 277–289 (1981)
10. H. Hirose, T. Sakumura, N. Tabuchi, T. Kiyosue, in *Proceedings of the International MultiConference of Engineers and Computer Scientists 2014, IMECS 2014*, Hong Kong, 12–14 Mar 2014. Lecture Notes in Engineering and Computer Science, pp. 668–672
11. H. Hirose, T. Sakumura, Foundation of mathematical deterioration models for the thermal stress. IEEE Trans. Electr. Insul., to appear
12. H. Hirose, T. Sakumura, N. Tabuchi, Optimum and semi-optimum life test plans of electrical insulation for thermal stress. IEEE Trans. Electr. Insul., to appear
13. H. Hirose, N. Tabuchi, T. Kiyosue, Mathematical deterioration model derivation for the thermal stress: generalized pareto distributions, in *6th Asia-Pacific International Symposium on Advanced Reliability and Maintenance Modeling* (2014)
14. T.J. Kielpinski, W. Nelson, Optimum censored accelerated life tests for normal and lognormal life distributions. IEEE Trans. Reliab. **24**, I310–I320 (1975)
15. W.Q. Meeker, A comparison of accelerated life test plans for Weibull and lognormal distributions and type I censoring. Technometrics **26**, 157–171 (1984)
16. W. Nelson, T.J. Kielpinski, Optimum accelerated life tests for normal and lognormal life distributions, in *General Electric Research and Development TIS Report 72-CRD-215* (1972)
17. W. Nelson, T.J. Kielpinski, Theory for optimum censored accelerated life tests for normal and lognormal life distributions. Technometrics **18**, 105–114 (1976)
18. Y. Zhang, W.Q. Meeker, Bayesian methods for planning accelerated life tests. Technometrics **48**, 49–60 (2006)
19. J.A. Nelder, R. Mead, A simplex method for function minimization. Comput. J. **7**, 308–313 (1965)

Charge Plasma Based Bipolar Junction Transistor on Silicon on Insulator

Sajad A. Loan, Faisal Bashir, Asim. M. Murshid,
Humyra Shabir, M. Rafat, M. Nizamuddin, Abdul Rahman Alamoud
and Shuja A. Abbasi

Abstract Charge plasma based devices are gaining interest due to various reasons, since these devices doesn't require conventional ways of creating different doping regions, therefore these devices are free from various doping related issues related, as random doping fluctuations, doping activations and the requirement of high temperature annealing are absent in these devices. In this work we put forward a novel lateral *pnp* bipolar transistor on silicon on insulator. Metals of different work function are used to induce *n* and *p* type charge plasma on undoped silicon to have emitter, base and collector regions. The 2D simulation study has revealed that a very high current gain is achieved in the proposed device in comparison to conventional *pnp* transistor. The charge plasma concept is very much appropriate in surmounting the doping related issues such as diffusion or ion implantation, random doping fluctuations and high thermal budget in current nano devices.

S.A. Loan (✉) · F. Bashir · M. Nizamuddin
Department of Electronics and Communication Engineering, Jamia Millia Islamia,
New Delhi 110025, India
e-mail: sajadiitk@gmail.com

F. Bashir
e-mail: faisalbashir4161@gmail.com

Asim.M. Murshid
Department of Computer Sciences, Kirkuk University, Kirkuk, Iraq
e-mail: asim.majeed2001@gmail.com

A.R. Alamoud · S.A. Abbasi
Department of Electrical Engineering, King Saud University, Riyadh, Saudi Arabia
e-mail: alamoud@ksu.edu.sa

S.A. Abbasi
e-mail: abbasi@ksu.edu.sa

H. Shabir
Department of Electronics Engineering, Jamia Millia Islamia, New Delhi 110025, India

H. Shabir · M. Rafat
Department of Applied Sciences, Jamia Millia Islamia, New Delhi 110025, India

© Springer Science+Business Media Dordrecht 2015
G.-C. Yang et al. (eds.), *Transactions on Engineering Technologies*,
DOI 10.1007/978-94-017-9588-3_17

Keywords Charge plasma · Current gain · Doping · Lateral BJT · SOI · Work function

1 Introduction to Charge Plasma Based Devices

Bipolar junction transistor possesses number of advantages such as controllable characteristics, high speed and low output resistance, such type of characteristics are essential for analog mixed signal design. In addition advantage of BJT over MOSFET and the isolation possible with the SOI devices has motivated to have SOI based bipolar-CMOS (BiCMOS), where both BJT and MOSFET can be fabricated on same substrate. But lateral *pnp* BJT shows very inferior performance relative to *npn* BJT, its current gain and cutoff frequency is poor. Even though optimized lateral *pnp* BJT for analog application exhibit poor speed and current gain [1, 2]. Thus requirement for high performance lateral *pnp* BJT is the need of hour. The increasing demand of higher performance portable devices has provided enough motivation to device designers to go for the system on chips (SoC). In a SoC, the entire system, which may a multi signal and/or multi technology, is fabricated on a single substrate. Among many technologies which are being used to realize a SoC efficiently, the bipolar-CMOS technology (BiCMOS) is potentially favorable and strong. BiCMOS is the only technology providing bipolar and CMOS domains together. Bipolar technology is best suited for the realization of analog circuitry, as it provides large output resistance (desirable for many analog applications), high transconductance and high speed [3–5]. For designing high performance push-pull circuits and active resistors for analog applications, a high performance *pnp* BJT is highly desirable. The *pnp* bipolar transistors can act as efficient drivers in the output stages of analog devices, like operational amplifiers, by reducing the supply current [6, 7]. Further, complementary bipolar technology using *npn* and *pnp* bipolar transistors have applications in amplifiers, feedback circuits, current mirrors and push pull circuits. The performance of BiCMOS can be significantly improved if it is realized on SOI substrates. The SOI-BICMOS possesses superior dc/ac isolation, reduces noise, higher speed, reduced cross talk etc. However, the technology is complex and costlier, as there are severe compatibility problems between vertical bipolar junction transistor (BJT) and CMOS devices. The problems of SOI-BiCMOS can be addressed by using lateral BJTs instead of vertical, as lateral BJT share fabrication scheme with CMOS and is more compatible with it. However, the lateral BJTs are inferior to vertical ones in terms of cutoff frequency (f_T) and current gain (β) due to large base resistance [8–13].

Keeping in mind the broad range of applications of *pnp* transistor, efforts have been made in this work to improve the performance of a lateral *pnp* transistor on SOI. The problem of poor current gain in a lateral *pnp* BJT has been addressed by proposing a new structure of lateral *pnp* transistor on SOI. The proposed device uses the concept of charge plasma [14–17] to realize emitter, base and collector regions and is being called as lateral *pnp* bipolar charge plasma transistor (BCPT). The p^+ emitter, n base and p collector regions have been creating not by using conventional ion implantation or diffusion techniques, however, by using metals of different work functions. These metals induce p and n type doping in a thin undoped silicon film. A significant improvement in current gain is achieved in the proposed device in comparison to the conventional *pnp* transistor. The 2D simulation study using Atlas device simulator [18] has shown a significant improvement (144 % increase) in the current gain in comparison to the conventional *pnp* transistor. However, cutoff frequency is poor in the proposed device and need to be improved to use the device for high frequency applications. Further, the proposed device is robust and reliable as doping related issues like doping fluctuations, doping activations and the requirement of high temperature annealing are absent in the proposed device.

2 Literature Survey

As the device dimension are scaled down, various issues arise, which degrade the performance of the device. An important issue which degrades the performance of the device is discreteness and randomness of the dopant charges. This doping issue in combination with the scaling of supply voltage has a significant effect on the functionality and overall performance of the device [19]. The doping related problem can be addressed by using charge plasma concept [19, 20]. The charge plasma concept involves realizing various doped regions in a device, like emitter, base, collector, source, drain etc. by using metals of different work function, not by using the conventional ways of ion implantation and diffusion. The charge plasma concept used for designing *pn* diode, shown in Fig. 1a, reveal a good rectifying behavior for well-chosen gate workfunctions and device dimensions [14].

The charge plasma concept has been used to realize various three terminal devices. A charge plasma *npn* bipolar charge plasma transistor (BCPT) demonstrated in [16] shows that device is free from doping related issues and possess very high current gain, however, it has poor cutoff frequency as compared to conventional BJT. In order to achieve higher cutoff frequency without degrading other performance parameters; SELBOX-BCPT [17] has been proposed. The SELBOX-BCPT device possesses high gain, high cutoff frequency and high breakdown voltage.

3 Schematics of the Conventional and the Proposed Devices

The schematic diagrams of the proposed lateral *pnp* BCPT and the conventional lateral BJT are shown in Fig. 2. The proposed and the conventional devices use the background p type doping of $N = 1 \times 10^{13}/cm^3$, the buried-oxide layer thickness

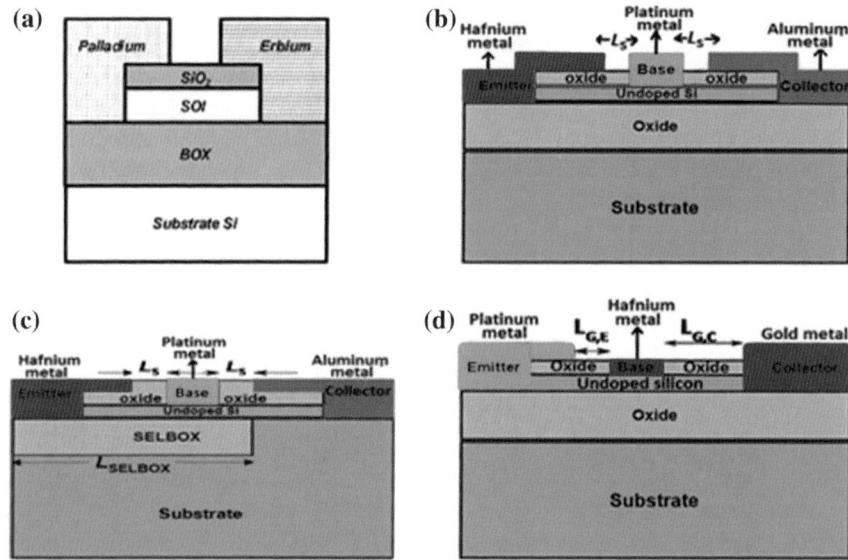

Fig. 1 Schematic diagrams of **a** *PN* Charge plasma diode **b** *npn* BCPT [16] **c** *npn* SELBOX-BCPT [17] **d** *PN* Schottky collector [5]

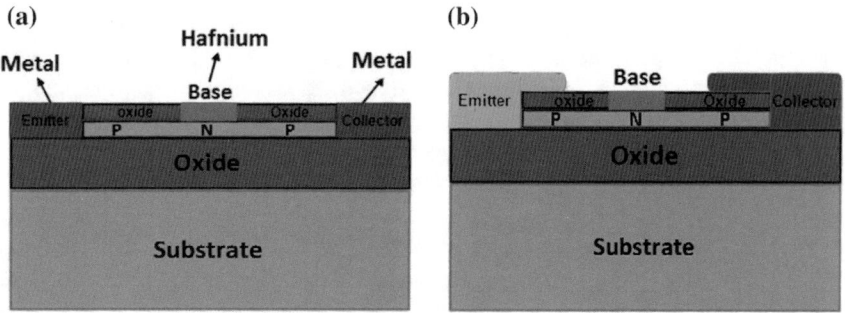

Fig. 2 Schematic diagrams of **a** conventional BJT and **b** proposed *pnp* BCPT

(t_{box}) of 375 nm, the gate-oxide thickness (t_{ox}) of 5 nm and the silicon film thickness (t_{Si}) of 15 nm. The electrode lengths of the emitter, base, and collector regions are chosen to be 0.2, 0.1, and 0.4 μm, respectively in the proposed device. For creating the emitter, by inducing holes in the undoped silicon body, platinum (work function $\Phi = 5.65$ eV) is employed as the emitter electrode metal. For inducing electrons to create the base region, hafnium (work function $\Phi = 3.90$ eV) is used. As the collector of p-n-p transistor needs to have a lower carrier concentration than the base region, to induce holes to create the collector region, Au (work function $\Phi = 5.0$ eV) is used. The separation between the electrodes (L_S) in the proposed device is taken as 100 nm. It is also important to choose an appropriate thickness for the silicon film to maintain uniform induced carrier distribution through-out the silicon thickness and it has to kept within the Debye length [14–17, 20].

A two dimensional (2D) device simulator Atlas [18] has been used to simulate the devices. To accurately design and simulate the devices, various models have been used. The models include *fldmob*, *consrh*, *auger* and *BGN* etc. The recombination effects are taken into account by using Klassens's model [21–23] for concentration dependent lifetimes for Shockley-Read-Hall (SRH) recombination with intrinsic carrier life times $n_{ie} = n_{ih} = 0.2$ μs [12, 15–17]. The other models used include Fermi-Dirac distribution, Philips unified mobility model, band gap narrowing (*BGN*) model [15–17], Selberherr impact ionization model (*selb*) [24] and Shirahata mobility model [25].

4 Simulation Results and Discussion

The important phenomenon in this work is the creation of charge plasma of different types (n and p) and different concentrations. This is being done by using metals of different work functions, as mentioned above. The induced charge plasma concentration in the emitter, base and collector regions, taken along a cross section at a distance of 2 nm, is shown in Fig. 3. It shows the induced charge concentration for both equilibrium conditions (with $V_{BE} = 0$ V and $V_{CE} = 0$ V) and non-equilibrium conditions like forward active mode (with $V_{BE} = 0.7$ V and $V_{CE} = 1$ V). The induced concentrations in the emitter, base and collector regions are $p_E = 2 \times 10^{18}/$ cm^3, $n_B = 1 \times 10^{20}/$cm^3 and $p_C = 2 \times 10^{17}/$cm^3 respectively [20].

Figure 4 shows the band diagrams of the proposed device under equilibrium and non equilibrium conditions. Figure 4a shows that the alignment of quasi Fermi levels under equilibrium conditions for both electrons and holes.

As expected, the quasi Fermi levels for electrons and holes under equilibrium conditions are aligned, as shown in Fig. 4a. Figure 4b shows the band diagram for

Fig. 3 Total carrier concentration in the proposed pnp BCPT structure for different bias conditions (under thermal equilibrium and under non-equilibrium conditions)

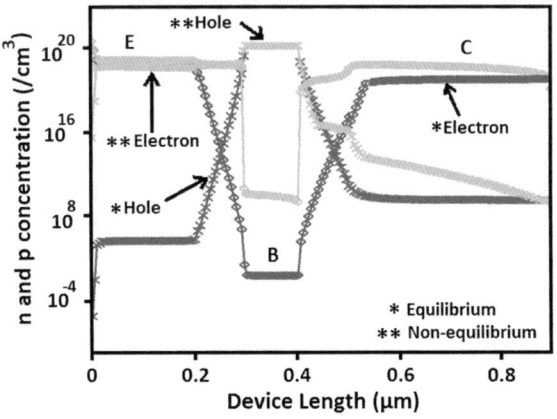

Fig. 4 Energy band diagrams of the PNP BCPT structure for different bias conditions **a** under thermal equilibrium **b** under non-equilibrium conditions

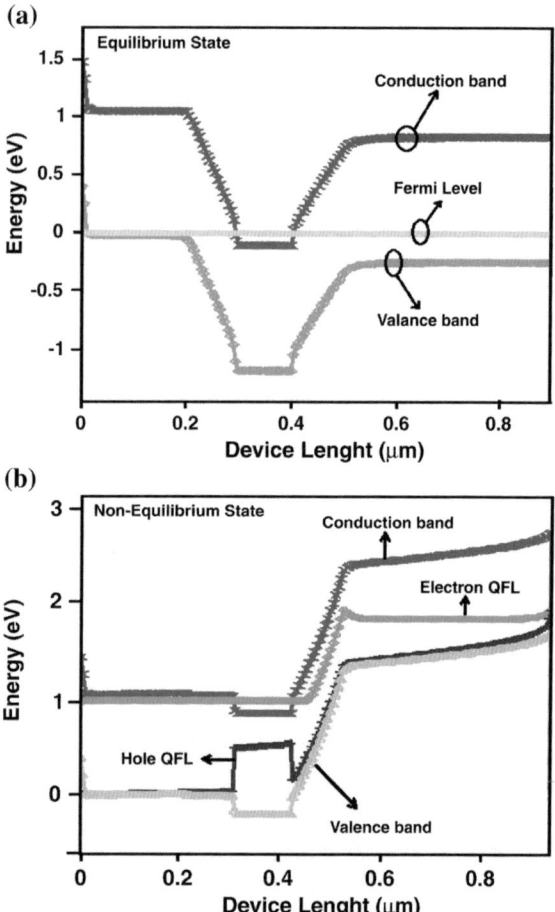

the proposed device under forward active mode. Under non-equilibrium conditions, the alignment breaks, there are different quasi Fermi levels for electrons and holes. This can be attributed to high plasma concentration on either side of the forward biased base-emitter junction.

The important characteristics of Gummel plots of the proposed and the conventional devices are shown in Fig. 5. It is clear that the base current of the proposed *pnp* BCPT device is significantly lower than the conventional BJT. This can be attributed to the presence of Surface Accumulation Layer Transistor (SALTran) effect [26, 27]. Figure 6 shows the comparison of the current gains of the proposed and the conventional devices. It is clear that the current gain is significantly higher in the proposed *pnp*-BCPT device. The higher current gain can be due to accumulation of holes at the surface. The accumulation of holes results in an electric field due to the concentration gradient from the metal–semiconductor interface toward the emitter–base junction. This electric field repels the minority

Fig. 5 Figure 4: Gummel plots for the conventional pnp BJT and the proposed pnp BCPT structures

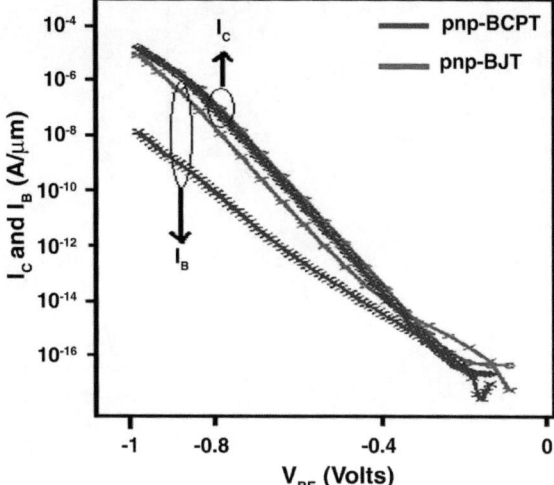

Fig. 6 Current gain comparison of the conventional pnp BJT, pnp BCPT structures

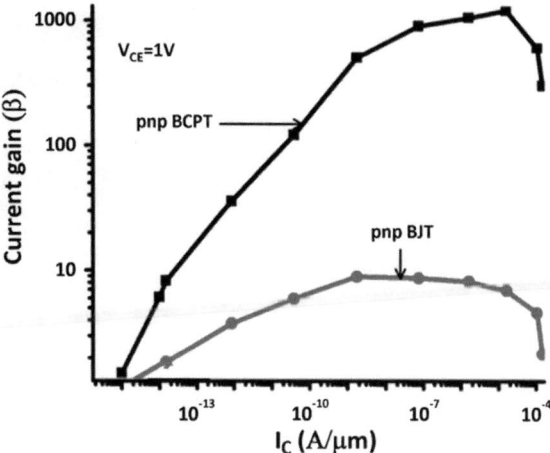

carrier electrons injected from the base into emitter and subsequently reduces the base current. Hence a large current gain is achieved in the proposed device in comparison to the conventional *pnp* transistor. Further it is observed that the peak current gain of proposed *pnp* BCPT is 1,450 and that of conventional *pnp* BJT is around 10. One of the biggest problem with the proposed device is its poor cutoff frequency in comparison to the conventional device. Figure 7 shows that the cutoff frequency is lower in the proposed device in comparison to the conventional device. Therefore, it cannot used efficiently for high frequency applications. Hence it is mandatory to address the f_T problem of the proposed device to increase its application domain.

Fig. 7 Cutoff frequency (f_T) comparison of the conventional PNP-BJT, PNP-BCPT structures

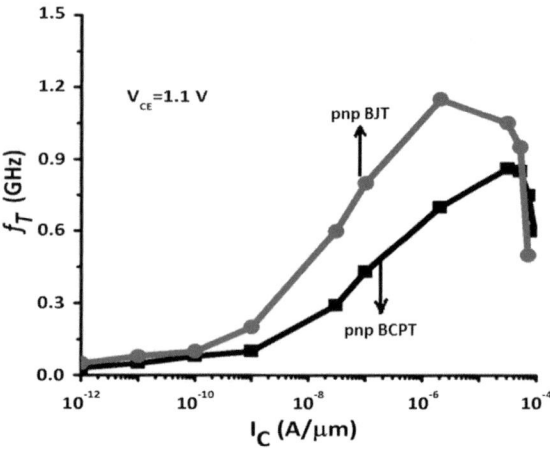

Fig. 8 Output characteristics of proposed pnp BCPT

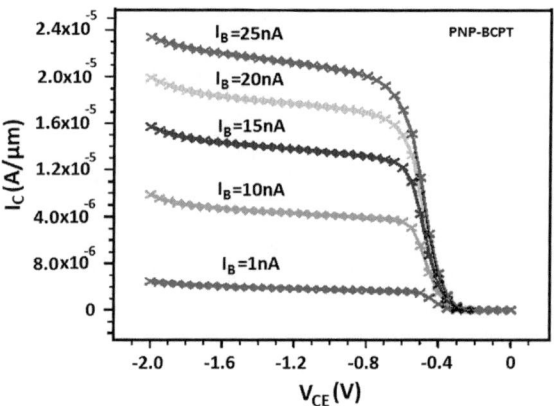

Figure 7 shows that the cutoff frequency of *pnp* conventional BJT is 1.15 GHz and that of the proposed *pnp* is 0.92 GHz. The lower cutoff frequency in the proposed device can be attributed to higher parasitic capacitance and the presence of gaps in the proposed device. The output characteristics of the proposed device are shown in Fig. 8. It is clear that the proposed device has reasonably good breakdown strength.

Figures 9 and 10 show the effect of emitter electrode work function variations on the current gain and the cutoff frequency of the proposed device. It is observed that work function variation significantly changes the current gain and the cutoff frequency of the proposed device. This can be attributed to the fact that the change in work function of emitter electrode changes the charge plasma concentration in the emitter, which changes emitter efficiency and in turn changes current gain and cutoff frequency. It is seen from Figs. 9 and 10 that the cutoff frequency and the current gain increase with the increase in emitter work function. However, increase cannot be linear and can saturate after some threshold emitter work function.

Fig. 9 Variation of current gain with emitter work-function

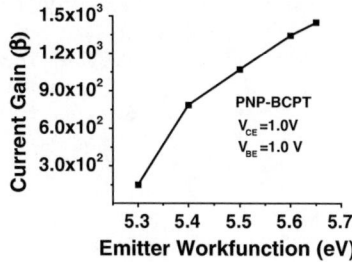

Fig. 10 Variation of cutoff frequency (f_T) with emitter work-function

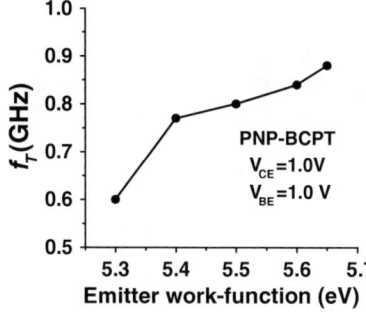

5 Conclusion and Future Work

In this paper, the charge plasma concept has been used to improve the performance of a lateral *pnp* transistor. Metals of different work functions has been used to induce *p* type and *n* type doping concentration in a thin silicon film of around 15 nm to realize emitter, base and collector regions. The simulation study has shown that a large current gain is achieved in the proposed device in comparison to the conventional device. Further, the proposed device is highly reliable as there are no doping related issues like doping fluctuations, doping activations and the requirement of high temperature. The major problem in the proposed device is the poor cutoff frequency, which need to be improved on priority.

Acknowledgments This work was supported by NSTIP strategic technologies programs number (11-NAN-2118-02) in the Kingdom Saudi Arabia.

References

1. A. Tamba, T. Someya, T. Sakagami, N. Akiyama, Y. Kobayashi, CMOS-compatible lateral bipolar transistor for BiCMOS technology. II. Experimental results. IEEE Trans. Electron Devices **39**, 1865–1869 (1992)
2. V.M.C. Chen, J.C.S. Woo, A low thermal budget, fully self-aligned lateral BJT on thin film SOI substrate for low power BiCMOS applications, in *Symposium on VLSI Technology Digest of Technical Papers*, June 1995, pp. 133–134

3. T.H. Ning, Why BiCMOS and SOI BiCMOS. IBM J. Res. Dev. **46**, 181–186 (2002)
4. K. Washio, SiGe HBT and BiCMOS technologies for optical transmission and wireless communication systems. IEEE Trans. Electron Devices **50**, 656–658 (2003)
5. S.A. Loan et al., A high performance charge plasma PN-Schottky collector transistor on silicon-on-insulator. Semicond. Sci. Technol. **29**(9), 095001, 9pp (2014).
6. R. Bashir et al., A complementary bipolar technology family with a vertically integrated PNP for high-frequency analog applications. IEEE Trans. Electron Devices **48**(11), 2525–2534 (2001)
7. M. Mitchell, S. Nigrin, F. Cristiano, P. Ashburn, P. Hemment, Characterization of NPN and PNP SiGe hetero-junction bipolar transistors formed by Ge+ implantation, in *Symposium on High Performance Electron Devices for Microwave Optoelectronics*, pp. 254–259 (1999)
8. Q. Quyang, J. Cai, T.H. Ning, J.B. Johnson, A simulation study on thin SOI bipolar transistors with fully or partially depleted collector. IEEE BCTM **1**, 28–31 (2002)
9. S.A. Loan, S. Qureshi, S.S.K. Iyer, A novel high breakdown voltage lateral bipolar transistor on SOI with multizone doping and multistep oxide. Semicond. Sci. Technol. **24**(2), 025017, 10pp (2009).
10. J. Cai, T.H. Ning, Bipolar transistor on thin SOI: concept, status and prospect, in *Proceedings of IEEE International Conference on Solid State and IC Technology*, pp. 2102–2107 (2004)
11. I.H.M. Sun, W.T. Ng, K. Kanekiyo, T. Kobayashi, H. Mochizuki, M. Toita et al., Lateral high speed bipolar transistors on SOI for RF SoC applications. IEEE Trans. Electron Devices **52**, 1376–1383 (2005)
12. H. Ni, T. Yamada, K. Inoh, T. Shino, S. Kawanaka, M. Yoshimi, Y. Katsumata, A novel lateral bipolar transistor with 67 GHz f(max) on thin-film SOI for RF analog applications. IEEE Trans. Electron Devices **47**(7), 1536–1541 (2000)
13. S.A. Loan, S. Qureshi, S.S.K. Iyer, A novel partial ground plane based MOSFET on selective buried oxide: 2D simulation study. IEEE Trans. Electron Devices **57**(3), 1–10 (2010)
14. B. Rajasekharan, C. Salm, R.J.E. Hueting, T. Hoang, J. Schmitz, The charge plasma p-n diode. IEEE Electron Device Lett. **29**(12), 1367–1369 (2008)
15. B. Rajasekharan, R.J.E. Hueting, C. Salm, T. van Hemert, R.A.M. Wolters, J. Schmitz, Fabrication and characterization of the charge-plasma diode. IEEE Electron Device Lett. **31** (6), 528–530 (2010)
16. M.J. Kumar, K. Nadda, Bipolar charge-plasma transistor: a novel three terminal device. IEEE Trans. Electron Devices **59**(4), 962–967 (2012)
17. S.A. Loan et al., A high performance charge plasma based lateral bipolar transistor on selective buried oxide. Semicond. Sci. Technol. **29**, 015011 (2014). (10pp)
18. ATLAS, *Device Simulation Software* (Silvaco Int, Santa Clara, 2010)
19. A. Asenov et al., Random dopant induced threshold voltage lowering and fluctuations in sub 50 nm MOSFETs: a statistical 3D atomistic' simulation study. Nanotechnology **10**, 153–158 (1999)
20. F. Bashir, S.A. Loan, M. Nizamuddin, H. Shabir, A.M. Murshid, M. Rafat, A.R.M. Alamoud, S.A. Abbasi, A novel high performance nanoscaled dopingless lateral PNP transistor on silicon on insulator, in *Proceedings of the International Multiconference of Engineers and Computer Scientists 2014, IMECS 2014*, Hong Kong, 12–14 Mar 2014. Lecture Notes in Engineering and Computer Science, pp. 679–682
21. D.B.M. Klassen, A unified mobility model for device simulation—I: model equations and concentration dependence. Solid State Electron. **35**(7), 953–959 (1992)
22. D.B.M. Klaassen, J.W. Slotboom, H.C. De Graaff, Unified apparent band-gap narrowing in n- and p-type silicon. Solid State Electron. **35**(2), 125–129 (1992)
23. D.B.M. Klassen, A unified mobility model for device simulation—II: temperature dependence of carrier mobility and lifetime. Solid State Electron. **35**(7), 961–967 (1992)
24. S. Selberherr, *Analysis and Simulation of Semiconductor Devices* (Springer, Wien, 1984)
25. M. Shirahata, H. Kusano, N. Kotani, S. Kusanoki, Y. Akasaka, A mobility model including the screening effect in MOS inversion layer. IEEE Trans. Comput. Aided Design Integr. Circuits Syst. **11**(9), 1114–1119 (1992)

26. M.J. Kumar, V. Parihar, Surface accumulation layer transistor (SALTran): a new bipolar transistor for enhanced current gain and reduced hot-carrier degradation. IEEE Trans. Device Mater. Rel. **4**(3), 509–515 (2004)
27. M.J. Kumar, P. Singh, A super beta bipolar transistor using SiGe base surface accumulation layer transistor (SALTran) concept: a simulation study. IEEE Trans. Electron Devices **53**(3), 577–579 (2006)

Carbon Nanotube Based Operational Transconductance Amplifier: A Simulation Study

Sajad A. Loan, M. Nizamuddin, Humyra Shabir, Faisal Bashir, Asim M. Murshid, Abdul Rahman Alamoud and Shuja A. Abbasi

Abstract MOS technology is convenient for implementing OTAs because MOSFETs are inherently voltage-controlled current devices. A variety of CMOS OTAs with different topologies have been developed for different purposes. In this chapter, design and simulation of novel operational transconductance amplifiers (OTAs) based on carbon nanotubes (CNT) has been performed. Two structures of CNT based OTAs have been proposed and have been compared with a conventional CMOS based OTA. The two CNT based OTAs include the one employing CNT based NMOS and conventional PMOS transistors, named as NCNT-PMOS-OTA and the other employing CNT based PMOS and conventional NMOS transistors, named as PCNT-NMOS-OTA. The proposed structures are designed using HSPICE and are based on 45 nm technology node. The key characteristics of the proposed devices, like DC voltage gain, average power, bandwidth and output resistance have been computed. It has been observed that CNT based OTAs result in high performance in comparison to CMOS-OTA. The DC gain has increased by 44.4 %

S.A. Loan (✉) · M. Nizamuddin · H. Shabir · F. Bashir
Department of Electronics and Communication Engineering, Jamia Millia Islamia, New Delhi 110025, India
e-mail: sajadiitk@gmail.com

M. Nizamuddin
e-mail: nizamelectro@rediffmail.com

H. Shabir
e-mail: humyraiitk@gmail.com

F. Bashir
e-mail: faisalbashir4161@gmail.com

A.M. Murshid
Department of Computer Sciences, Kirkuk University, Kirkuk, Iraq
e-mail: asim.majeeed@gmail.com

A.R. Alamoud · S.A. Abbasi
Department of Electrical Engineering, King Saud University, Riyadh, Saudi Arabia
e-mail: alamoud@ksu.edu.sa

S.A. Abbasi
e-mail: abbasi@ksu.edu.sa

© Springer Science+Business Media Dordrecht 2015
G.-C. Yang et al. (eds.), *Transactions on Engineering Technologies*,
DOI 10.1007/978-94-017-9588-3_18

in PCNT-NMOS OTA and 69.3 % in NCNT-PMOS OTA in comparison to CMOS-OTA. The average power has decreased by 24.18 % in PCNT-NMOS OTA and 14.98 % in NCNT-PMOS OTA in comparison to CMOS-OTA.

Keywords Carbon nanotube · CNTFET · DC gain · Graphene · MOSFET · OTA · Power consumption · Simulation

1 Introduction

Operational transconductance amplifiers (OTAs) are deemed to be promising to replace OP-AMPs as the building blocks. This new class of operational amplifier (OP-AMP) has not only all the advantages and applications of the conventional OP-AMP, however, is has some extra advantages and applications. Apart from differential inputs, it contains additional control terminal which enhance its flexibility and application domain [1–3]. The other advantages of OTAs include large bandwidth, large dynamic range, no excess phase issues, flexibility and tunability and realization of high integration level integrated circuits [4–6]. The realization of CMOS based OTA has resulted in low power and high performance, however, the era of CMOS is nearing its end. The further scaling of MOSFET below 60 nm is becoming difficult due to short channel effects and other reliability issues. Besides, use of highly scaled MOSFETs to realize OTA results in poor linearity and limited output resistance. Therefore, there is an immediate need to address the scaling problems of the conventional MOSFET and enhance the validity of Moore's law further. New efficient materials and devices need to be found to replace the existing silicon based MOSFETs [7–11].

One of the material and device of interest to researchers is CNT and CNTFET. CNT is being considered as a promising and is being projected to replace the widely used silicon [12–14]. It has unique properties, like high tensile strength more than steel, electrical conductivity more than the best conductor silver, thermal conductivity more than diamond. One of the important properties is the presence of nearly 1D ballistic transport capability in a CNT. This ballistic or near-ballistic transport is observed under low voltage bias because of the ultralong (~1 μm) scattering mean free path (MFP). The quasi-1-D structure provides better electrostatic control over the channel region than the 3-D (e.g., bulk CMOS) and 2-D devices (e.g., fully depleted SOI). These properties make carbon-nanotube field-effect transistor (CNTFET) one of the promising new devices to extend or complement traditional silicon technology. The CNTFET is a MOSFET like device with the channel replaced by parallel combination of CNTs. The CNT based channel results in very high mobility due to 1D ballistic transport of charge carriers and hence results in high drive capability to date [15, 16].

There has been a lot of work available in the literature on the digital applications of CNTFET but its analog applications have not been explored. This chapter explores in detail the performance and reliability of hybrid CMOS–CNTFET technologies to propose their suitability in a wide range of high performance analog circuit and system.

To cash the advantages of CNTFET, we have realized OTA based on N and P type CNTFETs. In this chapter, two types of CNT based OTAs have been designed and simulated. In one case CNT based NMOS and conventional PMOS transistors have been used to realize an OTA, named as NCNT-PMOS-OTA. The second type of OTA uses CNT based PMOS and conventional NMOS transistors, named as PCNT-NMOS-OTA. Both these devices are designed using HSPICE and are based on 45 nm technology node. The important features of the proposed OTAs, like DC voltage gain, average power, bandwidth and output resistance have been computed. The simulation study has shown high performance in CNT based OTAs in comparison to conventional CMOS-OTA. The DC gain has increased by 44.4 % in PCNT-NMOS OTA and 69.3 % in NCNT-PMOS OTA in comparison to CMOS-OTA. The average power has decreased by 24.18 % in PCNT-NMOS OTA and 14.98 % in NCNT-PMOS OTA in comparison to CMOS-OTA. The output resistance has also got significantly increased in CNT based OTA, however, has resulted in small bandwidth in comparison to conventional CMOS-OTA. It has been further observed that the performance of the CNT based OTAs can be improved further by optimizing the number of CNTs (N).

The rest of the chapter is divided into five sections. In Sect. 2, brief overview of CNT and CNTFET has been given. Section 3, discusses design and analysis of CNT based OTAs. Section 4 have results and discussion. Finally, Sect. 5 has concluded the chapter with future scope of work.

2 Carbon Nanotube (CNT) and Carbon Nanotube Field Effect Transistor (CNTFET)

Carbon nanotubes were discovered by Ijima of NEC Japan in 1993, while studying the surface of graphite electrode in an electric arc discharge. They are actually allotrope of carbon. They possess unique electrical, mechanical and thermal properties. They are highly conductive (electrical and thermal) and have very high aspect ratio. Because of these unique properties they have wide domain of applications, including field-emission displays, nanocomposite materials, nanosensors, and nanoelectronics. The carbon nanotubes exist in two forms: (i) Single wall carbon nanotube (SWCNT) and (ii) multiwall carbon nanotube as shown in Fig. 1a. SWCNT are actually tubes of graphite that are normally capped at the ends. They can be visualized as a layer of graphite rolled into a seamless cylinder. Their diameter is around 1 nm and length a few microns. They are superior to MWCNT, however, are costlier. MWCNT appear like a coaxial assembly of SWCNTs, like a coaxial cable. They diameter of MWCNT ranges from 5 to 50 nm and the inter

Fig. 1 **a** Different types
b different configurations of
SWCNT

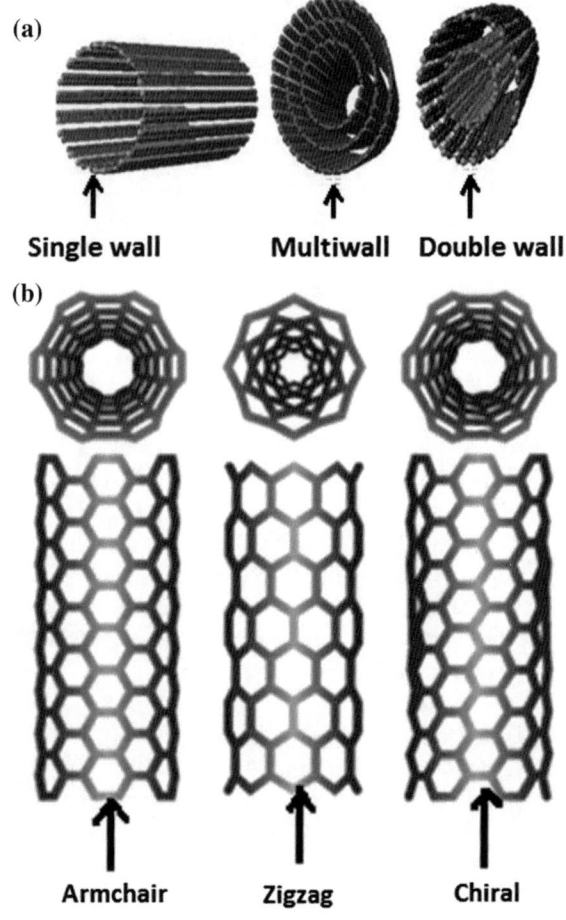

layer spacing is 3.4 Å. They are easy to produce in large quantity. However, the
structure is complex and the structural imperfections may diminish their unique
properties. The wrapping of graphite sheets in SWCNT can be represented by a pair
of indices (n, m), called as the chirality vector or roll-up vector, as shown in Fig. 1b.
There are three types of SWCNTs based on the chiral vector and chiral angle (θ).
SWCNT is arm chair type if n = m = 0 and θ = 30°, it is a Zig-Zag if $n = m = 0$ and
$\theta = 0°$ and a Chiral type if $n=/m=/0$ and θ lies between 0° and 30° [17–21].

Figure 2 shows Structure and energy band diagrams showing the principle of
operation of: (a) and (b) SB-CNTFET, (c) and (d) MOS-CNTFET, and (e) and (f)
T-CNTFET and (g) Schematic of Multi CNT-FET. Basically, three types of carbon
nanotube transistors (i. Schottky-barrier CNTFETs, ii. MOSFET-like CNTFET, and
iii. band-to-band tunneling CNTFETs) are being currently studied in research. First,
Schottky-barrier CNTFETs works on the principle of direct tunneling through a
Schottky barrier (SB) at the source–channel junction (Fig. 2a, b). The barrier width

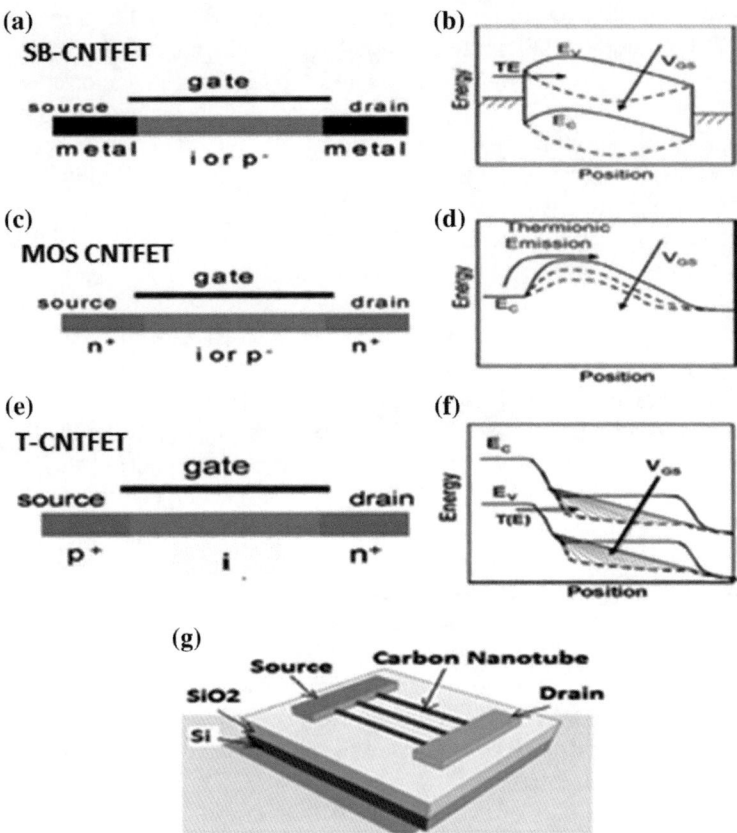

Fig. 2 Structure and energy bands **a, b** SB-CNTFET, **c, d** MOS-CNTFET, **e, f** T-CNTFET and **g** multi CNT-MOSFET

is modulated by the application of gate voltage and thus the transconductance of the device is dependent on the gate voltage. These devices are fabricated using direct contact of the metal with the semiconducting nanotube and consequently they have a SB at the metal nanotube junction. Two important aspects of these nanotube transistors are worth mentioning. (1) the energy barrier at the SB severely limits the transconductance of the nanotube transistors in the ON state and reduces the current delivery capability—a key metric to transistor performance. (2) Schottky-barrier CNTFETs (SB-CNTFETs) exhibit strong ambipolar characteristics and this constrains the use of these transistors in conventional CMOS logic families. It is worth mentioning, that SB-CNTFETs can be midgap (where the source–drain material is midgap) or band-edge (where the source–drain material is at for nFETs and at for pFETs). Band-edge SB-CNTFETs offer higher ON currents at the cost of higher leakage when compared with their midgap counterparts.

To overcome these handicaps associated with the SB-CNTFETs, there have been attempts to develop CNTFETs which would behave like normal MOSFETs. More recently, a tunable CNTFET with electrostatic doping has been demonstrated. This MOSFET-like CNTFET (Fig. 2c, d) operates on the principle of barrier height modulation by application of the gate potential. In MOS CNTFETs it is evident that: (1) the MOSFET-like CNTFETs have unipolar characteristics unlike SB-FETs; (2) the absence of SB reduces the OFF leakage current; (3) they have more scalability compared to their SB counterparts; (4) in the ON state the source–channel junction has no SB and hence, significantly higher ON current.

A third variety of CNTFETs, namely, the band-to-band tunneling CNTFETs (T-CNTFETs) (Fig. 2e, f) have also been demonstrated both experimentally as well as in theory. These devices have super cutoff characteristics (subthreshold slopes less than 60 mV/decade), low ON currents and can potentially be used for ultra-low-power applications. The MOSFET-like CNTFET device structure is used for the modeling because of both the fabrication feasibility and superior device performance of the MOSFET-like CNTFET as compared to the SB-controlled FET. In a MOSFET-like CNTFET, the channel is made up of parallel combination of SWCNTs. The source and drain regions are highly doped regions and the CNT channel is undoped. The important advantages of CNTFET include 1D ballistic transport of charge carriers, high mobility, large drive current and very low power consumption.

In carbon-nanotube field-effect transistor (CNTFET), the Chirality, Ch (i.e. the direction in which the graphene sheet is rolled) whose magnitude and relationship with CNT diameter (D_{CNT}) is given by Eqs. (1) and (2) respectively where 'a' is the grapheme lattice constant (0.249 nm) and n_1, n_2 are positive integers that specify the chirality of the tube

$$Ch = a\sqrt{\left(n_2^1 + n_2^2 + n_1 n_2\right)} \tag{1}$$

$$D_{CNT} = Ch/\pi \tag{2}$$

The CNT diameter, the width of the CNTFET transistor (W), number of CNTs in the channel of a CNTFET (N), inter-nanotube spacing (S) are related by Eq. (3). The Bandgap energy (Σg) and The threshold voltage (V_{th}) of the intrinsic CNT channel are related by Eq. (4) where 'acc' is the carbon-to-carbon bond distance (0.1412 nm for graphite), 'e' is the unit electron charge and 't' is the carbon-to-carbon bond energy (3.0 eV).

$$W = (N - 1)S + D_{CNT} \tag{3}$$

$$\sum g(eV) = 2 \, acct/D_{CNT} \, (nm) \tag{4}$$

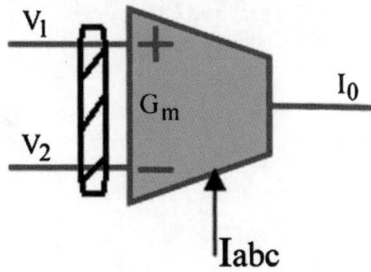

Fig. 3 Symbol of a CNT based OTA

An operational transconductance amplifier (OTA) is actually a voltage controller current source. OTA is similar to the conventional OP-AMP in many expect that an OTA has very high output impedance. An OTA posses an external bias current I_{abc}, which is responsible for the flexibility and tunability properties of an OTA. Figure 3 shows the symbol of a CNT based OTA. The OTA is best described in term of its transconductance (gm) rather than voltage gain. The output current of an OTA is given by the following equation

$$I_{OUT} = g_m(V_1 - V_2) \tag{5}$$

where g_m is the transconductance, V_1 and V_2 are the two voltages at the input of the OTA. The operational transconductance amplifier (OTA) consumes more power in analog integrated circuits in many applications. Low power consumption is becoming more important in handset devices, so it is a challenge to design conventional CMOS based low power OTA [22, 23]. Also, there is a tradeoff between speed, power, and gain for an OTA design because usually these parameters are contradicting parameters. CNT based OTA consumes very low power as compared to exist CMOS based OTAs, shown in Fig. 7.

3 Proposed CNT Based OTA

In this work, two CNT based OTA have been designed and compared with the conventional CMOS OTA. Figure 4 shows a conventional CMOS OTA. The OTAs designed are based on 45 nm technology node and have been designed using HSPICE. Figure 5 shows the circuit diagram of one of the proposed CNT based OTA. It uses N CNTFETs as sinks and conventional PMOS transistors as sources. It is being called as PMOS-NCNT_OTA. Similarly, another proposed CNT based OTA uses P CNTFETs as sinks and conventional NMOS transistors as sources.

Fig. 4 Conventional CMOS based OTA

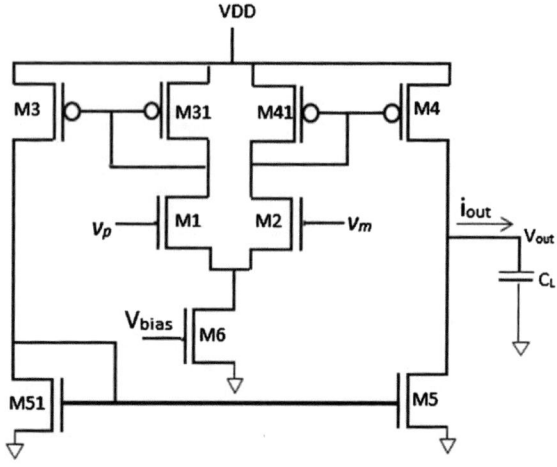

Fig. 5 Proposed PMOS-NCNT-OTA

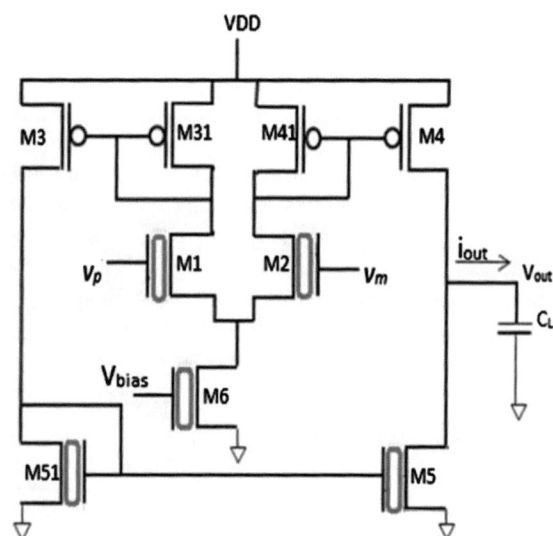

4 Simulation Results and Discussions

In this work, the effect of change in number of transistors (N) on the overall performance has been studied for the two proposed CNT based OTAs. It has been observed that the number of CNTs change the performance significantly and an optimized number of CNTs need to be used to have an optimum performance. Figure 6 shows that the increase in number of CNTs increases the DC gain in both the CNT based OTAs. However, there is a saturation in the DC gain when N > 15. It may

Fig. 6 Variation of DC gain
with N (V_{DD} = 0.9 V,
L = 45 nm, D_{CNT} = 1.5 nm,
S_{CNT} = 20 nm)

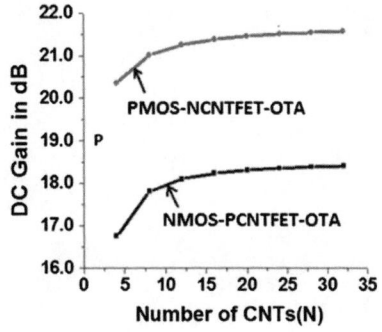

be due to screening effect due to large number of adjacent CNTs. Further, the DC gain
in PMOS-NCNTFET-OTA is more in comparison to NMOS-PCNTFET-OTA. This
can be attributed to large output resistance in PMOS-NCNTFET-OTA. Figure 7
shows that average power increases with the increase in number of transistors. This
can be due to increase in drive current by increasing N. Figure 8 shows that the
bandwidth increases with the increase in number of CNTs. Since increase in number
of CNTs, increases transconductance and hence driving capability, therefore,
bandwidth is bound to increase. The increase in bandwidth is more in NMOS-

Fig. 7 Variation of average
power with N

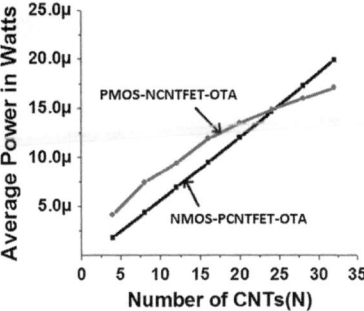

Fig. 8 Variation of 3-dB
bandwidth with N

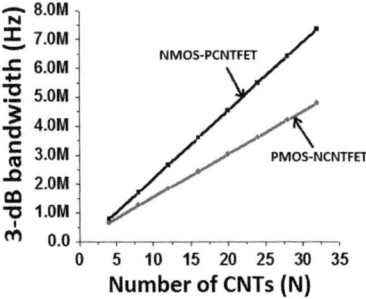

Fig. 9 Variation of output
resistance with the N

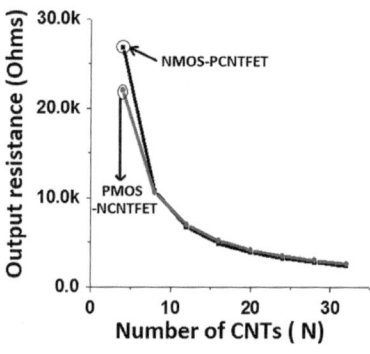

PCNTFET-OTA in comparison to PMOS-NCNTFET-OTA, due to low output
resistance in NMOS-PCNTFET.

Figure 9 shows that the output resistance decreases with the increase in number
of CNTs in the channel. Since increase in number of CNTs is actually equivalent to
increasing the width of a MOSFET, therefore, drive current increases and hence
output resistance decreases.

Figures 10, 11 and 12 show the plots between gain and frequency of CMOS-OTA,
NMOS-PCNTFET-OTA and PMOS-NCNTFET-OTA. It is seen that the gain is
highest in CNT based OTAs. However, the bandwidth is low in CNT based OTAs in
comparison to CMOS-OTA.

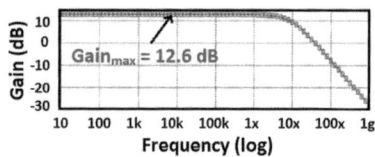

Fig. 10 DC gain plot of CMOS-OTA

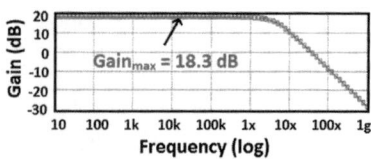

Fig. 11 DC gain plot of NMOS-PCNTFET-OTA

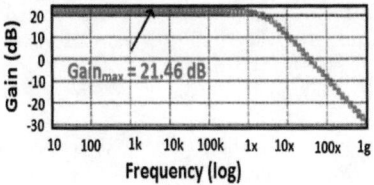

Fig. 12 DC gain plot of PMOS-NCNTFET-OTA

5 Conclusion and Future Work

The work presents the design and simulation of OTAs based on CNTFET. Three OTAs have been designed and compared: (i) conventional CMOS-OTA (ii) OTA using conventional PMOS as loads and N CNTFETs as drivers (iii) OTA uses conventional NMOS as loads and P CNTFETs as drivers. The simulation study has revealed that the use of CNT in the conventional MOSFET will improve the performance significantly. The power consumption will decrease drastically, the speed will increase significantly and the DC gain will also increase significantly. Further, it has been seen that the performance of the CNT based OTAs can be improved further by using an optimum number of CNTs (N). In future scope of the work, the proposed circuits based on Pure CNTFET can be designed.

Acknowledgments This work was supported by NSTIP strategic technologies programs, number (11_NAN_2118_02) in the kingdom of Saudi Arabia.

References

1. R. Jacob Baker, *CMOS Circuit Design, Layout and Simulation*, 2nd edn. (Nov. 2007)
2. J.O. Voorman, Transconductance Amplifier, U.S. Patent 4,723,110, 2 Feb 1988
3. H.S. Malvar, Electronically controlled active filters with operational transconductance amplifiers. IEEE Trans. Circuits Syst. **CAS-29**, 333–336 (1982)
4. T.H. Lin, C.K. Wu, M.C. Tsai, A 0.8-V 0.25-mW current-mirror OTA with 160-MHz GBW in 0.18 μm CMOS. IEEE Trans. Circuits Syst. II: Express Briefs **54**(2), 131–135 (2007)
5. X. Zhang, E.I. El-Masry, A novel CMOS OTA based on body-driven MOSFETs and its applications in OTA-C filters. IEEE Trans. Circuits Syst. I: Regul. Pap. **54**(6), 1204–1212 (2007)
6. A.R. Kim et al., Low-power class-AB CMOS OTA with high slew-rate, in *IEEE SoC Design Conference (ISOCC)*, 22–24 Nov 2009, pp. 313–316
7. S.A. Loan, S. Qureshi, S.S.K. Iyer, A novel partial-ground-plane-based MOSFET on selective buried oxide: 2-D simulation study. IEEE TED **57**(3), 671–680 (2010)
8. S.A. Loan, F. Bashir, M. Rafat, A.R. Alamoud, S.A. Abbasi, A high performance charge plasma based lateral bipolar transistor on selective buried oxide. Semicond. Sci. Technol **29**, 015011, 10 pp (2014)
9. Y. Tian, R. Huang, X. Zhang, Y. Wang, A novel nanoscaled device concept: quasi-SOI MOSFET to eliminate the potential weakness of UTB SOI MOSFET. IEEE Trans. Electron Devices **52**(4), 561–568 (2005)

10. T. Numata, S.I. Takagi, Device design for subthreshold slope and threshold voltage control in sub-100-nm fully depleted SOI MOSFETs. IEEE Trans. Electron Devices **51**(12), 2161–2167 (2004)

11. H.P. Wong, D.J. Frank, P. Solomon, C.J. Wann, J.J. Welser, Nanoscale CMOS. IEEE Proc. **87** (4), 537–570 (1999)

12. S. Iijima, Helical microtubules of graphitic carbon. Nature **354**, 56–58 (1991)

13. R. Saito, G. Dresselhaus, M. Dresselhaus, *Physical Properties of Carbon Nanotubes* (World Scientific Publishing Co. Inc., Singapore, 1998)

14. K. Yong-Bin, Integrated circuit design based on carbon nanotube field effect transistor. Trans. Electr. Electron. Mater. **12**(5), 175–188 (2011)

15. A. Javey et al., Self-aligned ballistic molecular transistors and electrically parallel nanotube arrays. Nano Lett. **4**, 1319–1322 (2004)

16. J. Appenzeller, Carbon nanotubes for high performance electronics (invited paper). Proc. IEEE **96**(2), 206 (2008)

17. J. Deng, H.S.P. Wong, A compact SPICE model for carbon nanotube field effect transistors including non-idealities and its application—part II: full device model and circuit performance benchmarking. IEEE Trans. Electron Devices **54**(12), 3195–3205 (2007)

18. S.A Loan, M. Nizamuddin, H. Shabir, F. Bashir, A.R.M. Alamoud, S.A. Abbasi, Design of a novel high gain carbon nanotube based operational transconductance amplifier. in *Proceedings of the International Multiconference of Engineers and Computer Scientists 2014, IMECS 2014*, Hong Kong, 12–14 Mar 2014. Lecture Notes in Engineering and Computer Science, pp. 797–800

19. A. Raychowdhury, K. Roy, Carbon nanotube electronics: design of high-performance and low power digital circuits. IEEE Trans. Circuits Syst. I: Regul. Pap. **54**(11), 2391–2401 (2007)

20. K. Navi et al., High speed capacitor-inverter based carbon nanotube full adder. Nanoscale Res. Lett. **5**, 859–862 (2010)

21. A. Keshavarzi et al., Carbon nanotube field-effect transistors for high-performance digital circuits—transient analysis, parasitics, and scalability. IEEE Trans. Electron Devices **53**(11), 2718–2726 (2006)

22. F. Ali Usmani, M. Hasan, Carbon nanotube field effect transistors for high performance analog applications: an optimum design approach. Microelectron. J. **41**, 395–402 (2010)

23. L. Tianwang, Y. Bo, J. Jinguang, A novel fully differential telescopic operational transconductance amplifier, IOP Science. J. Semiconductors **30**(8), 085002 (2009)

Analytical Study and Reduction of Harmonics Issued from LED Lamps in Lighting System

Chaiyan Jettanasen and Atthapol Ngaopitakkul

Abstract Harmonics are unavoidably generated in any non-linear electrical/electronic systems, and cause severe problems in terms of performance and operation. This paper focuses on analytical study and reduction of harmonics originated from LED lamps usually functioning with high switching-frequency driver. Since the driver is a switching device, it will be thus a direct harmonic and/or electromagnetic interference (EMI) source of the system. In order to suppress or reduce produced harmonics, a low-pass harmonic filtering technique is proposed and applied. The experimental results can reveal harmonic reduction effectiveness by comparing with a lighting standard, which is herein IEC 1000-3-2 (or EN 61000-3-2); this confirms finally the Electromagnetic Compatibility (EMC) of the system.

Keywords Harmonics · LED driver · LED lamp · Light emitting diode · Lighting standard · Low-pass filter

1 Introduction

Nowadays, Light Emitting Diode (LED) lamps become increasingly popular to be used in many applications, for example, inside and outside of the residence or office, street lights, building decoration, and vehicle application. The main purposes of using LED lamp are energy savings because of low energy consumption and overall efficiency augmentation. Furthermore, it has long lifetime, and is environmentally friendly because there is no composition of the toxic substance comparing to other types of lamps. Consequently, in the buildings, LEDs have replaced

C. Jettanasen (✉) · A. Ngaopitakkul
Faculty of Engineering, King Mongkut's Institute of Technology Ladkrabang,
Chalongkrung Rd., Bangkok, Thailand
e-mail: kjchaiya@kmitl.ac.th

A. Ngaopitakkul
e-mail: knatthap@kmitl.ac.th

© Springer Science+Business Media Dordrecht 2015
G.-C. Yang et al. (eds.), *Transactions on Engineering Technologies*,
DOI 10.1007/978-94-017-9588-3_19

incandescent lamps and fluorescent lamps, which have been usually used for many decades. Even though LED lamp has many advantages, it also has some disadvantages such as generation of harmonics or EMI in the system owing to the functioning of switching devices of the LED driver. The driver is essential for lightening the lamp.

Many research papers have focused on development of performance of LED lamps, lighting control, and illumination on the work surface [1–4], but there are few papers concerning side effects [5, 6] when employing this kind of lamp. To reduce or suppress harmonics generated in any electrical/electronic system, there are a number of traditional and innovative techniques [7–12]. However, in this paper, low-pass harmonic filtering approach will be studied and discussed. For high frequency related to EMI problem, it will not be focused herein because we mostly concern low-frequency harmonic current capable of flowing through other electric/electronic devices/equipments connected to the same electrical network; this can damage or malfunction the mentioned devices.

In this paper, the studied system and its experimental setup are first presented. Second, the harmonics measurement and its results are illustrated. Harmonic filter design is next carried out in order to overcome harmonics problem. Finally, the results when inserting harmonic filter at the input of LED lamp's driver are shown and compared with the lighting standard to reveal the effectiveness of passive filtering technique implemented in the studied system. This will further confirm the electromagnetic compatibility (EMC) of the considered system.

2 Studied System and Experiment Setup

The studied system is composed of an ac power source, and LED lamp and its driver set of different commercial brands. Three brands are studied in this paper; two brands (A and B) are constant voltage LED driver and one brand (C) is constant current LED driver. Various configurations are considered and carried out for each brand as presented below:

- 1 driver for 1 LED lamp
- 1 driver for 9 LED lamps
- 9 drivers for 9 LED lamps

The purpose is to know the effect of number of driver and number of LED lamp to generation of harmonics and/or EMI.

The example of LED lamp and its driver used in this study is shown in Fig. 1 and the overall experimental setup is illustrated in Fig. 2.

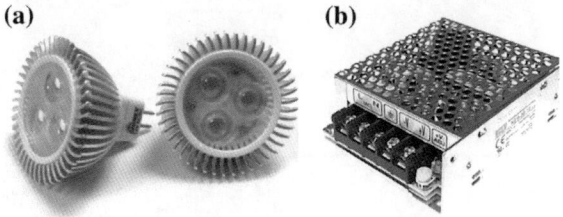

Fig. 1 LED lamps (**a**) and driver (**b**)

Fig. 2 Experimental setup of LED lamps

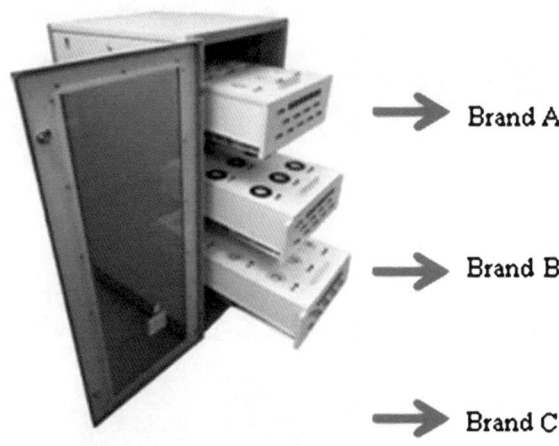

3 Harmonics Measurement and Results

For each configuration, the harmonic current is measured at the input of LED lamp using a power quality analyzer as depicted in Fig. 3. This measuring instrument can also provide a number of electrical quantities, such as values of power, power factor, and total harmonic distortion percentage.

The percentage of total harmonic distortion of current ($\%THD_i$) is compared for different configurations as presented in Table 1.

Note that LED lamp of brand C produces less harmonics than that of brand A and B because the current waveform is less distorted. Figures 4 and 5 show the current waveform of brand A and C, in time domain, respectively. However, it is clearly seen that these waveforms contain harmonic components by using Fast Fourier Transform (FFT) function of oscilloscope.

Moreover, harmonics magnitude ($\%f$) of each harmonics order is compared for each LED lamp brand and shown in Fig. 6.

According to the results, it is obvious that LED lamp of brand A has high $\%THD_i$; that is why, in this study, it is interesting to focus on this brand, and the low-pass harmonic filtering technique will be applied to improve power quality of current

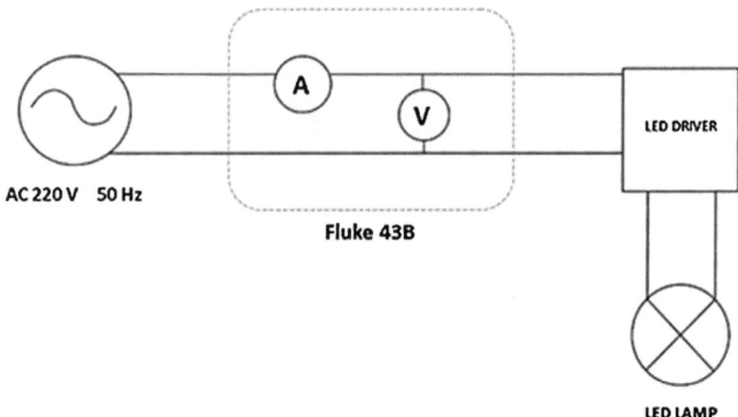

Fig. 3 Harmonics measurement in LED lamps system

Table 1 Percentage of total harmonic distortion of current for different configurations

LED lamps and drivers	%THD$_i$	
	1 driver	9 driver
LED lamp with driver of brand A	141.4	174.6
LED lamp with driver of brand B	75.6	76.3
LED lamp with driver of brand C	56.0	23.5

signal. The goal is try to design an effective harmonic filter with low cost in order to obtain the equivalent performance or power quality to the more expensive LED lamp driver.

4 Design of Low-Pass Harmonic Filter

LED lamp functioning with driver of brand A is chosen for harmonic filter design due to its highest harmonics generation. The acceptable level of harmonics is normally defined by a standard, which is herein IEC Std. 1000-3-2 (Group C); this ensures that the sensitive nearby electrical/electronic equipments or itself will be not affected by generated harmonics. Table 2 shows the limited harmonic current of lighting equipments according to the mentioned standard.

Note that for brand A and B, the harmonics level exceeds the maximum harmonic current permitted by the applied standard, whereas brand C rather respects to the standard.

The generated harmonics issued from LED lamp of brand A will be reduced by adding the harmonic filter at the input of the LED lamp. This filter is simply composed of one series inductor and one parallel capacitor. The resistor of the order

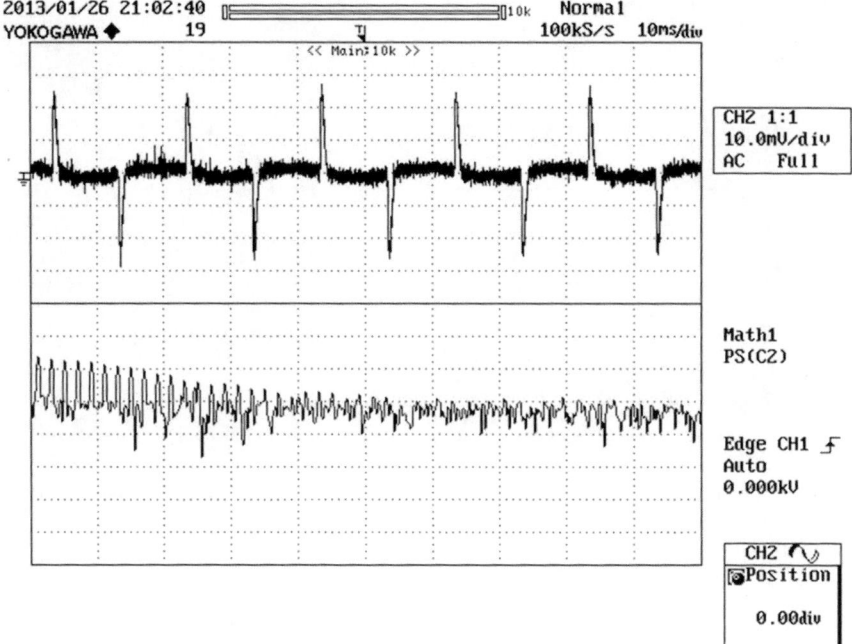

Fig. 4 Current waveform of LED lamp with 1 driver of brand A

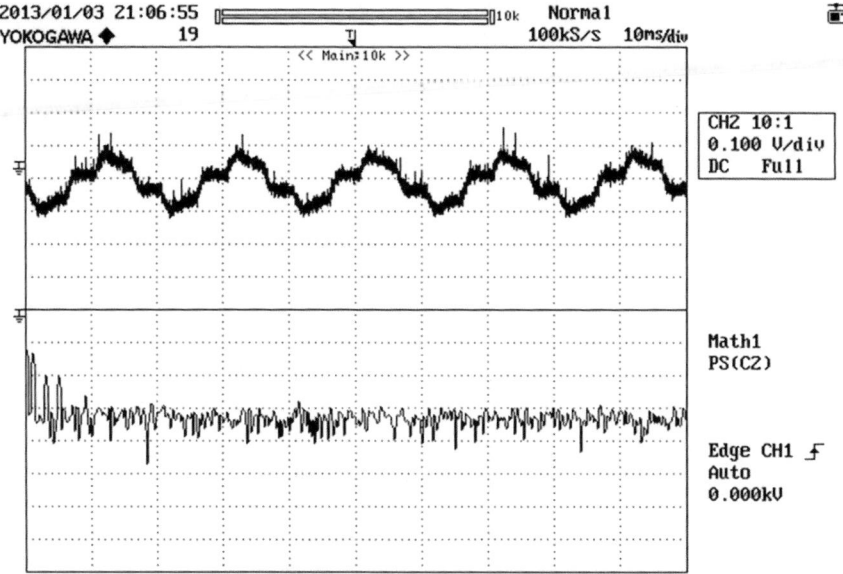

Fig. 5 Current waveform of LED lamp with 1 driver of brand C

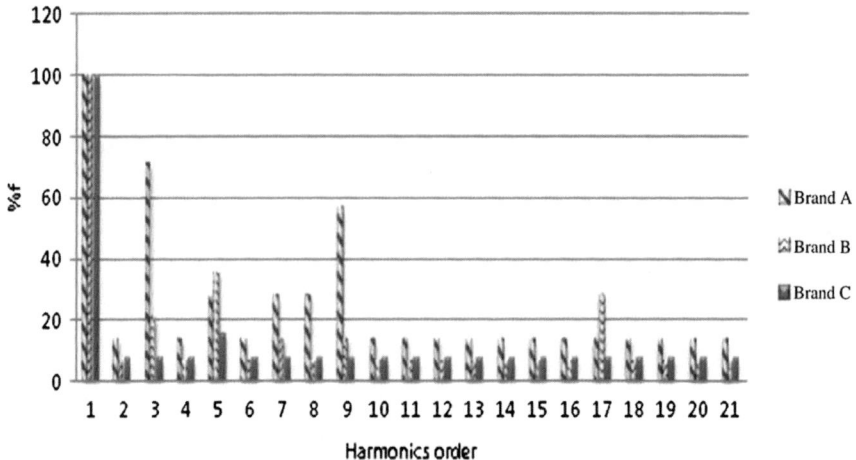

Fig. 6 Comparison of harmonic current magnitudes at different harmonic order for each brand (1 lamp and 1 driver)

Table 2 Limit of harmonic current of lighting equipments (group C) according to IEC 1000-3-2 standard

Harmonics order (n)	Maximum harmonic current permitted (calculated in percentage by comparing to the fundamental magnitude)
2	2
3	30 × (power factor)
5	10
7	7
9	5
11 ≤ n ≤ 39 (only odd order)	3

of MΩ can be added in parallel with the capacitor in order to discharge its electric charge and also for a reason of mechanical structure.

To reduce the harmonics magnitude, the cut-off frequency is a key parameter to be considered. Here, the cut-off frequency of harmonic filter is fixed at 150 Hz, and the value of capacitor is 11 μF, thus by using (1), the value of inductor will be 102.34 mH.

$$f_c = \frac{1}{2\pi\sqrt{LC}} \tag{1}$$

After inserting the experimentally designed filter as depicted in Fig. 7, the current waveform becomes more sinusoidal as shown in Fig. 8. Furthermore, the %THD$_i$ is improved, it is presently equal to 10.2 % (before insertion of harmonic filter, it was

Fig. 7 Experimentally designed filter used in this study

(a)

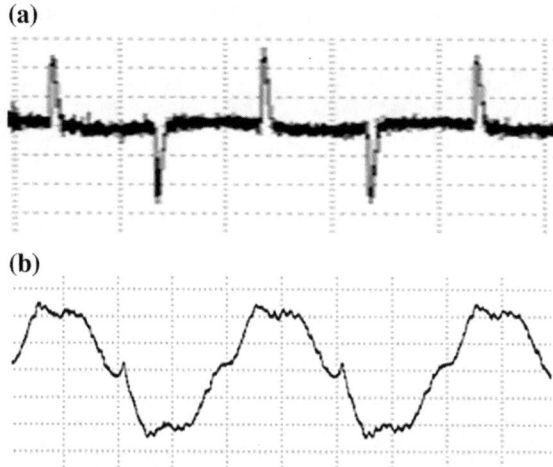

(b)

Fig. 8 Current waveform without harmonic filter (**a**), and with designed harmonic filter (**b**)

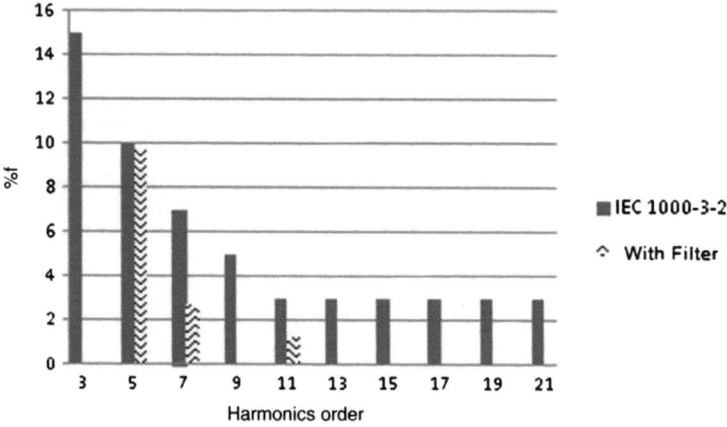

Fig. 9 Harmonics magnitude, obtained after insertion of harmonic filter, at different orders compared with IEC standard

equal to 141.4 %). The spectrum of current waveform with harmonic filter is also determined. The result is shown in Fig. 9. The results are obviously shown that the harmonics level is now conformed to the lighting equipment standard.

5 Conclusion

The analytical study of harmonics generated by LED lamp driver for lighting applications or in lighting systems has been conducted in this paper. Since the harmonics level exceeds the applied IEC 1000-3-2 standard, it must be reduced by an attenuation approach. The passive low-pass harmonic filtering technique was proposed and applied in this study, and with this filter, the level of harmonics respects satisfactorily to the standard. Finally, this confirms the Electromagnetic Compatibility of the overall system.

Acknowledgments The authors wish to gratefully acknowledge financial support for this research sponsored by the Faculty of Engineering, King Mongkut's Institute of Technology Ladkrabang (KMITL), Thailand.

References

1. A. Pandharipande, D. Caicedo, Daylight integrated illumination control of LED systems based on enhanced presence sensing. Energy Build. **43**(4), 944–950 (2011)
2. W.R. Ryckaert, K.A.G. Smet, I.A.A. Roelandts, M. VanGils, P. Hanselaer, Linear LED tubes versus fluorescent lamps: an evaluation. Energy Build. **49**(8), 429–436 (2012)
3. M. Ali, M. Orabi, M.E. Ahmed, A.E.L Aroudi, Design considerations of a single-stage LED lamp driver with power factor correction, in *2nd International Conference on Electric Power and Energy Conversion Systems (EPECS)*, pp. 1–6 (2011)
4. T.M. Roffi, I. Idris, K. Uchida, S. Nozaki, N. Sugiyama, H. Morisaki, F.X.N. Soelami, Improvement of high-power-white-LED lamp performance by liquid injection, in *International Conference on Electrical Engineering and Informatics*, pp. 1–6 (2011)
5. S. Uddin, H. Shareef, A. Mohamed, M.A. Hannan, An analysis of harmonics from LED lamps, in *Asia-Pacific Symposium on Electromagnetic Compatibility (APEMC)*, pp. 837–840 (2012)
6. S. Uddin, H. Shareef, A. Mohamed, M.A. Hannan, An analysis of harmonics from dimmable LED lamps, in *IEEE International Conference on Power Engineering and Optimization Conference (PEDCO)*, pp. 182–186 (2012)
7. H.K. Channi, H.S. Sohal, Power quality innovation in harmonic filtering. Int J Res Eng Appl Sci (IJREAS) **2**(2), 518–528 (2012)
8. M. Arias, D.G. Lamar, F.F. Linera, D. Balocco, A.A. Diallo, J. Sebastián, Design of a soft-switching asymmetrical half-bridge converter as second stage of an LED driver for street lighting application. IEEE Trans. Power Electron. **27**(3), 1608–1621 (2012)
9. S.Y.R. Hui, L.M. Lee, H.S.H. Chung, Y.K. Ho, An electronic ballast with wide dimming range, high PF, and low EMI. IEEE Trans. Power Electron. **16**(4), 465–472 (2001)
10. F.L. Tomm, A.R. Seidel, A. Campos, M.A.D. Costa, R.N. do Prado, HID lamp electronic ballast based on chopper converters. IEEE Trans. Industr. Electron. **59**(4), 1799–1807 (2012)

11. T.J. Liang, H.K. Liao, J.F. Chen, C.M. Huang, H.T. Lin, C.A. Cheng, A two-stage electronic ballast for HID lamp with flyback PFC, in *7th International Power Electronics and Motion Control Conference (IPEMC)*, pp. 192–198 (2012)
12. C. Jettanasen, A. Ngaopitakkul, Analytical study of harmonics issued from LED lamp driver, in *Proceedings of the International MultiConference of Engineers and Computer Scientists 2014, IMECS 2014*, Hong Kong, 12–14 Mar 2014. Lecture Notes in Engineering and Computer Science, pp. 683–686

Simulation Study of a Novel High Performance Oxide Engineered Schottky Collector Bipolar Transistor

Sajad A. Loan, Humyra Shabir, Faisal Bashir, M. Nizamuddin, Asim M. Murshid, Abdul Rahman Alamoud and Shuja A. Abbasi

Abstract The metal source/drain (MSD) Schottky barrier MOSFET offers several benefits enabling the scaling of MOSFET below 30 nm gate lengths. MSD Schottky barrier MOSFET possesses low parasitic S/D resistance, abrupt junctions that enable the scaling of the device to sub-10-nm gate lengths, superior control of leakage current due to the intrinsic Schottky potential barrier, and elimination of parasitic bipolar action. In this work, we propose a novel lateral Schottky Collector Bipolar Transistor (SCBT) employing multi zone base and multi step buried oxide has been proposed and simulated. The proposed device is simulated by using a 2D numerical simulator MEDICI. The simulation study has revealed that the proposed device with two base doping zones has ∼30 % higher breakdown voltage than the conventional device. The breakdown voltage increases further and is ∼75 % higher, when the number of zones in the proposed devices is increased to three. The increase in breakdown voltage can be attributed to the creation of extra electric field peaks in extended base region by multi doping zones in the base.

S.A. Loan (✉) · H. Shabir · F. Bashir · M. Nizamuddin
Department of Electronics and Communication Engineering, Jamia Millia Islamia,
New Delhi 110025, India
e-mail: sajadiitk@gmail.com

H. Shabir
e-mail: humyraiitk@gmail.com

F. Bashir
e-mail: faisalbashir4161@gmail.com

M. Nizamuddin
e-mail: nizamelectro@rediffmail.com

A.M. Murshid
Department of Computer Sciences, Kirkuk University, Kirkuk, Iraq
e-mail: asim.majeed2001@gmail.com

A.R. Alamoud · S.A. Abbasi
Department of Electrical Engineering, King Saud University, Riyadh, Saudi Arabia
e-mail: alamoud@ksu.edu.sa

S.A. Abbasi
e-mail: abbasi@ksu.edu.sa

© Springer Science+Business Media Dordrecht 2015
G.-C. Yang et al. (eds.), *Transactions on Engineering Technologies*,
DOI 10.1007/978-94-017-9588-3_20

Keywords Bipolar transistor · Breakdown voltage · Electric field · Multi zone · Oxide engineering · Schottky collector · Silicon-on-insulator

1 Introduction to Schottky Barrier Devices

The International Technology Roadmap for Semiconductors (ITRS) states that the scaling of the semiconductor device further and further in nanoscale is extremely difficult and semiconductor industry has already "entered the era of material-limited device scaling" [1]. The basic materials for the conventional MOSFET device— silicon, silicon dioxide, doped silicon, and polysilicon—have been extended to their performance limitations [1]. New materials, such as high-k gate dielectrics, strained-silicon substrates, metal gates, and metal source/drains (S/Ds) (to replace doped silicon S/Ds), are the materials of choice and can be the future materials.

The bipolar junction transistors used in switching circuits usually face the problem of storage charge due to minority carrier injection into the base and collector. This results in the degradation of switching speed of the device. May [2] fabricated a Schottky Barrier Collector Transistor (SBCT), which possess Schottky barrier for the collector-base junction and a p-n junction for the emitter-base junction. In such a device, no injection of minority carriers take place from the metal to semiconductor, while carriers injected into the metal have practically zero life time. Hence SBCT does not have a significant storage time in a saturated switch mode. Further, such a device possesses negligible collector series resistance, as the collector is metallic [3, 4].

The idea of completely replacing doped source and drain regions in a conventional MOSFET with metals was first proposed by Nishi in 1966, when he submitted a Japanese patent [5]. In 1968, Lepselter and Sze [6] proposed a structure of PMOS bulk device employing PtSi for the S/D regions. However, the proposed device fabricated by Lepselter and Sze showed poor performance with room-temperature drive current ten times lower than that of a conventional MOSFET.

Figure 1 shows the schematic cross sections of (a) conventional MOSFET and (b) Schottky barrier MOSFET. In a Schottky barrier MOSFET, metal source and metal drain have been used instead of doped source and doped drain regions of a

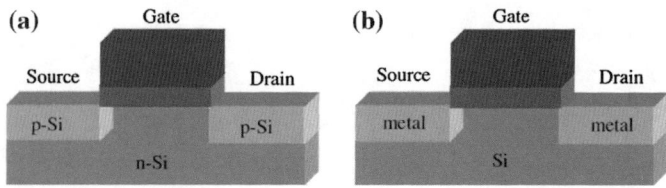

Fig. 1 Schematic cross-section of **a** conventional MOSFET and **b** Schottky barrier MOSFET

conventional MOSFET. The advantages Schottky barrier MOSFET possess includes lower sheet resistance, smaller junction depth, low temperature processing and realization of more integration levels. The issues associated with a Schottky barrier MOSFET include limited drive current because of Schottky barrier height and higher leakage current because of tunneling current between Si the substrate and the metal silicide layer.

The SCBT on SOI offers the advantages of ultimate isolation, reduced crosstalk and substrate noise. There is also an enhancement in the operating speed of devices, due to reduction in parasitic capacitances. However, there is a problem in the integration of the vertical SCBT with SOI-CMOS technology. The vertical bipolar device requires the SOI layer to be thicker for the collector region. This gives rise to additional cost of SOI wafers and compatibility problems with the thin film SOI [4].

An attractive alternative to vertical SCBT is the lateral SCBT on SOI. The lateral SCBT on SOI possess low parasitic capacitance and promise low power consumption. Besides, it allows tuning of the SOI layer to optimize CMOS device performance without degradation of the bipolar device performance. The lateral SBCT on SOI has promising applications in high speed analog and mixed signal circuit designing and in non-saturating VLSI circuits in BiCMOS technology. However, the major drawback of the SBCT on SOI is its extremely low breakdown voltage (VCE ≤ 3 V). This can be attributed to the presence of an accumulated or depleted space charge region, producing high electric field at the Schottky collector-base interface. Further, the field induced barrier lowering effect and the image force increases the reverse leakage current, which causes early breakdown of the device [7–11].

2 Schematics of the Conventional and the Proposed Devices

The schematic cross-sectional views of the three devices simulated in this paper are shown in Fig. 2. They are conventional, two zone and three zone proposed devices. Table 1 shows various device and process parameters used in simulation. In this paper, we performed the numerical simulation study of a multi doping zone base Schottky collector transistor on buried oxide multistep [12] using MEDICI [9]. Two types of structures have been simulated and studied, one with two doping zones and

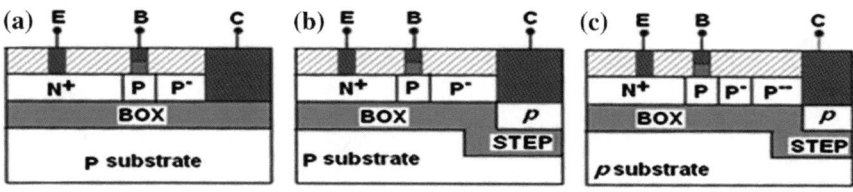

Fig. 2 Schematic cross-section of **a** conventional **b** two zone and **c** three zone SBCT on BODS

Table 1 Simulation parameters used

Parameter	2ZC	2ZP	3ZP
Substrate doping (cm^{-3})	1×10^{15}	1×10^{15}	1×10^{15}
Emitter doping (cm^{-3})	5×10^{19}	5×10^{19}	5×10^{19}
Base doping			
B$_1$ (cm^{-3})	7×10^{17}	7×10^{17}	7×10^{17}
B$_2$ (cm^{-3})	5×10^{16}	5×10^{16}	5×10^{16}
B$_3$ (cm^{-3})	–	–	1×10^{16}
Emitter length (μm)	3.8	3.8	3.8
Base length			
B$_1$ (μm)	1.2	1.2	1.2
B$_2$ (μm)	1	1	1
B$_3$ (μm)	–	–	0.8–1
Collector length (μm)	4	4	3
SOI thickness (μm)	0.2	0.2	0.2
T_{BOX} (μm)	0.2–1.0	0.2–1.0	0.2–1.0
Field oxide thickness (μm)	1	1	1

other with three doping zones in the base and both of them use buried oxide double step (BODS) [7] wafer. They have been named as two zone proposed (2ZP) and three zone proposed (3ZP) devices. The simulations have shown that the 2ZP and 3ZP devices have 30 and 75 % higher common emitter breakdown voltage (BV_{CEO}) than the conventional two zone base device. The multizone base creates additional electric field peaks which reduce the electric field at the metal-base interface and hence increases the breakdown voltage of the device. The breakdown voltage is further increased by using buried oxide double step (BODS) under the metal collector. The silicon island under the collector and BODS share more potential and hence increases the breakdown voltage-further.

3 Simulation Results and Discussions

To increase the accuracy of simulations, various models representing various physical phenomena have been incorporated. These models include *analytic*, *prp-mob*, *fldmob*, *consrh*, *auger* and *BGN* [9]. Figures 3 and 4 show the Gummel plot and the current gain of conventional and 3ZP devices. As can be seen from these figures, the plots are overlapping for most of the base voltage. This can be attributed to the same base area and base current in both the devices. The reason of same base current and base area is that the buried oxide is fully covering base in both the devices, hence keeping base area and base current constant. By choosing proper doping concentration in base, the current gain has been chosen to be 42 in all

Fig. 3 Gummel plot of
conventional and three zone
proposed (*3ZP*) device

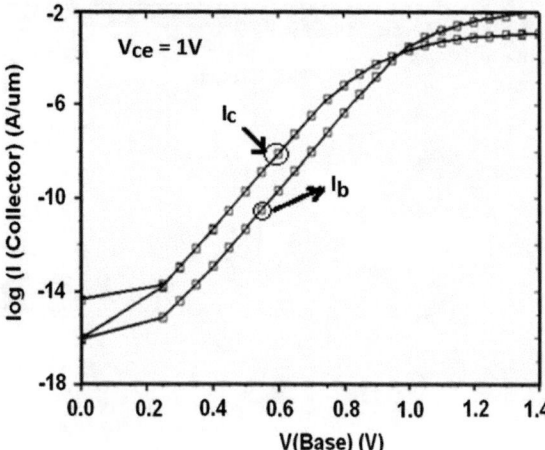

Fig. 4 Current gain versus
collector current of
conventional and SCBT 3ZP
device

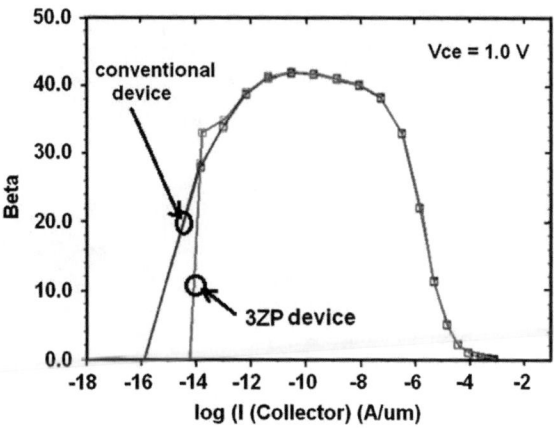

devices for better comparison of results. The output characteristics of conventional, 2ZP, conventional device with three zone base and 3ZP devices are shown in Figs. 5 and 6.

For each device the common emitter breakdown voltage (BV_{CEO}) is calculated at a collector current of 1×10^{-6} A/μm. It is observed from the analysis that the 3ZP device possesses about 40 % more breakdown voltage than the conventional device with three doping zone base. On comparing 3ZP device with 2ZP device, we see that the breakdown voltage is 59 % more in 3ZP device. The comparison of 3ZP and 2ZP devices with the conventional devices shows an increase of 75 and 30 % in breakdown voltage respectively. The uniformity of the lateral surface electric field is increased by using the multi zone base, as it creates additional electric field peaks in the base which reduce the electric field at the metal-base interface, thus increasing the breakdown voltage of the device. The buried oxide double step

Fig. 5 Common–emitter I–V
characteristics of the 3ZP
device and conventional
devices with 3Z base

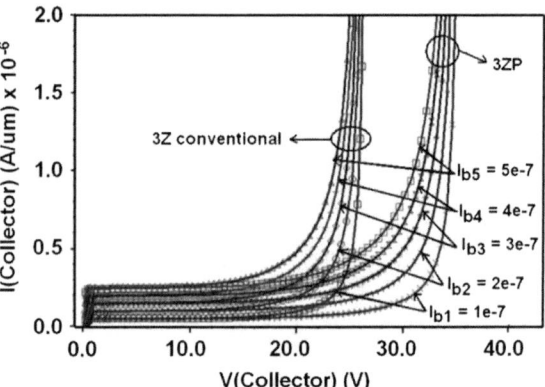

Fig. 6 Common emitter I–V
characteristics of the 2Z
conventional, 3Z
conventional, 2ZP and 3ZP
devices

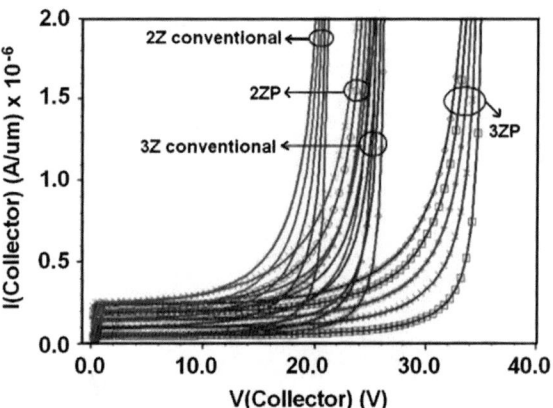

(BODS) under the metal collector also helps in increasing BV_{CEO} as it shares some
applied potential along with a substrate island under the collector. The potential
contours in the conventional and in the 3ZP device are shown in Fig. 7. As can be
seen from Fig. 7b, the potential contours crowd along the oxide layer and hence the
critical field at which breakdown occurs is obtained at a lower voltage of ∼20 V.
On the other hand, in 3ZP device, shown in Fig. 7a, the critical field at which the
breakdown occurs results at a higher voltage of ∼25 V. Although, the crowding of
potential contours is still there in oxide, however, the depletion region width is
more, which can sustain more electric field and hence results in higher breakdown
voltage.

Further, the low doping at the collector side and the high doping at the emitter
side is chosen so that it gets fully depleted in the absence of externally applied bias.
This helps in improving the breakdown voltage by reducing the electric field at the
metal-base interface.

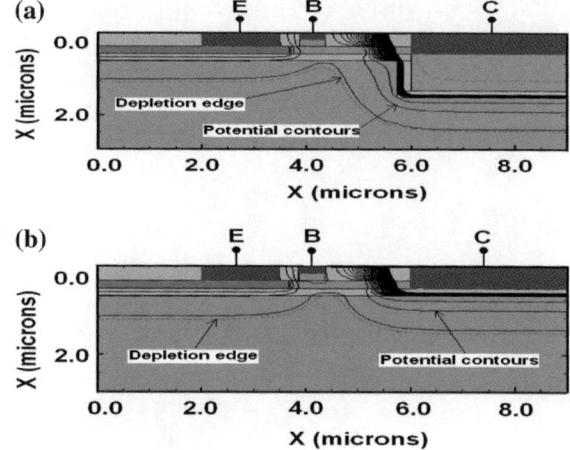

Fig. 7 Potentials contours in a 3ZP SBCT and in b conventional SBCT device

Figure 8 shows the electric field profile in the conventional, 2ZP and the 3ZP devices. It is clear that the electric field peak at the metal-base junction in conventional device is more than that of 2ZP device. The magnitude of electric field peak is least in the 3ZP device. This reduced electric field results in an increase in breakdown voltage of 3ZP device. Figure 9 shows the effect of increasing the thickness of step oxide on breakdown voltage. It is seen that increasing T_{STEP} in upward direction towards collector does not change the breakdown voltage significantly. As we go on increasing the T_{STEP} in the upward direction, an SCBT on extended box [4] can result. Such a structure has more or less same breakdown voltage as that of SCBT on BODS. However, SCBT on BODS is thermally efficient in comparison to extended box structure, due to thin oxide and a silicon island under collector.

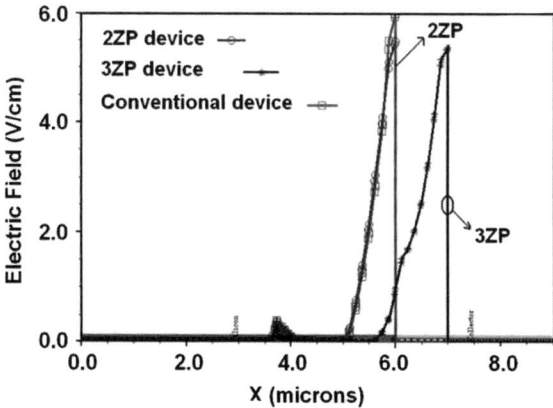

Fig. 8 Electric field profile along the silicon film in 3ZP, 2ZP and in conventional SBCT devices

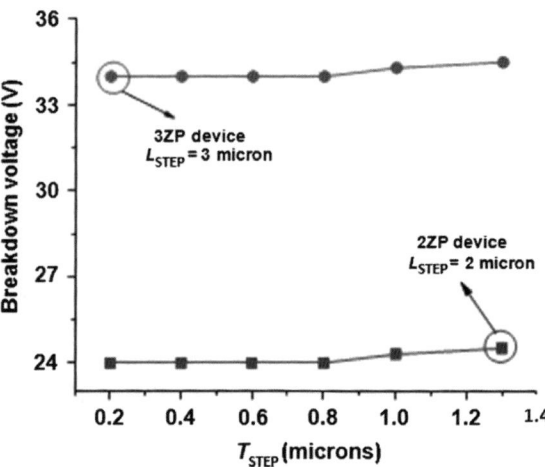

Fig. 9 Effect of change in TSTEP on breakdown voltage in 2ZP and 3ZP devices

4 Energy Band Diagram of the Proposed Device

The ohmic or Schottky behavior of a metal semiconductor junction depends on the Schottky barrier height, Φ_b, of the junction. In case of Schottky contact, Φ_B is significantly higher than the thermal energy kT, the semiconductor is depleted near the metal. For an ohmic contact, barrier height is lower and the semiconductor is not depleted near the metal and hence an ohmic contact is formed. Figure 10 is the band diagram of a metal and semiconductor contact at zero bias (equilibrium). It also shows the graphical definition of the Schottky barrier height, Φ_B, for an n-type semiconductor as the difference between the interfacial conduction band edge E_C and Fermi level E_F. The energy band diagrams of the proposed structure has been obtained for both equilibrium and active mode conditions, as shown in Fig. 11. It clearly shows a barrier at metal semiconductor interface which is responsible for negligible reverse saturation current.

Fig. 10 Band diagram for Schottky barrier under equilibrium conditions

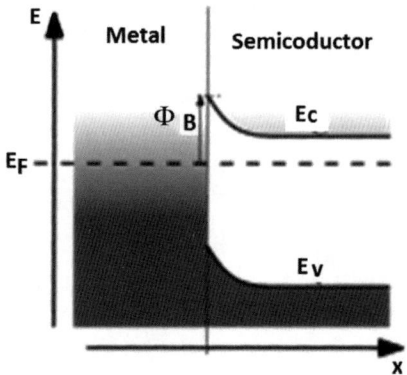

Fig. 11 Energy band
diagram **a** under equilibrium
and **b** forward active mode

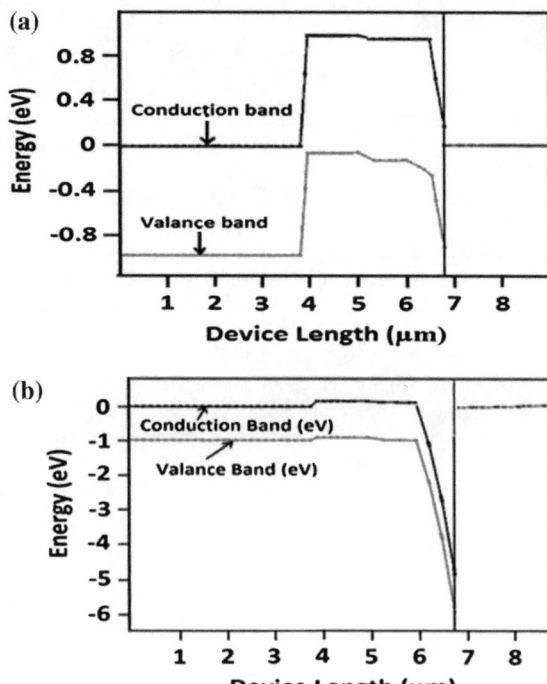

5 Conclusion and Future Work

The important feature of Schottky barrier MOS technology is the use of metallic source and drain regions, instead of the doped source and drain regions of the conventional MOSFET. Metallic S/Ds result in lower resistance and atomically abrupt, providing a long-term scalability advantage. The formation of Schottky barrier between metal S/D junction to the channel results in improved I_{OFF} leakage control. It further enables lower doping in the channel region, which results in higher channel mobility.

In this work, high performance multi doping zone base Schottky collector bipolar transistor on buried oxide double step has been proposed. The 2D numerical simulations have revealed that the 2ZP and 3ZP devices have 30 and 75 % higher common emitter breakdown voltage (BV_{CEO}) than the conventional SCBT device. The increase in breakdown voltage is attributed to the generation of additional electric field peaks which reduce the electric field at the metal-base interface, thus increasing the breakdown voltage of the device. The BODS under the metal collector and the silicon island under the collector share some potential and hence increase the breakdown voltage further.

Acknowledgments This work was supported by NSTIP strategic technologies programs number (11-NAN-2118-02) on the Kingdom of Saudi Arabia.

References

1. *International Technology Roadmap for Semiconductors 2003 Edition* (Semiconductor Industry Association, Austin, TX)
2. G.A. May, The Schottky barrier collector transistor. Solid State Electron. **11**(6), 613–619 (1968)
3. S. Akbar, N. Anantha, C. Hsieh, J. Walsh, Method of fabrication of Schottky bipolar transistor. IBM Technol. Discl. Bull. **33**, 11 (1991)
4. M.J. Kumar, S.D. Roy, A new high breakdown voltage lateral Schottky collector bipolar transistor on SOI: design and analysis. IEEE Trans. Electron Devices **52**(2), 496–501 (2005)
5. Y. Nishi, Insulated gate field effect transistor and its manufacturing method. Japan Patent 587,527 (1970)
6. M.P. Lepselter, S.M. Sze, SB-IGFET: an insulated-gate field-effect transistor using Schottky barrier contacts for source and drain. Proc. IEEE **56**(8), 1400–1402 (1968)
7. S.A. Loan, S. Qureshi, S.S.K. Iyer, A novel high breakdown lateral BJT on SOI with multizone doped and multistep oxide: a numerical simulation study. Semicond. Sci. Technol. **24**(2), 025017 (2009)
8. M.J. Kumar, D.V. Rao, A new lateral PNM Schottky collector bipolar transistor on SOI for nonsaturating VLSI logic design. IEEE Trans. Electron Devices **49**(6), 1070–1072 (2002)
9. TMA MEDICI 4.2. (Technology Modeling Associates Inc., Palo Alto)
10. J.M. Larson, J.P. Snyder, Overview and status of metal S/D Schottky-barrier MOSFET technology. IEEE Trans. Electron Devices **53**(5), 1048–1058 (2006)
11. S.A. Loan, H. Shabir, F. Bashir, M. Rafat, A.R.M. Alamoud, S.A. Abbasi, A high performance charge plasma PN-Schottky collector transistor on silicon-on-insulator. Semicond. Sci. Technol. **29**(095001), 1–9 (2014)
12. S.A. Loan, H. Shabir, F. Bashir, M. Rafat, A.R.M. Alamoud, S.A. Abbasi, High performance oxide engineered lateral Schottky bipolar transistor, in *Proceedings of the International Multi Conference of Engineers and Computer Scientists 2014, IMECS 2014*, Hong Kong, 12–14 Mar 2014. Lecture Notes in Engineering and Computer Science, pp. 701–704

Two Area Load Frequency Control of Hybrid Power System Using Genetic Algorithm and Differential Evolution Tuned PID Controller in Deregulated Environment

Gargi Konar, Kamal Krishna Mandal and Niladri Chakraborty

Abstract Load frequency control (LFC) of interconnected power system ensures zero steady state error in frequency dynamics and proper sharing of load by generators. Use of PID controllers in LFC enhances smooth and efficient control of area control error (ACE). For the sake of sustainable and environment friendly power generation, use of renewable energy sources along with non-renewable one are accentuated to generate power in open market scenario these days. In this work, PID controllers tuned with Genetic Algorithm (GA) and Differential Evolution (DE) are used for two area LFC with thermal, hydro and diesel generators. The incorporation of PID controller gives desired power system dynamic responses. It is observed that frequency reaches the steady state value within reasonable time and generators of connected areas share the tie line power according to their participation factors. MATLAB codes are developed for GA and DE based tuning of PID controller. The optimized values of PID gains are used to study the power system dynamics due to change in loads using Simulink.

Keywords Area control error · Differential evolution · DISCO · GENCO · Genetic algorithm · Hybrid power generation · PID controller · Tie line power

G. Konar (✉) · K.K. Mandal · N. Chakraborty
Power Engineering Department, Jadavpur University Saltlake Campus,
LB-8, Sector-III, Kolkata 700098, India
e-mail: gargi_konar@yahoo.co.in

K.K. Mandal
e-mail: kkm567@yahoo.com

N. Chakraborty
e-mail: chakraborty_niladri2004@yahoo.com

© Springer Science+Business Media Dordrecht 2015
G.-C. Yang et al. (eds.), *Transactions on Engineering Technologies*,
DOI 10.1007/978-94-017-9588-3_21

1 Introduction

In recent years the power utilities have gone through a major change from monopoly to competition. In that competitive environment, deregulation in power sector has been initiated all over the world. As an effect, the electricity consumers have gained the opportunity to choose among several energy providers. With these changes, the electricity generation, transmission and distribution systems are needed to follow new strategies owing to deregulation [1–3].

Reliable and good quality power transfer should be maintained in an interconnected power system under deregulated environment through proper choice of automatic generation control components [4]. Sudden change in load introduces frequency fluctuations and tie-line power exchange. Suitable load frequency control with the consideration of bilateral contracts between participating areas nowadays became mandatory. Optimal output feedback, linear feedback, Kalman estimator [2, 5–7] are such few control strategies adopted elsewhere to accomplish the same.

These days some soft computing techniques have achieved popularity over the classical control strategies in designing load frequency control. Several optimization techniques like Genetic algorithm, Particle Swarm Optimization, Bacterial Foraging are currently being applied for the automatic generation control in multi-area system under deregulation [8–10]. Such optimization techniques are also used for automatic generation control of interconnected power system without deregulation [11–13]. These techniques are used either to tune the different types of controllers or to set the parameters for power system stabilizers. These actions enable operators to improve the frequency deviation situation and restoration of the tie line power fluctuations quickly. In deregulated environment participation contract between two or more areas are regulated by an 'independent system operator' (ISO) [14]. Contract violation and its effects are also important in these situations [2].

As the current era of civilization is facing acute energy crisis, attentions are being drawn to use renewable energy resources for power generation. Thus the incorporation of hydro, wind, solar, biomass power generating system in interconnected power system has become a familiar situation. The load frequency control of such systems along with conventional power generation is being focused these days. Such a two area interconnected power system having a combination of thermal, hydro and dual fuel diesel generators are considered in this work.

In this paper, two area automatic generation control in a deregulated environment is considered. The power generating companies (GENCO) can sell electricity at competitive prices to the distribution companies (DISCO) in open market under deregulation. Two GENCOs and one DISCO are considered in each area under study. GENCOS share loads of its own area as well as that of the other area as demanded by the DISCOs. This participation is based on the contract made between the two systems as per the corresponding DISCO Participation Matrix (DPM) in restructured environment [14]. The PID controller is used to nullify the effect of frequency and tie-line power deviations in both the areas. MATLAB code has been developed to achieve genetic algorithm and differential evolutional based PID controller tuning. PID controller tuning ensures the improvements in the system response in terms of settling time, rise time, overshoot and steady state value. Studies are made for different contract conditions. The results obtained from two different optimization techniques are compared for sudden changes in load in both areas. This situation is studied for two cases. In the first case all generators are active, but in the other, the hydro generator is not generating power. In both cases the load sharing by all generators are studied. The block diagrams of two area load frequency control under deregulation and hybrid power generation scenario are drawn in Simulink and effects of sudden load change on different GENCOs are noted.

2 Two Area Load Frequency Control in Restructured Power System

Restructuring of power sector initiated lots of challenges in power system operation and control. Load frequency control being one the major concerns of power system study, requires modifications in respect to study of the effects of bilateral contracts between participating GENCOs and DISCOs on the system dynamics. To improve the dynamical transient response proper choice of PID controller gains are achieved in this work through soft computing techniques.

2.1 Two Area Power System in Restructured Environment

Restructured power system consists of Generation companies—GENCOs, Transmission companies—TRANSCOs, Distribution companies—DISCOs, and independent system operators ISO. In this paper, the power system comprises two areas having one DISCO and two GENCOs in each area as shown in Fig. 1. The corresponding DISCO Participation Matrix (DPM) is shown in Eq. 1. Area 1 consists of one thermal (GENCO1) and one hydro generator (GENCO2) whereas area 2 consists of one thermal (GENCO3) and one dual fuel diesel generator (GENCO4).

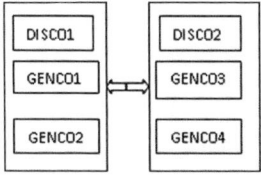

Fig. 1 Configuration of the power system

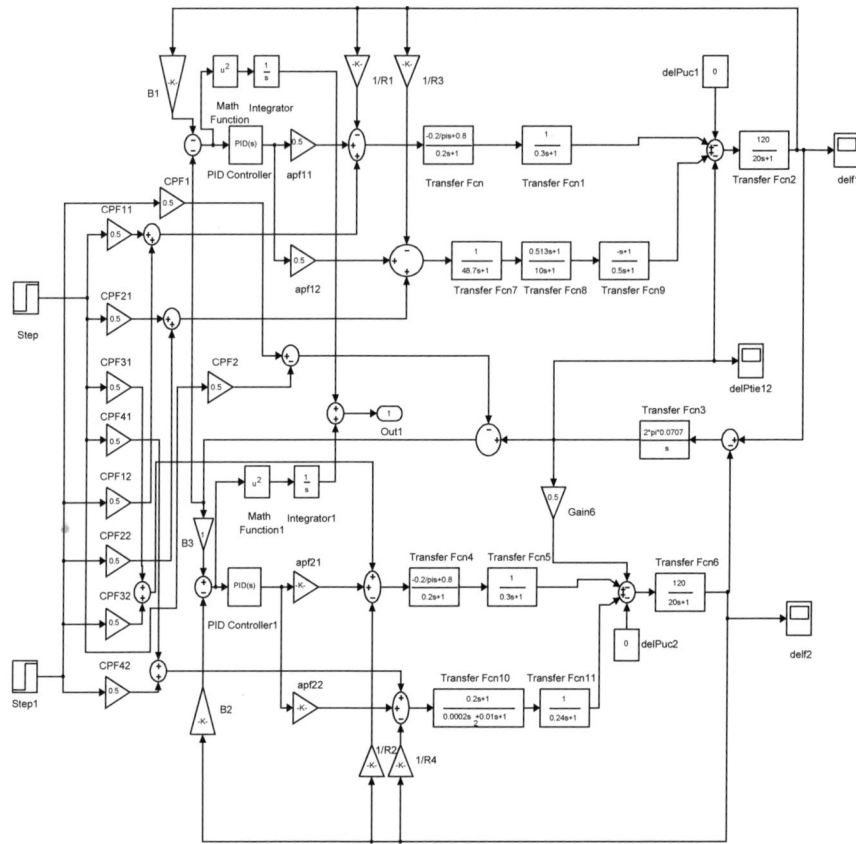

Fig. 2 Complete block diagram of two area load frequency control in restructured environment

Each DISCO can buy power from all four GENCOs according to the fractions assigned to the elements of the corresponding column of DPM matrix. Thus a GENCO sells a fraction of total load to a DISCO as per contract made between them. Thus a DISCO buys the total required from all the GENCOs as per the contract made, i.e. sum of all the elements in each column of DPM is unity i.e. $\sum_i cpf_{ij} = 1$, where *cpf* is the contract participation factor [14].

Figure 2 depicts the block diagram of the two area load frequency control under deregulated environment. In this model the control vectors u_1 and u_2 are determined from the knowledge of the area control error (ACE) and frequency deviations of area 1 and 2 respectively. Equations 2 and 3 represent the ACEs of both areas. ACEs are calculated based on the frequency deviation (ΔF_1) and the tie-line power deviation i.e. the difference between the scheduled power deviation ($\Delta P_{tie12sch}$) and the actual power deviation ($\Delta P_{tie12actual}$). The latter two are represented through Eqs. 4 and 5.

$$DPM = \begin{bmatrix} cpf_{11} & cpf_{12} \\ cpf_{21} & cpf_{22} \\ cpf_{31} & cpf_{32} \\ cpf_{41} & cpf_{42} \end{bmatrix} \tag{1}$$

$$ACE_1 = B_1 \Delta F_1 + \Delta P_{tie12error} \tag{2}$$

$$ACE_2 = B_2 \Delta F_2 + \Delta P_{tie12error} \tag{3}$$

$$\Delta P_{tie12sch} = cpf_{12} \Delta P_{L2} - cpf_{21} \Delta P_{L1} \tag{4}$$

$$\Delta P_{tie12error} = \Delta P_{tie12actual} - \Delta P_{tie12sch} \tag{5}$$

Here $\Delta P_{tie12error}$ is the tie-line power error. Using ACE participation factor (apf), ACE signal is distributed among the GENCOs. These ACE signals are then fed into PID controller.

2.2 PID Controller Design

This is the most important part of the Automatic Generation Control (AGC). The choice of proportional-integral-derivative (PID) controller than proportional plus integral (PI) controller ensures better system response in terms overshoot and settling time [15]. The ACE signals are controlled using the PID controller to produce control vectors for the AGC. In this work, the PID controller tuning is done through Genetic Algorithm (GA) and Differential Evolution (DE). The proportional (k_p), integral (k_i) and derivative (k_d) gains are set using GA and DE. The transfer function of the PID controller (Eq. 6) used for both the areas are considered to be identical factor [14].

To get the optimized values of the PID gains, suitable objective function is developed here. However the values of the gains are appropriately chosen. This objective function (OB) can be defined as the sum of the squares of the area control errors (ACE_1 and ACE_2) in each area as shown in Eq. 7.

$$G_c(s) = k_p + \frac{k_i}{s} + k_d s \tag{6}$$

$$OB = \int_0^\infty \sum_{i=1}^2 (ACE_i)^2 dt \tag{7}$$

The optimization problem is based on the minimization of the Objective Function subject to the conditions that the PID gains k_p, k_i and k_d of both the controllers will lie within the minimum and the maximum limits as shown in Eq. 8.

$$k_p^{\min} \le k_p \le k_p^{\max}, \quad k_i^{\min} \le k_i \le k_i^{\max}, \quad k_d^{\min} \le k_d \le k_d^{\max} \tag{8}$$

Thus the PID controller parameters are optimized using Eqs. 7 and 8 with the help of two soft computing techniques—Genetic Algorithm and differential evolution as discussed in the next section.

2.3 PID Controller Parameter Tuning Using Genetic Algorithm

A combination of Darwinian Survival of the fittest principle with genetic operation popularly known as Genetic Algorithm became an effective method of optimization. This global optimization technique involves stochastic search algorithm. Since the gains of PID controllers, k_p, k_i and k_d, are to be optimized using GA, three binary strings are assigned to each member of the population in this problem. To accommodate the entire range of possible solutions, large value of population size (100) is chosen. The implementation of GA starts with parameter encoding [16]. This is done with great care so that the link between the objective function and the strings are maintained properly. The decimal integers of binary strings are obtained following Eq. 9.

$$y_j = \sum_{i=1}^{l} 2^{i-1} b_{ij} \ (j = 1, 2, \ldots L) \tag{9}$$

where
y_i is the decimal coded value of the binary string
b_{ij} is the ith binary digit of the jth string
l is the length of the string
L is the population size

Following a fixed mapping rule, the continuous variable x_j (Eq. 10) is found in the search space where $x_{\min}$ and $x_{\max}$ are the minimum and the maximum values of the variable x_j.

$$x_j = x_{\min} + \frac{x_{\max} - x_{\min}}{2^l - 1} y_j \ (j = 1, 2, \ldots L) \tag{10}$$

In the next step the most challenging task is done i.e. the evaluation of the best values of PID controller gains to minimize the objective function. This task ensures smallest overshoot, fastest rise time and quickest settling time.

Another important step in GA is to select the highly fit strings in population as the parents and a mating pool is formed. The probability [16] for selecting the ith string is

$$p_i = \frac{f_i}{\sum_{j=1}^{L} f_j} \quad (11)$$

This step is followed by crossover operation where new strings are generated by exchanging the information among the strings of the mating pool. The mutation operator is also introduced to bring variations. Here mutation rate is chosen to be 0.5.

This newly tuned PID gains are used to form the PID controller transfer function. The controller transfer function is then used to simulate the overall system response of two area Load Frequency Control in deregulated environment for a given step input. The main objective is to achieve the minimum overshoot, quickest rise time and the settling time for frequency deviation and tie-line power characteristics.

2.4 PID Controller Parameter Tuning Using Differential Evolution

This optimization technique has came up as most promising evolutionary algorithm to fulfill the requirements of handling the non-differentiable, non-linear and multimodal objective function, good convergence properties, less computation time etc. in minimization problem [17]. It differs with GA in mutation operation.

In the initialization process [17], it generates a vector V of size N_v which is fixed. Evolution of vector V over generation G is the objective of this search process. Each entity X_i is a vector. The components of X_i are nothing but the problem decision variables. Hence,

$$P^{(G)} = \left[x_i^{(G)}, \ldots X_{N_V}^{(G)} \right] \quad (12)$$

$$X^{(G)} = \left[x_{1,i}^{(G)}, \ldots X_{D,i}^{(G)} \right]^T, \quad i = 1, 2, \ldots N_v \quad (13)$$

The complete search area is covered while choosing the initial population with the help of uniform probability distribution function. Thus the initialization of all random variables are done as

$$X_{j,i}^{(0)} = x_j^{\min} + \sigma_j \left(X_j^{\max} - X_j^{\min} \right), \quad i = 1, 2, \ldots N_v \quad and \quad j = 1, 2, \ldots D \quad (14)$$

Here, $X_j^{\min}$ and $X_j^{\max}$ are the lower and the upper limits of the jth decision variables, $\sigma_j \in [0, 1]$ and $X_{j,i}^{(0)}$ is the jth parameter of the ith individual of the initial population and D is the number of decision variable.

The next step is nothing but the mutation operation which generates mutant vectors V_i according to Eq. (15).

$$V_i^{(G)} = X_k^{(G)} + f_m(X_l^{(G)} - X_m^{(G)}) \tag{15}$$

Here, X_k, X_l and X_m are randomly chosen vectors. The mutation factor lies between [0, 2] to control the perturbation size in mutation operation. In this problem the mutation factor is chosen to be 0.65. This operation is followed by crossover operation [17] to generate trial vectors U_i using Eq. (16).

$$U_{j,i}^{(G)} = \begin{cases} V_{j,i}^{(G)}, & \text{if } n_j \leq C_R \text{ or } j = q \\ X_{j,i}^{(G)}, & \text{otherwise} \end{cases} \tag{16}$$

where $i = 1, 2, \ldots N_v$; $j = 1, 2, \ldots D$; $X_{j,i}^{(G)}$, $V_{j,i}^{(G)}$ and $U_{j,i}^{(G)}$ are the jth parameter of the ith target vector, mutant vector and the trial vector respectively at Gth generation; n_j is a uniformly distributed random number in the range of [0, 1] and $C_R(\in[0, 1])$ is the user defined cross over factor. The mutation operation is followed by the selection of better offsprings. If the fitness function of the off spring is better compared to the parent, then the parent is replaced by the offspring; otherwise the parent is retained in the next generation. Mathematically,

$$X_i^{(G+1)} = \begin{cases} U_i^{(G)}, & \text{iff } (U_i^{(G)}) \leq f(X_i^{(G)}) \\ X_i^{(G)}, & \text{otherwise} \end{cases} \tag{17}$$

This process is repeated for several generations allowing the individuals to inherit better fitness values in search of optimal solution.

3 Results and Discussions

The genetic algorithm (GA) and differential evolution (DE) based MATLAB code is developed in this work for PID controller tuning. The values of PID controller gains k_p, k_d and k_i thus obtained are used in PID blocks of two area LFC block diagram (Fig. 2) drawn in MATLAB/Simulink. This interconnected power system having thermal, hydro and dual fuel diesel generators are simulated with GA and DE tuned PID gains for several cases. The power system data used here are given in Table 1 and the optimized values of the PID gains based on GA and DE optimizations are shown in Table 2.

The DISCOs of this problem take power from the GENCOs according to the DPM. Here it is assumed that the each element of DPM has a value of 0.5. At the same time each GENCO participates in automatic generation control according to the area participation factors $apf_1 = 0.5$ and $apf_2 = 0.5$.

Table 1 Power system data

Power system data	Values
Steam generator time constants T_{g1}, T_{g2}	0.2 s
Steam turbine time constants T_{t1}, T_{t2}	0.3 s
Speed governor time constant for hydro turbine	48.7 s
Servomotor time constant of hydro turbine	10 s
Hydro turbine time constants T_{w1}, T_{w2}	1 s, 1 s
Diesel engine speed governor time constants T_{d1}, T_{d2}, T_{d3}	0.2, 0.0021, 0.008
Diesel power generating system time constant	0.024 s
Power system gains K_{p1}, K_{p2}	120 Hz/pu MW
Power system time constants T_{p1}, Tp_2	20 s
B_1, B_2	0.425 pu MW/Hz
Speed regulation of governors R_1, R_2, R_1, R_2	2.4 Hz/pu MW

Table 2 Values of PID gains

Area	k_p		k_i		k_d	
	GA	DE	GA	DE	GA	DE
1	0.015	2.467	0.1204	0.468	0.009	1.94
2	0.224	2.072	0.414	2.748	0.231	0.383

3.1 Study of Effects of Sudden Change of Load in Both Areas

The transient behavior of actual tie line power deviations due to sudden change of load in area1 (0.2 pu) and area2 (0.1 pu) are observed (Figs. 3 and 4). As per contracts made in deregulated environment, the scheduled tie line power deviation due to the said load changes in area1 and area2 is 0.05 pu which is met by both the systems at steady state condition. Though the GA based system has lower over-shoots (8 %), the DE based system gives better response in terms of settling time and rise time. Also the numbers of oscillations are quite less in case of latter system. Table 3 depicts the comparative study of transient responses of two systems.

Figures 5 and 6 shows the frequency deviation plots with respect to time. As per desired response, the frequency deviation is nullified in case of both areas. The DE based system has lower peak overshoot compared to the other and lesser settling times (Table 3). It is observed that the transient plot of frequency deviation of area2 less number of oscillations compared to area1. Therefore the emphasis can be given on the use of diesel generator when operated with biomass for clean power generation.

Transient behavior of changes in power generations in all the generators are also observed due to the corresponding sudden changes in load in area1 and area2. The generation capacities of all generators are assumed to be equal. Figures 7 and 8 shows the transient response of changes in generations of area1 using DE and GA

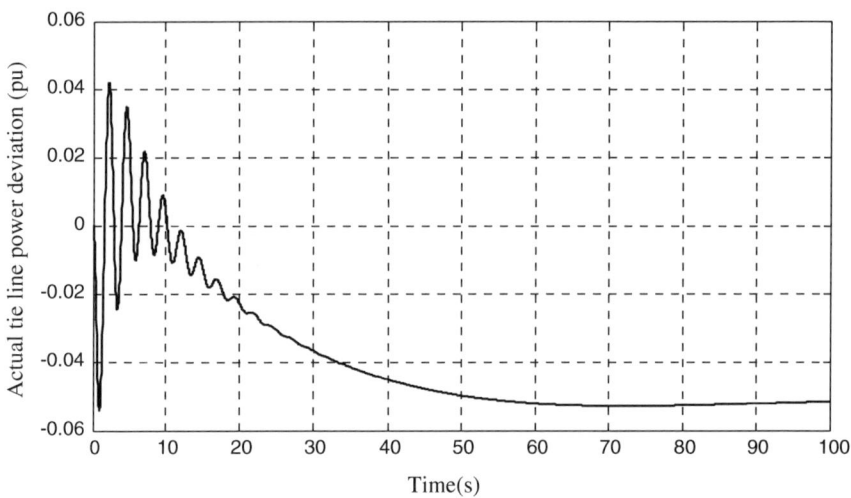

Fig. 3 Change in tie line power with respect to time using GA based PID tuning

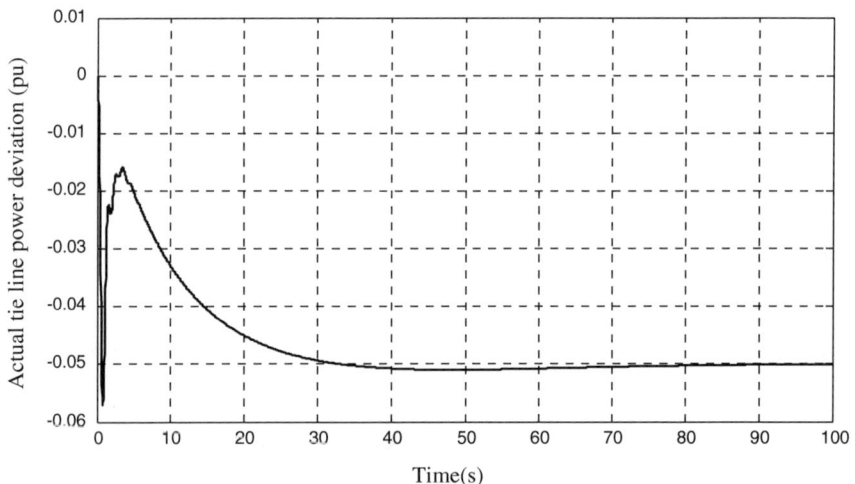

Fig. 4 Change in tie line power with respect to time using DE based PID tuning

based PID tuning. DE based PID tuning of area 1 and 2 ensures better response because the thermal generators have to generate less compared to the hydro generator. This fact encourages more generations from renewable sources i.e. hydro power compared to the thermal. Similarly Figs. 9 and 10 show the changes in power generations in area2 for GA and DE based PID tuning respectively. This is a favorable situation in terms of sustainable generation. The negative signs indicates that the power generation has reduced and vice versa. To meet the demand of 0.2 pu

Table 3 Performance study

Parameters	ΔF_1		ΔF_2		$\Delta P_{tie12actual}$	
	GA	DE	GA	DE	GA	DE
Settling time (s)	41.5	16	41.8	17	50	30
Rise time (s)	0.8	0.72	0.3	0.16	0.4	0.35
Peak over-shoot (%)	62	30	40	25	8	14

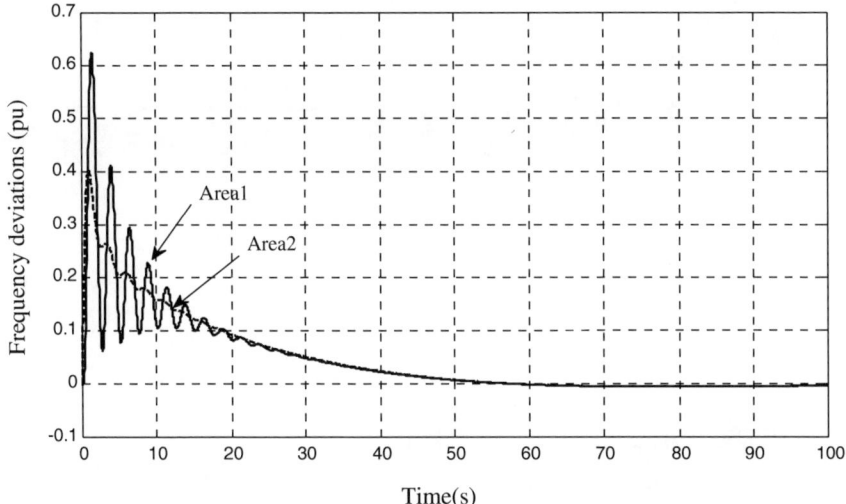

Fig. 5 Transient behavior of changes in frequency in area1 and area2 using GA based system

increment in area1 and 0.1 pu increment in area2, thermal generators and hydro generators may decrease their generation whereas the diesel generator operating in dual fuel mode may increase their generation level. Here dual fuel mode signifies the power generation from diesel generator using biomass i.e. sustainable power generation. This situation ensures environment friendly power generation.

3.2 Study of Effects of Sudden Change of Load in Both Areas with the Hydro Plant Disconnected

The power generation scenario is also observed for a sudden change in load as per Sect. 3.1. The only difference is that the Hydro generator is disconnected due to lack of water reserve. Since the area participation factors remain same, the load increments are met in area1 through the thermal generation (Fig. 11) only. But with the change in area participation factors in area2 the generation level of diesel

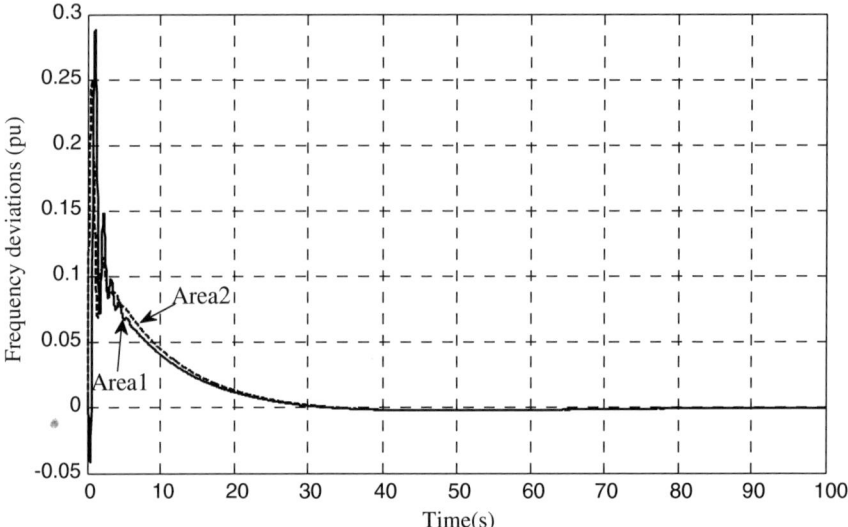

Fig. 6 Transient behavior of changes in frequency in area1 and area2 using DE based system

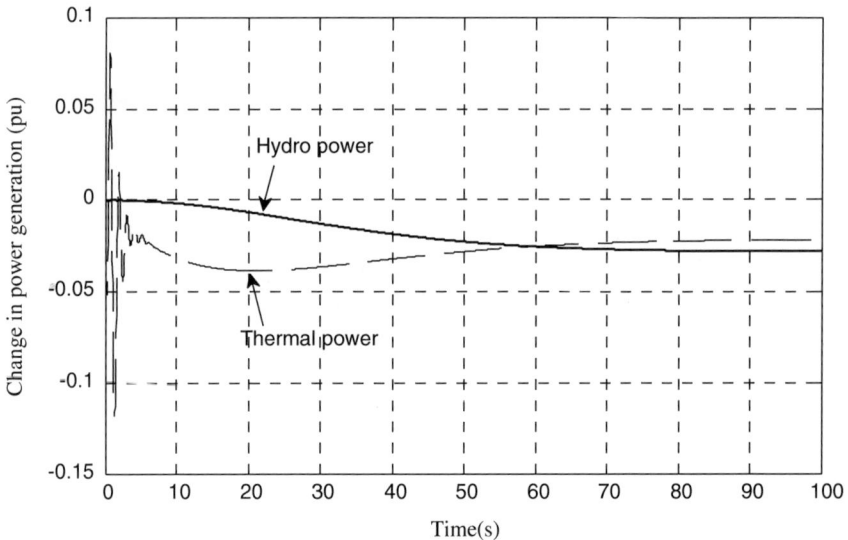

Fig. 7 Transient behavior of changes in thermal and hydro power generations of area1 using DE based system

generator with biomass resource is increased and the thermal generation is decreased. This situation helps to satisfy the deficiency in generation caused by the disconnection of hydro generator in area1 (Fig. 12). This clearly strengthens the use of renewable energy for electrical power generation.

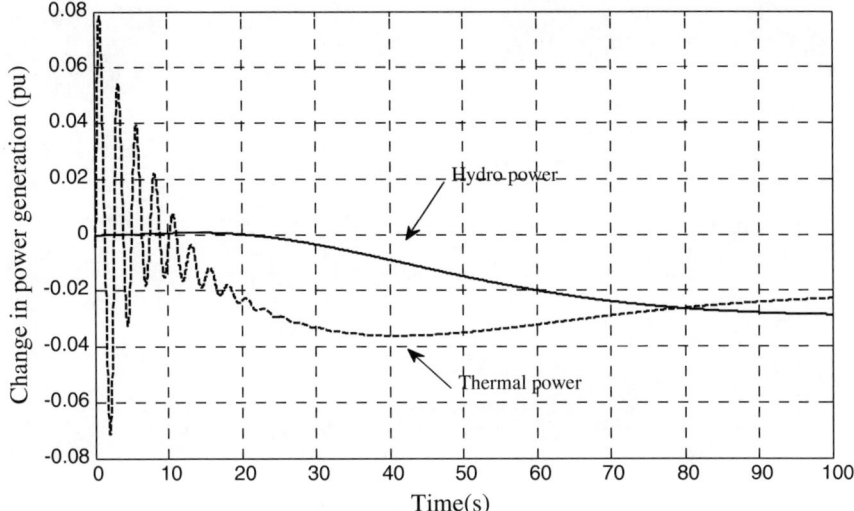

Fig. 8 Transient behavior of changes in thermal and hydro power generations of area1 using GA based system

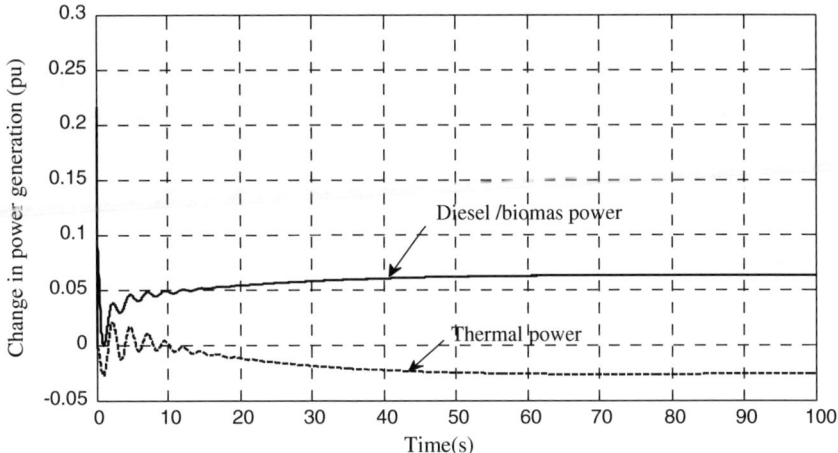

Fig. 9 Transient behavior of changes in thermal and diesel power generations of area2 using GA based system

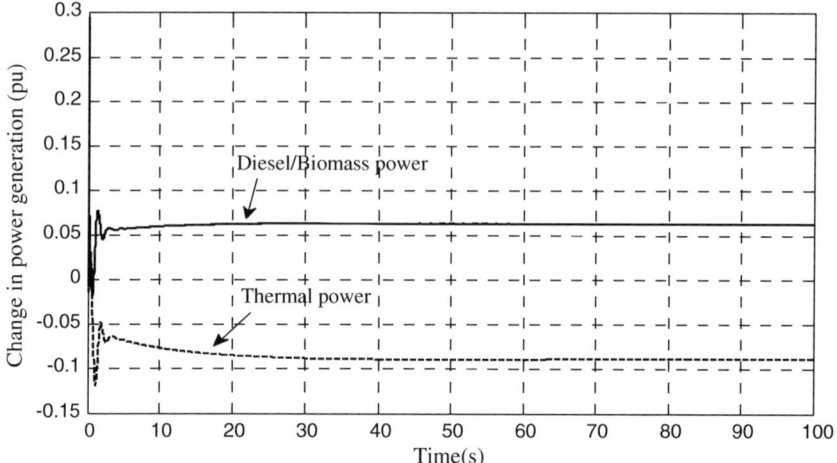

Fig. 10 Transient behavior of changes in thermal and diesel power generations of area2 using DE based system

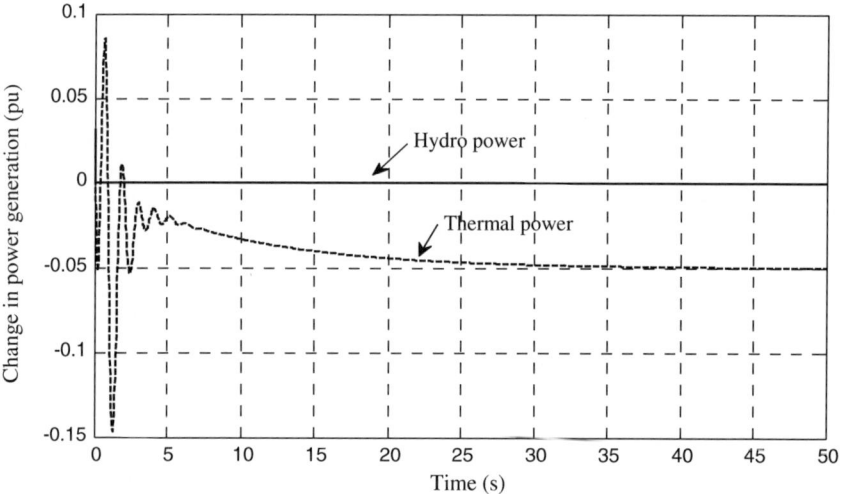

Fig. 11 Transient behavior of changes power generation area 1 using DE based system with hydro generator disconnected

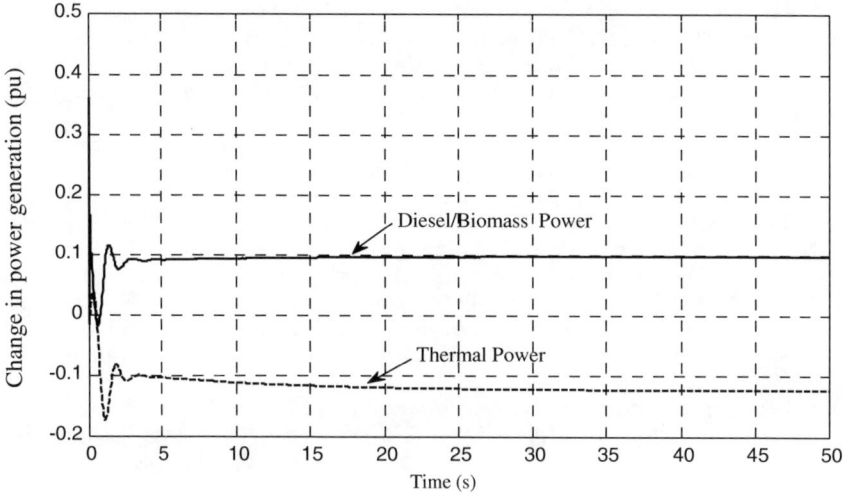

Fig. 12 Transient behavior of changes in power generation in area2 DE based system with hydro generator disconnected and changed area participation factors

4 Conclusion and Future Work

In this work, two area load frequency control is established under deregulation. The PID controllers which are used to bring the system dynamics within comfortable limits is tuned with the help of genetic algorithm and differential evolution techniques. With the variations of load in both areas, the generators participate as per their contracts. Use of renewable power generation in such a case found interesting. The lack of generation due to disconnection of hydro generator is fulfilled by the generators of the other area with changed participation factor i.e. overloading of the same area generator can be avoided.

Acknowledgments This work was supported by Departmental Research Scheme (DRS) of University Grant Commission of Government of India awarded to the Department of Power Engineering, Jadavpur University in the year 2003.

References

1. A.G. Kagiannas, D.T. Askounis, J. Psarras, Power generation planning: from monopoly to competetion. Electr. Power Energy Syst. **26**, 413–421 (2004)
2. J. Sadeh, E. Rakhshani, in *Multi-area Load Frequency Control in a Deregulated Power System Using Optimal Output Feedback Control.* Proceedings of 5th International Conference on European Electricity Market 2008, EEM 2010, 28–30 May 2008, Lisboa, pp. 1–6
3. W. Tan, H. Zhang, M. Yu, Decentralised load frequency controlling a deregulated environment. Electr Power Energy Syst. **41**, 16–26 (2012)

4. H. Sadat, *Power System Analysis* (Tata McGraw Hill, NewDelhi, 2002)
5. V. Donde, M.A. Pai, L.A. Hiskens, Simulation and optimization in an AGC system after deregulation. IEEE Trans. Power Syst. **16**(3), 481–489 (2001)
6. F. Liu, Y.H. Song, J. Ma, S. Mei, Q. Lu, Optimal load-frequency control in restructured power systems. IEE Proc. Gen. Trans. Distrib. **150**(1), 87–95 (2003)
7. E. Rakhshani, J. Sadeh, in *Simulation of Two-area AGC System in a Competitive Environment Using Reduced-Order Observer Method*. Proceedings of 5th International Conference on European Electricity Market 2008, EEM 2010, Lisboa, 28–30 May 2008, pp. 1–6
8. G. Konar, K.K. Mandal, N. Chakraborty, in *Two Area Load Frequency Control Using GA Tuned PID Controller in Deregulated Environment*. Lecture Notes in Engineering and Computer Science: Proceedings of International MultiConference of Engineers and Scientists 2014, IMECS 2014, Hong Kong, 12–14 Mar 2014, pp. 752–757
9. J. Nanda, S. Mishra, L.C. Saikia, Maiden application of bacterial foraging based optimization technique in multi-area automatic generation control. IEEE Trans. Power Syst. **24**(2), 602–609 (2009)
10. S. Debbarma, L.C. Saikia, in *Bacterial Foraging Based FOPID Controller in AGC of an Interconnected Two-Area Reheat Thermal System Under Deregulated Environment*. International Conference On Advances In Engineering, Science And Management (ICAESM-2012), India, 30–31 March 2012, pp. 303–308
11. E.S. Ali, S.M. Abd-Elazim, BFOA based design of PID controller for two area load frequency with nonlinearities. Electr. Power Energy Syst. **51**, 224–231 (2013)
12. R.K. Sahu, S. Panda, U.K. Rout, DE optimized parallel 2-DOF PID controller for load frequency control of power system with governor dead-band nonlinearity. Electr. Power Energy Syst. **49**, 19–33 (2013)
13. A.M. Jadhav, K. Vadirajacharya, Performance verification of PID controller in an interconnected power system using particle swarm optimization. Energy Procedia **14**, 2075–2080 (2012)
14. D.P. Kothari, I.J. Nagrath, *Power System Engineering*, 2nd edn. (Tata McGraw Hill, New Delhi, 2008)
15. K. Ogata, *Modern Control Engineering*, 5th edn. (Pearson Education Limited, New Delhi, 2010)
16. D.P. Kothari, J.S. Dhillon, *Power System Optimization*, 2nd edn. (PHI Learning Private Limited, New Delhi, 2011)
17. An introduction to differential evolution, Kelly Fleetwood. Available at http://www.maths.uq.edu.au/MASCOS/Multi-Agent04/Fleetwood.pdf

Selection of Proper Activation Functions in Back-Propagation Neural Network Algorithm for Transformer and Transmission System Protection

Atthapol Ngaopitakkul and Sulee Bunjongjit

Abstract This paper presents an analysis on the selection of an appropriate activation function used in neural networks for fault diagnosis decision algorithm in transformer and transmission line protection scheme. A decision algorithm based on a combination of discrete wavelet transform (DWT) and back-propagation neural networks (BPNN) is developed. The discrete wavelet transform is employed for extracting the high frequency component contained in the fault signals. The training process for the neural network and fault diagnosis decision are implemented using toolboxes on MATLAB/Simulink. The activation functions in each hidden layers and output layer have been varied to find out and to select the best activation function for fault diagnosis decision algorithm. It is found that the use of Hyperbolic tangent-function for the hidden layers, and linear activation function for the output layer gives the most satisfactory accuracy in these particular case studies.

Keywords Activation functions · Back-propagation neural network · Fault · Protection · Transmission system · Transformer · Wavelet transforms

1 Introduction

Power System normally contains a large number of transmission lines, busbars, generations, transformers, and protective devices. Occurrence of a fault in the power system network may endanger the operation of many power system and

A. Ngaopitakkul (✉)
Faculty of Engineering, King Mongkut's Institute of Technology Ladkrabang,
Chalongkrung Road, Ladkrabang, Bangkok 10520, Thailand
e-mail: knatthap@kmitl.ac.th

S. Bunjongjit
Faculty of Engineering, Rajamangala University of Technology Rattanakosin,
Salaya Phutthamonthon, Nakhon Pathom 73170, Thailand
e-mail: kbsulee@yahoo.com

© Springer Science+Business Media Dordrecht 2015
G.-C. Yang et al. (eds.), *Transactions on Engineering Technologies*,
DOI 10.1007/978-94-017-9588-3_22

279

potentially lead to charges outages. Fault diagnosis involves identification of the fault location and fault type (sometimes call "fault classification"). It may not be possible for a fault diagnosis system to identify the exact fault location and fault type when the fault event is complex. This can happen when the protective devices malfunction such as wrong tripping zone, failure to trip, misoperation, improper operating time, wrong direction of fault, and etc. During the course of recent years, the development of fault diagnosis techniques for the protection scheme in power system has been progressed with the several technique and algorithms [1–15]. Many research reports have been developed to prevent unnecessary operation of the protective equipment under different non-fault conditions.

In [5], the variation of maximum coefficients from discrete wavelet transform (DWT) extracted from post fault current signals, can be used as an input for the training process of an artificial neural network in a decision algorithm to classify fault types in electrical transmission system. Although this decision algorithm can give more satisfactory results, but the effects of loop structure of the transmission network have not been yet taken into account. The development and hardware implementation of the transient recognition scheme were presented in [7] by using wavelet transforms for the extraction of features, and the HMM was used for distinguishing the transients originating from faults from other types of transients. In [8], a transient signal analysis with discrete wavelet transform (DWT) was based on a ratio index quantified in a certain window of analysis for identifying and correctly differentiating the inrush current from incipient internal faults. The ratio index was defined as the relationship between the maximum coefficient from the first detail of the DWT decomposition and the spectral energy of the other frequency components presented in the same detail.

Artificial neural networks (ANNs) techniques have been proposed in some approaches in the literature to improve protective relay due that these algorithms can give precise results. Back-propagation neural network (BPNN) is a kind of neural networks, which is widely applied today owing to its effectiveness to solve almost all types of problem. Therefore, a decision algorithm, used for fault diagnosis in the power system to decrease complexity and duration of maintenance time, is required. However, the activation function is a key factor in the artificial neural network structure. Back-propagation neural networks support a wide range of activation functions such as sigmoid function, linear function, and etc. In previous research works [16], by considering the fault classification in underground cable system, the comparison of the activation functions is considered while the activation functions have been varied to identified the fault diagnosis in power system [17]. The result shown that the choice of activation function can change the behavior of the BPNN considerably. There is no theoretical reason for selecting a proper activation function. Hence, the objective of this paper is to investigate an appropriate activation function for the fault diagnosis decision algorithm. The activation functions in each hidden layers and output layer are varied, and the results obtained from the decision algorithm are studied.

The decision algorithm is a part of a transformer and transmission line protective scheme proposed in this paper. It is noted that the discrete wavelet transform is employed for extracting the high frequency component contained in the fault signals. The construction of the decision algorithm is detailed and implemented with various case studies based on Thailand electricity transmission and distribution systems.

2 Simulation

Artificial neural network requires fault signal samples from simulations to training and test processes but fault signal in power system hardly occurs, so various fault signals pattern will be obtained from simulation. The ATP/EMTP program is employed in simulating the transients of fault signals, at a sampling rate of 200 kHz.

2.1 Internal Fault in Transformer Windings

A 50 MVA, 115/23 kV two-winding three-phase transformer was employed in simulations with all parameters and configuration provided by a manufacturer [18]. The single line diagram of the simulated system is shown in Fig. 1. In this study, the scheme under investigations is a part of Thailand electricity transmission and distribution system. It can be seen that the transformer as a step down transformer with a star-star connection, is connected between two subtransmission sections. To implement the transformer model, simulations were performed with various changes in system parameters as follows:

- The angles on phase A voltage waveform for the instants of fault inception were 0°–330° (each step is 30°).
- Internal faults type at the transformer windings (both primary and secondary) which is winding to ground fault was investigated.
- The fault position designated on any phases of the transformer windings (both primary and secondary) was varied at the length of 10, 20, 30, 40, 50, 60, 70, 80 and 90 % measured from the line end of the windings.
- Fault resistance was 5 Ω.

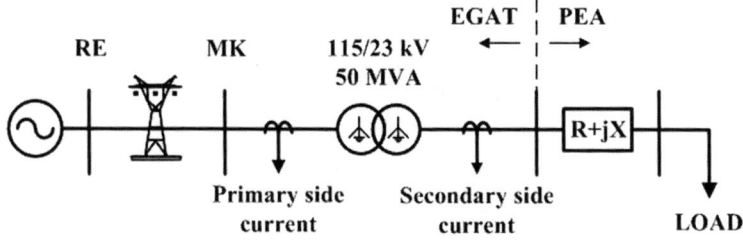

Fig. 1 The system used in simulations studies [17]

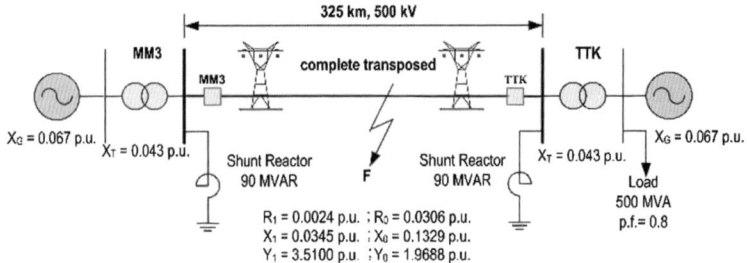

Fig. 2 The transmission system used in simulations studies [17]

2.2 Fault in Transmission Line

The fault in transmission system is chosen based on the Thailand's transmission system as shown in Fig. 2. Transmission line with frequency dependent parameters can be calculated by supporting routing line cable constants (LCC) in ATP/EMTP. The LCC model is based on the geometrical and material data for an overhead line including the corresponding electrical data. Fault patterns in the simulation are performed with following changes of the system parameters:

- Fault types under consideration, namely: single phase to ground (SLG: AG, BG, CG), double-line to ground (DLG: ABG, BCG, CAG), line to line (L-L: AB, BC, CA) and three-phase fault (3-P: ABC).
- Fault locations on each transmission line were at the distance of 10, 20, 30, 40, 50, 60, 70, 80 and 90 %, measured from the sending end.
- Inception angle on a voltage waveform was varied between 0° and 330°, with the increasing step of 30°. Phase A was used as a reference.
- The fault resistance is 10 Ω.

3 Fault Diagnosis Decision Algorithm and Results

The fault signals generated using ATP/EMTP were interfaced with MATLAB/ Simulink for a construction of fault diagnosis process. The fault diagnosis decision algorithm can be divided into two processes. For the first process, the fault detection decision algorithm must be detected using the discrete wavelet transform (DWT) to classify between fault condition and normal condition. For the next process, after fault condition is detected, the fault classification decision algorithm will be classifying the fault type using the back-propagation neural networks (BPNN).

3.1 Fault Detection Using DWT

After simulating the fault signals, fault detection decision algorithm [9, 18] is processed using the positive sequence current signals. The clark transformation matrix was used to calculate the positive sequence and the zero sequence of currents. The obtained current signals are imported to extract the high frequency transient components to several scales using mother wavelet daubechies4 (db4) in the wavelet toolbox. After applying the DWT to the positive sequence currents, the comparison of the coefficients from each scale is considered. Coefficients obtained using DWT of signals are squared so that the abrupt change in the spectra can be clearly found. This sudden change is used as an index for the occurrence of faults. By performing many simulations, it has been found that the coefficient in scale 1 (50–100 kHz) from DWT seems enough to indicate the fault condition, so it is unnecessary to use other coefficients from higher scales in this algorithm.

3.2 Fault Classification Using BPNN

After the fault detection process, the coefficients detail of scale 1, which is obtained using the DWT, is used for training and test processes of the back-propagation (BPNN). In this paper, a structure of a BPNN consists of three layers which are an input layer, two hidden layers, and an output layer as shown in Fig. 3. Each layer is connected with weights and biases. In addition, the activation function is a key factor in the BPNN structure. The choice of activation function can change the behavior of the BPNN considerably. Hence, the activation functions in each hidden layers and output layer are varied as illustrated in Table 1 in order to select the best activation function. A training process was performed using neural network toolboxes in MATLAB which can be summarized in Fig. 4.

By observing Fig. 3, before the fault classification decision algorithm process, a structure of the BPNN consists of four inputs as illustrated in Fig. 5. The maximum coefficients detail (phase A, B, C and zero sequence of post-fault current) of DWT

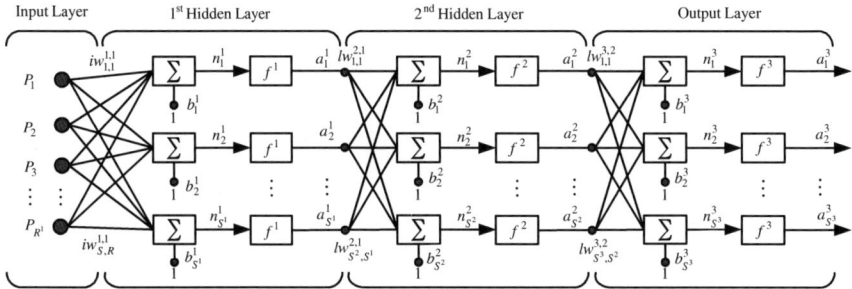

Fig. 3 Example of an ATP/EMTP simulated fault signal for an AG fault

Table 1 Activation functions in all hidden layers and output layers for training neural networks

Activation function in		
First hidden layer	Second hidden layer	Output layer
Hyperbolic tangent sigmoid function	Logistic sigmoid function	Linear function
		Logistic sigmoid
		Hyperbolic tangent sigmoid
	Hyperbolic tangent sigmoid function	Linear function
		Logistic sigmoid
		Hyperbolic tangent sigmoid
Logistic sigmoid function	Logistic sigmoid function	Linear function
		Logistic sigmoid
		Hyperbolic tangent sigmoid
	Hyperbolic tangent sigmoid function	Linear function
		Logistic sigmoid
		Hyperbolic tangent sigmoid

at the first peak time that can detect fault, is performed as input variables. The output variables of the BPNN for identifying fault types are designated as either 0 or 1, corresponding to phases A, B, C and ground (G). If each output value of BPNN is less than 0.5, no fault occurs on each phase; conversely, if this output value of BPNN is more than 0.5, a fault does occur.

During the training process stage 1, the initial number of neurons in each hidden layer are 2 neurons for the first hidden layer and 1 neuron for the second hidden layer, respectively. Each neuron in input layer broadcasts input signal to each neuron in first hidden layer and the weight and biases were random. Each neuron in first hidden layer is interconnected by weights to compute input signal in first hidden layer. It computed its activation and sent its output to second hidden layer. Next, each neuron in the second hidden layer computed its activation and sent its output to output layer. Finally, each neuron in output layer computed its activation to the output (response) of BPNN for given input pattern [9, 16, 18].

For stage 2 of training process, mean absolute percentage error (MAPE) used as an index for efficiency determination of the BPNN is computed. The error between output of BPNN and its target output was calculated. Error is used to distribute the information on the error and feedback to all neurons in the previous layer. It is also used (later) to adjustment of weights and biases. It is not necessary to propagate the error back to the input layer, but error is used to adjust weights and biases between the first hidden layer and the input layer.

For stage 3 of training process, all errors have been determined, the weights and bias for all layers are adjusted simultaneously. The adjustment to the weight and bias will update layer by layer (from output layer to input layer). Its weights and bias

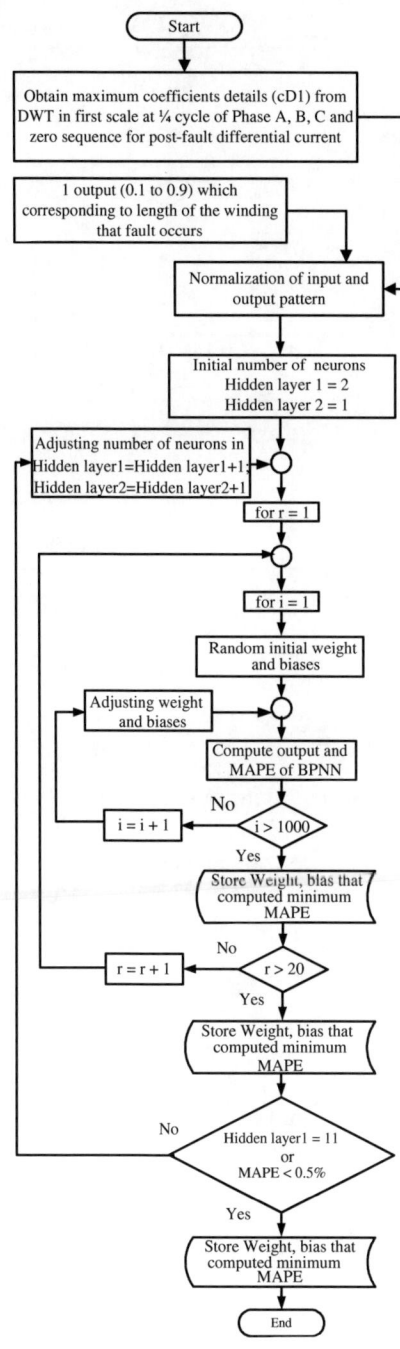

Fig. 4 Flowchart for the training process

Fig. 5 Wavelet transform
from scale 1 to 5 for the zero
sequence of the current
signals

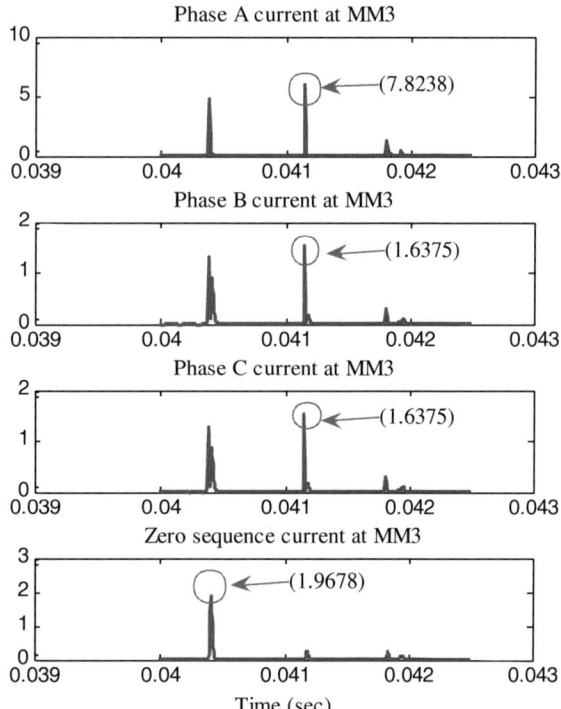

adjustment try to match its output to the target output, adjusted by using the Leven-
berg–Marquardt. Next, input values are propagated again, to calculate new output
(response) of BPNN and compare with target output. This process was repeated for
20,000 iterations in order to compute the best value of MAPE or until the goal error is
reached (the MAPE of test sets was less than 0.5 %).

4 Results

After the training process, case studies were varied so that the algorithm capability
can be verified. The system under consideration is shown in Figs. 1 and 2 for
transformer protection and transmission line protection respectively.

4.1 Internal Fault Type

After training process, the decision algorithm is employed to identify the internal
fault types in the transformer winding. The total number of the case studies is 810.

Fig. 6 Comparison of MAPE of fault type for various activation functions between each transformer winding

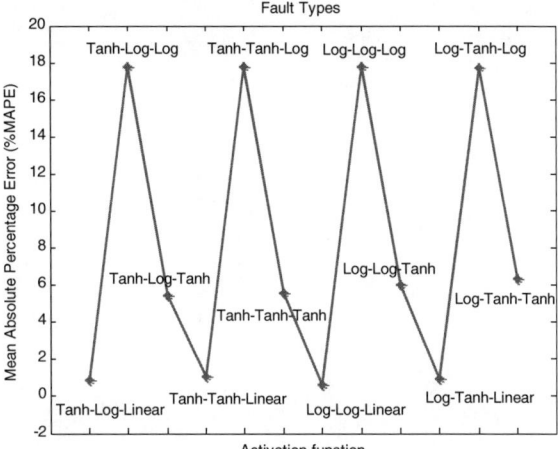

Case studies are performed with various types of fault at each position in the transformer including the variation of fault inception angles. The result obtained from various activation functions of case studies both high voltage and low voltage winding is shown in Fig. 6. From Fig. 6, it can be seen that there are four cases of activation functions with average error less than 5 % as follows:

1. Hyperbolic tangent—Logistic—Linear.
2. Hyperbolic tangent—Hyperbolic tangent—Linear.
3. Logistic—Logistic—Linear.
4. Logistic—Hyperbolic tangent—Linear.

The average accuracy of various types of internal fault in each phase of high voltage winding and low voltage winding is shown in Fig. 7. It can be seen that Hyperbolic tangent—Hyperbolic tangent—Linear as activation function in each layer, is tested with various fault types on both high voltage and low voltage windings of the three-phase transformer. The accuracy of internal fault type from the classification of the decision algorithm is highly satisfactory.

4.2 Transmission Line Fault Type

After training process, the decision algorithm is employed to identify the fault types in the transmission line. Case studies were performed with various fault types at each location along the transmission lines and varying fault inception angles and

Fig. 7 Comparison of
average accuracy when
identifying the types of
internal fault among various
activation functions. **a** High
voltage, **b** low voltage

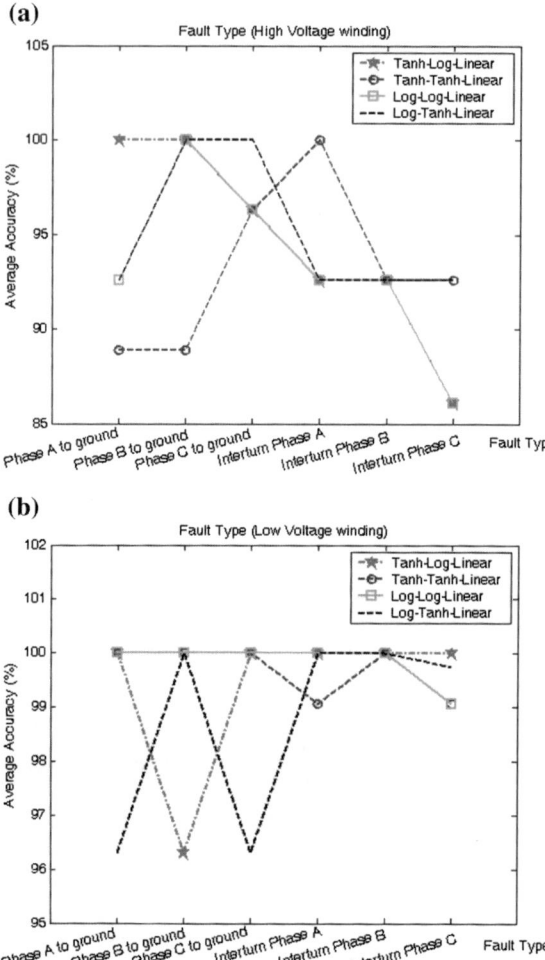

locations along each transmission line, as shown in Fig. 8. By observing Fig. 8, the
comparison between an average accuracy in fault identifications obtained from the
various activation functions is shown.

From Fig. 8, it is shown that when the case studies are tested with various
locations at each transmission line, the accuracy of fault type from the BPNN
algorithm is highly satisfactory. In addition, the proposed algorithm Hyperbolic
tangent—Hyperbolic tangent—Linear as activation function in each layer can give
a better performance in predicting the fault types.

Fig. 8 Comparison of average accuracy when identifying the fault type at various lengths of the transmission line among various activation functions. **a** Hyperbolic in first hidden layer, **b** logistic in first hidden layer

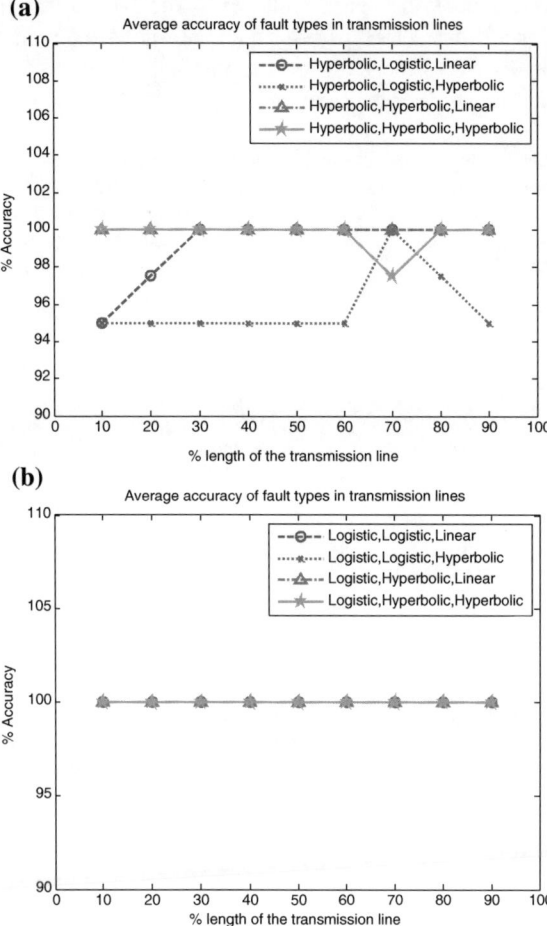

(a)

Average accuracy of fault types in transmission lines

Legend:
- – –⊖– – Hyperbolic,Logistic,Linear
- ·····◄···· Hyperbolic,Logistic,Hyperbolic
- –·△·– Hyperbolic,Hyperbolic,Linear
- ——✳—— Hyperbolic,Hyperbolic,Hyperbolic

Y-axis: % Accuracy
X-axis: % length of the transmission line

(b)

Average accuracy of fault types in transmission lines

Legend:
- – –⊖– – Logistic,Logistic,Linear
- ·····◄···· Logistic,Logistic,Hyperbolic
- –·△·– Logistic,Hyperbolic,Linear
- ——✳—— Logistic,Hyperbolic,Hyperbolic

Y-axis: % Accuracy
X-axis: % length of the transmission line

5 Conclusion

In this paper, study of an appropriate activation function for the decision algorithm used in the transformer and trans-mission line protection has been discussed. The fault diagnosis decision algorithm can be discriminated between faulty condition and normal condition using the discrete wavelet transform (DWT). After fault condition can be detected, the fault classification decision algorithm will classify the fault type and fault location using the back-propagation neural networks (BPNN). The maximum coefficient from the first scale at ¼ cycle of phase A, B, and C of post-fault current signals and zero sequence current obtained by the DWT have been used as an input variable in a decision algorithm. The activation functions in each hidden layer and output layer have been varied, and the results obtained from the decision algorithm have been investigated with the variation of fault inception angles, fault types, and fault locations. The results have illustrated that the use of Hyperbolic tangent sigmoid function in the first and the second layers with Linear function in the output layer is the most appropriate scheme for the transformer and transmission line protection scheme.

Acknowledgment The authors wish to gratefully acknowledge financial support for this research from the King Mongkut's Institute of Technology Ladkrabang Research fund, Thailand. The authors would like also to thank for partially supported by the Faculty of Engineering, Rajamangala University of Technology Rattanakosin Research fund.

References

1. O.A.S. Youssef, Discrimination between faults and magnetizing inrush currents in transformers based on wavelet transforms. Electr. Power Syst. Res. **63**(2), 87–94
2. M.E. Hamedani Golshan, M. Saghaian-nejad, A. Saha, H. Samet, A new method for recognizing internal faults from inrush current conditions in digital differential protection of power transformers. Electr. Power Syst. Res. **71**(1), 61–71
3. A.A. Yusuff, A.A. Jimoh, J.L. Munda, Determinant-based feature extraction for fault detection and classification for power transmission lines. IET Gener. Transm. Distrib. **5**(12), 1259–1267
4. F.E. Perez, R. Aguilar, E. Orduna, G. Guidi, High-speed non-unit transmission line protection using single-phase measurements and an adaptive wavelet: zone detection and fault classification. IET Gener. Transm. Distrib. **6**(7), 593–604
5. P. Chiradeja, A. Ngaopitakkul, Identification of fault types for single circuit transmission line using discrete wavelet transform and artificial neural networks, in *Proceedings of the International Multiconference of Engineers and Computer Scientists 2009, IMECS2009*, 18–20 Mar 2009, Hong Kong. Lecture Notes in Engineering and Computer Science, pp. 1520–1525
6. D. Barbosa, U. Chemin Netto, D.V. Coury, M. Oleskovicz, Power transformer differential protection based on clarke's transform and fuzzy systems. IEEE Trans. Power Deliv. **26**(2), 1212–1220
7. N. Perera, A.D. Rajapakse, Development and hardware implementation of a fault transients recognition system. IEEE Trans. Power Deliv. **27**(1), 40–412
8. S. Seyedtabaii, Improvement in the performance of neural network-based power transmission line fault classifiers. IET Gener. Transm. Distrib. **6**(8), 731–737
9. A. Ngaopitakkul, C. Jettanasen, Combination of discrete wavelet transform and probabilistic neural network algorithm for detecting fault location on transmission system. Int. J. Innovative Comput. Inf. Control **7**(4), 1861–1874
10. J. Upendar, C.P. Gupta, G.K. Singh, Fault classification scheme based on the adaptive resonance theory neural network for protection of transmission lines. Electr. Power Compon. Syst. **38**(4), 424–444
11. A. Abu-Siada, S. Islam, A novel online technique to detect power transformer winding faults. IEEE Trans. Power Deliv. **27**(2), 849–857
12. Z. Gajic, Use of standard 87T differential protection for special three-phase power transformers part I: theory. IEEE Trans. Power Deliv. **27**(3), 1035–1040
13. M. Moscoso, S. Hosseini, G.J. Lloyd, K. Liu, Operation and design of a protection relay for transformer condition monitoring, in *Proceedings of 11th International Conference on Developments in Power System Protection, DPSP2012*, pp. 1–6
14. A. Ashrafian, M. Rostami, G.B. Gharehpetian, Hyperbolic S-transform-based method for classification of external faults, incipient faults, inrush currents and internal faults in power transformers. IET Gener. Transm. Distrib. **6**(10), 940–950

15. O.A.S. Youssef, A wavelet-base technique for discrimination between faults and magnetizing inrush currents in transformers. IEEE Trans. Power Deliv. **18**(1), 170–176

16. A. Ngaopitakkul, C. Pothisarn, Selection of proper activation functions in back-propagation neural networks algorithm for fault classification in underground cable. IAENG Trans. Eng. Technol. **7**(1), 308–319

17. N. Suttisinthong, B. Seewirote, A. Ngaopitakkul, C. Pothisarn, Selection of proper activation functions in back-propagation neural network algorithm for single-circuit transmission line, in *Proceedings of the International MultiConference of Engineers and Computer Scientists 2014, IMECS2014*, 12–14 Mar 2014, Hong Kong. Lecture Notes in Engineering and Computer Science, pp. 758–762

18. A. Ngaopitakkul, A. Kunakorn, Internal fault classification in transformer windings using combination of discrete wavelet transforms and back-propagation neural networks. Int. J. Control Autom. Syst. (IJCAS) **4**(3), 365–371

EOS Protection for USB2 Transceiver

Suhas Vishwasrao Shinde

Abstract This paper presents CMOS integrated electrical over stress (EOS) protection circuit for USB2 transceiver. A unique full speed SE0 state in USB2 standard protocol produces overshoot and undershoots in presence of an on board choke. The purpose of choke is to minimize EMI for USB2 differential signaling. This works fine as long as data signaling is differential. In the presence of SE0 state, single transition occurs on either Dp or Dn pad causing high inductive kick back from choke resulting in excessive overshoot and undershoots. The electrical over stress on cascode devices leads to device degradation resulting in violation of USB2 output driver impedance spec of 45 Ω ±5 %. This in turn increases 'defects per million' count over time. To mitigate this reliability issue, EOS protection circuit is proposed. This circuit is designed in Intel-22 nm process and evaluated by computer simulations across all PVT conditions. Proposed EOS protection circuit reduces respective overshoot and undershoot of 4.2 V and −0.6 to 3.7 V and −0.34 V. For the growing EOS concern of deep submicron devices in 14, 10 nm and future technologies, this design scheme becomes an attractive choice.

Keywords CMOS integrated circuits · EOS · EOP · Mealy state m/c · SE0 state · USB2

1 Introduction

An USB2 standard was developed ∼20 years prior without much consideration about future manufacturing processes. The USB2 low/full speed signaling is 3.3 V hence 3.3 V supply is commonly used for transceiver circuits implementing low/full speed as well as high speed signaling modes. The advanced manufacturing processes such as 22, 14 and 10 nm uses devices with reduced threshold voltages and

S.V. Shinde (✉)
Intel Mobile Communications, GmbH, Am Campeon 10-12, 85579 Neubiberg, Germany
e-mail: suhas.v.shinde@intel.com

© Springer Science+Business Media Dordrecht 2015   293
G.-C. Yang et al. (eds.), *Transactions on Engineering Technologies*,
DOI 10.1007/978-94-017-9588-3_23

electrical stress limits. Circuit implementation in these technologies with 3.3 V supply is almost impossible unless process offers a flavor of high voltage devices such as 1.8 V with higher electrical stress limit. However, this contradicts the trend of advanced process technology where devices are scaled down. Their threshold voltages and stress limits are reduced. This necessitates some innovations in process as well as in design to mitigate EOS issue. Due to cost reason Intel 22 nm process does not support 3.3 V devices so output driver is always designed with stacked 1.8 V devices to minimize electrical over stress. However large overshoot/undershoot may violate stress limits of these devices hence alternative approach in circuit design was an urgency to deal with this EOS issue. This motivates this work on developing novel EOS protection circuit for USB2 transceiver in Intel 22 nm process [1]. This scheme is based on charge injection and drain from pad subjected to undershoot and overshoot respectively. This helps neutralizing the pad partially to reduce overshoot and undershoot. It uses scheme of detection of SE0 condition and single transition on either Dp or Dn pad and accordingly inject or drain charge from pad to reduce overshoot and undershoot. Although this paper discusses EOS issue in reference to USB2, it can be used for USB3 and other standards where such OS/US scenario occurs.

This paper [2] is divided into six sections. Section 2 describes SE0 signaling protocol in USB2 standard. Section 3 explains EOS issue specifically to USB2 application. Section 4 explains principle of suppressing overshoot and undershoots. Section 5 discusses detailed circuit implementation. Section 6 gives simulation results validating proposed EOS protection circuit. Finally, some concluding remarks in conclusion section.

2 USB2-SE0 Protocol

The Single Ended Zero (SE0) state in USB2 standard communication protocol is used to indicate end-of-packet (EOP). This EOP is signalled by driving Dp and Dn pads to logical low state for two bit times followed by driving the lines to the J state for one bit time. USB2 defines two states of differential signaling, J state and K state. The J state corresponds to logical high and logical low driven on pads Dp and Dn respectively. On the contrary, K state is defined as logical low and logical high driven on Dp and Dn pads respectively. The transition from SE0 state to the J state defines the end of data packet at the receiver. The J state is asserted for one bit time and then both the Dp and Dn output drivers are placed in high impedance state. All USB2 devices attach to USB through ports on specialized USB devices known as hubs. The width of EOP is twice the bit duration and can be as low as one bit duration. USB2 has three transfer rates, high speed is 480 Mbps, full speed is 12 Mbps and low speed is 1.5 Mbps. From these transfer rates we find bit duration for full speed to be 83 ns and low speed 666 ns. Considering timing variations due to differential buffer delay, rise and fall time mismatches, noise and other random effects, EOP width can be between 160 and 175 ns for full speed and between 1.25

(a)

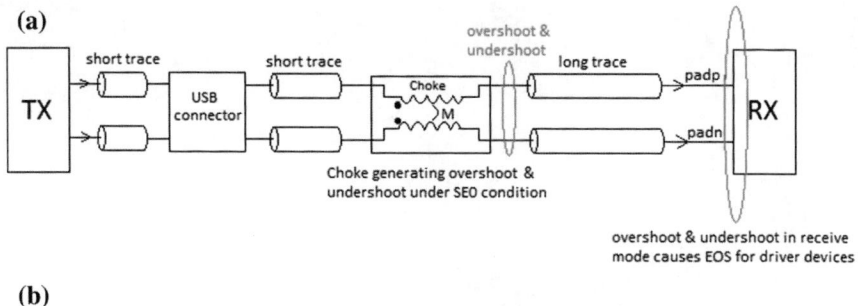

(b)

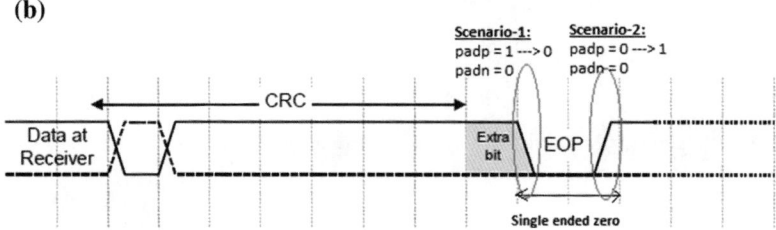

Fig. 1 **a** Application platform of USB2 and **b** SE0, EOP condition with highlighted scenario-1 and scenario-2

and 1.5 µs for low speed. As shown in Fig. 1b it can be one bit duration where it may reduce as little as 82 ns for full speed and 670 ns for low speed. Our point of interest here is EOP since differential transitions are missing. In addition we will be focusing only on full speed mode since it will have 3.3 V signaling with rise and fall times as low as 4 ns. The next section will make this intent clear.

3 An EOS Issue

One of the techniques to deal with EMI issue in serial interfaces is using mutually-coupled inductor choke. Choke placement is closer to the output of transmitter following USB2 connector as shown in Fig. 1a. The differential signaling through choke is adjusted for 50 % cross-over minimizing common mode noise. This in turn reduces common mode noise induced electromagnetic interference. However, when single ended transition occurs on differential lines as shown in Fig. 1b, the same choke generates overshoot and undershoots due to inductive kick back. The effect is severe for high slew rates. In USB2 standard communication protocol single edge transition occurs in SE0 state as shown in Fig. 1b. The SE0 state in high speed does not generate overshoot and undershoot to stress the devices because its small swing signaling, 400 mV. However full speed is 3.3 V signaling with highest possible slew rate of 4 ns hence generates excessive overshoot and undershoots. The low speed is also 3.3 V signaling but fastest slew rate is 20 ns hence overshoot and undershoots are not generated. The generated overshoot and undershoots propagate

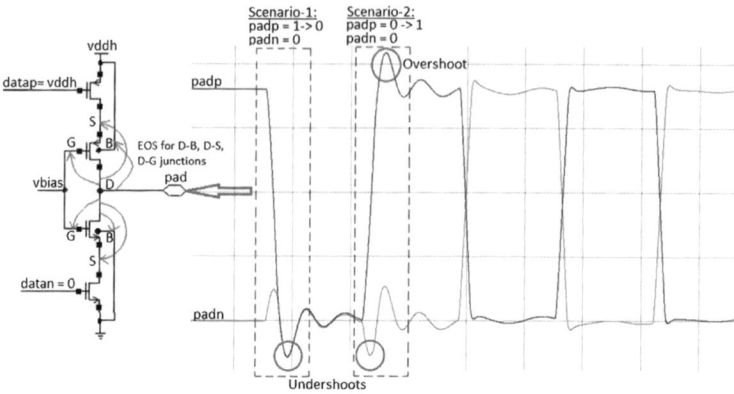

Fig. 2 On RX side, over stressed tri-stated driver due to subjected overshoot and undershoots

down the line to receiver. At receiving end, transmitter is tri-stated and sees these overshoot and undershoots. The drain-bulk, drain-source and drain-gate junction of 1.8 V cascode transistor experience electrical over stress as depicted in Fig. 2. As activity factor increases, the transistor devices see OS/US very often and degrade quite rapidly. This leads to change in device parameters like threshold voltage, transconductance and thereby finally affects the transmitter output impedance. USB2 defines transmitter output impedance specification to be 45 Ω ±5 %. This specification may get violated over time leading to increased defects per million (DPM) number. To mitigate this reliability issue, EOS protection circuit is designed in Intel 22 nm process.

This EOS issue is divided into two scenarios as shown in Fig. 2. Scenario-1 is defined as padp making transition from high to low when padn is at logic low level. This scenario corresponds to undershoot on both padp and padn. Scenario-2 is defined as padp making transition from logic low level to logic high level and padn is still at logic low level. This scenario corresponds to overshoot on padp and undershoot on padn. These two scenarios occur due to tightly coupled two inductors of 200 nH present in choke. Observed overshoot may go as high as 4.2 V and undershoot as low as −0.6 V. These voltage levels are high enough to violate the EOS limits of cascode devices of transmitter. Although not shown in Fig. 2, there could be other structures connected to pads like receiver circuit, ESD protection circuit. These other devices/components connected to pad also undergo electrical stress and may degrade over time. This requires reliability simulations like "Aging" to determine amount of degradation over time and caused defects per million. Removing choke solves this EOS issue but results into increased EMI. Instead other solutions to suppress EMI are not applicable to this USB2 application. For example Spread spectrum clocking for data bit stream may solve both EMI and EOS issue but violates the 500 ppm clocking requirement of USB2.

4 Principle of Overshoot/Undershoot Suppression

This design scheme [1] detects overshoot and undershoot scenarios ahead of time and injects the charge into pad when undershoot occurs and drains the charge from pad when overshoot occurs. The basic idea of charge injection and drain is illustrated in this section. Here we will be considering two different scenarios to address the issue. Scenario-1 is when padp makes transition from logic high to logic low and padn is at logic low level during EOP window. Scenario-2 is when padp makes transition from logic low level to logic high level and padn is at logic low level during EOP window.

Scenario-1: (padp = 1 → 0, padn = 0)

In SE0 state (FS mode) of USB2 protocol padn is low and padp transitions from high to low for every end of packet (EOP). This scenario causes single edge transition and thereby undershoot on padp and padn. Two different low threshold detectors are used to detect signals padp and padn going low. The inverted outputs of these detectors are labeled as padp' and padn'. The OR logic of these two signals results in logic low level when padn = 0 and padp makes transition from high to low (crosses low threshold). The XNOR logic of this signal with its delayed version creates a window for charge injection. The charge injection and drain circuit is enabled in this window injecting a controlled amount of charge into padp and padn to neutralize the pads partially to lower undershoot. The total amount of charge injected can be found by integrating charge density function. This can also be given as charge distribution function with respect to time. The final steady state value of this function is the total amount of charge injected as shown in Fig. 3a. The total amount of charge injected can be controlled by peak current limiter and charge injection window length. The exact nature of charge density is controlled by controlling rise and fall slew at input of current starved devices. This input is exponential in nature and derived from output of RC circuit. The charge injection mechanism is as illustrated in Fig. 4a. The constant current sources limits the peak charge density for controlled power consumption. The input to devices is slew controlled for rising edge to generate the required charge density profile. This rising slew is exponential in nature and generated by RC circuit. Thus injected charge into pads lowers undershoot.

Scenario-2: (padp = 0 → 1, padn = 0)

As shown in Fig. 3b the FSM part of this scheme detects the SE0 condition of USB2 protocol where single edge transition occurs at the beginning of packet. Upon SE0 state detection it asserts start signal high. This start signal enables charge injection and drain circuit. The charge injection and drain circuit detects rising edge on padp and asserts end signal high. The controlled amount of charge is drained from padp and injected in padn lowering the overshoot and undershoot respectively. Once charge injection and drain action is over it asserts end signal low. This in turn causes start signal to go low disabling the charge injection and drain circuit. The total amount of charge injected/drained can be found by integrating charge density

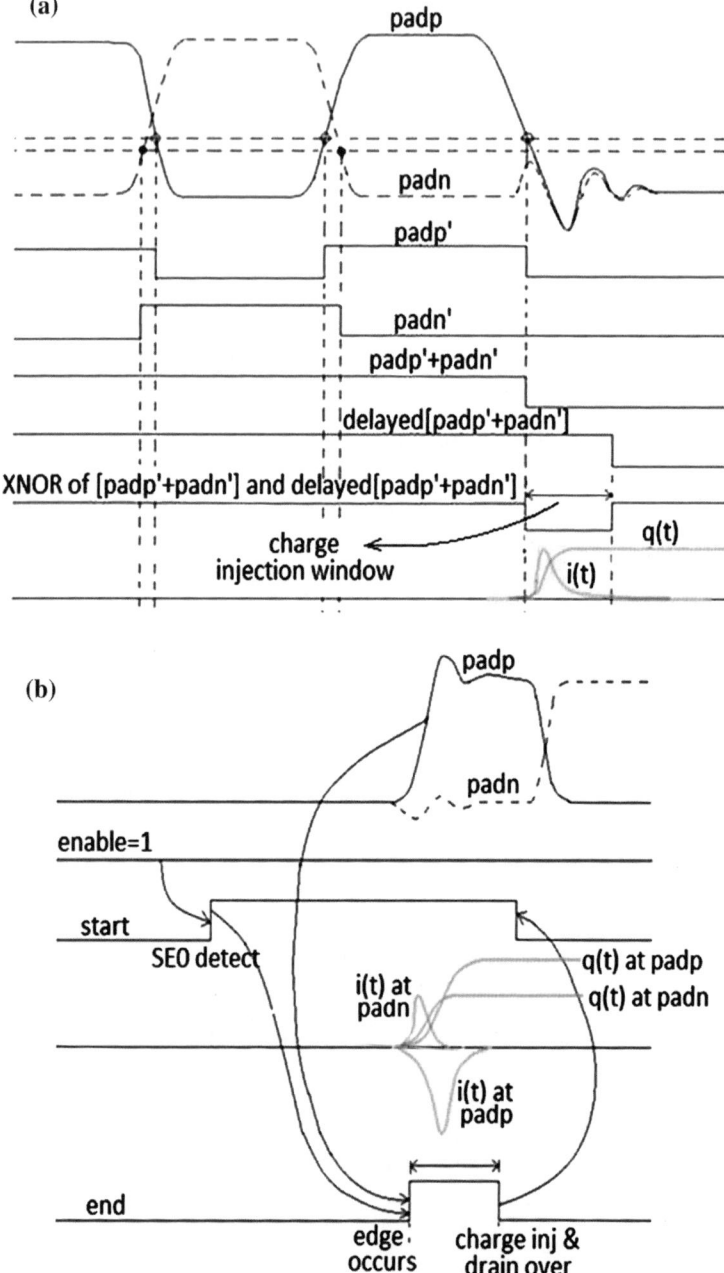

Fig. 3 **a** Scenario-1 solution and **b** scenario-2 solution

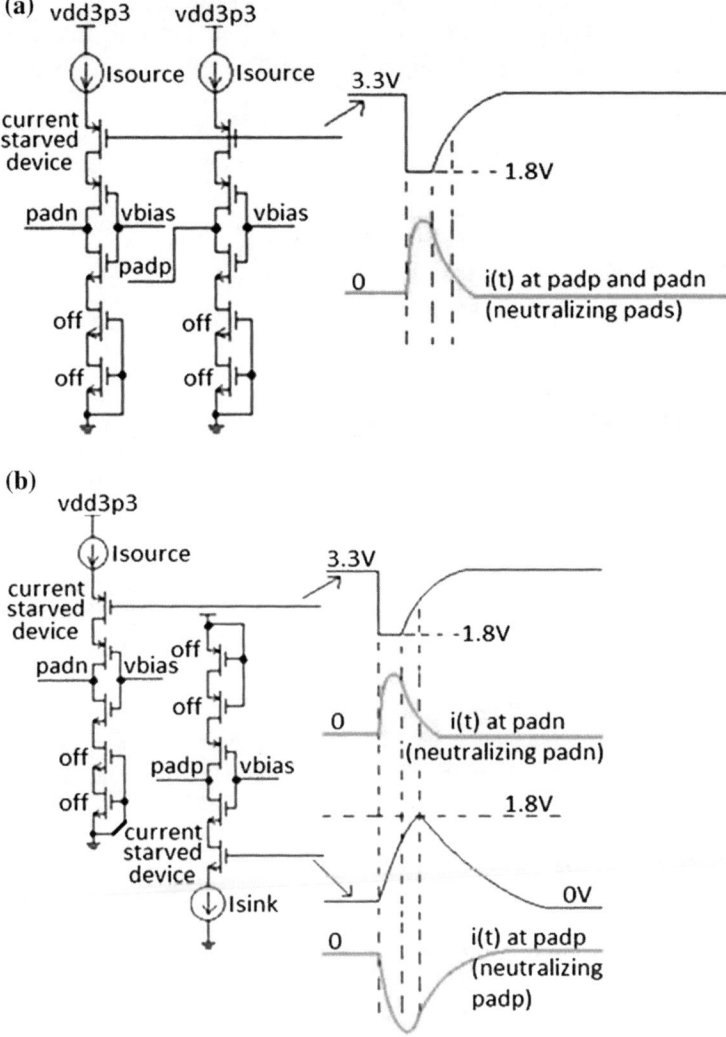

Fig. 4 Implementation of **a** scenario-1 and **b** scenario-2 circuit

function. This can also be given as charge distribution function with respect to time. The final steady state value of this function is the total amount of charge injected/ drained as shown in Fig. 3b. The total amount of charge injected/drained can be controlled by peak current limiter and charge injection window length. The exact nature of charge density is controlled by controlling rise and fall slew at input of current starved devices. This input is exponential in nature and derived from output of RC circuit. The charge injection and drain mechanism is as illustrated in Fig. 4b where Isource and Isink are the constant current source and sink respectively. These sources limit the peak charge density for controlled power consumption. The input signals slew for current starving devices are exponential in nature and are adjusted by RC charging and discharging circuits. The slew timing is designed such a way that required charge injection and drain profile can be generated to partially neutralize the pad.

Under both scenarios the charge density and distribution function can be given by,

$$\text{Charge density function, } i(t) = \frac{dq}{dt} \quad \text{and}$$

$$\text{Charge distribution function, } q(t) = \int\limits_{0}^{t} (t)dt = \int\limits_{0}^{t} dq$$

The total amount of charge injected/drained can be computed as

$$\text{Total Charge injected/drained} = \int\limits_{0}^{\infty} (t)dt = \int\limits_{0}^{\infty} dq$$

Since the controlled amount of charge is injected/drained it results in controlled power consumption. This technique addresses the issue exactly where it is needed (i.e. circuit not active on every edge) hence it provides optimum power solution.

It should be noted that the conventional solutions like clamping driver output to deal with overshoot and undershoot do not work due to more complex transmission line load. Any sudden clamp or on-off of devices leads to more ringing and much severe overshoot and undershoot. For this reason slow ON/OFF of devices with controlled amount of charge injection and drain is necessary to reduce overshoot and undershoot.

5 EOS Protection Circuit

The top level block diagram of EOS protection circuit is shown in Fig. 5a. It contains charge injection/drain block and finite state machine block.

Further charge injection/drain block is expanded in Fig. 5b where it contains sub-blocks, edge detection, logic circuit, charge injection/drain legs, RC based slew control and peak current limiter. The finite state machine block is sub divided into SE0 detector, digital glitch filter and Mealy type finite state machine as shown in Fig. 5c. On the top hierarchy, padp and padn signals go to charge injection/drain block and clock and enable signals act as input to finite state machine block. The output "end" from charge injection/drain block goes as input to finite state machine block and finite state machine output "start" feeds back to charge injection/drain block. The detailed circuit schematic is as shown in Fig. 7. The circuit operation is as explained below in reference to two different scenarios.

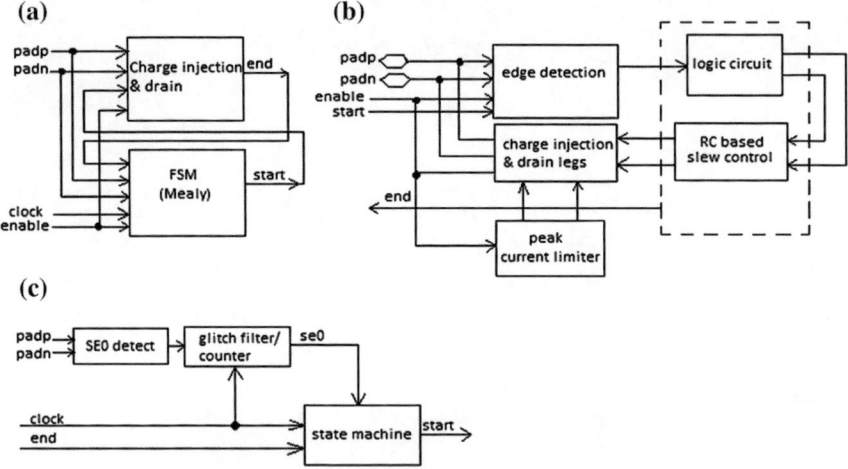

Fig. 5 **a** Top level block diagram of EOS protection circuit, **b** block diagram of charge injection/drain circuit, and **c** SE0 detect and finite state machine block diagram

Scenario-1: (padp = 1 → 0, padn = 0)

In this scenario padn signal is low and padp signal transitioning from high to low is detected by low threshold detectors. The low threshold detectors are basic inverters tuned to threshold values which are very low. The detector on padp has threshold of Vt1 and detector on padn has threshold value of Vt2. Here Vt2 is designed to be less than Vt1. The low threshold is achieved by making PMOS device weaker than NMOS device. Vt2 < Vt1 is ensured by making PMOS of detector-2 much weaker than PMOS of detector-1 and NMOS of detector-2 much stronger than NMOS of detector-1. Output of these detectors go low when signal goes above these threshold values. padp' and padn' are the inverted output of these signals as shown in Fig. 3a. The OR logic of these signals padp' and padn' goes low when padn is low and padp also goes low (crosses Vt1). XNOR logic of this signal with its delayed version gives output low. This is the window where charge injection and drain circuit is enabled and charge injection into padp and padn is done to partially neutralize the pads. This in turn lowers undershoot. Here very precise detection of threshold values (∼400 mV) is not required and hence low power inverter based detectors are used.

Scenario-2: (padp = 0 → 1, padn = 0)

The FSM part of this scheme detects the SE0 state in USB2 protocol by detecting padp = padn = 0 and asserts start signal high enabling charge injection and drain circuit. Charge injection and drain circuit detects the rising edge and creates a required window to inject/drain the charge in to/from padn/padp. This window is defined by end signal which goes high when lower threshold of edge is detected and goes low when charge injection and drain action is over. Charge injection and drain circuit asserts end signal low after injecting and draining the charge. This signal

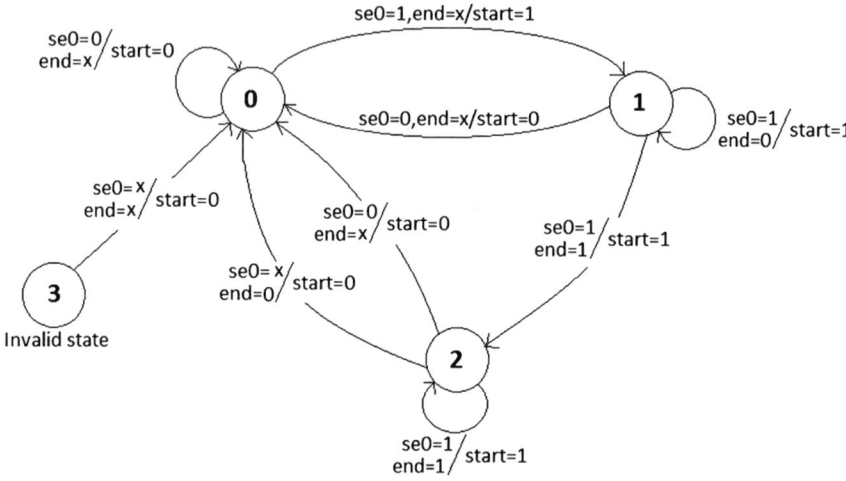

Fig. 6 State diagram of mealy type finite state machine

goes as input to FSM which in turn de-asserts the start signal to charge injection circuit disabling it. The timing diagram is as shown in Fig. 3b. The state machine design is based on state diagram described in Fig. 6. This state machine has two inputs, 'se0' and 'end' and total 4-states with one invalid state. The output of state machine, 'start' depends upon present state and present inputs (Mealy type) [3]. There is a possibility of state machine entering into invalid state due to glitch/noise or power ramp up. Under such circumstances the state machine brings itself back to valid state from invalid state as shown in Fig. 6. This state machine asserts start signal high only when it detects SE0 state and de-asserts start signal when end signal goes low after charge injection and drain. As shown in Fig. 7 the finite sate machine takes in stepped down padp and padn signals and detects SE0 state by detecting both padp = padn = 0 levels. The purpose of state machine is to detect SE0 state so that charge injection/drain on every falling/rising edge during normal data transmission can be avoided. Thus built in time out signal generated by threshold detector is more meaningful. The two bit counter is used to count clock pulses until count reaches to $(11)2$. This ensures that padp = padn = 0 for long enough time and indeed its SE0 state. This removes glitches and avoids false SE0 detection hence can be called as digital glitch filter. When Charge injection and drain circuit is enabled, the comparator output goes high when rising edge occurs on padp. This comparator is folded cascode architecture with differential inputs and single ended output [4, 5]. The stepped down padp and padn signals goes as input to this comparator. The input devices are purposefully skewed so that output switches high when padp signal goes above padn signal by ~100 mv. The

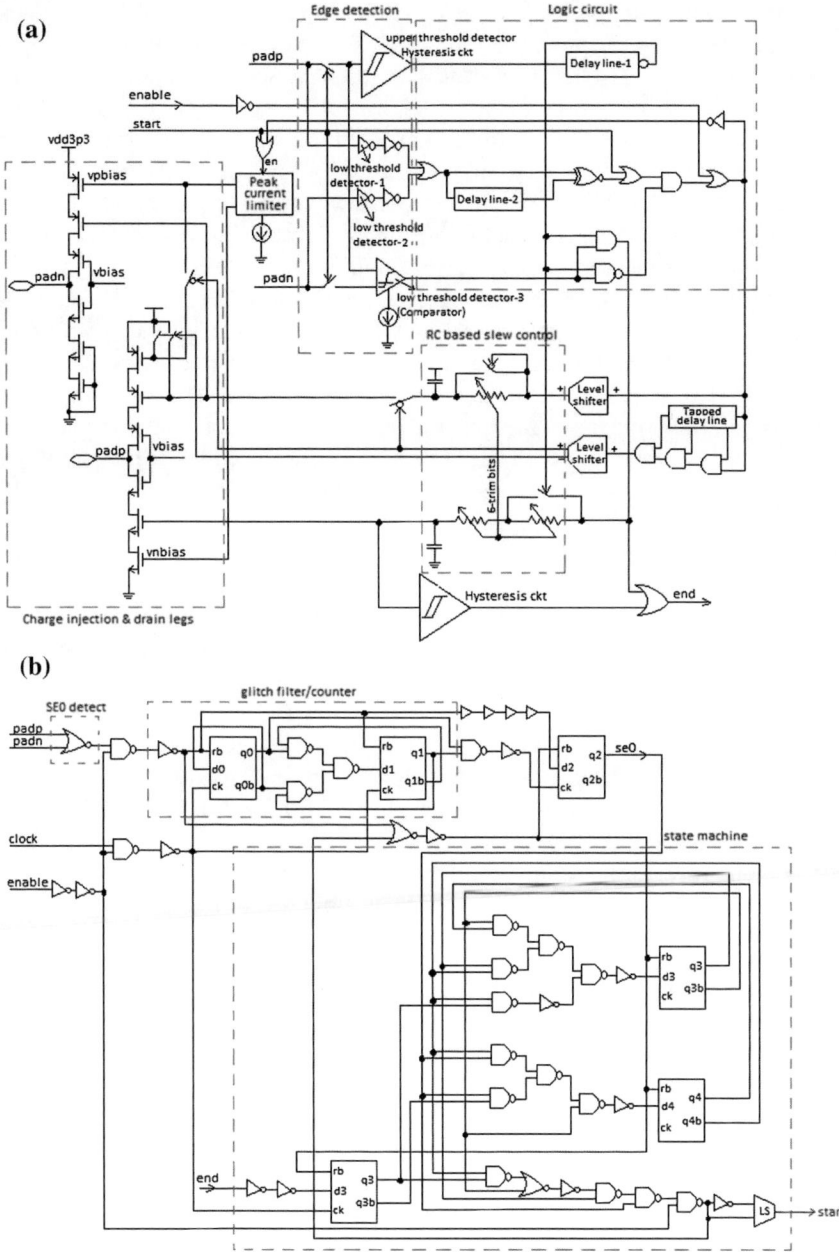

Fig. 7 **a** Charge injection and drain circuit and **b** mealy type finite state machine

differential input comparator architecture is used to make design more robust against input common mode noise. The upper threshold of rising edge on padp is detected at ∼2.5 V by hysteresis circuit. The upper threshold detection is necessary due to large range of rise time specification in Full speed (FS) mode of USB2 protocol. The rise time can be from 4 to 20 ns in FS mode of USB2. These threshold circuits help to create window for charge injection and drain. Inverter based delay line is used to stretch this window further by some margin since overshoot lasts for ∼5 ns on padp after upper threshold detection. The charge injection and drain mechanism is as illustrated in Fig. 4 where Isource and Isink are the constant current source and sink respectively. These sources limit the peak charge density. These current sources/sinks further allow designers to get rid of calibration on these devices. The input signal slew for current starving devices is adjusted by RC based charging and discharging circuits. The slew timing is designed such a way that required charge injection and drain profile can be generated to partially neutralize the pad. The capacitors in RC circuit are implemented as MOS caps and resistors by poly. The poly resistor variation is minimized by six bit calibration code. This improves precision of charging and discharging time bases generated by RC circuits. The peak injected and drained charge density becomes very high under fast corner and damps the overshoot to very low value. This unnecessarily increases power consumption and results in overdesign. This scheme limits the peak current optimizing design for power consumption and reliability. The peak current limit circuit is basic current mirror circuit which takes in constant current and mirrors to charge injection and drain legs so that peak current never exceeds certain value under fast process corner. The charge injection and drain circuit uses level shifter to shift input signaling levels from 0–>1.8 V to 1.8–>3.3 V for PMOS devices comprising charge injection leg. This level shifting is necessary because the devices used are 1.8 V devices in 3.3 V domain. This unique combination of analog and digital components and intelligence (FSM) makes this architecture unique. The added intelligence (FSM) saves significant amount of power since it enables circuit only when it's required.

6 Simulation Results

A test set up shown in Fig. 1a is prepared with appropriate SPICE models for different components [6] and simulations are run. The test set up is comprised of TX modeled as pulse generator and 45 Ω impedance with ±5 % tolerance, Short traces modeled by transmission line model, USB connector model, Choke model, Long, 14-inch trace modeled by transmission line model, RX modeled by 10 pF in parallel with 15 KΩ, and EOS Protection Circuit in Intel 22 nm process.

Figure 8a shows overshoot and undershoot caused by on board choke during SE0 EOP window in full speed mode. The observed overshoot is 4.2 V and undershoot is −0.6 V. Figure 8b, c shows reduced overshoot and undershoot by proposed EOS protection circuit. In typical conditions overshoot and undershoot is

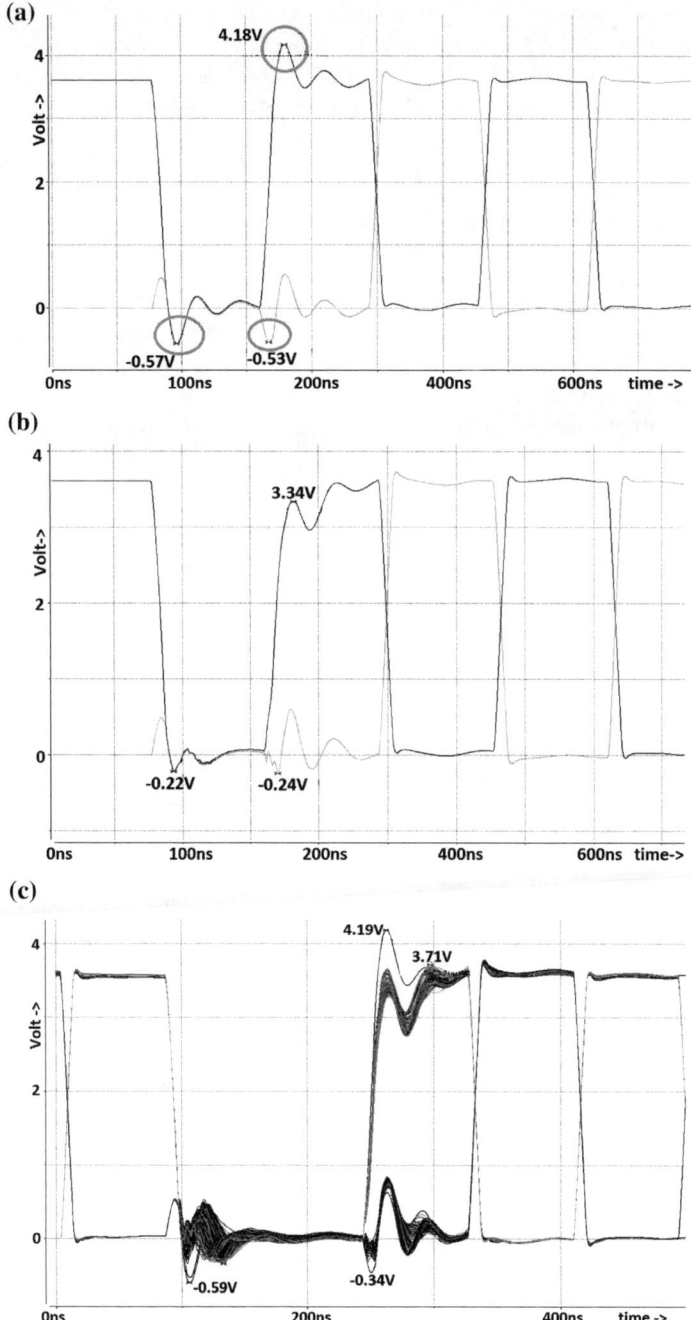

Fig. 8 **a** Typical simulation result without EOS protection circuit, **b** typical simulation result with EOS protection circuit, and **c** all PVT simulation result with EOS protection circuit

reduced to 3.3 and −0.24 V respectively. Under worst case, across all PVT conditions overshoot and undershoot is reduced as low as 3.7 and −0.34 V respectively. PVT conditions consider all 15 corners, ±10 % variation on 1.8 and 3.3 V supply, and Temperature variation from −40 to 125 °C.

7 Conclusion and Future Work

This paper demonstrates a design solution to address issues such as EMI and reliability with advanced sub-micron process technology. It uses some intelligence to detect the EOS issue and addresses exactly where and when it's needed. This reduces power dissipation and improves reliability. This EOS protection circuit can be used for future submicron technologies like 14 and 10 nm.

Acknowledgment The author wishes to thank Intel Microelectronics, HIP USB2 circuit team, Malaysia for providing necessary access to tools and Intel 22 nm process technology. Author also would like to thank IMC WPRD ETS DSI IPR PHY group, Germany for their encouragement and support on this work.

References

1. Inventor: S.V. Shinde, Patent application filed, PCT/US2011/059229
2. S.V. Shinde, PVT insensitive reference current generation, in *Proceedings of the International MultiConference of Engineers and Computer Scientists 2014, IMECS 2014*, 12–14 Mar 2014, Hong Kong. Lecture Notes in Engineering and Computer Science, pp. 777–784
3. M. Morris Mano, *Digital Design* (Prentice Hall, New Jersey, 2002)
4. B. Razavi, *Design of Analog CMOS Integrated Circuits* (McGraw-Hill, New York, 2001)
5. P.E. Allen, D.R. Holberg, *CMOS Analog Circuit Design* (Holt, Reinhart, and Winston, Inc., New York, 1987)
6. K.S.D. Oh, X.C.C. Yuan, *High speed signaling: Jitter modeling, analysis, and budgeting.* Prentice Hall modern semiconductor design series (2012)

An Embedded SIP-VoIP Service in Enhanced Ethernet Passive Optical Network

I-Shyan Hwang, AliAkbar Nikoukar, Lamarana Jallow
and Andrew Tanny Liem

Abstract Voice-over-Internet Protocol (VoIP) is an infrastructure for the delivery of voice communications over IP networks by enlisting a group of hardware and software technologies. The VoIP regulation (channel setup, signaling, digitization and encoding) is similar to public switch telephone network (PSTN). The VoIP packetize the digital data and send it over packet-switch network as IP packet, however, the PSTN transmits the data over the circuit-switched network. The VoIP offers different services (i.e. text messaging, call conferencing and video call) although a major concern to deliver such services to the end user is bandwidth availability with very low latency. Ethernet passive optical network (EPON) is regarded as the best solution in access networks in deployment, cost, bandwidth and user equipment friendly to satisfy the VoIP requirements. In this paper, we propose an embedded VoIP server in the Optical Line Terminal (OLT). The traffic classifier and SIP-VoIP components are added to OLT to enhance the EPON to handle VoIP service. The optical network unit (ONU) separates the VoIP traffic based on class of service (CoS) and Type of Service and assign a high priority to this kind of traffic. The DBA is designed to handle four types of traffic (EF, voip, AF and BE) and satisfy the VoIP requirements. Our simulation results show that the proposed embedded VoIP server improves the overall VoIP quality of service (QoS) in terms of mean packet delay, jitter and packet loss. Moreover, the proposed DBA improves the system throughput in advance.

I.S. Hwang (✉) · A. Nikoukar · L. Jallow
Department of Computer Science and Engineering, Yuan-Ze University, Chung-Li 32003, Taiwan
e-mail: ishwang@saturn.yzu.edu.tw

A. Nikoukar
e-mail: nikoukar@yu.ac.ir

L. Jallow
e-mail: s1006071@mail.yzu.edu.tw

A.T. Liem
Department of Computer Science, Klabat University, Manado 95371, Indonesia
e-mail: andrew_heriyana@yahoo.com

© Springer Science+Business Media Dordrecht 2015
G.-C. Yang et al. (eds.), *Transactions on Engineering Technologies*,
DOI 10.1007/978-94-017-9588-3_24

Keywords Embedded server · EPON · PSTN · QoS · SIP · VoIP

1 Introduction

The current number of telephone line has exceeded over 1 billion users and about 6.8 billion mobile-cellular subscriptions in the world [1]. The Public Switch Telephone Network (PSTN) is still where the majority of voice calls are carried. The well-engineered PSTN uses circuit switching technology, which reserves resources along the entire communication channel for the duration of a given call. This has brought about the quality that it provided for the past 100 years. In recent years, we have seen many developments and new services coming to market, services that the PSTN cannot fully handle. Services like voice email, soon, telephony service providers will move to networks, based on open protocols known as voice over Internet protocol (VoIP) [2]. Operational cost will be reduced, with the provisioning of multiple service types in a single network with a common IP network infrastructure. As voice and data are in the same network, this will be easier to run, maintain and manage [3]. VoIP also known as Internet telephony, is the ability to transmit voice communication over the packet-switched IP networks. VoIP has revolutionized telephone communication, as it is cheaper for the end users. Other additional features which the users previously had to pay for are offered for free, features like caller ID, call waiting, call forwarding and call conferencing. Moreover, extended services also come up with establishment of VoIP, which includes emailed voicemail and easy management of phone contacts.

Service providers can offer some added services by including VoIP telephony for the users, with which they can manage the network that can comprise of voice, video and data.

The integration of all traffic types onto a single network may seem nifty, but a few problems are realized. While cost reduction, new functionality and increased mobility are realized. With the introduction of VoIP to the existing networks, this will result in bad voice quality for VoIP when compared to PSTN. VoIP requires packets to be delivered with strict timing; low latency, jitter, packet loss and sufficient bandwidth.

To set up and tear down calls, Signaling Protocols are needed to establish the telephone calls over the internet. The role of signaling protocols can be broken down into four functions—User Location, Session Establishment, Session Management and Call Participant Management. H.323, Session Initiation Protocol (SIP), Media Gateway Control Protocol (MGCP), SKYPE are some protocols architectures that can be used as signaling protocols. The debate is adopting a protocol that has quality as close to as the PSTN and deployment might differ from market to market. Due to simplicity, scalability and low overhead, SIP can be implemented in networks of any size. SIP is more flexible in the sense that it covers intentionally only subsets of functionality needed for VoIP Telephony and is characterized with the ability to be used with different transport and other protocols [4]. SIP performed better when compared with H.323 under extreme traffic congestion and different queuing policies. And a higher percentage of successful call establishments were achieved with SIP when compared to H.323 [5].

The Session Initiation Protocol (SIP) is the IETF protocol for VOIP and other text and multimedia sessions, widely used for controlling multimedia communication sessions such as voice and video calls over Internet Protocol (IP). SIP is used for creating, modifying and terminating of sessions and also how a call is established, how voice data is transferred during call with one or more participants, and also the termination of the session. SIP is similar to how HyperText Transfer Protocol (HTTP) functions, and shares some of its design principles. It typically adopts client-server architecture (request/response). Client generates requests and sends to the server. After the request is processed, the response is then sent to the server. Similar to HTTP, SIP is based on text-based messages. The messages are exchanged between clients and the servers as either requests or responses. The INVITE request is the most important type of request and is used to invite a user to a call. The ACK request which the caller sends to the callee to simply acknowledge the receipt of the latter's response. The BYE request is used to terminate the session between two users Fig. 1 explains the basic component of SIP calling mechanism.

There are a number of available SIP based Open Source VoIP soft-switch platforms, providing rich telephony services [6]. It offers a wide range of features to end users (call forwards, voicemail, conferencing, call blocking, click-to-dial, call-lists showing near-real-time accounting information, etc.), which can be configured by them using the customer-self-care web interface. A web-based administrative panel is provided for operators, allowing them to configure users, peering, billing profiles, etc., and also the possibility of viewing real-time statistics of the system.

Deploying passive optical networks (PON) from the service providers to be as close as possible to the customers will give the required bandwidth for customers various multimedia service need, at a very cost efficient and flexible infrastructure

Fig. 1 Basic call flow of SIP

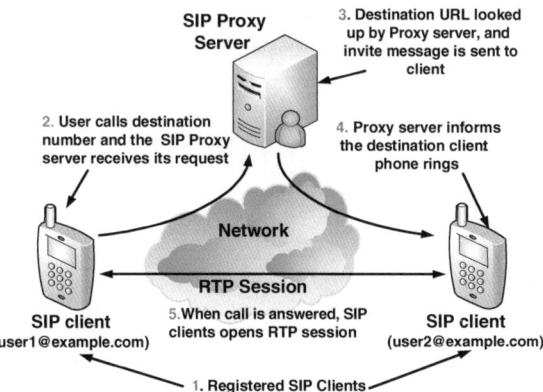

from both parties. A PON is a form of fiber optic access network. Access networks provide end-user connectivity. They are placed in close proximity to end users and deployed in large volumes. Access networks exist in many different forms for various practical reasons. In an environment where legacy systems already exist, carriers tend to minimize their capital investment by retrofitting existing infra-structure with incremental changes. Compared to traditional copper-based access loops, optical fiber has virtually unlimited bandwidth (in the range of tera-hertz or THz of usable bandwidth). Deploying fiber all the way to the home therefore serves the purpose of future proofing capital investment.

PON offers many system architectures, the EPON is regarded as the best solution for access networks due to its simplicity, high data rate, and low cost compared to the other PONs [7]. EPON is based on Ethernet standard, which makes it easy to deploy from the central office to customer premises. EPON is generally deployed in a tree like topology, using passive (non-active electronics within the access net-work) 1:N splitter. The optical line terminal (OLT) is located at the central office and N number of optical network units (ONUs) is close to the customer premises. In EPON, multi-point control protocol (MPCP) is used to control the P2MP fiber networks. The MPCP is implemented at the medium access to control layer to perform the bandwidth allocation, auto-discovery process, and ranging. Two con-trol messages—Gate message, which carries the granted bandwidth information from the OLT to ONU in the downward direction; and REPORT message, which is used by ONU to report its bandwidth request (local queue length) to the OLT in upstream direction [7]. Due to the many traffic types the ONU is capable of sup-porting up to eight priority queues. After each ONU sends there report message based on queue state information, to the OLT. The OLT executes the Dynamic Bandwidth Allocation (DBA), which calculates and allocates the transmission timeslot for each ONU based on their queue state information. The DBA plays a major role in providing an efficient bandwidth allocation scheme for the ONUs to share the network resources accordingly, and also to provide better Quality of Service (QoS) for the end users.

In this paper, we propose an OLT architecture that contains the VoIP server (VoS) by using the Field-programmable gate array (FPGA) [8]. The rest of the paper is organized as follows. Section 2 talks about the related work, the proposed architecture is introduced in Sect. 3. The performance evaluation is explained in Sect. 4, followed by the conclusion in Sect. 5.

2 Related Work

If the VoIP is to be adopted in every organization there should be a mechanism in place to handle as many calls at a time. Bandwidth and architecture is a major concern for network operators and carriers. Reference [9] shows a maximum of 12 calls can be achieved in wireless Local Area Networks. As to lease line traffic, circuit emulation servicer over packets (CESoP) is a simple and cost effective solution for "tunneling" TDM circuits through a packet-switched network. Although CESoP also supports voice application, it is more costly and complex compared to the cost of implementing a VoIP based solution [10].

The token bucket (TB) is used as policing unit, which monitors the traffic entering the network. Being between a VoIP host and an ingress node, it ensures that the generated traffic conforms to a certain pre-determined profile. In case the traffic violates this profile, the TB policer drops incoming packets in a way to make the outgoing traffic fit the given profile [11]. In this case it will be observed a significant amount of packets will be dropped causing impossible voice communication, when there is a number of simultaneous calls.

With VoIP servers typically located outside our reach, Skype in particular, which is also proprietary. Having the VoIP server closer to the users will decrease some of the packet delays usually caused by the long distance calls.

With all the pretty features of VoIP, there are so many problems that arise. VoIP promised so many fancy features which are offered for free of service to consumers. Bandwidth has been a major concern ever since the inception of the internet. Applications like IPTV, video on demand, VoIP, online gaming and so on are huge consumers of bandwidth.

For the past 100 years, the quality of the call service provided by the PSTN has always been on point. A tremendous amount of work has to be done for VoIP to equate its quality with the PSTN. Service providers should come up with services that have quality equal to the PSTN. Delays of less than 150 ms is acceptable by most people; when the delays get to 250 ms then speech is annoying but comprehensible; when delay reaches 600 ms then speech becomes unintelligible and incoherent [8]. Packet loss is a major concern for VoIP systems, as a delivery ratio of more than 99 % is required for VoIP. A VoIP voice call is considered acceptable only when packet loss rate is less than 2 % [12].

Delay jitter can also be a major problem in voice communication. At the sending side, voice packets are usually transmitted at the constant rate, while at the other end packets may arrive or received at uneven rat. When the jitter becomes large, that will cause the delay packets to be cast aside, which will result in audible gaps in the communication. If a number of sequential packets are dropped, then voice becomes unintelligible. With all the problems stated above, broadband access network is the best solution to provide the required bandwidth, minimum delays and avoid packet losses.

3 Proposed Architecture

In the proposed architecture, a fully functioning telephony system is built in the OLT using the FPGA. The in-built VoS will be able to perform the calling needs for any of the registered users within the EPON from here on called VoIP-OLT. Usually the VoS are located outside the system in this case we have it inside the EPON architecture. In most cases, VoIP servers are usually located somewhere in

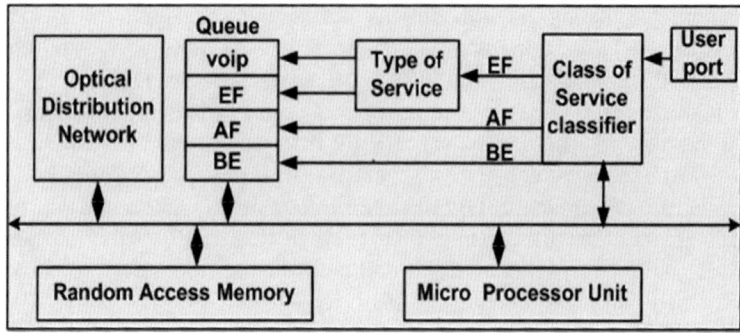

Fig. 2 Detailed ONU structure

the Internet, and that adds the delays to the packets traveling as every hop that it passes, adds to the delay. In this architecture, the VoS component in the OLT is called SIP-VoIP.

3.1 Architecture and Operation of ONU

The first place where major effects happen is at the ONU. When the ONU receives all the packet types from the users, it passes all the packets to the Class of Service (COS) classifier, which separates the assured forwarding (AF) and best effort (BE) from the expedited forwarding (EF) traffic type. The COS then passes the EF to the Type of Service (TOS) classifier using the Diffserv mechanism. The TOS will identify the different services if they are ordinary EF traffic or they are from VoIP originating devices (voip) traffic type, and then separate them accordingly to EF queue and voip queue. The TOS then sends the two queues to the ONU buffer which now contains voip, EF, AF and BE queues. Figure 2 shows us the ONU structure.

3.2 Architecture and Operation of OLT

After the ONU receives its grant message, it sends its buffer to the OLT. The packets are first sent to the COS for identifications and separations. The COS will be able to identify the voip packets as they have already been separated by the ONU.

The COS then sends the voip packets to the SIP-VoIP in the OLT as shown in Fig. 3. All voip packet processing are done in the SIP-VoIP. The SIP-VoIP function is one of the added components of the OLT. It uses the FPGA mechanism to have the VoS server in built. The firmware in the SIP-VoIP will host the functions and algorithms necessary for the operation of the VoS. The central process unit (CPU) is

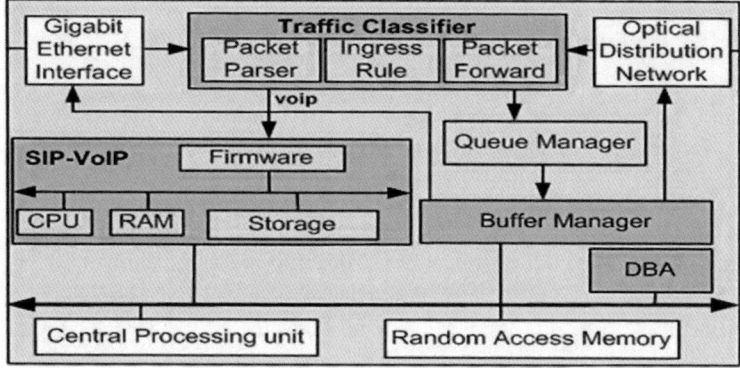

Fig. 3 Detailed OLT structure

for processing of voip packets and it uses the functions of the firmware to do the operation. The random access memory (RAM) will be used for buffering, and the storage will be used to store the databases and as well the registered users.

The operation of the SIP-VoIP component is given in Fig. 4. It is separated into two parts, where the public part can be accessed by the public users, and the private part which can only be accessed by the authorized users. The SIP Load-Balancer is used for protecting the underlying elements of the SIP server. As the name implies is can also be used as a load balancer if the system is scaled to more than just one pair of servers. SIP Proxy/Registrar is the busy bee of the SIP server. It handles user registrations, the authentication of end points. The Proxy Server also knows the location of Callee if a call is made to certain number, and check if it is in the same domain or else send it to the right proxy Server. SIP Back to Back User Agent (B2BUA) uncouples the call-leg 1 (i.e. from the caller to the SIP Server) from call-leg 2 (which is from the SIP server to the callee). It also does topology hiding; it hides caller information where necessary, and outbound authentication. The SIP App-Server will be used for other voice applications like voicemail, and can also be

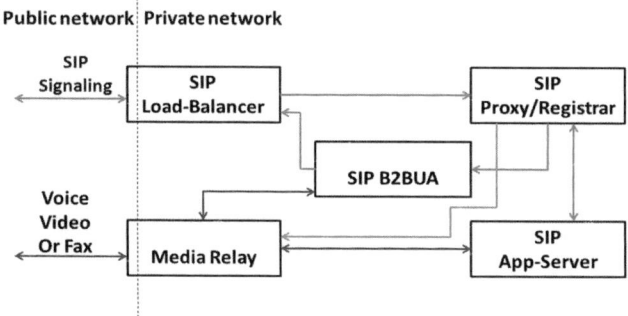

Fig. 4 Operation of the SIP-VoIP

used for reminder calls. The Media Relay is controlled by the SIP proxy. After the calls are authenticated and connected the media relay is called on for packet relay.

3.3 VoIP Dynamic Bandwidth Allocation (VoIP-DBA)

A new DBA is proposed called VoIP-DBA, which is designed to handle traffic allocation. When the OLT receives the REPORT message, it identifies the four different queues. The VoIP-DBA first checks its available bandwidth and the required bandwidth for the voip request and allocates all the requested bandwidth by the voip traffic. The VoIP-DBA then checks its remaining bandwidth and allocates bandwidth to the EF traffic as indicated by the report. After the voip and EF traffic has been allocated bandwidth it checks its remaining bandwidth and allocates the bandwidth to the requested AF traffic, and if there is any remaining bandwidth, then the BE traffic will also be allocated. In this case, priority is given to Voip then EF then AF and last is BE. The buffer manager which usually uses FIFO by the OLT for downstream transmission is redesigned to a priority queuing mechanism, and the highest priority is always given to the VoIP packets, followed by EF then AF and BE.

3.4 VoIP Call Handling

One of the most important functions in VoIP is the INVITE function. The INVITE is a SIP method that specifies the action that the requester (Calling Party) wants the server (Called Party) to take. The INVITE request contains a number of header fields. Header fields are named attributes that provide additional information about a message. The ones present in an INVITE include a unique identifier for the call, the destination address, Calling Party Address, and information about the type of session that the requester wishes to establish with the server.

Figure 5 shows the flowchart of the voip call handling in the EPON. The caller initiates an INVITE function, which is received by the ONU. The COS in the ONU will handle the packet as EF traffic and send it to the TOS for further classification. The TOS will then separate it accordingly to voip or EF. Then the voip packet is sent to the buffer. After the ONU has been granted time for transmission, the voip packet is then sent to the OLT, which will separate the voip packet from the other packets by using the COS and is sent to the SIP-VoIP for processing.

When SIP-VoIP receives the invite packet, the message first passes through the load balance the to the SIP proxy server, which will check if the user is authorized in the callers authentication credentials. The SIP proxy confirms if the called party is a local user or in another domain; in this case, all calls are in the same EPON architecture. If it is a registered user, the Proxy will replace the Request URI with the URI of the registered contact. When the proxy has completed the necessary

Fig. 5 Operation of the SIP-VoIP

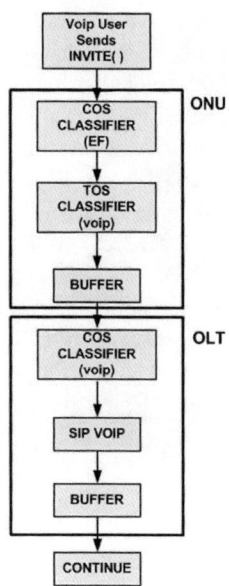

calling features, and the destination of the called party is determined, it invokes the Media Relay and the copy of the INVITE message is sent to the SIP B2BUA. The function of the SIP B2BUA will be to copy only explicitly allowed headers to a new INVITE message another function can be to force audio calls to use specific codec. The new message is then sent to the callee through the SIP load-balancer. The callee sends its replies through the same elements. The media relay is called on again to prepare the recommended ports for media stream. Once the message has been sent from the callee to the caller, and all other features are negotiated, the end points can now send traffic to each other either through the media relay or end-to-end. When the call ends, either of the parties can end call by sending a BYE message, the SIP proxy will inform the media relay to close the ports. After the processing, the packet is then sent to the buffer, where it is then sent to the right destination in the PON.

4 Performance Evaluation

To evaluate the performance of the proposed scheme in terms of mean packet delay, system throughput, jitter and packet loss, the system model is set up in the OPNET simulator with one OLT and 32 ONUs. The data rate of upstream and downstream direction is 1 Gbps. The ONU buffer size is 10 Mb. The distance from ONU to OLT is uniform from 10 to 20 km. The self-similarity and long-range dependence (LRD) which is chosen as the network traffic model for AF and BE. This model generates highly burst AF and BE traffics with Hurst's parameter of 0.8. High-priority traffic (i.e., EF, voip traffic) is modeled using the Poisson distribution. The EF and voip packets size is fixed to 70 and 160 bytes respectively. The AF and BE packet size is uniformly distributed between 64 and 1,518 bytes. The maximum transmission cycle time is 2 ms. The simulation parameter is shown in Table 1. In order to simulate the performance of the proposed mechanism uniform traffic load of ONUs is considered. We compared the performance of normal Interleave Polling with Adaptive Cycle Time (IPACT) DBA [13] and the proposed DBA which is called VoIP. Three significant scenarios are designed and analyzed with various voip, AF and BE service proportions to show the effectiveness of the voip traffic management. Table 2 shows the traffic proportions of EF, AF and BE for IPACT DBA in different scenarios. The EF traffic portion of the IPACT includes part of voip traffic. Table 3 shows the traffic proportions of EF, voip, AF and BE for IPACT DBA in different scenarios [11].

Table 1 Simulation parameters

Simulation environment	Value
Number of OLT	1
Number of ONUs	32
Up/down link capacity	1 Gbps
OLT-ONU distance (uniform)	10–20 km
Max cycle time	2 ms
Guard time	5 μs
DBA computation	10 μs
Control message length	0.512 μs
ONU buffer size	10 Mb
AF and BE packet size (bytes)	Uniform (64, 1,518)
EF packet size (bytes)	Constant (70)
VoIP packet size (bytes)	Constant (160)

Table 2 IPACT traffic proportion in different simulation scenario

	EF (%)	AF (%)	BE (%)
Scenario 1	20	40	40
Scenario 2	30	30	40
Scenario 3	40	40	20

Table 3 voip Traffic proportion in different simulation scenario

	EF (%)	voip (%)	AF (%)	BE (%)
Scenario 1	10	10	40	40
Scenario 2	10	20	30	40
Scenario 3	10	30	40	20

4.1 Mean Packet Delay

The packet delay is composed of queuing delay, the polling delay and granting delay. The queue delay is the delay from the beginning of the time-slot until the beginning of frame transmission. The polling delay is a fundamental delay, and defines as the time between packet arrival and the next request sent by the ONU; granting delay is the time interval from an ONU's request for a transmission window until the beginning of the time-slot in which this frame is to be transmitted. Figure 6 shows the mean packet delays for the various traffic types in the different scenarios versus the offered load. The EF delay of the proposed VoIP DBA in all scenarios is less than the EF delay in the IPACT. The reason is that the original IPACT handles just three kinds of traffic which voip traffic is part of EF traffic, and it causes the heavy load for EF traffic thus the delay in the IPACT is increased. However, the VoIP DBA can handle voip traffic separately and reduce the EF traffic load. The mean EF delay is shown in Fig. 6a. Figure 6b describes the mean AF delay of the IPACT and VoIP DBA in different scenarios versus traffic load. The Af delay of the VoIP and IPACT almost is the same when the traffic is under 50 %. The AF delay of the VoIP is less than the IPACT in highly load especially in scenario3. Figure 6c depicts the mean BE delay of the IPACT and VoIP DBA in different scenarios versus traffic load. The BE delay of the

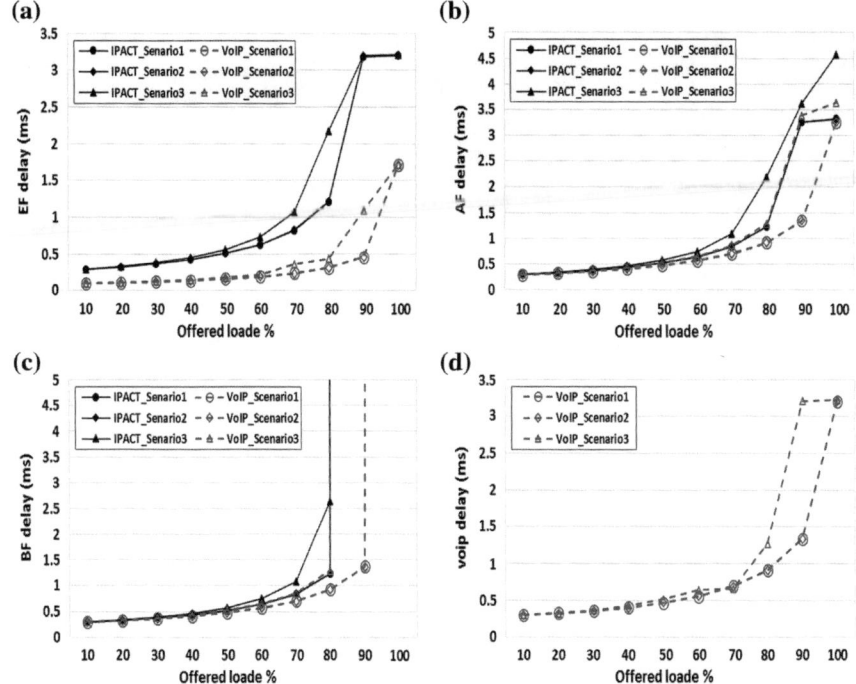

Fig. 6 Mean packet delay

VoIP and IPACT in the lightly load is the same; however, the BE delay of VoIP DBA is better than the IPACT but in both DBA the BE is saturated after 90 % traffic load. The voip delay of the VoIP DBA is shown in Fig. 6d. The voip delay in all traffic loads is under acceptable range, and it is below 3.5 ms.

4.2 Jitter

The delay variance σ^2, jitter, is calculated as $\sigma^2 = \sum_1^N \left(d_i^{voip} - \overline{D} \right)^2 / N$, where d_i^{voip} is the delay time of voip packet i, $\overline{D}$ is the average delay time of the voip traffic and N is the total number of received voip packets. Jitter or delay variance has a significant impact on voice quality when a smaller jitter value is required to deliver a high-quality voice transmission. The same was also done EF packets. Figure 7 shows the jitter for high-priority traffic (i.e. EF and voip) traffic of IPACT and VoIP DBA in different scenarios versus the offered load.

The EF jitter of the VoIP DBA in the lightly load is similar to the IPACT and in the highly load both DBA have the similar trend, and their values is almost same. The EF jitter of the VoIP DBA and IPACT are shown in Fig. 7a. The voip jitter is shown in Fig. 7b. The voip jitter in the lightly load is very less and in the highly load the max value in different scenarios are similar with EF jitter.

4.3 Packet Loss

The cycle time and the buffer size are two major factors, which can affects to the packet loss. Choosing the large cycle time will reduce the packet loss because each ONU has more time-slots to send its queues; however, it leads to increase packet delay. The other way to reduce the packet loss is to increase the queue size. Nevertheless, it causes

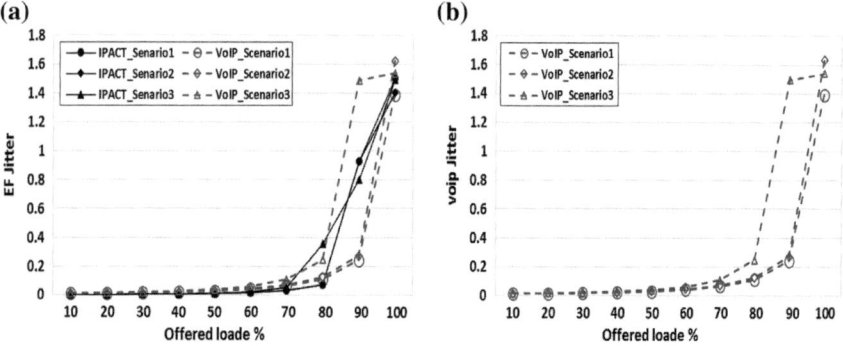

Fig. 7 High-priority traffic jitter

Fig. 8 Packet loss ratio

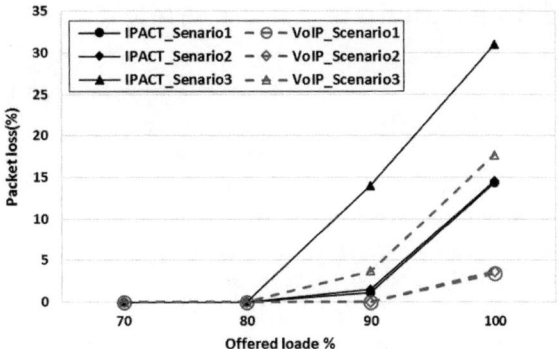

higher delay. Figure 8 shows the packet loss ratio for the various traffic types in the different scenarios versus the offered load. The packet loss of EF, voip and AF for VoIP and IPACT DBA in all scenarios are zero. The packet loss happens for the BE traffic. The voip packet loss in all scenarios is less than the IPACT.

4.4 System Throughput

The system throughput is calculated as sum of the traffic between the OLT and ONUs. Although the EPON rate is 1 Gbps, the system throughput depends on the combined efficiency. The scheduling overhead and the encapsulation overhead are two major components of the combined efficiency. The maximum net throughput of the EPON in upstream direction is 897.6 Mbps where the upstream encapsulation and scheduling overheads are 7.42 and 5.76 % (i.e., 32 ONUs with 1.5 ms cycle time), respectively.

Figure 9 describes the system throughput of VoIP and IPACT DBA for the different traffic types in the different scenarios versus the offered load. The system throughput of VoIP DBA is higher than the IPACT in all scenarios and traffic loads. It is because of the usage of the unused remainder, guard time, of every cycle by the voip packets.

Fig. 9 System throughput

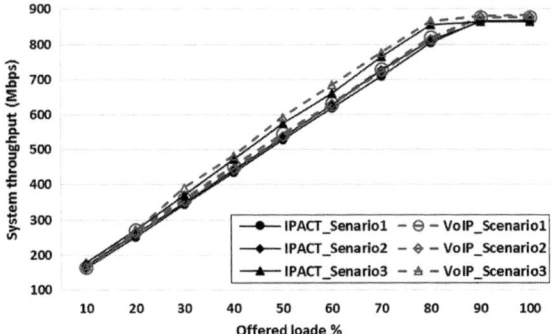

5 Conclusion and Future Work

This paper has proposed an embedded VoIP Server in the OLT and the mechanism to handle VoIP traffic over enhanced EPON. The ONU separates the VoIP traffic and reports it to the OLT. The proposed VoIP-DBA will guarantee the VoIP service requirements such as bandwidth and low latency of the voip packets. This will eliminate the usual delay that occurs when VoIP servers are located outside the EPON. The simulation results show that the proposed architecture can remove the limitation of concurrent call. Moreover, the overall QoS metrics in terms of mean packet delay, jitter, packet loss and system throughput is improved. To decrease cost, network and service providers can incorporate the SIP-VoIP component in the OLT. This architecture can be further implemented to support direct call between users without sending the data to the feeder fiber as local traffic.

References

1. Available online: http://www.itu.int/en/ITU-D/Statistics/Pages/stat/default.aspx
2. P.V. Mockapetris, Telephony's next act. IEEE Spectr. **43**(4), 28–32 (2006)
3. K. Salah, On the deployment of VoIP in ethernet networks: methodology and case study. Comput. Commun. **29**(8), 1039–1054 (2006)
4. I. Basicevic, M. Popovic, D. Kukolj, Comparison of SIP and H.323 protocols, in *The 3rd International Conference on Digital Telecommunications*, pp. 162–167 (2008)
5. B.S. De, P.P. Joshi, V. Sahdev, D. Callahan, End-to-end voice over IP testing and the effects of QoS on signaling, in *Proceedings of the 35th Southeastern Symposium on System Theory Morgantown*, pp. 142–147 (2003)
6. Available online: http://www.sipwise.com/doc/2.8/handbook-ce.pdf
7. I.S. Hwang, A. Nikoukar, C.H. Teng, K.R. Lai, Scalable architecture for VOD service enhancement based on a cache scheme in an ethernet passive optical network. IEEE/OSA J. Opt. Commun. Netw. **5**(4), 271–282 (2013)
8. L. Jallow, I.S. Hwang, A. Nikoukar, A.T. Liem, *A SIP-based VoIP Application in enhanced ethernet passive optical network architecture*, Lecture Notes in Engineering and Computer Science: Proceedings of The International MultiConference of Engineers and Computer Scientists, 12–14 Mar 2014, Hong Kong, pp. 617–622
9. J.S. Li, H.C. Kao, W.H. Lin, Achieving maximal VoIP calls in 802.11 wireless networks. Comput. Commun. **33**(11), 1296–1303 (2010)
10. G. Wu, D. Liu, Y. Chang, C. Zhang, The implementation of TDM service in EPON system, in *SPIE 6784 Network Architectures, Management, and Applications V*, 67843J (2007). doi:10.1117/12.743418
11. S. Sharafeddine, A. Riedl, J. Glasmann, J. Totzke, On traffic characteristics and bandwidth requirements of voice over IP applications, in *8th IEEE International Symposium on Computers and Communication*, pp. 1324–1330 (2003)
12. L. Cai, Y. Xiao, X. Shen, J.W. Mark, VoIP over WLAN: voice capacity, admission control, QoS, and MAC. Int. J. Commun Syst **19**(15), 491–508 (2006)
13. G. Kramer, B. Mukherjee, G. Pesavento, IPACT a dynamic protocol for an Ethernet PON (EPON). IEEE Commun. Mag. **40**(2), 74–80 (2002)

Extracting Naming Concepts by Analyzing Recipes and the Modifiers in Their Titles

Shoko Wakamiya, Yukiko Kawai, Hidetsugu Nanba
and Kazutoshi Sumiya

Abstract On user-generated recipe-sharing sites such as Rakuten recipe, various modifiers such as "Kid-friendly" and "Simple" are often used in the titles of the recipes to signify their characteristics. Although a modifier is used in a number of recipes' titles, the underlying concepts utilized vary. In this paper, we propose a system which extract and present Naming Concepts for recipes based on modifiers in their titles. Specifically, the system obtains typical ingredients and cooking utensils by summarizing the recipes for a dish to extract the differences between the elements of recipes and the typical elements in terms of addition, deletion and exchangeability and extracts additional information from procedures. Then, it identifies Naming Concepts for the recipes by extracting feature patterns based on the differences extracted and grouping them on the basis of the patterns. Finally, it presents recipes with granted Naming Concepts for readers. In the experiment, we extract the Naming Concepts of given recipes with a real recipe dataset.

Keywords Cooking utensils · Ingredients · Modifiers · Naming concepts · Recipe features · Typical · User-generated recipes

S. Wakamiya (✉) · Y. Kawai
Kyoto Sangyo University, Kyoto, Japan
e-mail: shokow@cc.kyoto-su.ac.jp

Y. Kawai
e-mail: kawai@cc.kyoto-su.ac.jp

H. Nanba
Hiroshima City University, Hiroshima, Japan
e-mail: nanba@hiroshima-cu.ac.jp

K. Sumiya
University of Hyogo, Himeji, Japan
e-mail: sumiya@shse.u-hyogo.ac.jp

© Springer Science+Business Media Dordrecht 2015
G.-C. Yang et al. (eds.), *Transactions on Engineering Technologies*,
DOI 10.1007/978-94-017-9588-3_25

1 Introduction

Cooking is one of the most important creative activities in daily life. Nowadays, we can obtain large numbers of recipes from cooking websites. For example, Rakuten recipe [1] provides over 863,000 user-generated recipes written in Japanese. COOKPAD [2], another famous cooking website, provides more than 1,800,000 recipes. Although these websites have recipes that meet a wide variety of users' demands, they are difficult to find because numerous recipes for any particular dish are available on the sites. For example, when we searched for recipes on the Rakuten recipe website using the query "carbonara," we were presented with about 1,500 different recipes. Thus, in order to find recipes that meets users' demands, invariably, clear distinctions must be made among recipes. In this paper, we propose a method to extract "Naming Concept," which can identify the features of each recipe [3].

Each recipe on the Rakuten recipe website comprises a title, dish categories, an ingredient list, and a procedure that gives step-by-step instructions on how to cook the dish. Here, recipe titles are typically represented in the form "modifier + dish name." For example, in the two titles "Simple! carbonara" and "kid-friendly omelette rice," "Simple!" and "kid-friendly" are modifiers and "carbonara" and "omelette rice" are dish names. The modifiers are assigned after considering the features of each recipe. In addition, the same modifier might be used in different ways. Figure 1 shows Naming Concepts for recipes whose titles include the

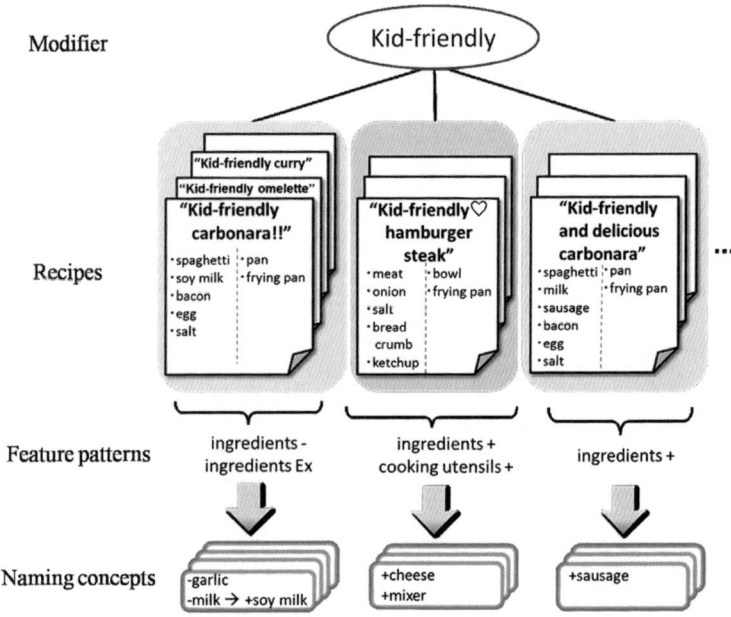

Fig. 1 A modifier of recipes based on naming concepts

modifier "Kid-friendly." In the figure, there are three types of Naming Concepts: "Kid-friendly" is used in the title of a recipe in the first type of Naming Concept because soy milk, which is considered to be preferred by many children, is used as an ingredient in carbonara instead of garlic and milk. In the second type of Naming Concept, the same modifier is used from a procedure in each recipe because a mixer is additionally used to mince the ingredients and enable children to eat without difficulty.

In this paper, we develop a system which extracts Naming Concepts for recipes by identifying the characteristic ingredients, cooking utensils, and procedure in each recipe. For this, the system extracts the typical ingredients and cooking utensils for the dish. Next, it extracts the differences between the typical elements of the dish and the elements of a recipe for the same dish and extracts tips as additional information. Then, it groups recipes with the same modifier by feature patterns of the differences. By using this system, a user can find an appropriate recipe by checking recipes with their Naming Concepts in the list of search results.

The remainder of this paper is organized as follows: Sect. 2 discusses related work. Section 3 presents our procedure for extracting Naming Concepts from recipes in detail. Section 4 shows the Naming Concepts for some recipes and discusses experimental results obtained. Finally, we conclude this paper and outline future work in Sect. 5.

2 Related Work

Lots of researches related to cooking have been conducted. Even among them, researches targeting recipes are gathering attentions of many researchers such as recipe recommendation [4–7]. Ueda et al. [5] proposed a method that recommends personalized recipes by measuring each user's food preference based on ingredients extracted from the user's recipe browsing and cooking history. Tsukuda et al. [8] proposed a method that enables users to browse from the current recipe to a desired recipe by adding one element into it or deleting one element from it. These methods just analyze addition and deletion of elements. On the other hand, we analyze both those factors along with element exchangeability.

In fact, it is important to extract typical ingredients and procedures in recipes for understanding features of each recipe. Tsukuda et al. [9] analyzed the typicality of an object in terms of target of analysis and type of typicality. Yamakata et al. [10] proposed a method that creates a typical cooking procedure from multiple recipes by converting each recipe text into recipe trees and by integrating them. Then, it extracts features of each recipe by comparing with the typical one. Comparing with these methods, we attempt to extract typical elements and identify differences by comparing the elements of a particular recipe with the typical elements used for the dish. Although [10] is similar to ours in terms of its attempt to define typicality and to extract recipe features on the basis of typicality, our work differs in that our aim is to extract recipe features based on modifiers.

Interestingly, data structure of recipes is different from the one of general Web pages. Therefore, by investigating the structure, we will be able to grasp intentions of recipe writers. Chung [11] proposed an efficient method that finds related words in a recipe domain using a data structure. Interestingly, their investigations revealed that people usually write the main ingredient in the first position of the ingredients list of each recipe and that such an ingredient is strongly related to the categories to which the recipes belong. Nanba et al. [12] constructed a recipe ontology based on the method proposed by Chung [11] and distributional similarity [13, 14] which they used for multi-recipe summarization. We utilize this ontology to extract Naming Concepts. Approaches focusing on modifiers include the method proposed by Takahashi et al. [15] to measure relevancy between a web text and modifiers in its title by extracting suitable words and conflicting words. The method determines whether modifiers are relevant to the contents or not by measuring information credibility. By contrast, our assumption is that recipes' Naming Concepts based on modifiers in their titles can be interpreted from multiple perspectives.

3 Extraction of Naming Concepts Based on Modifiers in Recipe Titles

3.1 Our Proposed Approach

In this work, we define Naming Concepts as features of recipes that present concepts of modifiers. We assume that a Naming Concept can be extracted by considering the differences in the various recipes for a dish and the patterns of differences in the recipes of a modifier. On the right side of Fig. 2, the arrows indicate the differences between typical elements of a dish and the elements of each recipe. For example, in the analysis focused on the recipe with the title "Kid-friendly carbonara!!," we can extract atypical ingredients and cooking utensils by analyzing the recipes for "carbonara." Focusing on the modifiers of the recipes, we extract the elements that can possibly signify the features of the recipes such as deletion of ingredients or exchange of cooking utensils. On the left side of Fig. 2, we extract the pattern of differences between a set of recipes based on a modifier. Consequently, our proposed method extracts Naming Concepts by extracting elements that are typically used in the dish from recipes for the same dish, extracting the elements that are different from the typical elements in each recipe, extracting additional information from the procedures, and grouping the recipes using feature patterns.

We extract Naming Concepts as follows:

1. Extracting elements that are typical in the recipes
2. Extracting the differences of recipes
3. Extracting additional information
4. Grouping recipes based on feature patterns

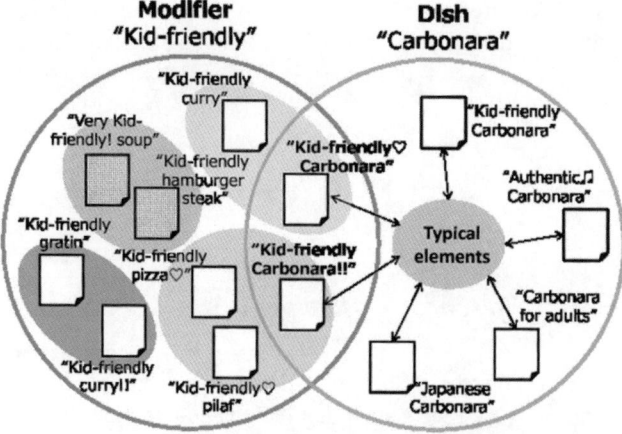

Fig. 2 Extracting typical elements and naming concepts based on relations between modifier and dish

The detailed steps for extracting Naming Concepts are described from Sects. 3.2 to 3.5.

3.2 Extracting Typical Elements in Recipes

In this work, R is constructed from a set of recipes separated by dish categories. A recipe r_{j_k} belongs to a dish category j consisting of M_{j_k}, I_{j_k}, and U_{j_k}. M_{j_k} is a set of modifiers included in the title of the recipe r_{j_k}. I_{j_k} is a set of ingredients of the recipe r_{j_k}, and U_{j_k} is a set of cooking utensils in the procedure of the recipe r_{j_k}. Then, M, I, and U are universal sets of modifiers, ingredients, and cooking utensils, respectively.

$$R = \{R_1, \ldots R_j, \ldots R_m\}, \quad R_j = \{r_{j_1}, \ldots r_{j_k}, \ldots r_{j_n}\},$$
$$r_{j_k} = (M_{j_k}, I_{j_k}, U_{j_k}) \quad M_{j_k} \subset M, \; I_{j_k} \subset I, \; U_{j_k} \subset U$$
$$M = \{m_1, m_2, \ldots\}, \quad I = \{i_1, i_2, \ldots\}, \quad U = \{u_1, u_2, \ldots\}$$

In this paper, we assume that the Naming Concepts of modifiers are detected in the differences between a recipe and its typical elements. Therefore, we extract typical elements of a recipe for a dish. One recipe's data is input, and we extract ingredients $t_j.I'$ and cooking utensils $t_j.U'$. Here, t_j is a set of recipe R_j of a category that one recipe belongs to. In this work, we consider ingredients and cooking utensils that are frequently used as typical ingredients and cooking utensils for a dish. Therefore, we extract ingredients and cooking utensils from a set of recipes R_j and calculate recipe frequency (*RF*) of each element. Thus, a set of typical elements

t_j consists of a set of ingredients I', which means that RF is α and over and a set of cooking utensils U' which means that RF is β and over, as follows:

$$t_j = (I', U'), \quad t_j \in T$$
$$t_j.I' = \{i_l | RF(i_l, R_j) > \alpha, \ i_l \in I_j\},$$
$$t_j.U' = \{u_o | RF(u_o, R_j) > \beta, \ u_o \in U_j\}$$

3.3 Extracting Differences of Recipes

Next, we extract the differences between elements of a recipe r_{j_k} and the typical elements t_j in the category R_j to which that recipe belongs. First, we extract a set of additional ingredients I_{add}, a set of deleted ingredients I_{del}, a set of additional cooking utensils U_{add}, and a set of deleted cooking utensils U_{del} as differences.

$$I_{add} = r_{j_k}.I - t_j.I', \quad I_{del} = t_j.I' - r_{j_k}.I$$
$$U_{add} = r_{j_k}.U - t_j.U', \quad U_{del} = t_j.U' - r_{j_k}.U$$

We determine the relation between the differences of the extracted ingredients and the cooking utensils, because in some cases one different element influences another different element, or different elements are independent of each other, making it irrelevant. Next, the element included in a set of additional elements I_{add} and U_{add} is represented by +, and then in a set of deleted elements I_{del} and U_{del} is represented by −. For example, when we compare the ingredients of a recipe with a title such as "healthy sweet and sour pork" that has typical ingredients "healthy sweet and sour pork," we extract differences $I_{add} = \{chicken, bambooshoot\}$, $I_{del} = \{pork, liquor\}$. Then, we use "chicken" instead of "pork," so we consider that these correspond. Conversely, "bamboo shoot" and "liquor" are only added and deleted; therefore, they have no relation. Thus, in a scenario where an element + is included in I_{add} or U_{add} and an element − is included in I_{del} or U_{del} correspond mutually, we consider that they have an exchangeable relation. Otherwise, we consider that there is no relation between their differences.

In order to determine relations, we calculate the degree of co-occurrence between different elements. In general, when two elements + and − are exchangeable, we consider that they do not co-occur. Therefore, a recipe that contains element + does not contain element −, and a recipe that contains element − does not contain element +. Thus, when the frequency of co-occurrence is low, we consider that elements + and − are exchangeable. Then, we pair elements + and − and extract the pair as differences in order to determine their relations. Consequently, we calculate the degree of co-occurrence of the various pairs of elements. Next, we treat element − (which is included as a typical element) and element − (which is included in one recipe) as denominators. In cases where only the degree of co-occurrence based on a typical element and one recipe are lower than the threshold amount, we determine that the different elements in the pair have an exchangeable relation. Therefore, a pair comprising different elements is added to a set of pairs comprising exchangeable ingredients I_{ex} or a set of pairs comprising exchangeable cooking utensils U_{ex}, and the elements that fall under I_{add} and I_{del}, or U_{add} and U_{del} are deleted. Then, the exchangeability relation of the different elements are determined and signified as "$-$** $\rightarrow$ + **" using the arrow. For example, we represent the recipe called "easy carbonara" that contains "+microwave" and "$-$pan" in cooking utensils as "$-$pan $\rightarrow$ + microwave." On the other hand, when the degree of co-occurrence of different elements is higher than the threshold amount, we consider that the different elements in the pair are independent, and therefore have no relations. Thus, we determine that elements + are additional elements and elements $-$ are deleted elements.

3.4 Extracting Additional Information

We consider that recipes' features of modifiers can be extracted from cooking procedures. For example, recipes for "Kid-friendly" may have features such as "small cut" and "easy-to-eat size." Therefore, we extract these features by analyzing the cooking procedures for the recipes. In order to extract them, we use word segmentation on the cooking procedures. Next, the cooking procedures are associated with word classes. We then extract additional information by dependency parsing.

3.5 Grouping Recipes Based on Feature Patterns

We define the feature patterns of recipes $P_{r_{j_k}}$ based on viewpoints grouped by the relations between different elements. More specifically, we define them based on a set of six viewpoints: additional ingredients I_{add}, deleted ingredients I_{del}, exchangeable ingredients I_{ex}, additional cooking utensils U_{add}, deleted cooking utensils U_{del}, and exchangeable cooking utensils U_{ex}. Then, to simplify, we represent patterns using binary vectors: when the element count is one or more, we present "1"; when the element count is zero, we present "0."

$$P_{r_{j_k}} = [bi(|I_{add}|), bi(|I_{del}|), bi(|I_{ex}|), bi(|U_{add}|), bi(|U_{del}|), bi(|U_{ex}|)]$$

We group recipes that use the same modifier by the feature patterns of viewpoints that have up to 64 ($=2^6$) patterns as Naming Concepts of modifiers.

4 Experiment

4.1 Dataset

We conducted an experiment using a recipe dataset provided by Rakuten Data Release from Rakuten Institute of Technology. From the dataset, we selected 192 recipes whose titles included the modifier "kid-friendly." We then extracted ingredients and cooking utensils as recipe elements by considering inconsistent spelling using a recipe ontology [12] that we constructed by integrating methods for extracting related words of a given word based on the data structure of user-generated recipes [11] and summarizing multi-documents. In order to extract the typical ingredients and cooking utensils for a dish, we calculated *RF* (Recipe Frequency) of all the recipes used in ten recipes for the dish. In the experiment, we set the values of thresholds α and β at 0.5. Table 1 presents the typical elements extracted for three dishes: hamburger steak, carbonara, and curry.

4.2 Result: Naming Concept Extraction

We extracted different elements by comparing the elements in a recipe with the typical elements for that recipe. Examples of the recipes used to extract Naming Concepts are shown in Table 2, and examples of the elements that differ from typical elements are shown in Table 3. In order to determine the relations between different elements, we made all possible pairs of elements of + and − from the set of different elements.

Next, we calculated the ratio of co-occurrence of each pair + and − in all the recipes for a dish. In the experiment, we set the value of the threshold at 0.1. Then, when the degree of co-occurrence was lower than the threshold, we determined the relation of the pair to be exchangeable. Conversely, when the degree of co-occurrence was higher

Table 1 Typical elements in recipe categories

	Hamburger steak	Carbonara	Curry
I	Onion	Egg	Curry powder
	Ground meat	Pepper	Onion
	Bread crumb	Salt	Water
	Salt	Bacon	Rice
	Pepper	Spaghetti	Carrot
	Egg	Cheese	Cooking oil
	Nutmeg	Water	Salt
	Worcester sauce	Cooking oil	Butter
	Ketchup	Fresh cream	
	Milk	Garlic	
	Cooking oil		
U	Frying pan	Frying pan	Pan
	Bowl		

Table 2 Example of recipes

	Kid-friendly, excellent and simple cheese in hamburger steak	Kid-friendly carbonara of tuna and corn	Beans curry can eat with children
I	Ground meat Onion Butter Bread crumb Soy milk Egg Salt Pepper Wine Sauce Ketchup Cheese Tomato	Spaghetti Cheese Egg Milk Non-dairy creamer Pepper Cooking oil Garlic Tuna Corn Water	Onion Garlic Tomato Soy bean Ground meat Curry powder Soy milk Water Cooking oil
U	Bowl Frying pan	Bowl Frying pan	Mixer Frying pan

Table 3 Elements that differ from typical elements

	Kid-friendly, excellent and simple cheese in hamburger steak	Kid-friendly carbonara of tuna and corn	Beans curry can eat with children
I_{add}	+Butter +Soy milk +Wine +Cheese +Lettuce +Tomato	+Milk +Non-dairy creamer +Tuna +Corn	+Garlic +Tomato +Soy bean +Ground meat +Soy milk
I_{del}	−Nutmeg −Milk −Cooking oil	−Salt −Bacon −Fresh cream	−Rice −Carrot −Salt −Butter
U_{add}		+Bowl	+Mixer +Frying pan
U_{del}			−Pan

than the threshold, we determined that no relation existed between the pair. Table 4 presents the results of the determination of the exchangeability relation using the typical elements in Table 1 and target recipes in Table 2 and calculating the degree of co-occurrence. We calculated the number of recipes containing elements + and − as the denominator, and determined the exchangeability relation when the degree of co-occurrence was lower than the threshold. We present the results for the extracted differing elements and features in the three recipes. Then, we present the different elements and viewpoints in Table 5.

We grouped 150 recipes into feature patterns of differences extracted from the experimental recipes, and extracted Naming Concepts for the modifier "kid-friendly."

Table 4 Results of calculation of confidence for judging the relation

Recipe titles	Pairs to determine relations	Degree of co-occurrence		Exchangeable
		+	−	
Kid-friendly, excellent and simple cheese in hamburger steak	+Butter, −nutmeg	0.12	0.16	
	+Butter, −cooking oil	**0.06**	**0.05**	T
	+Soy milk, −milk	**0.00**	**0.00**	T
	+Wine, −milk	**0.00**	0.18	
Kid-friendly carbonara of tuna and corn	+Milk, −fresh cream	0.24	0.17	
	+Non-dairy cream, −fresh cream	**0.00**	**0.00**	T
	+Tuna, −bacon	**0.00**	**0.00**	T
	+Corn, −bacon	**0.00**	0.25	
Beans curry can eat with children	+Garlic, −salt	0.13	0.23	
	+Garlic, −butter	0.10	**0.06**	
	+Tomato, −carrot	**0.04**	0.13	
	+Soy bean, −carrot	**0.00**	**0.09**	T
	+Ground meat, −carrot	**0.00**	0.15	
	+Mixer, −pan	0.70	**0.00**	
	+Frying pan, −pan	**0.00**	**0.00**	T

The recipes were grouped into 21 patterns from the 64 patterns in Sect. 3.5. In Table 6, we present the feature patterns which the number of grouped recipes is more than a threshold. In this experiment, we empirically set the threshold as 5.0 %. We emphasized the values of the degree of co-occurrence which were less than the threshod shown in bold in Table 4. As a result, eight patterns included the feature I_{ex} and five patterns included the feature I_{del}.

Table 7 presents recipe titles grouped into the top-three patterns. The recipes grouped by feature patterns do not become unbalanced because of the dishes. Therefore, recipes for the same dish are grouped by various patterns.

4.3 Discussion

In the experiments, typical cooking utensils that are normally included in typical elements were not included. For example, when cooking carbonara, a pan is typically used to boil spaghetti. However, in the experiment, it was not included in the typical elements because there were few recipes that had "pan" expressly written. When we extracted difference elements, cooking utensils were included in the difference elements that were normally included in typical elements because cooking utensils were used in pictures, but not included in the procedures. In the

Table 5 Results of extracted difference elements and features in the three recipes

Recipe titles	Different elements	Viewpoints
Kid-friendly, excellent and simple cheese in hamburger steak	+Wine	−
	+Cheese	−
	+Lettuce	I_{add}
	+Tomato	I_{del}
	−Cooking oil	I_{ex}
	−Nutmeg	−
	+Butter	−
	−Milk	−
	+Soy milk	−
Kid-friendly carbonara of tuna and corn	+Milk	−
	−Salt	I_{add}
	−Fresh cream	I_{del}
	+Non-dairy cream	I_{ex}
	−Bacon	U_{del}
	+Tuna	−
	−bowl	−
Beans curry can eat with children	+Garlic	−
	+Tomato	−
	+Ground meat	I_{add}
	+Soy milk	I_{del}
	−Rice	I_{ex}
	−Salt	U_{add}
	−Butter	U_{ex}
	−Carrot	−
	+Soy bean	−
	+Mixer	−
	−Pan	−
	+Frying pan	−

future, we need to improve the extraction of typical elements and determine the exchangeability relations. Therefore, we consider that we can make up for cooking utensils by inferring the cooking utensils used from the actions in the procedures. For example, we can infer that "boil" means "pan" and "fry" means "frying pan," and so on.

When we determined the exchangeability relation and compared our results with the correct data, we found that a number of pairs had appropriately determined the relation. However, with regard to cooking utensils, there are many recipes in which only actions are written (e.g. "boil" and "fry"). Therefore, cooking utensils that would not normally determine the relation of exchangeability here, do so because of the lowness of their degree of co-occurrence, resulting from them not being cooking utensils are not expressly written in recipes.

Table 6 Naming concepts for "kid-friendly"

	I_{add}	I_{del}	I_{ex}	U_{add}	U_{del}	U_{ex}	Appearance ratio (%)
Pattern 1	0	0	1	0	0	0	18
Pattern 2	0	1	1	0	0	0	18
Pattern 3	0	0	1	0	1	0	8
Pattern 4	0	1	1	1	1	0	7
Pattern 5	1	1	1	0	0	0	6
Pattern 6	0	1	1	1	0	0	5
Pattern 7	1	1	1	0	1	0	5
Pattern 8	1	0	1	0	0	0	5

Table 7 Recipe titles (translated from Japanese) grouped into the top-three patterns

Pattern	Recipe titles
Pattern 1	Kid-friendly! Sweet and sour pork of pork loin
	Smile for children 3 kinds of hamburger steaks
	Kid-friendly hamburger steaks of fish
	Healthy, big and kid-friendly hamburger steak with vegetable
	Kid-friendly the curry with being grated up vegetable
	Chocolate in the curry for children
	Kid-friendly! Corn in the curry for children
	The curry for children
	The curry eat with children
	Kid-friendly milk curry
	The curry for children with a milk
Pattern 2	Tomato in the sweet and sour pork kid-friendly!
	The hamburger steak in Denmark! Kid-friendly!
	Kid-friendly hamburger steaks of tohu
	Kid-friendly hamburger steaks! Not use flour and egg
	Kid-friendly hamburger steaks with many kinds of vegetables
	For children don't like vegetables! Hamburg steak of green pepper
	Kid-friendly! Curry with pork cutlet and mushroom
	Kid-friendly cheese hamburg steak
	The curry for children!
Pattern 3	Kid-friendly sweet and sour pork with potato
	Kid-friendly and soft hamburg steak
	Kid-friendly! Hamburg steak with hijiki
	Kid-friendly! Raisin in dried curry
	Simple and kid-friendly curry in Keema
	Kid-friendly curry of Japanese radish and tohu
	Kid-friendly mochi in the curry

In Table 5, we consider that the Naming Concepts for "kid-friendly" are addition, deletion, and exchange of ingredients. For example, in the recipe for carbonara in Table 5, we could extract the exchangeability relation in which "tuna" is used instead of "bacon" as the feature. Thus, there are also many recipes that include cooking utensils as Naming Concepts. However, there are some extracted cooking utensil elements for which it is difficult to consider that the elements are Naming Concepts. For example, in the recipe for curry in Table 5, we extracted and used "frying pan" instead of "pan." However, it is difficult to consider these elements as Naming Concepts. Therefore, we need to consider whether extracted elements really present the concepts of modifiers or noise.

In the results of grouped recipes based on feature patterns and extracted Naming Concepts, we found that recipes included in the same category have various Naming Concepts such as those listed in Table 7. In this paper, we extracted Naming Concepts for "kid-friendly." However, we need to extract more Naming Concepts for various modifiers. By comparing the difference between modifiers, we

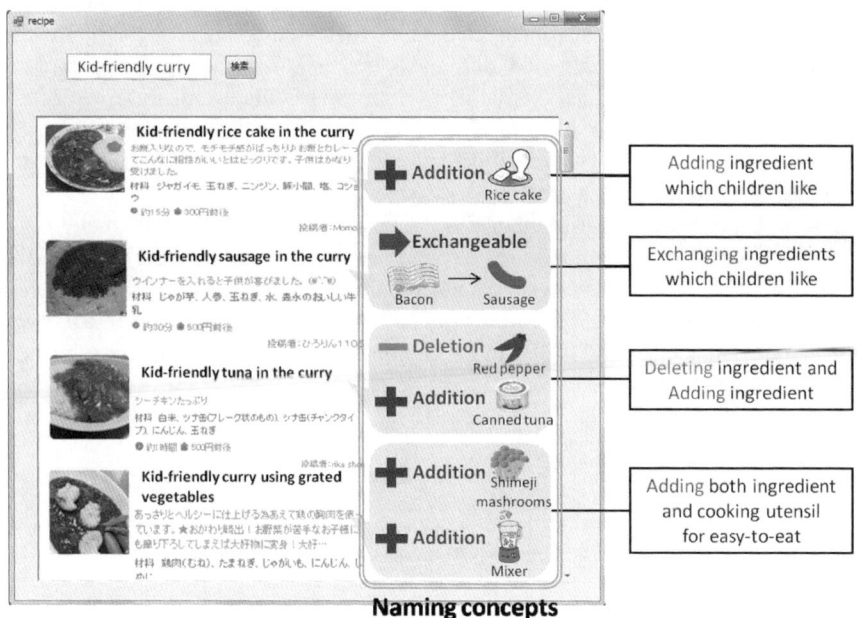

Fig. 3 User interface of our proposed system

could analyze the relations between modifiers. For example, when the trends for Naming Concepts are similar between modifiers, they have similar relations.

Figure 3 depicts the user interface of our proposed system that presents the Naming Concepts of recipes. In this system, when a user searches for recipes containing modifiers such as "Kid-friendly" and "Simple", it is difficult to understand the features of the recipes because a list of search results often show only titles and pictures for the recipes. Therefore, if users want to know the features of the recipes, they look into the details of each recipe. However, it is a really daunting task and determining its features is a time-consuming process. Therefore, our system enables users to comprehend the features of the recipes solely by checking the list of search results.

5 Conclusion

In this work, we proposed a system that extracts and presents Naming Concepts for recipes, which are defined as characteristic elements summarized by modifiers in the recipes' titles. We extracted different elements of ingredients and cooking utensils, determined the relations between them by calculating their degree of co-occurrence and extracted Naming Concepts by grouping the recipes based on feature patterns. Further, we experimented with a real recipe dataset to extract the Naming Concepts of given recipes.

In future work, we plan to enable the inferring of cooking utensils that are not written expressly in procedures because we found that typical cooking utensils tend to be omitted in the procedures of recipes. For instance, although we generally use a "frying pan" when "frying" something, it is often omitted because users/readers can easily associate the action "fry" with the cooking utensil "frying pan" in the recipes. Therefore, we plan to infer cooking utensils from procedures by considering actions in procedures.

Acknowledgments This research was supported in part by Strategic Information and Communications R&D Promotion Programme (SCOPE), the Ministry of Internal Affairs and Communications of Japan and JSPS KAKENHI Grant Number 26280042. The experimental Rakuten recipe dataset was provided by Rakuten Data Release from the Rakuten, Inc.

References

1. Rakuten recipe, http://recipe.rakuten.co.jp/
2. COOKPAD, http://cookpad.com/
3. A. Tachibana, S. Wakamiya, H. Nanba, K. Sumiya, Extraction of naming concepts based on modifiers in recipe titles, in *Proceedings of the International MultiConference of Engineers and Computer Scientists 2014, IMECS 2014*, 12–14 Mar 2014, Hong Kong. Lecture Notes in Engineering and Computer Science, pp. 507–512

4. C.-Y. Teng, Y.-R. Lin, L.A. Adamic, Recipe recommendation using ingredient networks, in *Proceedings of the 4th Annual ACM Web Science Conference (WebSci '12)*, pp. 298–307 (2012)

5. M. Ueda, M. Takahata, S. Nakajima, User's food preference extraction for cooking recipe recommendation, in *Proceedings of the 2nd Workshop on Semantic Personalized Information Management: Retrieval and Recommendation*, pp. 98–105 (2011)

6. M. Ueda, S. Asanuma, Y. Miyawaki, S. Nakajima, Recipe recommendation method by considering the user's preference and ingredient quantity of target recipe, in *Proceedings of International MultiConference of Engineers and Computer Scientists 2014 (IMECS 2014)*, pp. 519–523 (2014)

7. A. Yajima, I. Kobayashi, Easy cooking recipe recommendation considering user's conditions, in *Proceedings of the 2009 IEEE/WIC/ACM International Joint Conference on Web Intelligence and Intelligent Agent Technology—Volume 03 (WI-IAT '09)*, vol. 3, pp. 13–16 (2009)

8. K. Tsukuda, T. Yamamoto, S. Nakamura, K. Tanaka, Plus one or minus one: A method to browse from an object to another object by adding or deleting an element, in *Proceedings of the 21st International Conference on Database and Expert Systems Applications*, pp. 258–266 (2010)

9. K. Tsukuda, S. Nakamura, T. Yamamoto, K. Tanaka, Typicality analysis of an object and its application to search, in *WebDB Forum 2011*, 2G-1-2 (2011) (in Japanese)

10. Y. Yamakata, S. Imahori, Y. Sugiyama, S. Mori, K. Tanaka, Feature extraction and summarization of recipes using flow graph, in *Proceedings of the 5th International Conference on Social Informatics, LNCS 8238*, pp. 241–254 (2013)

11. Y.-J. Chung, Finding food entity relationships using user-generated data in recipe service, in *Proceedings of the 21st ACM International Conference on Information and Knowledge Management*, pp. 2611–2614 (2012)

12. H. Nanba, Y. Doi, M. Tsujita, T. Takezawa, K. Sumiya, Summarization of multiple cooking recipes, in *Proceedings of the 5th Symposium on Wisdom of Crowds, NLC2013-41*, vol. 113, no. 338, pp. 39–44 (2013) (in Japanese)

13. L. Lee, Measures of distributional similarity, in *Proceedings of the 37th Annual Meeting of the Association for Computational Linguistics*, pp. 25–32 (1999)

14. D. Lin, Automatic retrieval and clustering of similar words, in *Proceedings of the 36th Annual Meeting of the Association for Computational Linguistics and the 17th International Conference on Computational Linguistics*, pp. 768–774 (1998)

15. R. Takahashi, S. Oyama, H. Ohshima, K. Tanaka, Evaluating truthfulness of modifiers attached to web entity names, in *Proceedings of the 11th International Conference on Web-Age Information Management*, pp. 429–440 (2010)

Searching Comprehensive Web Pages of Multiple Sentiments for a Topic

Shoko Wakamiya, Yukiko Kawai, Tadahiko Kumamoto,
Jianwei Zhang and Yuhki Shiraishi

Abstract We have developed a novel system for searching comprehensive Web pages by focusing on multiplicity of sentiments of writers for a topic. Recently, lots of studies and services based on sentiment analysis have been conducted, since it is still difficult to search and summarize information satisfying users' needs by text analysis only. In this paper, we propose a system for searching and visualizing comprehensive Web pages in terms of sentiments by extracting multiple sentiments of Web pages on a query topic and re-retrieving Web pages using sub-topic keywords. Specifically, this system extracts sentiment features of each Web page using a sentiment dictionary consisting of three sentiment dimensions; "Happy ⟺ Sad," "Glad ⟺ Angry," and "Peaceful ⟺ Strained." Next, in order to conduct a re-retrieval, it extracts sub-topic keywords from Web pages of maximum (or minimum) sentiment features on three sentiment dimensions, respectively. Then, it re-retrieves Web pages using the query topic keyword and the extracted sub-topic keywords. Then, it plots them on sentiment graphs based on their sentiment features. By using the graphs, we can grasp not only sentiment tendency but also comprehensive sentiments for a query topic. In the experiment, we evaluate our proposed method using the developed system.

S. Wakamiya (✉) · Y. Kawai
Kyoto Sangyo University, Kyoto, Japan
e-mail: shokow@cc.kyoto-su.ac.jp

Y. Kawai
e-mail: kawai@cc.kyoto-su.ac.jp

T. Kumamoto
Chiba Institute of Technology, Narashino, Japan
e-mail: kumamoto@net.it-chiba.ac.jp

J. Zhang · Y. Shiraishi
Tsukuba University of Technology, Tsukuba, Japan
e-mail: zhangjw@a.tsukuba-tech.ac.jp

Y. Shiraishi
e-mail: yuhkis@a.tsukuba-tech.ac.jp

© Springer Science+Business Media Dordrecht 2015
G.-C. Yang et al. (eds.), *Transactions on Engineering Technologies*,
DOI 10.1007/978-94-017-9588-3_26

337

Keywords Information retrieval · Multiple sentiments · Re-retrieval · Sentiment analysis · Sentiment dictionary · Sentiment tendency

1 Introduction

General search engines such as Google, Yahoo!, and Bing have become main tools to obtain information on a topic from the WWW. These engines typically search Web pages concerning a given query and rank them by considering relevancy and popularity between each Web page and the query. Generally, users only look through Web pages in the top-ranked search results and may often find the information satisfying their search needs. However, existing search engines sometimes fail to give user-desired pages at top ranks due to the diversity of users' search intentions.

Previous work have been devoted to improve search results in three main directions: (i) ranking pages in initial search results, (ii) suggesting query expansion or reformulation for re-retrieving new pages, and (iii) analyzing sentiments of pages for opinion mining. As for (i), many aspects have been utilized to achieve effective ranking of search results, such as query logs [1], authorship [2], social tags [3], re-finding [4], and multiple pairwise relationships between pages [5]. As for (ii), query expansion or reformulation involves expanding or revising the search query to match additional or new pages by means of some techniques and information such as global analysis [6] and social annotation [7]. As for (iii), sentiment analysis and opinion mining [8, 9] have attracted a lot of research interests, which study sentiments and its related concepts such as opinions and attitudes. Interestingly, the role of sentiment in information retrieval has been investigated in some researches [10]. Especially, researches on sentiment retrieval or opinion retrieval [11–13] aim to provide a general opinion search service. These are similar to traditional Web search in the way that both of them try to find pages relevant to a query. On the other hand, these are different from the latter in the way that sentiment retrieval needs further determinations whether the pages express opinions on the query topic and whether their polarities are positive or negative.

In this paper, in order to attempt to grasp multiple sentiments for a topic, we develop a comprehensive Web search system based on multiple sentiments [14]. For this, we adopt more diverse sentiments such as $dim1$: Happy ⇔ Sad, $dim2$: Glad ⇔ Angry, and $dim3$: Peaceful ⇔ Strained, not restricted to positive-negative sentiment. In addition, users can specify not only the three sentiment features, but also strengths of the sentiments to rank and retrieve pages. These features are also utilized to perform a re-retrieval to obtain pages with various sentiments. Specially, we propose a system consisting of the following parts:

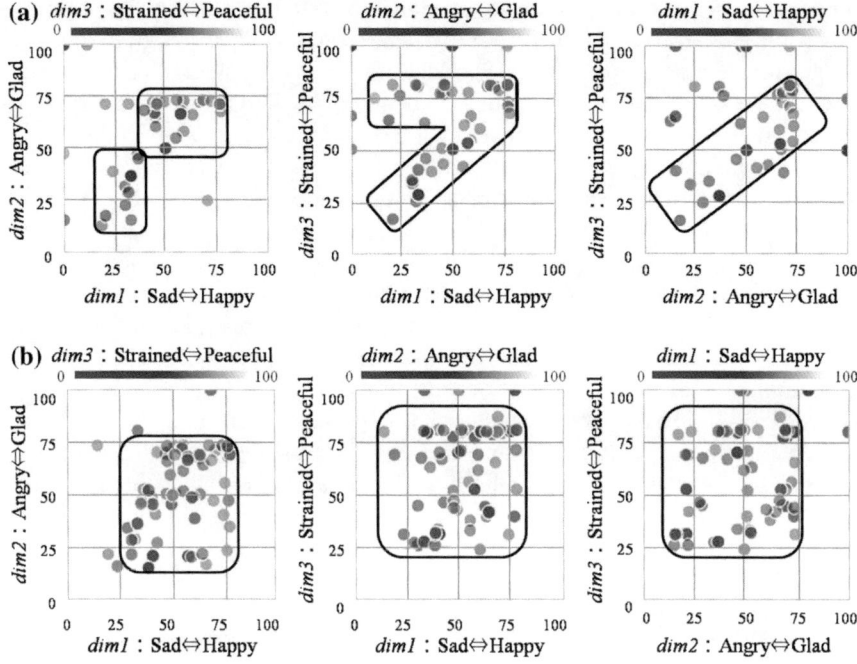

Fig. 1 Sentiment distribution of search results. **a** Search results by existing search engine (Google). **b** Search results by our proposed method

- extracting multiple sentiment features of Web pages on a given query topic using a sentiment dictionary [15] consisting of words and their sentiment features on three sentiment dimensions,
- extracting sub-topic keywords related to respective sentiments for expanding the query topic keyword, and
- re-retrieving comprehensive Web pages in terms of sentiments.

Figure 1 shows sentiment distributions of search results for a query topic 'child benefit' by plotting the search results based on their sentiment features on sentiment graphs. In these graphs, two of three sentiment dimensions are represented by the horizontal and vertical axes, and the remaining sentiment dimension is represented by graduations of color of plots. If a Web page expresses negative sentiments such as "Sad" on dim1, "Angry" on dim2, and "Strained" on dim3, its values on the horizontal and vertical axes are close to 0 and its color becomes deeper. On the other hand, if a Web page expresses positive sentiments such as "Happy" on dim1, "Glad" on dim2, and "Peaceful" on dim3, its values on the horizontal and vertical axes are close to 100 and its color becomes lighter. Especially, Fig. 1a is the sentiment distribution of Web pages searched by Google, and Fig. 1b is the one of Web pages searched by our proposed method. The query topic 'child benefit' is a law introduced in Japan about distributing social security payment to parents of children. In Fig. 1a, the sentiments of Google's search results on this topic are biased towards a little sad, a little angry, and a little strained on three sentiment graphs, because most of Web pages introduce that many people decline this offer or parents may abuse this grant. By the sentiment distribution of the Web pages, we can grasp sentiment tendency of writers for the topic. However, it is not possible to know comprehensive sentiments of writers. On the other hand, in Fig. 1b, the search results are more evenly distributed on three sentiment graphs. Therefore, we retrieve Web pages by considering comprehensive sentiments for a topic by our proposed method. When a user wants to read Web pages with other sentiments, s/he just clicks a point plotted on a sentiment graph.

The rest of this paper is structured as follows. Section 2 provides an overview of our proposed system. Section 3 describes how to extract sentiment features of Web pages using a sentiment dictionary. Section 4 explains methods to extract sub-topic keywords and to re-retrieve comprehensive Web pages with multiple sentiments. Section 5 shows experimental results and evaluates the effectiveness of our proposed system. Section 6 reviews related work. Finally, we conclude the paper with future work in Sect. 7.

2 System Overview

Figure 2 shows the overview of the proposed system. Given a query topic keyword by a user, the system performs the following process:

Step 1. Web pages concerning the query topic are returned as initial search results by Google Web Search API.[1] In order to analyze sentiments of Web pages, the system obtains their titles and snippets, respectively.

Step 2. The system extracts sentiments of each Web page using a sentiment dictionary proposed in our previous work [15]. In the dictionary, each word has its sentiment features on three sentiment dimensions, $dim1$: Happy $\Leftrightarrow$ Sad, $dim2$: Glad $\Leftrightarrow$ Angry, and $dim3$: Peaceful $\Leftrightarrow$ Strained. In order to extract the sentiment features of each Web page, the system extracts and averages the sentiment features of words in the title and snippet of a Web page. The details of sentiment extraction are described in Sect. 3.

Step 3. In order to present Web pages with their sentiments in the search results, the system generates three sentiment graphs consisting of two axes based on two sentiment dimensions and plots whose polarities are based on the other sentiment dimension. Then, it plots the Web pages based on their sentiments on the graphs, respectively.

Step 4. The system extracts sub-topic keywords related to each sentiment for expanding the query topic keyword. For this, it finds a representative Web

---

[1] Google Web Search API: https://developers.google.com/web-search/docs/.

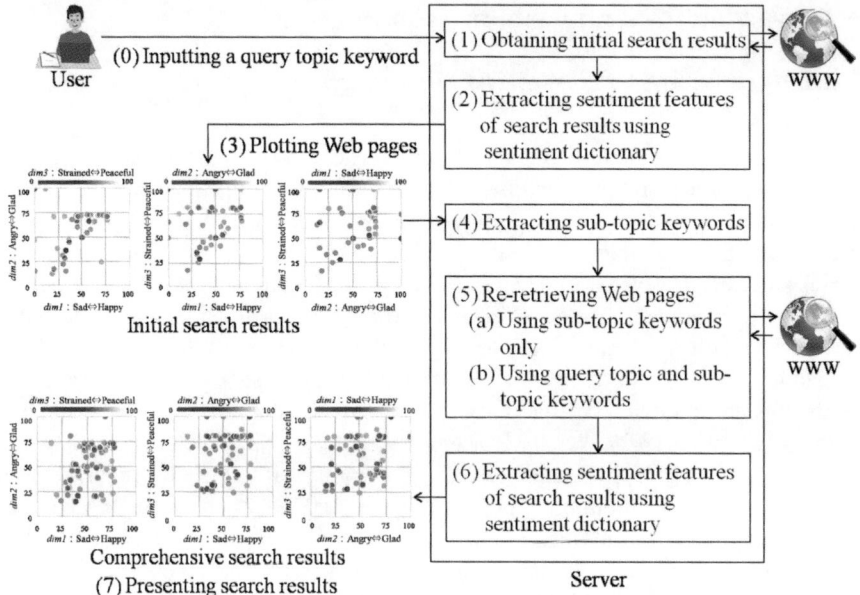

Fig. 2 Overview of comprehensive web pages search system based on multiple sentiments

page for each sentiment which is defined as a Web page whose sentiment features are maximum or minimum on two sentiment dimensions. Then, it extracts important words from the Web page based on term frequency. The details of a method to extract sub-topic keywords are described in Sect. 4.2.

Step 5. The system re-retrieves Web pages by the following two methods.

(a) it retrieves Web pages using the sub-topic keywords only, and extracts pages which include the query topic keyword from the search results, and

(b) it retrieves Web pages using both the sub-topic keywords and the query topic keyword.
The details of methods to re-retrieve comprehensive Web pages with multiple sentiments are described in Sect. 4.1.

Step 6. Similar to Step 2., the system extracts sentiments of the re-retrieved Web pages.

Step 7. Finally, the re-retrieved Web pages are presented to the user. Similar to Steps 3., the system plots them based on their sentiment features on the sentiment graphs, respectively.

3 Extraction of Sentiments in Web Page

3.1 Construction of Sentiment Dictionary

We construct a sentiment dictionary, in which each entry indicates the correspondence of a word and its sentiment features on three dimensions. The three-dimension sentiments are $dim1$: Happy ⇔ Sad, $dim2$: Glad ⇔ Angry, and $dim3$: Peaceful ⇔ Strained, that are formed based on a statistical analysis and a clustering analysis in our previous work [15]. A sample of the sentiment dictionary is shown in Table 1. Sentiment feature $s(w)$ of a word w on each dimension is a value between 0 and 1. The values close to 1 mean the sentiments of the words are close to "Happy," "Glad," or "Peaceful," while the values close to 0 mean the words' sentiments are close to "Sad," "Angry," or "Strained." For example, the sentiment feature of the word 'prize' on $dim1$: Happy ⇔ Sad is 0.862, which means the word 'prize' conveys a "Happy" sentiment. On the other hand, the sentiment feature of the word 'deception' on $dim2$: Glad ⇔ Angry is 0.075, which means 'deception' conveys an "Angry" sentiment.

For each of the three dimensions, we set two opposite sets (OW_L and OW_R) of original sentiment words (Table 2). The basic idea of sentiment dictionary construction is that a word expressing a left sentiment on a dimension often occurs with the dimension's OW_L, but rarely occurs with its OW_R. For example, the word 'prize' expressing the sentiment "Happy" often occurs with the words "Happy," "Enjoy," "Enjoyment," and "Joy," but rarely occurs with the words "Sad," "Grieve," "Sadness," and "Sorrow." We compare the co-occurrence of each target word with the two sets of original sentiment words for each dimension by analyzing the news articles published by a Japanese newspaper YOMIURI ONLINE between 2002 and 2006.

First, for each dimension, we extract the set S of news articles including one or more original sentiment words in OW_L or OW_R. Then, for each news article, we

Table 1 A sample of the sentiment dictionary

Word w	$s(w)$ on Happy ⇔ Sad	$s(w)$ on Glad ⇔ Angry	$s(w)$ on Peaceful ⇔ Strained
'prize'	0.862	1.000	0.808
'cooking'	1.000	0.653	0.881
'deception'	0.245	0.075	0.297
'death'	0.013	0.028	0.000

Table 2 Original sentiment words for three dimensions

Dimension	OW_L	OW_R
$dim1$: Happy ⇔ Sad	Happy, enjoy, enjoyment, joy	Sad, grieve, sadness, sorrow
$dim2$: Glad ⇔ Angry	Glad, delightful, delight	Angry, infuriate, rage
$dim3$: Peaceful ⇔ Strained	Peaceful, mild, primitive, secure	Tense, eerie, worry, fear

count the numbers of the words that are included in OW_L and in OW_R. The news articles, in which there are more words included in OW_L than in OW_R, constitute the set S_L. Inversely, the news articles, in which there are more words included in OW_R than in OW_L, constitute the set S_R. N_L and N_R represent the numbers of the news articles in S_L and S_R, respectively. For each word w occurring in the set S, we count the number of news articles including w in S_L and mark it as $N_L(w)$. Similarly, we count and mark the number of news articles including w in S_R as $N_R(w)$.

Sentiment feature $s(w)$ of a word w is calculated as follows:

$$s(w) = \frac{P_L(w) \times weight_L}{P_L(w) \times weight_L + P_R(w) \times weight_R},$$

where $weight_L$ and $weight_R$ are variables calculated by $\log_{10} N_L$ and $\log_{10} N_R$, respectively. Then, functions $P_L(w)$ and $P_R(w)$ calculate corrected conditional probabilities by dividing $N_L(w)$ and $N_R(w)$ into N_L and N_R, respectively.

3.2 Extracting Sentiments for Individual Pages and Summarizing Sentiments of Search Results

The sentiment features of a Web page are calculated by looking up sentiment features of respective words in the page from the sentiment dictionary and averaging them.[2] In this way, a page has sentiment features ranging from 0 to 1, since the sentiment features of the words in the sentiment dictionary range from 0 to 1. Considering the comprehensibility, each sentiment feature of a page is multiplied by 100 and converted to a value ranging from 0 to 100. Then, we show sentiment features of search results in response to a query as sentiment polarity and graduations of color of plots on sentiment graphs as shown in Fig. 1a, b.

4 Comprehensive Web Search Based on Multiple Sentiments

4.1 Sub-topic Keywords Extraction Based on Sentiments

After calculating sentiment features from initial search results, the sub-topic keywords with correlation and diverse sentiments are detected for re-retrieving Web pages. The system generates a sentiment query vector $V_q = (v_{q1}, v_{q2}, v_{q3})$ using the sentiment feature on each dimension as its element. For each Web page in the list of

---

[2] Since the title and snippet of a Web page summarize the content of the page and their text is shorter than the full page, the system actually calculates the sentiment features using the text of the title and snippet for each page so as to shorten the response time.

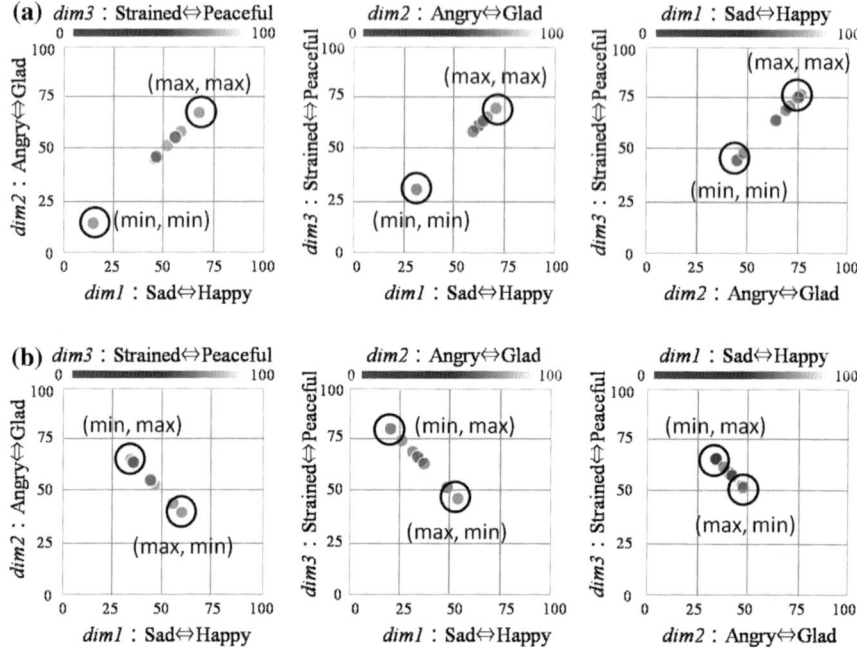

Fig. 3 Extracting representative web pages concerning sentiments. **a** Web pages mapped to $y = x$. **b** Web pages mapped to $y = -x$

initial search results, sentiment page vector $V_p = (v_{p1}, v_{p2}, v_{p3})$ $(p = 1, \ldots, N,$ where N is the number of initial search results) is determined, the elements of which are the sentiment features on three dimensions of each Web page.

We extract sub-topic keywords from representative Web pages concerning respective sentiments as follows:

1. The sentiment features on the three sentiment dimensions with respect to the query topic (sentiment summary of initial search results) are represented as a sentiment query vector $V_q = (v_{q1}, v_{q2}, v_{q3})$. Each page in the list of initial search results is also represented as a sentiment page vector $V_p = (v_{p1}, v_{p2}, v_{p3})$.

2. Each Web page is mapped to $y = x$ and $y = -x$ on pairs of sentiment features $((v_{p1}, v_{p2}), (v_{p2}, v_{p3}), (v_{p3}, v_{p1}))$ as shown in Fig. 3a, b, respectively. Then, we extract 12 Web pages whose pairs of sentiment features are maximum or minimum on respective sentiment graphs.

3. Keywords are extracted from the detected Web pages by calculating an importance of respective words based on term frequency. If the term frequency of a keyword is larger than a threshold, it is determined as a candidate of sub-topic keyword.

4. The system looks up sentiment features of respective candidate sub-topic keywords from the sentiment dictionary, converts their sentiment features to the

scale ranging from 0 to 100, and forms sentiment sub-topic keywords vector $V_s = (v_{s1}, v_{s2}, v_{s3})$.

The importance of a word w in a Web page of search results for a topic query q, $imp_q(w)$, is calculated based on tf-idf (term frequency times inverse of document frequency) as follows:

$$imp_q(w) = score(w) \times \log \frac{count(q)}{count(q \wedge w)},$$

where $score(w)$ is a score based on the term frequency ($0 \leq score(w) \leq 100$) and means relative importance between terms. Then, $count(q)$ is the number of Web pages searched by a query topic keyword q and $count(q \wedge w)$ is the number of Web pages searched by both words q and w. When the importance $imp_q(w)$ of a word is higher than a threshold, we regard it as an important word related to each sentiment. In the experiment described in Sect. 4.2, we empirically set a threshold as 40.

4.2 Re-retrieval of Web Pages Using Sub-topic Keywords and Query Topic Keyword

Keywords with high importance are determined as sub-topic keywords related to sentiments and are utilized to expand the initial query topic for re-retrieval of comprehensive Web pages. In order to re-retrieve Web pages written with more multiple sentiments, we propose two methods (a) and (b) as follows:

(a) Using sub-topic keywords only: In this method, the system re-retrieves Web pages using sub-topic keywords only. Then, it obtains Web pages whose snippets include the query topic keyword. By this, the sentiment distribution of the extracted Web pages is based on not the one of Web pages concerning the query topic keyword, but the one of respective sub-topic keywords.
(b) Using both query topic keyword and sub-topic keywords: The system can directly obtain Web pages which have high relevance with the query keyword and include all sentiments based on the three sentiment dimensions.

5 Evaluation

5.1 Analyzing Multiplicity of Sentiments of Web Pages

In order to verify the effect of our proposed method, we searched Web pages on a topic with its sub-topic keywords and plotted the Web pages based on their sentiment features on sentiment graphs as described in Sect. 4.1. In this experiment, we

Table 3 Query topic keywords and sub-topic keywords (translated from Japanese)

	Query topic keywords	Sub-topic keywords
1	'Liberal democratic party'	Microsoft, opinions, we, MSN Sankei news, president, Kafesuta, Chiyoda-ku, Tokyo, liberal democratic party, topics by
2	'Election'	Koizumi theater midst, candidates database, notification order, liberal democratic party official, all domestic, ballot box, kawasaki, Japan's largest, candidate poster board, politician, representative
3	'Special secret protection law'	Freedom of information act, Yosuke Isozaki, much ado about nothing, leakage, night watchman, good balance, prime minister aide, the ruling and opposition parties, anonymous, haver, protection law
4	'Tax increase'	Price, consumption tax act, public works, tax increase issue, monetary base, bank of Japan, Asahi Shimbun published, government budget, revenue, asset purchase policy, market, glossary
5	'TPP'	Commitment, ratchet provisions, agriculture, forestry and fisheries, year agreement, WTO Doha round, ISDS provisions, trans-pacific partnership agreement, Australian government, Amari minister, economic partnership agreement
6	'Democratic party'	Verbal abuse, democratic party, opinions, Turkmenistan, Thumbnail26, republican party, mental abuse, liberal democratic party, party platform, two-party system, December 2013 Ohata secretary-general briefing, president
7	'Territorial issue'	Parties to each other, partners, list of shops, territorial sea base, former soviet union, border issue, four islands conclusion of a peace treaty negotiations and northern, international court of justice, claim, Ichiran-ya, U.S. senate Senkaku problem, Mongolia

used seven keywords of query topics; there are four trend terms such as 'special secret protection law,' 'tax increase,' 'TPP', and 'territorial issue' and three general terms such as 'liberal democratic party,' 'election,' and 'democratic party.' Table 3 shows the query topics and their extracted sub-topic keywords. We compare the sentiment distributions of search results by an existing method (Google) with those of search results by proposed methods by using the query and sub-topic keywords. The number of search results are 96 Web pages, respectively.

Table 4 shows standard deviations of search results by the three types of searches for the seven query topic keywords. As for the search types, Proposed (a) means a search using sub-topic keywords only as described in Sect. 4.2(a) and Proposed (b) means a search using both query topic keyword and sub-topic keywords as explained in Sect. 4.2(b). In the experiment, we could confirm that the standard deviations of results of proposed methods become higher than the one of results of existing search engine for almost all query topics. Furthermore, the overall averages of each sentiment feature for all query topics become more diverse than the existing search engine.

Table 4 Re-ranking effect

Query topic keywords	Search type	Standard deviation			Average
		dim1 Happy⇔Sad	*dim2* Glad⇔Angry	*dim3* Peaceful⇔Strained	
1 'liberal democratic party'	Proposed (a)	15.48	15.52	17.25	16.08
	Proposed (b)	18.02	18.39	20.39	**18.93**
	Existing	18.18	19.47	18.89	18.84
2 'election'	Proposed (a)	17.51	15.14	16.63	**16.43**
	Proposed (b)	15.08	16.03	13.69	14.93
	Existing	14.04	13.03	15.26	14.11
3 'special secret protection law'	Proposed (a)	10.58	15.41	14.99	13.66
	Proposed (b)	16.73	21.43	22.13	**20.10**
	Existing	16.31	17.47	16.97	16.91
4 'tax increase'	Proposed (a)	18.03	12.37	12.78	14.39
	Proposed (b)	14.76	16.17	14.93	15.28
	Existing	17.62	19.37	20.12	**19.04**
5 'TPP'	Proposed (a)	17.32	21.30	17.00	18.54
	Proposed (b)	16.89	20.22	20.25	19.12
	Existing	17.32	19.29	21.46	**19.36**
6 'democratic party'	Proposed (a)	19.13	18.54	20.00	19.22
	Proposed (b)	18.78	19.85	22.02	**20.22**
	Existing	18.64	19.85	20.79	19.76
7 'territorial issue'	Proposed (a)	23.32	11.84	8.63	14.60
	Proposed (b)	20.21	19.02	15.01	**18.08**
	Existing	18.24	18.22	11.95	16.14
Average	Proposed (a)	17.34	15.73	15.32	16.13
	Proposed (b)	17.21	18.73	18.35	**18.09**
	Existing	17.19	18.10	17.92	17.74

However, for example, for the query topics 4 ('tax increase') and 5 ('TPP'), some sentiments of the search results by the existing method became higher than those of proposed methods. In that case, we considered that these query topics were controversial and had a tendency to be written with multiple sentiments. On the other hand, as for the query topic 'tax increase,' it was caused since some sub-topic keywords were lower relevance with the topic and sentiments. Therefore, we need to improve the method to extract meaningful sub-topic keywords.

5.2 Effect of Re-retrieval

In order to quantitatively evaluate the effect of re-retrieving Web pages, we compared sentiment distributions of 96 Web pages extracted by the initial search in Fig. 4a and by the re-retrieval in Fig. 4b, respectively. In Fig. 4, we show sentiment distributions of Web pages searched by the query topic 'territorial issue.' The topic means in conflict with economic interests such as resources between nations. The sentiments on this query topic that initial search results reflect were a little sad, a little angry, and peaceful (Fig. 4a), because most of Web pages in the initial search list introduce that territorial issue takes place to prioritize the interests of their own country. On the other hand, in Fig. 4b, the sentiments of Web pages searched by

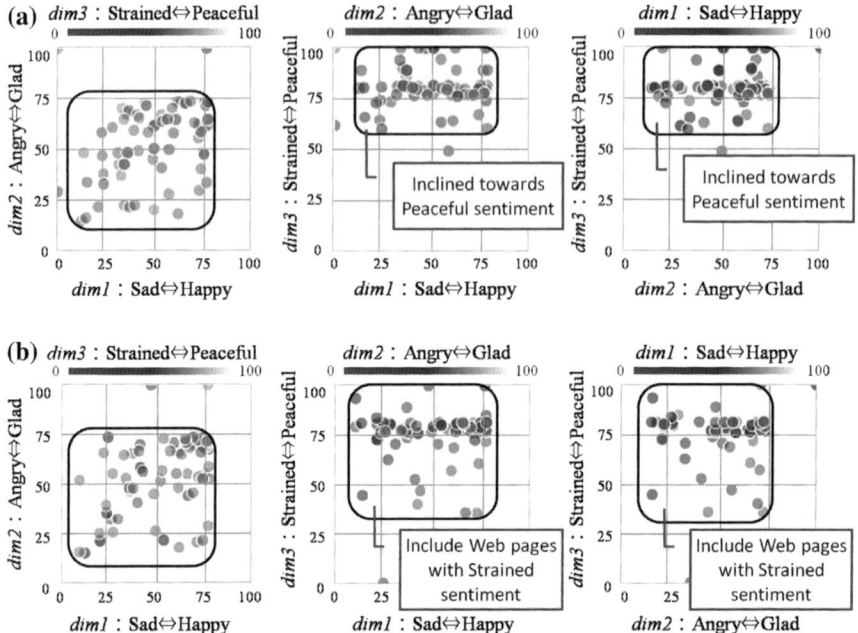

Fig. 4 Sentiment graphs. **a** Initial search results. **b** Re-retrieval results

re-retrieval became more multiple and some Web pages written with strained sentiment were searched. For example, we could find Web pages arguing about 'territorial issue' which is a tense area of conflict between countries express sad, angry, and strained sentiments.

5.3 Evaluations of Sentiment Features and Topical Precision of Search Results

In order to evaluate a topical precision between a query topic and each Web page, we conducted an experiment with participants. The participants consist of 10 students of a university. We first present 21 Web pages (=seven query topic keywords × 3 pages) searched by existing search engine and our proposed method, respectively. Here, as our proposed method, we utilized the method Proposed (b) which used both query topic keyword and sub-topic keywords as explained in Sect. 4.2b, because the multiplicity of sentiments of Web pages based on the standard deviation was higher than the method Proposed (a) in Sect. 5.1. In this experiment, the participants selected sentiment values of each Web page by seven levels; -3.0, -2.0, -1.0, 0, $+1.0$, $+2.0$, $+3.0$. In order to compare the sentiment features using the sentiment dictionary with the sentiment values by the participants, we computed the averages of absolute difference of them. For this, as for the

Table 5 Averages of differences of sentimental features in search results

	dim1	dim2	dim3	Average
Existing	13.28	15.59	28.37	19.08
Proposed	12.01	15.36	24.63	17.33

Table 6 Topical precision in search results

	Topical precision (%)
Existing	91.00
Proposed	89.41

sentiment values by the participants, we averaged sentiment values on three sentiment dimensions of each Web page and converted the values between 0 and 100. Table 5 shows the results. From this table, we could confirm that the difference for results based on the proposed method became larger than the one for results based on the existing method. In addition, the difference on the dimension 3 (*dim3*) became larger than the average in both existing method and proposed method.

Next, in order to evaluate a topical precision of search results, we conducted an experiment with the same participants. In this experiment, we asked the participants whether they could associate the searched Web pages with respective topics. If not, they answered keywords about topics that they associated with the Web pages. Table 6 shows the averages of topical precisions. From this result, we could confirmed that the topical precision of results based on our proposed method was slightly lower than the one of results based on the existing method. However, both averages were about 90 %, so we considered the topical precisions were adequately good.

When checking search results of both methods, we could find that search results by the existing method included Web pages such as dictionaries explaining meanings of a topic and official pages, and there were many Web pages whose contents were similar. On the other hand, in search results by the proposed method, we could find that lots of news articles and blogs about a topic were included. Because a query topic was regarded as sub-topic in many Web pages, the number of Web pages whose main topics were exactly same as a query topic was decreased. Finally, we could extract and present comprehensive Web pages related to a query topic and written with multiple sentiments.

6 Related Work

There are a number of studies on re-ranking search results considering various aspects. Zhuang and Cucerzan [1] proposed a Q-Rank method to refine the ranking of search results by constructing the query context from query logs. Bogers and van den Bosch [2] presented a passage-based approach to leverage information about the centrality of the document passages with respect to the initial search results list.

Yan et al. [3] proposed a Query-Tag-Gap algorithm to re-rank search results based on the gap between search queries and social tags. Tyler et al. [4] utilized the prediction of re-finding (finding the pages that users have previously visited) to re-rank pages. Kang et al. [5] improved search results by using demographical contexts such as gender, age, and income.

Another research direction for improving retrieval accuracy is to re-retrieve new search results based on query expansion or reformulation. Xu and Croft [6] utilized the hints such as query logs, snippets and search result documents from external search engines to expand the initial query. Lin et al. [7] proposed an automatic diagnosis of term mismatch to guide interactive query expansion or create conjunctive queries.

Similar to these researches we also aim to obtain a user-desired re-ranking and re-retrieval search results. Different from them we improve a personalized Web search using the sentiment features. The pages with sentiments similar to users' can be re-ranked to the top rank and the pages with opposite sentiments can be re-retrieved based on the extraction of opposite sub-queries.

On the other hand, sentiment analysis and opinion mining [8, 9] are one of the hottest research areas that extract sentiments (or opinions, attitudes) from text such as movie reviews, book reviews, and product evaluations. Some researches have applied sentiment knowledge to information retrieval and its relevant research areas. Arapakis et al. [10] investigated the role of sentiment features in collaborative recommendation and their experimental results showed the sentiment features extracted from movie reviews were capable of enhancing recommendation effectiveness.

Specially, sentiment retrieval or opinion retrieval is a newly developed research subject, which requires documents to be retrieved and ranked according to opinions about a query topic. Eguchi and Lavrenko [11] proposed several sentiment retrieval models based on probabilistic language models, assuming that users both input query topics and specify sentiment polarity. Similar methods proposed in [13] and [12] unified topic relevance and opinion relevance respectively based on a quadratic combination and a linear combination combined topic-sentiment word pairs in a bipartite graph to effectively rank the documents. Opinion retrieval from User Generated Content (UGC) such as blogs [16] and Twitter [17] also yields comparable retrieval performance.

Different from the above researches mainly focusing on review documents and positive-negative sentiments [18, 19], we consider any Web pages and more diverse sentiments. As we showed in the experiments, there are Web pages that express both positive sentiment and negative sentiment in different dimensions.

7 Conclusions

In this paper, we proposed a novel system for searching comprehensive Web pages written with multiple sentiments on a query topic by extracting sub-topic keywords related to each sentiment and re-retrieving Web pages. Specifically, in order to extract sentiments of Web pages, we used a sentiment dictionary consisting of words and their sentiment features on three sentiment dimensions, $dim1$: Happy ⇔ Sad, $dim2$: Glad ⇔ Angry, and $dim3$: Peaceful ⇔ Strained. Then, we developed a system which can present initial search results and comprehensive search results using sentiment graphs. By presenting initial search results, it could provide sentiment tendency for a query topic. Search results by our proposed methods could show more diverse Web pages in terms of their sentiments.

In future work, we plan to improve our method to extract Web pages with more multiple sentiments by repeatedly and efficiently re-retrieving Web pages. In addition, we need to improve the method to extract sub-topic keywords for re-retrieving Web pages. Furthermore, we will discuss a design of user interface for visualizing sentiment distribution of search results.

Acknowledgments This research was supported in part by Strategic Information and Communications R&D Promotion Programme (SCOPE), the Ministry of Internal Affairs and Communications of Japan, and JSPS KAKENHI Grant Numbers 24780248, 26280042, 26330347, 26330351, and 26870090.

References

1. Z. Zhuang, S. Cucerzan, Re-ranking search results using query logs, in *Proceedings of the 15th ACM International Conference on Information and Knowledge Management, CIKM '06*, pp. 860–861 (2006)
2. T. Bogers, A. van den Bosch, in *Authoritative Re-ranking of Search Results*, ed. by M. Lalmas, A. MacFarlane, S. Rüger, A. Tombros, T. Tsikrika, A. Yavlinsky. Advances in Information Retrieval, Lecture Notes in Computer Science, vol. 3936, pp. 519–522 (2006)
3. J. Yan, N. Liu, E.Q. Chang, L. Ji, Z. Chen, Search result re-ranking based on gap between search queries and social tags, in *Proceedings of the 18th International Conference on World Wide Web, WWW '09*, pp. 1197–1198 (2009)
4. S.K. Tyler, J. Wang, Y. Zhang, Utilizing re-finding for personalized information retrieval, in *Proceedings of the 19th ACM International Conference on Information and Knowledge Management, CIKM '10*, pp. 1469–1472 (2010)
5. C. Kang, X. Wang, J. Chen, C. Liao, Y. Chang, B. Tseng, Z. Zheng, Learning to re-rank web search results with multiple pairwise features, in *Proceedings of the Fourth ACM International Conference on Web Search and Data Mining, WSDM '11*, pp. 735–744 (2011)
6. J. Xu, W.B. Croft, Query expansion using local and global document analysis, in *Proceedings of the 19th Annual International ACM SIGIR Conference on Research and Development in Information Retrieval, SIGIR '96*, pp. 4–11 (1996)
7. Y. Lin, H. Lin, S. Jin, Z. Ye, Social annotation in query expansion: A machine learning approach, in *Proceedings of the 34th International ACM SIGIR Conference on Research and Development in Information Retrieval, SIGIR '11*, pp. 405–414 (2011)
8. B. Liu, *Sentiment Analysis and Opinion Mining* (Morgan & Claypool Publishers, Colorado, 2012)
9. B. Pang, L. Lee, Opinion mining and sentiment analysis. Found. Trends Inf. Retr. **2**(1–2), 1–135 (2008)
10. I. Arapakis, J.M. Jose, P.D. Gray, Affective feedback: An investigation into the role of emotions in the information seeking process, in *Proceedings of the 31st Annual International*

ACM SIGIR Conference on Research and Development in Information Retrieval, SIGIR '08, pp. 395–402 (2008)

11. K. Eguchi, V. Lavrenko, Sentiment retrieval using generative models, in *Proceedings of the 2006 Conference on Empirical Methods in Natural Language Processing, EMNLP '06*, pp. 345–354 (2006)

12. X. Huang, W.B. Croft, A unified relevance model for opinion retrieval, in *Proceedings of the 18th ACM Conference on Information and Knowledge Management, CIKM '09*, pp. 947–956 (2009)

13. M. Zhang, X. Ye, A generation model to unify topic relevance and lexicon-based sentiment for opinion retrieval, in *Proceedings of the 31st Annual International ACM SIGIR Conference on Research and Development in Information Retrieval, SIGIR '08*, pp. 411–418 (2008)

14. K. Minami, S. Wakamiya, N. Hata, Y. Kawai, T. Kumamoto, J. Zhang, Y. Shiraishi, Comprehensive web search based on sentiment features, in *Proceedings of the International MultiConference of Engineers and Computer Scientists 2014, IMECS 2014*, 12–14 Mar 2014, Hong Kong. Lecture Notes in Engineering and Computer Science, pp. 483–488

15. J. Zhang, Y. Kawai, T. Kumamoto, S. Nakajima, Y. Shiraishi, Diverse sentiment comparison of news websites over time, in *Proceedings of the 6th KES International Conference on Agent and Multi-Agent Systems: Technologies and Applications, KES-AMSTA'12*, pp. 434–443 (2012)

16. W. Zhang, C. Yu, W. Meng, Opinion retrieval from blogs, in *Proceedings of the Sixteenth ACM Conference on Conference on Information and Knowledge Management, CIKM '07*, pp. 831–840 (2007)

17. Z. Luo, M. Osborne, T. Wang, Opinion retrieval in twitter, in *Proceedings of the International AAAI Conference on Weblogs and Social Media* (2012)

18. S. Chelaru, I.S. Altingovde, S. Siersdorfer, W. Nejdl, Analyzing, detecting, and exploiting sentiment in web queries. ACM Trans. Web **8**(1), 6:1–6:28 (2013)

19. G. Demartini, S. Siersdorfer, Dear search engine: What's your opinion about...?: Sentiment analysis for semantic enrichment of web search results, in *Proceedings of the 3rd International Semantic Search Workshop, SEMSEARCH '10*, pp. 4:1–4:7 (2010)

Analyzing Early Mentioning of Past Buzzwords for Determination of Bloggers' Buzzword Prediction Ability

Seiya Tomonaga, Shinsuke Nakajima, Yoichi Inagaki,
Reyn Nakamoto and Jianwei Zhang

Abstract The goal of our research is to discover factors which predict which words will become buzzwords—terms representing topics that have become popular—within the blogosphere. In this paper, we propose a method which evaluates bloggers' buzzword prediction ability by analyzing how early bloggers mentioned past buzzwords. We do so by measuring how early a buzzword is first mentioned until the buzzword's peak in popularity. We describe this method and also report the evaluation on buzzword classification.

Keywords Blog analysis · Blog big data analysis · Blog mining · Buzzword prediction · Category classification · Trend analysis

S. Tomonaga · S. Nakajima (✉)
Faculty of Computer Science and Engineering, Kyoto Sangyo University, Motoyama,
Kamigamo, Kita-ku, Kyoto-City 603-8555, Japan
e-mail: nakajima@cse.kyoto-su.ac.jp

S. Tomonaga
e-mail: i1358068@cse.kyoto-su.ac.jp

Y. Inagaki · R. Nakamoto
Yoichi Inagaki Kizasi Company, Inc., 6th Floor Tomoe Nihonbashi Building,
20-14 Hakozaki-Cho, Nihonbashi, Chuo-ku, Tokyo 103-0015, Japan
e-mail: inagaki@kizasi.jp

R. Nakamoto
e-mail: reyn@kizasi.jp

J. Zhang
Faculty of Industrial Technology, Tsukuba University of Technology,
Amakubo 4-3-15, Tsukuba-City, Ibaraki 305-8520, Japan
e-mail: zhangjw@a.tsukuba-tech.ac.jp

© Springer Science+Business Media Dordrecht 2015
G.-C. Yang et al. (eds.), *Transactions on Engineering Technologies*,
DOI 10.1007/978-94-017-9588-3_27

353

1 Introduction

Discovering buzzword before they become popular is very difficult task. However, it is very beneficial to do so, especially from say a marketing point of view. Therefore, we are working on developing methods to discover buzzwords early based on analyzing consumer generated media such as blogs and social media. By analyzing these such sources, we believe it is possible to discover future buzzword candidates.

There has been several related works in regards to predicting which topics would become buzzwords. Nakajima et al. [1, 2] implemented a system which predicted what topics would likely become buzzwords in the future and found moderate success in our experiments. In the paper, we proposed a method to discover bloggers who frequently mentioned buzzwords before they became popular and used these bloggers to find new buzzword candidates. Here, bloggers who were good predictors of buzzwords in a certain field or category were assumed to continue being a good predictor of buzzwords in the future.

In our own previous research [3–5], we looked into measuring a blogger's buzzwords prediction ability. However, we did not look into automatic detection of past buzzwords and additionally, we did not take into account the time from the first mention of a buzzword until the peak popular of a buzzword.

In this paper, we propose automatic detection method of past buzzwords and a method for estimation of the time from the first appearance to the peak of popularity. We utilize not only the buzzword itself, but also consider the related co-occurrence words, which captures the depth and richer information of the topic.

For example, in the case of the iPhone 5, we would value a blogger who mentions the specs and features of iPhone5 over a blogger who just mentions that they want a iPhone 5. The blogger who mentioned the specs would most likely be more knowledgeable and therefore a better predictor of future buzzwords.

We measure the growth period of buzzword, and then we rate a blogger's buzzword predictive ability by rating how early a blogger mentions a buzzword and it's related co-occurrence words. With this, we can provide a method to find bloggers with high buzzword predictive ability. In addition, in order to identify areas of bloggers' buzzword prediction ability, we categorize buzzword candidate.

Note that this paper are extended and revised of the paper has been submitted in IMECS 2014 [6].

In Sect. 2, we describe related work. In Sect. 3, we describe the measurement of bloggers' buzzword prediction ability based on analyzing frequency of early mentions of past buzzwords. In Sect. 4, we describe the evaluation about category classification of buzzword candidate. In the last Section, we describe our future work as well as our conclusions.

2 Related Work

Research for discovering buzzwords and analyzing of trend diffusion by analyzing blog are as follows. Okumura [7] proposed a system for measure how popular a keyword has become by looking at the changes in frequency of a keyword. Furukawa et al. [8] looked into determining the most important words by tracking how a topic propagated. Kim et al. [9] analyzed trends of news diffusion in news media/SNS/blog based on crowd phenomena.

There also existing systems for detecting buzzwords or trends, such as Yahoo! Buzz Index [10], BlogPulse [11]. Yahoo! Buzz Index [10] calculates a topic's buzz score based on the percentage of Yahoo! users searching for that subject on a given day, and identifies "leaders" (subjects with the highest buzz scores) and "movers" (subjects with the highest percentage increase in buzz scores from one day to the next). BlogPulse [11] extracted key phrases and key people from blog entries by calculating the ratio of the frequency of occurrence of a phrase or a person name to its average frequency over the past two weeks.

These systems analyze the number of bloggers who write a word as well as the change in frequency in order to extract buzzwords. In contrast, our proposed method first discovers bloggers with good buzzword prediction ability and then extracts buzzwords candidates from their blogs.

3 Measurement of Bloggers' Buzzword Prediction Ability

We extract seed words (buzzwords of past) by analyzing past blog articles and then bloggers who mention these words early are determined to have high buzzword predictive ability. Our data set comes from kizasi.jp, a company which mines blog data. As of September 6, 2013, the data set consisted of 12,103,387 bloggers and 172,018,786 entries.

The steps of our method is as follows and will be explained in each section:

1. Classification of bloggers into knowledge groups (Sect. 3.1)
2. Extraction of seed words by analyzing blog set (Sect. 3.2)
3. Estimation of growth period of seed words (Sect. 3.3)
4. Calculation of bloggers' buzzword prediction score for individual entries (Sect. 3.4)
5. Calculation of bloggers' buzzword prediction ability (Sect. 3.5)

3.1 Classification of Bloggers into Knowledge Groups

Our purpose is to discover bloggers who pickup and discuss new topics before they popular. In order to do so, it is determine which areas each blogger is knowledgeable in. We assume that a blogger who possess a high level of knowledge about a certain field or category should be a good predictor of future buzzwords.

In the same line of thinking, a blogger who is knowledgeable about one category would not necessarily be a good predictor of buzzword in another category. For example, a blogger who frequency writes about internet would not necessarily be a good predictor of buzzwords in the topic of the economy.

Therefore, we classify each blogger into knowledge groups and then rank the bloggers within that group according to the knowledge level.

We use knowledge groups which were determined from our previous research. This system grouped bloggers into knowledge groups and then ranked their knowledge within that particular group.

3.2 Extraction of Seed Words by Analyzing Blog Set

3.2.1 Extraction of Seed Word Candidates

We first extract seed word candidates.

We start with using a keyword ranking system provided kizasi.jp [12] which rankings popular keywords written in blogs. This list shows the top keywords written in blogs for that day.

We first take the top hundred keywords for each day over a two year period. Next, we exclude repeated words, common words, certain seasonal keywords, and lastly infrequently occurring words. Seasonal keywords are words that appear regularly depending on the season. For example, "New Year's" or "Summer Festival". Also, every four years, there would keywords such as "Olympic", "FIFA World Cup" and "WBC" etc. The remaining keywords become our seed word candidates.

3.2.2 Category Classification of Seed Word Candidates

Since it is necessary to classify buzzwords into categories, we must determine which categories each seed word candidate belongs to. In addition, in order to efficiently find bloggers who are good buzzword predictors, we focus on bloggers who are knowledgeable in the categories in which these seed words belong.

Thus, it is necessary to find the category which matches the each keyword the best. For categories, we use the knowledge groups which will be described in Sect. 3.1.

In order to classify a seed word, we first calculate the semantic proximity of a seed word candidate and knowledge groups. We do so by calculating the similarity of co-occurrence word set of seed word candidate and the co-occurrence word set of a knowledge group. Each co-occurrence word set of a seed word candidate consists of the top 400 co-occurring words in all blog posts in the data set. Each co-occurrence word set for a knowledge group is the top 400 co-occurring words in blog posts which were written by bloggers who belong to the knowledge group. We considered using three different measures for calculation similarity between these word sets: Jaccard similarity measure, Simpson diversity Index, and lastly, cosine similarity.

3.2.3 Recognition of Seed Words Based on Degree of Influence

Since we are trying to extract bloggers who mentioned past buzzword before they become popular, it is necessary to focus on influential seed words.

To calculate a seed word's influence, we use the total number blog posts of a word during period T after the peak of the number of blog entries containing the word.

Here are the steps for calculating the degree of influence: First, we investigate the number of posts of each of the seed word candidates. Second, we calculate the moving average of the number of posts for each seed word candidate for the past two years. We then confirm the peak of the number of posts for that seed word candidate.

We assume that the peak is the point of highest recognition by the general population. If the number of posts are very low during period T after the peak, we assume the seed word candidate is quickly forgotten and is of low influence. If the number of posts does not decrease during period T following the peak, that keyword's influence is high. These words become seed words (Fig. 1).

In addition, we try to find several knowledge groups highly relevant to each seed word candidate. As the result, we can identify if the domains of the seed words can be used as learning data.

3.3 Estimation of Growth Period of Seed Words

We now describe the process for the estimation of the growth period of a seed word. The growth period is the period from when a seed word first starts to be posted in blogs until the peak of a seed word's popularity.

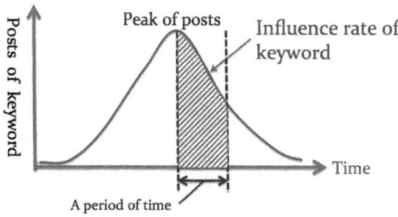

Fig. 1 Conceptual view of influence rate

3.3.1 Calculating the Similarity Between Word Set of Peak and Word Sets Before the Peak

In order to estimate the growth period of the seed words, we first find when a seed word starts to be mentioned. In doing so, it is necessary to confirm that the topic contents at that time matches the content of when it is at its peak.

For example, if "iOS5" is the seed word, the peak time would contains keywords about the specifications or features the new iOS. If a blog post just asking when iOS 5 is to be released rather than talking about specifics, it would not match the peak time and thus, we cannot consider the blogger to be have good predictive ability. In other words, to use only the seed word "iOS5" and not write any specific or useful information is not considered to be a early mention of a buzzword.

On the other hand, if a blogger posts that the "voice assistant feature named Siri will be included on the iOS5", we can considered that the blogger has some prediction ability about the related category.

We estimate the time when the topic of a seed word begins including the nuances of the peak of the epidemic by calculating the similarity between the co-occurrence word set of the seed word before its peak to the co-occurrence word set of its peak. For example, here is "iOS5" (Fig. 2).

In addition, the co-occurrence word set of seed words are top 400 co-occurring words after removal of generic terms during the topics peak popularity. Words that appear more frequently as we approach the peak of popularity are of particular importance.

In order to discover specific keywords at the time of peak popularity, we use Spearman's rank correlation to determine if there is an increasing number of blog posts containing a co-occurrence word.

Fig. 2 Calculating the similarity between word set of peak and word sets before the peak

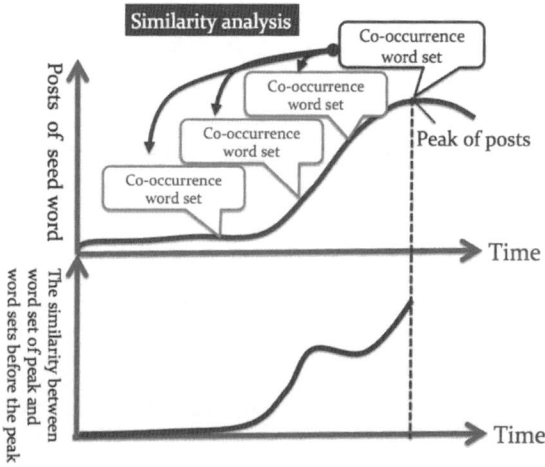

If there is a correlation of fairly strong positive (more than 0.6 coefficient of Spearman's rank correlation), this co-occurrence word has increasing number of posts as we approach the peak and it is considered to be a important keyword at the time of its peak popularity.

3.3.2 Determination of the Growth Period of Seed Word

It is possible to examine the change in content from a specific period to the peak period by calculating similarity between topics of that specific period to the peak period.

If there is a rise in similarity in the vicinity of the peak, we consider this point to be when the seed word began.

However, there are no guarantee that the similarity is zero during other points. We much consider points where there is some, but low similarity to be background noise.

Therefore, we try to apply the following two methods to determine the beginning of the growth period of a seed word.

1. Find the x-intercept of the first-order approximation line of the similarities between the word set of each period and the word set of the peak period. The x-intercept is regarded as the start point of the growth period of the seed word (Fig. 3).
2. Set a threshold θ for the similarity between the word set for each period versus the word set of the peak period. Then, the point that the similarity exceeds the threshold θ for the first time is regarded as the start of the growth period (Fig. 4).

In this way, we try to find the start point of the growth period. The growth period of each seed word is determined by both the start point and the popularity peak.

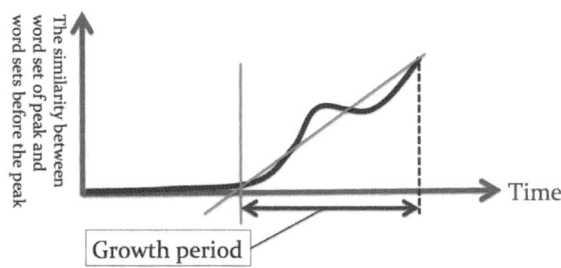

Fig. 3 The method for determining growth period of seed words (using x-intercept)

Fig. 4 The method for
determining growth period of
seed words (using threshold)

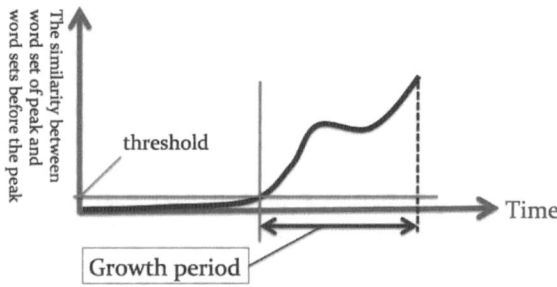

Growth period

3.4 Calculation of Bloggers' Buzzword Prediction Score for Individual Entries

In this section, we explain the method for calculating bloggers' buzzword prediction score for blogger entries.

The bloggers' buzzword prediction score is calculated for each blog article posted in the growth period of the seed word. Highest scores are given to entries near the beginning of the growth period. The lowest scores are given to entries at the end of the growth period, i.e. the peak.

The calculation for a blogger's buzzword prediction score $PredictionScoreEnrty_i$ for entry $entry_i$ is shown in formula 1.

$$PredictionScoreEntry_i = \frac{(entry_{all} + 1) - order_i}{entry_{all}} \tag{1}$$

$entry_{all}$ is the number of entries that a blogger posted about the seed word during the growth period of seed word.

$Order_i$ is the value representing how early a blogger early posted about the seed word during the growth period. It is normalized value from 0 to 1 based on the particular entry order in that blogger's posts during that period.

For example, say the number of entries $entry_{all}$ is 100. $order_i$ would be assigned to entries 1–100 as $1, 0.99, \ldots, 0.02, 0.01$.

3.5 Calculation of Bloggers' Buzzword Prediction Ability

In this section, we explain calculation method of bloggers' buzzword prediction ability of each seed word.

Below, we show conditions which reflect a high prediction ability for blogger's buzzword prediction.

Fig. 5 Similarity analysis of blog posts for bloggers' buzzword prediction ability

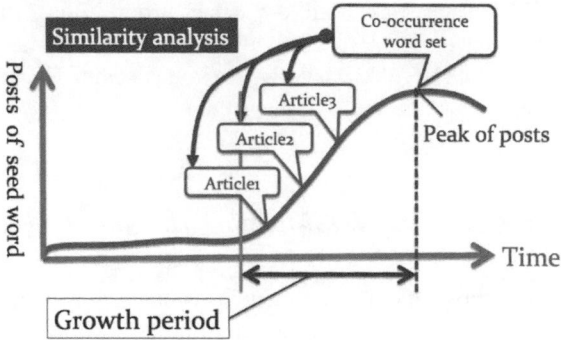

- Posting articles related to the target seed word early in the growth period.
- Posting many articles related to the target seed word in the growth period (Fig. 5).

The calculation of bloggers' buzzword prediction ability $PredictionScore_{(A,x)}$ for a seed words A is show in formula 2

$$PredictionScore_{(A,x)} = \sum_{k=1}^{N} Sim(D_A, entry_k) \times PredictionScoreEntry_k \quad (2)$$

N is the number of articles that blogger x posted in the growth period. D_A is the co-occurrence word set at the peak of seed word A. $Sim(D_A, entry_k)$ is the similarity between the topic of peak of seed word A and topic of article $entry_k$. $PredictionScoreEntry_k$ is the bloggers' buzzword prediction score for entry $entry_k$ as described in Sect. 3.4.

4 Category Classification Evaluation Experiment of Buzzword Candidates

In this section, we discuss our empirical evaluation for category classification of seed word candidates as described in Sect. 3.2.2.

It is necessary to categorize seed word candidates in order to understand categories related to each buzzword. In addition, in order to efficiently discover bloggers who are good buzzword predictors, we focus on bloggers who are knowledgeable in topics related to the seed word candidates. From this reason, it is necessary that we determine which knowledge groups are semantically similar to each seed word candidate.

The categories used are taken from "Blog Ranking Based on Bloggers' Knowledge Level for Providing Credible Information" [13]. However, in that paper, many of the categories are too specific. Therefore, we use only the top 120 categories from that.

4.1 Category Classification Method of Seed Words

The purpose of our experiment is to confirm the method can accurately classify seed words into appropriate categories. In other words, the goal is to evaluate if we can assigned a strongly related group to each seed word.

The relevance measure, which represents how semantically similar a seed word and a knowledge group is, calculated by the similarity between the co-occurrence word set of the seed word candidate and the knowledge group. After that, we take the knowledge group which has the high similarity. Similarities measures used in this evaluation experiment are jaccard coefficient (3), simpson coefficient (4), cosine similarity (5), and co-occurrence ranking points weighted cosine similarity (6).

$$jaccard(S, C) = \frac{|S \cap C|}{|S \cup C|} \tag{3}$$

$$simpson(S, C) = \frac{|S \cap C|}{\min(|S|, |C|)} \tag{4}$$

$$cosSim(S, C) = \frac{|S \cap C|}{\sqrt{|S|^2}\sqrt{|C|^2}} \tag{5}$$

$$cosSim_{Rank}(S, C) = \frac{\sum_{i=1}^{N}(r_{Si} \cdot r_{Ci})}{\sqrt{\sum_{i=1}^{N} r_{Si}^2}\sqrt{\sum_{i=1}^{N} r_{Ci}^2}} \tag{6}$$

Here, S is the word set of the top 400 words co-occurring with the seed word candidate with in the set of crawled blog entries C is the word set of the top 400 words co-occurring with the knowledge group's representative keyword.

In addition, r_S is ranking score from 400 to 1 that we allocate to each keyword S in order of co-occurrence degree. r_C is ranking score from 400 to 1 that we assign to each keyword C in order of their co-occurrence degree. N is number of keywords to be compared—in this experiment, it is 400.

Using each of the four methods, we determine which seed words and knowledge groups have high similarities, and then determine which of the four methods gave the best results.

As an additional baseline comparison, we used a method from "ratios of bloggers in a community who talked about the target seed word to all the bloggers in the community [1, 2]". The method is based on the idea that if the majority of knowledge group P mentions seed words Q, that word is related to group P.

4.2 Method of Experiment and Evaluation

Seed words which were used this experiment are "AKB48", "Android", "Facebook", "Joshikai", "K-POP" and "Smartphone". These words were popular keywords in the past. The goal of this experiment is to investigate that proposed method can automatically determine the correct knowledge group for each keyword. We first manually created a ranking of the top ten related categories for each of the six seed words. We call the rankings "ideal knowledge group ranking". The similarity method that produced categories most similar to our ideal knowledge group would be the best method.

Blog posts to be used in this experiment were separated into three month periods from July 2012 to July 2013 as such: "2012.07.01–2012.09.30 (period 1)", "2012.09.30–2012.12.31 (period 2)", "2012.12.31–2013.04.01 (period 3)" and "2013.04.01–2013.07.02 (period 4). Then, co-occurrence lists were created for each period for each knowledge group. The reason for having multiple periods is that we assume that over time a seed words related categories would change, and thus, we wished to confirm the changing of related categories. In addition, we use $nDCG$ and DCG as comparative evaluation measure. DCG is discounted cumulative gain, and it evaluates whether a correct ranking, including order, can be reproduced. In other words, it is a measure in which not only whether the ideal ranking is included, but also considers whether ranking order is correct as well. Additionally, $nDCG$ is normalized discounted cumulative gain, and its values are additionally normalized to 1.

We show the formula of DCG and $nDCG$ below.

$$DCG_p = rel_1 + \sum_{i=2}^{p} \frac{rel_i}{\log_2(i)} \tag{7}$$

$$nDCG_p = \frac{DCG_p}{IDCG_p} \tag{8}$$

rel_i is a item's score of the ith item in the target ranking. p shows the number of items in the target ranking. In this experiment, $p = 10$. In addition, $IDCG_p$ shows the ideal value of DCG_p. Also, it is possible to add a higher weight to the top items in a ranking, and thus, the value of DCG is higher if a method can rank a top item of ranking higher. In our experiment, we used 10, 9, 8, ..., 1 for our ideal ranking and then calculated DCG. We regard this ideal ranking's DCG as $IDCG_p$.

4.3 Results of Evaluation of Knowledge Group Ranking Based on Relevance

Here are the results of our experiment. Tables 1, 2, 3, 4, 5 and 6 shows the *nDCG* of previous studies as well as each method. The maximum value in each period in each table became bold (including the same rate). *A* is simpson's coefficient, *B* is jaccard's coefficient, *C* is cosine similarity, and *D* is weighted cosine similarity based on co-occurrence point.

Looking at Tables 1, 2, 3, 4, 5 and 6, ranking result is exactly the same result; however, each score of method *A*, *B* and *C* is different. Therefore, the value of *nDCG* is also the same value. On the other hand, value of *nDCG* of method *D* is different value of *nDCG* of method *A*, *B*, and *C*. By contrast, results of ranking of Tables 6 and 4 are better than method *A*, *B*, and *C*. Lastly, the results of ranking other seed words are lower than method *A*, *B*, and *C*.

At first, we had expected that method *D* to produce the best results; however, it did not always do so. In the future, we will consider tuning method *D* and further

Table 1 nDCG of each method to ideal ranking (AKB48)

Period	Proposed method				Previous method
	A	B	C	D	
Period1	**0.617**	**0.617**	**0.617**	0.549	0.091
Period2	0.605	0.605	0.605	**0.627**	0.144
Period3	**0.549**	**0.549**	**0.549**	0.300	0.104
Period4	**0.572**	**0.572**	**0.572**	0.547	0.102
Average	**0.586**	**0.586**	**0.586**	0.506	0.110

Table 2 nDCG of each method to ideal ranking (android)

Period	Proposed method				Previous method
	A	B	C	D	
Period1	0.834	0.834	0.834	**0.838**	0.584
Period2	**0.880**	**0.880**	**0.880**	0.819	0.584
Period3	**0.806**	**0.806**	**0.806**	0.806	0.579
Period4	**0.813**	**0.813**	**0.813**	0.765	0.531
Average	**0.833**	**0.833**	**0.833**	0.807	0.569

Table 3 nDCG of each method to ideal ranking (facebook)

Period	Proposed method				Previous method
	A	B	C	D	
Period1	**0.476**	**0.476**	**0.476**	0.403	0.327
Period2	0.384	0.384	0.384	**0.402**	0.327
Period3	0.384	0.384	0.384	**0.401**	0.327
Period4	**0.491**	**0.491**	**0.491**	0.455	0.327
Average	**0.433**	**0.433**	**0.433**	0.415	0.327

Table 4 nDCG of each method to ideal ranking (joshikai)

Period	Proposed method				Previous method
	A	B	C	D	
Period1	**0.391**	**0.391**	**0.391**	0.373	0.033
Period2	0.302	0.302	0.302	**0.397**	0.034
Period3	0.443	0.443	0.443	**0.469**	0.033
Period4	0.333	0.333	0.333	**0.385**	0.033
Average	0.367	0.367	0.367	**0.406**	0.033

Table 5 nDCG of each method to ideal ranking (K-POP)

Period	Proposed method				Previous method
	A	B	C	D	
Period1	**0.829**	**0.829**	**0.829**	0.683	0.486
Period2	0.676	0.676	0.676	**0.708**	0.506
Period3	**0.658**	**0.658**	**0.658**	0.618	0.503
Period4	**0.689**	**0.689**	**0.689**	0.645	0.550
Average	**0.713**	**0.713**	**0.713**	0.663	0.511

Table 6 nDCG of each method to ideal ranking (smartphone)

Period	Proposed method				Previous method
	A	B	C	D	
Period1	0.775	0.775	0.775	**0.909**	0.050
Period2	0.678	0.678	0.678	**0.869**	0.071
Period3	0.553	0.553	0.553	**0.754**	0.071
Period4	0.676	0.676	0.676	**0.700**	0.071
Average	0.670	0.670	0.670	**0.808**	0.066

investigating other seed words. We plan to looking to find which method works best in differing conditions.

In any case, the accuracy of the proposed method is higher than accuracy of previous studies. Therefore, we consider the basis of the method to be effective. We experimented with four periods of each three month. However, as the value of *nDCG* differed, the ranking which was calculated for each period had differing results. The implication is that blog content differ for each period.

Also, we were able to rank the top ten relevant knowledge group using each of the proposed methods, and we particular found success with the top one or two relevant rankings. However, our proposed method still has areas that need improving. If we consider limiting the ranking to the top few, we believe the method is quite effective.

5 Conclusion

Our goal is develop to method for discovering bloggers with buzzword prediction ability, and then using them to detect buzzword candidates from those bloggers' contents. In this paper, we proposed a method for determining a bloggers' prediction ability by analyzing content and usage period for past buzzwords. Also, we experimented with buzzword candidate category classification and we were able to get positive results.

In the future, we will implement and evaluate "the estimation of growth period of seed words (Sect. 3.3)", "the calculation of bloggers' buzzword prediction score for individual entries (Sect. 3.4)", "the calculation of bloggers' buzzword prediction ability (Sect. 3.5)". We are also working towards implementing a system of practical use.

Acknowledgments This work was supported in part by the MEXT Grant-in-Aid for Scientific Research(C) (#23500140, #26330351).

References

1. S. Nakajima, J. Zhang, Y. Inagaki, R. Nakamoto, Early detection of buzzwords based on large-scale time-series analysis of blog entries, in *23rd ACM Conference on Hypertext and Social Media (ACM Hypertext 2012)*, June 2012, pp. 275-284
2. S. Nakajima, J. Zhang, Y. Inagaki, R. Nakamoto, Early detection of gradual buzzwords based on large-scale time-series analysis of blog entries, in *Information Processing Society of Japan: Database (TOD56)* (2013)
3. S. Nakajima, A. Jatowt, Y. Inagaki, R. Nakamoto, J. Zhang, K. Tanaka, Finding good predictors in blogsphere based on temporal analysis of posting patterns. DBSJ J **10**(1), 13–18 (2011)
4. S. Tomonaga, S. Nakajima, A. Jatowt, Y. Inagaki, R. Nakamoto, J. Zhang, K. Tanaka, Investigating predictability levels in blogosphere based on temporal analysis of blog articles, in *Proceedings of SoC2012*, June 2012, pp. 79–84
5. S. Tomonaga, S. Nakajima, J. Zhang, Y. Inagaki, R. Nakamoto, Method for measuring bloggers' buzzword prediction ability based on analysis of past buzzword mention frequency, in *DEIM Forum 2013*, C1-2, Mar 2013
6. S. Tomonaga, S. Nakajima, Y. Inagaki, R. Nakamoto, T. Ogura, J. Zhang, Measurement of bloggers' buzzword prediction ability based on analyzing frequency of early mentions of past buzzwords, in *Proceedings of the International Multiconference of Engineers and Computer Scientists 2014, IMECS 2014*, 12–14 Mar 2014, Hong Kong. Lecture Notes in Engineering and Computer Science, pp. 462–467
7. M. Okumura, Blog mining: Towards trend and sentiment analysis on the web(<special issue> web interaction in the era of social network). Japanese Soc. Artif. Intell. **21**(4), 424–429 (2006)
8. T. Furukawa, Y. Matuo, I. Ohmukai, K. Uchiyama, M. Ishizuka, Keyword discrimination using topic diffusion process in weblogs. J. Japan Soc. Fuzzy Theor. Intell. Inf. **21**(4), 557–566 (2009)

9. M. Kim, D. Newth, P. Christen, Trends of news diffusion in social media based on crowd phenomena, in *WWW Companion '14 Proceedings of the Companion Publication of the 23rd International Conference on World Wide Web Companion*, pp. 753–758
10. Yahoo! Buzz Index. http://buzzlog.yahoo.com/overall/
11. N.S. Glance, M. Hurst, T. Tomokiyo, BlogPulse: Automated trend discovery for weblogs, in WWW 2004 workshop
12. kizasi.jp -From blog, to know the topic, to find the sign. http://kizasi.jp/
13. S. Nakajima, J. Zhang, Y. Inagaki, T. Kusano, R. Nakamoto, Blog ranking based on bloggers' knowledge level for providing credible information, in *Proceedings of the 10th International Conference on Web Information Systems Engineering, WISE2009*, Oct 2009, pp. 227–234

An E-commerce Recommender System Using Measures of Specialty Shops

Daisuke Kitayama, Motoki Zaizen and Kazutoshi Sumiya

Abstract The use of online shopping sites, such as Amazon and Rakuten, has increased in recent years. Many stores now offer online shops, with the categories of the shop items representing various intended uses. For example, a flashlight may be used for camping or emergencies, so this item may be listed under categories such as "Outdoors" or "Emergency Supplies" in the shop. In this paper, we aim to build a recommender system for specialty shops based on the viewpoints of items browsed by users. We first extract the viewpoints of browsed items by using the category structures of online shops. Thereafter, we analyze the category structures and the selection of goods to determine specialty shops.

Keywords Category structure · Degree of specialty · E-commerce · Evaluation measure · Online shopping · Recommender system

1 Introduction

Online shopping has increased dramatically in recent years, with many stores, such as Amazon.com,[1] Yahoo! Shopping,[2] and Rakuten.com[3] offering online shopping. These sites not only have their own categories for all items, but also many specific

---

[1] http://www.amazon.com.

[2] https://shopping.yahoo.com.

[3] http://www.rakuten.com.

D. Kitayama (✉)
Kogakuin University, 1-24-2 Nishi–shinjuku, Shinjuku, Tokyo 163-8677, Japan
e-mail: kitayama@cc.kogakuin.ac.jp

M. Zaizen
Micware Co., Ltd, 1-1-3 Higashikawasaki-cho, Chuo-ku, Kobe, Hyogo 650-0044, Japan
e-mail: lazy3147@gmail.com

K. Sumiya
University of Hyogo, 1-1-12 Shinzaikehoncho, Himeji, Hyogo 670-0092, Japan
e-mail: sumiya@shse.u-hyogo.ac.jp

© Springer Science+Business Media Dordrecht 2015
G.-C. Yang et al. (eds.), *Transactions on Engineering Technologies*,
DOI 10.1007/978-94-017-9588-3_28

category structures in the participating shops. In other words, Yahoo! Shopping and Rakuten.com, for example, have a general category structure because they may handle general requests from end users and participating shops. However, participating shops may use the individual category structures, which we consider to be the shops' specialties. For example, "Flashlights" and "Retort-packed food" are in the category "Outdoor Gear" in a department store. However, the same items are listed in the "Emergency Supplies" category in a specialty shop for survival goods. Therefore, if users are looking for these items as survival goods, we would recommend other items in the category "Emergency Supplies". We deem these categories to represent some of the item's intended uses. In this study, we recommend participating shops and items in these shops based on viewpoints of browsed items, taking into account these shops' viewpoint specialties [1]. For example, we recommend specialty shops with "Outdoors" and "Emergency Supplies" categories for flashlights and retort-packed food.

2 Our Approach

2.1 A Recommender System for Specialty Shops

In this study, we use specific category structures in participating shops to infer viewpoints among browsed items. Specifically, we use categories with the lowest common ancestors that include browsed items in participating shops, because these categories are likely to represent the intended uses of items. For example, when the browsed items are flashlights and retort-packed food, and the categories with the lowest common ancestor that have these items in participating shops are "Outdoor Gear" or "Emergency Supplies," we can infer that the viewpoints among these browsed items are outdoor leisure and disaster preparedness. Thereafter, we analyze the category structures in participating shops, in order to identify and recommend specialty shops based on these viewpoints. Figure 1 shows an example of a recommendation. A user has browsed "Flashlight" and "Retort-packed food." We present the categories with the lowest common ancestor, namely, "Outdoor Gear" and "Emergency Supplies" as viewpoints for these items. Then, we present child categories such as "Outdoor Gear > Camping" or "Emergency Supplies > Foods" underneath the categories of the lowest common ancestor . When choosing a category, a user can browse recommended items that are suitable for the user's purposes. Therefore, we decided that the recommended categories should be found in specialty shops. We determine shops' specialties by using the following measures.

- The degree of a shop's specialty based on the classification method of items. Specialty shops can classify items into appropriate categories. Therefore, we define specialty shops as shops that have a detailed category structure and uniformed granularity.

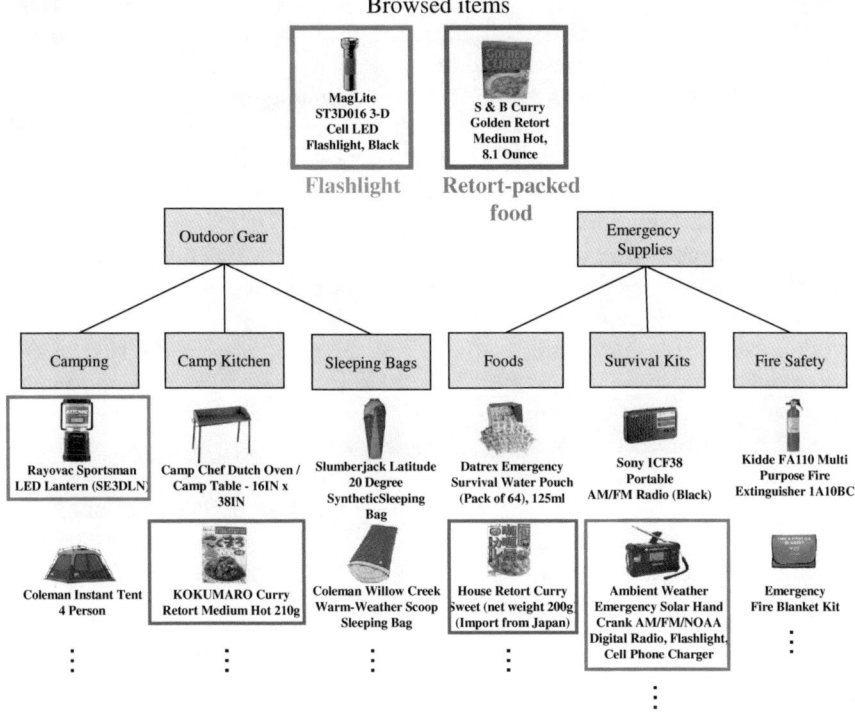

Fig. 1 An E-commerce recommender system based on specialties in online shops. The category structure shown on the *left* is the specialty shop for "emergency supplies". The category structure shown on the *right* is the specialty shop for "outdoor gear"

- The degree of a shop's specialty based on the selection of items. We consider the selection of items to determine the degree of the shop's specialty. Specialty shops may comprehensively deal in items of a specific category. Furthermore, these shops may offer unique items. Therefore, we define specialty shops as shops have a large coverage of items as well as many unique items.
- The degree of a shop's specialty based on the main target. We consider that all of the items of a specialty shop match a specific viewpoint. In other words, shops that have only items corresponding to the viewpoints of browsed items are specialty shops for the particular user. We consider that a shop's specialty can be represented by the ratio of main target items.

2.2 Related Work

Online shopping sites recommend various items based on users' item browsing histories, using collaborative filtering methods [2–6] or hybrid approaches with content-based filtering methods [7–13]. However, with such methods, other items from the same category as the item browsed by a user are often recommended, although items have viewpoints. In other words, conventional recommender systems may not treat user's viewpoint in current session.

Cao and Li [14] proposed a method for recommending suitable items for a user's purposes. This is similar to our work on recommending suitable items for a user's viewpoint, but our method differs in that it considers specialties when recommending shops based on a user's viewpoint.

Hijikata et al. [15] proposed a discovery-oriented collaborative filtering method for users that was based on a recommendation process. Our method is also able to identify specialty shops that are unfamiliar to users; therefore, the above-mentioned method may complement ours.

Shoji and Hori [16] experimented with an interface to support decision making during online shopping. The authors targeted users who decided not to purchase items, and proposed a communication model based on sales clerks. Our method, which is based on this communication model, aims to present shop information related to browsed items.

Online shopping sites can rank participating shops according to opening day, number of items, and number of reviews. However, such sites do not consider shop specialties. We are not aware of any studies that have evaluated the specialty of a shop.

3 Definition of Measures for Specialty Shops Based on Viewpoints

In this section, we define the measures used for determining specialty shops, and explain three methods for calculating the degree of a shop's specialty regarding viewpoints among browsed items. We define a shop's specialty as the result of a calculation using the item classification method, item selection, and the main target category of items in the shop.

3.1 The Degree of a Shop's Specialty Based on the Item Classification Method

We consider an item classification method to determine the degree of a shop's specialty. Specialty shops can classify items into appropriate categories; therefore, we define specialty shops as those with detailed category structures and uniformed granularity. First, we describe the degree of detail of a category structure, where a detailed category structure is a deep and wide structure. Therefore, the degree of detail of a category structure in a shop i based on browsed items X is calculated using the following expression:

$$Detail(X, i) = W(X, i) \times D(X, i) \tag{1}$$

where W and D are the number of leaf categories as the width and the number of layers in the category structure as the depth, respectively under the lowest common ancestor category with items X in a shop i.

Next, we describe the degree of uniformed granularity. We assume that well classified category structures have uniformed granularity. Therefore, the degree of uniformed granularity in a shop i based on browsed items X is calculated using the following expression:

$$Uniformity(X, i) = \frac{1}{1 + \sigma} \tag{2}$$

$$\sigma = \sqrt{\frac{1}{p} \sum_{l=1}^{p} \left(C(X, i, l) - \frac{\sum_{l=1}^{p} C(X, i, l)}{p} \right)^2} \tag{3}$$

where C is the number of items belonging to category l. This category l is one of the leaf categories under the lowest common ancestor category with items X in shop i. σ is the standard deviation of the number of items belonging to leaf categories under the lowest common ancestor category with items X in shop i. Intuitively, *Uniformity* is the degree of uniformity in the quantities of items belonging to the end categories.

Finally, we calculate the degree of specialty based on the classification method *Classification(X, i)* by using the following expression:

$$\begin{aligned} Classification(X, i) = {}& \alpha \times Detail(X, i) \\ & + (1 - \alpha) \times Uniformity(X, i) \end{aligned} \tag{4}$$

Figure 2 is an example of shops' degrees of specialty based on the item classification method. Shop A's category structure of "C^1" with a width of six and a depth of three is more detailed than shop B's category structure of "C^1" with a width of four and a depth of three. Shop A has greater uniformity in the quantities of items belonging to categories than shop C. Therefore, shop A has the most detailed specialty category structure of viewpoint "C^1".

3.2 The Degree of a Shop's Specialty Based on the Selection of Items

We use the selection of items to determine the degree of a shop's specialty. Specialty shops may sell a comprehensive range of items in a specific category. Furthermore, these shops may deal in unique items. Therefore, we define specialty

Fig. 2 Classification by category structures based on viewpoints. Shop A and shop B have a high degree of uniformity, as well as a high degree of detail

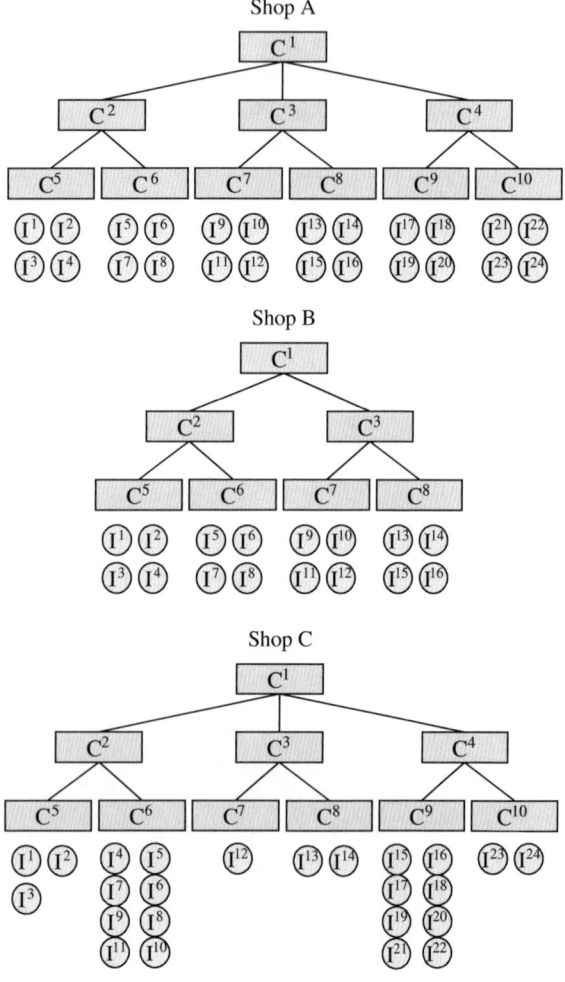

shops as those with a large coverage of items and many unique items. We first describe the degree of item coverage. It is assumed that specialty shops have most of the items of a specific category. Therefore, the degree of item coverage in a shop i based on browsed items X is calculated using the following expression:

$$Cover(X, i) = \frac{|G(X, i)|}{|\bigcup_{n=1}^{m} G(X, n)|} \tag{5}$$

where G is a set of items of a viewpoint based on items X in shop i. The function *Cover* returns the ratio of the number G of a viewpoint based on items X in shop i to the number G of the viewpoint based on items X in all shops. Then, *Cover* is defined as the degree of item quantities of a certain viewpoint based on browsed items X.

Figure 3 shows examples of shops' degrees of specialty based on *Cover*. Because Shop D's "C^1" category has more items than Shop E's "C^1" category, Shop D is more specialized for the viewpoint "C^1" based on *Cover* than Shop E.

Next, we describe the degree of item uniqueness. We assume that specialty shops have many unique items that are not sold in other shops. Therefore, the degree of item uniqueness in a shop i, based on browsed items X, is calculated using the following expression:

$$Unique(X, i) = \sum_{o \in G(X,i)} U(X, o) \tag{6}$$

$$U(X, o) = \begin{cases} e^{-\log\left(\frac{|S(X \cup o)|}{|S(X)|}\right) - 1} & \left(\frac{|S(X \cup o)|}{|S(X)|} \leq \delta\right) \\ 0 & (other) \end{cases} \tag{7}$$

where G is a set of items of a viewpoint based on items X in shop i. o shows an item that is contained in $G(X, i)$ and S is a set of shops having the argument items. Using S, we determine an item o's uniqueness $U(X, o)$. When the number of $S(X \cup o)$ is δ ($0.0 \leq \delta \leq 1.0$) the number of $S(X)$ and above, $R(X, o)$ is 0. δ means the threshold of a unique item.

Figure 4 shows examples of the rarity of items. Because items I^1, I^2, I^3, and I^4 belong to Shop A, Shop B, and Shop C, they are not unique items. However, item I^5 is a unique item, because it belongs only to Shop A. From this, *Uniqueness* means the degree of the selection of hard-to-find items that most shops do not have.

Finally, we calculate the degree of uniqueness based on the selection of items, *Selection(X, i)*, which is calculated using the following expression:

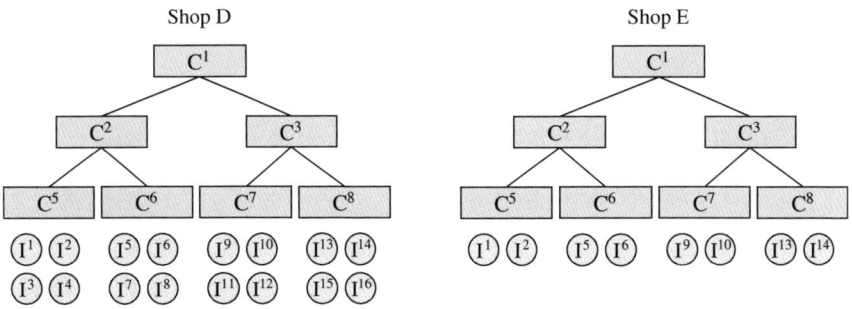

Fig. 3 Coverage of items based on a viewpoint. Shop D has a high degree of coverage

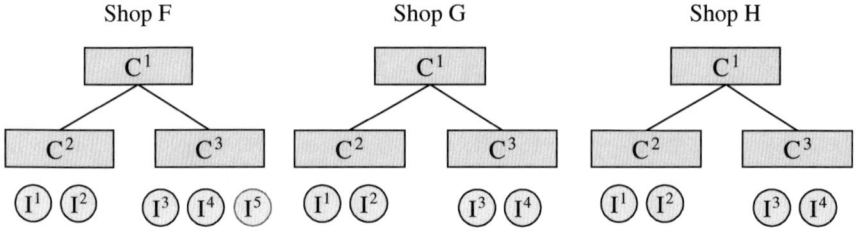

Fig. 4 Uniqueness of items based on a viewpoint. Shop F has a high degree of uniqueness

$$Selection(X, i) = \beta \times \left(Cover(X, i) - \frac{\sum_{n=1}^{m} Cover(X, n)}{m} \right)^2$$
$$+ \gamma \times \left(Unique(X, i) - \frac{\sum_{n=1}^{m} Uniqueness(X, n)}{m} \right)^2 \qquad (8)$$

$$\beta = \begin{cases} 1 & \left(Cover(X, i) \geq \frac{\sum_{n=1}^{m} Cover(X, n)}{m} \right) \\ -1 & (other) \end{cases} \qquad (9)$$

$$\gamma = \begin{cases} 1 & \left(Unique(X, i) \geq \frac{\sum_{n=1}^{m} Uniqueness(X, n)}{m} \right) \\ -1 & (other) \end{cases} \qquad (10)$$

where the function *Cover* returns the degree of the quantities of all items, and the function *Unique* returns the degree of the selection of hard-to-find items of a viewpoint based on browsed items X in shop i.

3.3 The Degree of a Shop's Specialty Based on the Main Target

We assume that all of the items of specialty shop match a specific viewpoint. In other words, shops having only items of a viewpoint based on browsed items are specialty shops for the particular user. We consider that a shop's specialty can be determined by the ratio of main target items. Then, the degree of specialty based on the main target of a shop is calculated using the following expression:

$$Precision(X, i) = \frac{\log |G(X, i)|}{\log |N(i)|} \qquad (11)$$

where G is a set of items of a viewpoint based on items X in shop i, and N is a set of items in shop i. The function *Precision* returns the ratio of the number of G to the number of N.

4 Experiment

4.1 Experimental Setting

We calculated the degree of specialty of shops in rakuten.co.jp based on the three viewpoints "Emergency Supplies," "Outdoor," and "Kitchen and Dining" (see Table 1). We selected 50 shops for each viewpoint by using the Rakuten Item Search API. In the experiment, we calculated the degrees of these shops' specialty based on viewpoints (see Tables 2, 3, and 4). We present the calculation results in Tables 2, 3 and 4, showing *Detail, Uniformity, Cover, Unique, Precision, Classification*, and *Selection*. In these tables, we include shops that have a high degree of *Precision* (>0.9), *Classification* (>0.07) or *Selection* (>0.01).

4.2 Results of the Shops' Specialty Based on the Item Classification Method

Table 2 shows that shop anzenlife and shop bousaikan have high degrees of specialty based on the item classification method (*Classification*). Because these shops have a relatively detailed category structure for "Emergency Supplies" and each leaf categories are classified by similar granularity, these shops are specialty shops in terms of the classification method. Therefore, it is believed that the *Classification* values of these shops are reasonable. In contrast, shop bousaiss, shop bousai-web and other shops have a low degree of *Classification*, and most shops are thus not included in the table.

Shop maxshare's and shop royal3000's category structure[4] for "Emergency Supplies" are each composed of only one category, so these shops are not specialty shops in terms of the classification method. Shop bousaianshin has the most detailed category structure. However, this shop effectively with this category structure. In detail, a few leaf categories have a lot of items and other leaf categories have a few items. In other words, the category structure of this shop does not classified well; that is, this shop's *Uniformity* is low. Therefore, this shop is not a specialty shop in terms of the classification method, and its low degree of *Classification* is reasonable.

---

[4] These shops are not included in the table. Because these Precision, Classification and Selection values are low.

Table 1 Experimental data: shop data

Viewpoint	Number of shops
Emergency supplies	50
Outdoor	50
Kitchen and dining	50

Results of the calculations using our method, shown in Tables 3 and 4, exhibit the same tendencies of those shown in Table 2.

4.3 Results of the Shops' Specialty Based on the Selection of Items

Shop bousaianshin has a high degree of specialty based on its selection of goods (*Selection*), and its *Cover* is very high. Thus, this shop has a great deal of items of the viewpoint "Emergency Supplies." In addition, the selection of hard-to-find items (*Unique*) in this shop is very large.

We suspect that there is a tendency for *Unique* to be high if *Cover* is high. The correlation coefficient between *Unique* and *Cover* reached 0.99 or more and was very high in all of the results. However, this is an undesirable outcome. If a shop has many items, the degree of *Unique* must be low when the shop does not have many hard-to-find items. We suspect that the cause of this problem lies in the method used for determining whether an item in a shop is found in other shops. The Rakuten Item Search API cannot retrieve Japanese Article Numbers (JANs); therefore, instead of using JANs, we use item names qualified by each shop to determine whether an item in a shop is found in other shops. As a result, the *Unique* method is not very accurate, and we therefore need to modify the method of item identification.

4.4 Results of the Shops' Specialty Based on the Main Target

Some of the shops with a high degree of *Precision* are bousaiss, bouhanbousai, and bousai-web. All items in these shops are of the viewpoint "Emergency supplies," so on the whole they offer emergency supplies. Results of the calculations using our method shown in Tables 3 and 4 indicate the same tendency of the results shown by Table 2. Therefore, we believe that the *Precision* values of these shops are reasonable. However, it should be noted that *Precision* does not guarantee quality when confirming items in shops.

Table 2 Experimental data and results: shop data based on "Emergency supplies"

Shop's name	Number of items	Number of items based on the viewpoint	Category structure depth and width	Detail	Uniformity	Cover	Uniqueness	Precision	Classification	Selection
anmakuya	1,149	1,097	5, 108	0.5562	0.0326	0.0123	0.0222	0.9934	0.0045	−0.0001
anzenlife	405	403	5, 146	0.7523	0.0918	0.0089	0.0119	0.9992	0.0173	−0.0004
be-kan	714	707	5, 173	0.8916	0.0349	0.0079	0.0141	0.9985	0.0078	−0.0004
bousaianshin	13,324	13,286	5, 194	1.0000	0.0062	0.6642	1.0000	0.9997	0.0015	1.3570
bousaikan	1,190	259	5, 70	0.3602	0.2064	0.0107	0.0153	0.7847	0.0186	−0.0003
bousaiss	441	441	3, 48	0.1476	0.0587	0.0222	0.0277	1.0000	0.0022	0.0000
ganpon	511	433	5, 75	0.3860	0.0707	0.0170	0.0244	0.9734	0.0068	0.0000
saibou	487	238	4, 77	0.3168	0.1094	0.0099	0.0126	0.8843	0.0087	−0.0004
bouhanbousai	14	14	2, 8	0.0155	0.4617	0.0007	0.0006	1.0000	0.0018	−0.0012
bousai-web	237	237	3, 53	0.1631	0.0343	0.0110	0.0153	1.0000	0.0014	−0.0003
Average	13,457	426	2.18, 22.04	0.0988	0.2028	0.0197	0.0296	0.5692	0.0015	0.0264

Table 3 Experimental data and results: shop data based on "Outdoor"

Shop's name	Number of items	Number of items based on the viewpoint	Category structure depth and width	Detail	Uniformity	Cover	Uniqueness	Precision	Classification	Selection
yamalab	475	283	4, 27	0.1044	0.0869	0.0103	0.0286	0.9160	0.0023	−0.0004
hikers	212	211	5, 56	0.2722	0.1275	0.0020	0.0055	0.9991	0.0087	−0.0020
fieldlife	499	497	7, 123	0.8390	0.0401	0.0056	0.0144	0.9994	0.0084	−0.0012
basecamp	1,492	1,485	6, 171	1.0000	0.0382	0.0614	0.1464	0.9994	0.0096	0.0117
hab	8,519	6,288	4, 123	0.4790	0.0082	0.3715	1.0000	0.9664	0.0010	1.0322
Average	11,015	344	2.14, 14.72	0.0647	0.1551	0.0184	0.0474	0.5578	0.0009	0.0201

Table 4 Experimental data and results: shop data based on "Kitchen and dining"

Shop's name	Number of items	Number of items based on the viewpoint	Category structure depth and width	Detail	Uniformity	Cover	Uniqueness	Precision	Classification	Selection
good-choice	6,633	4,339	4, 139	1.0000	0.0347	0.1691	0.7362	0.9518	0.0087	0.4608
neoleaf	60,945	9,118	4, 35	0.2505	0.0042	0.2954	1.0000	0.8276	0.0003	0.9334
livingearth	518	62	4, 15	0.1063	0.2857	0.0012	0.0047	0.6603	0.0076	-0.0051
joyfullmart	1,041	669	5, 29	0.2595	0.0322	0.0101	0.0439	0.9364	0.0021	-0.0010
zaka-mmc	4,992	951	5, 70	0.6288	0.0692	0.0329	0.1212	0.8053	0.0109	0.0024
Average	10,118	578	2.74, 17.28	0.1103	0.1448	0.0190	0.0742	0.6182	0.0017	0.0260

4.5 Discussion

Overall, the results show that shops with a high degree of specialty do not necessarily have a large number of items, but have higher category structure depths and widths than the average. We may efficiently narrow down items by using measures of the degree of specialty.

We confirmed that our method for calculating the degree of specialty of shops based on the item classification method (*Classification*) and the main target of a shop (*Precision*) could determine specialty shops. *Detail* and *Uniformity* are independent measures and represent the specialty of *Classification*.

In contrast, our method for calculating the degree of specialty of shops based on the selection of items (*Selection*) could not determine specialty shops. However, we suspect that *Selection* can determine specialty shops if hard-to-find items are not considered (*Unique*).

5 Conclusion

In this paper, we proposed a method for determining the degree of shops' specialty based on a viewpoint. These values are then extracted using the category structures of online shops to build a recommender system for specialty shops, based on the viewpoints of items browsed by users. In addition, to verify our method, we calculated the degree of specialty of shops in rakuten.co.jp based on the viewpoint "Emergency Supplies," "Outdoor," and "Kitchen and Dining"

In future work, we intend to repeat the experiment after modifying the method for determining whether an item in a shop is also found in other shops, in order to verify the method for calculating *Unique*. Furthermore, we intend to evaluate the usability of the recommendations of specialty shops determined by our method, to confirm that the system can match user viewpoints.

Acknowledgments This research was supported in part by the Grant-in-Aid for Young Scientists (B) 24700098 from the Ministry of Education, Culture, Sports, Science, and Technology of Japan.

References

1. M. Zaizen, D. Kitayama, K. Sumiya, An E-commerce recommender system based on degree of specialties in online shops, in *Proceedings of the International Multiconference of Engineers and Computer Scientists 2014, IMECS 2014*, 12–14 Mar 2014, Hong Kong. Lecture Notes in Engineering and Computer Science, pp. 492–496
2. J.S. Breese, D. Heckerman, C. Kadie, Empirical analysis of predictive algorithms for collaborative filtering, in *Proceedings of the 14th Conference on Uncertainty in Artificial Intelligence, UAI'98*, San Francisco, CA, USA (Morgan Kaufmann Publishers Inc., Massachusetts, 1998), pp. 43–52

3. J.L. Herlocker, J.A. Konstan, A. Borchers, J. Riedl, An algorithmic framework for performing collaborative filtering, in *SIGIR*, pp. 230–237 (1999)
4. P. Resnick, N. Iacovou, M. Suchak, P. Bergstrom, J.Riedl, Grouplens: An open architecture for collaborative filtering of netnews, in *Proceedings of the 1994 ACM Conference on Computer Supported Cooperative Work, CSCW'94*, New York, NY, USA (ACM, 1994), pp. 175–186
5. B. Sarwar, G. Karypis, J. Konstan, J. Riedl, Item-based collaborative filtering recommendation algorithms, in *Proceedings of the 10th International Conference on World Wide Web, WWW'01*, New York, NY, USA (ACM, 2001), pp. 285–295
6. M. Sun, G. Lebanon, P. Kidwell, Estimating probabilities in recommendation systems. J. Roy. Stat. Soc.: Ser. C (Appl. Stat.) **61**(3), 471–492 (2012)
7. C. Basu, H. Hirsh, W. Cohen, Recommendation as classification: Using social and content-based information in recommendation, in *Proceedings of the 15th National Conference on Artificial Intelligence* (AAAI Press, 1998), pp 714–720
8. B.M. Kim, Q. Li, Probabilistic model estimation for collaborative filtering based on items attributes, in *Proceedings of the 2004 IEEE/WIC/ACM International Conference on Web Intelligence, WI '04*, Washington, DC, USA (IEEE Computer Society, 2004), pp. 185–191
9. P. Melville, R.J. Mooney, R. Nagarajan, Content-boosted collaborative filtering for improved recommendations, in *18th National Conference on Artificial Intelligence*, Menlo Park, CA, USA (American Association for Artificial Intelligence, 2002), pp. 187–192
10. D.M. Pennock, E. Horvitz, S. Lawrence, C. Lee Giles, Collaborative filtering by personality diagnosis: A hybrid memory and model-based approach, in *Proceedings of the 16th Conference on Uncertainty in Artificial Intelligence, UAI '00*, San Francisco, CA, USA (Morgan Kaufmann Publishers Inc, Massachusetts, 2000), pp. 473–480
11. D.Y. Pavlov, D.M. Pennock, A maximum entropy approach to collaborative filtering in dynamic, sparse, high-dimensional domains, in *Proceedings of Neural Information Processing Systems* (MIT Press, Massachusetts, 2002)
12. X. Su, R. Greiner, T.M. Khoshgoftaar, X. Zhu, Hybrid collaborative filtering algorithms using a mixture of experts, in *Proceedings of the IEEE/WIC/ACM International Conference on Web Intelligence, WI'07*, Washington, DC, USA. (IEEE Computer Society, 2007), pp. 645–649
13. C.-N. Ziegler, G. Lausen, L. Schmidt-Thieme, Taxonomy-driven computation of product recommendations, in *Proceedings of the 13th ACM International Conference on Information and Knowledge Management, CIKM'04*, New York, NY, USA (ACM, 2004), pp. 406–415
14. Y. Cao, Y. Li, An intelligent fuzzy-based recommendation system for consumer electronic products. Expert Syst. Appl. **33**(1), 230–240 (2007)
15. Y. Hijikata, T. Shimizu, S. Nishida, Discovery-oriented collaborative filtering for improving user satisfaction, in *Proceedings of the 14th International Conference on Intelligent User Interfaces, IUI'09*, New York, NY, USA (ACM, 2009), pp. 67–76
16. H. Shoji, K. Hori, S-conart: an interaction method that facilitates concept articulation in shopping online. AI Soc. **19**(1), 65–83 (2005)

Cooking Recipe Recommendation Method Focusing on the Relationship Between User Preference and Ingredient Quantity

Mayumi Ueda and Shinsuke Nakajima

Abstract There are numerous websites on the Internet that recommend cooking recipes. However, these websites order recipes according to date of submission, access frequency, or user ratings for recipes, and therefore, they do not reflect a user's personal preferences. In this paper, we propose a recipe recommendation method based on the user's culinary preferences. We employ the user's recipe browsing and cooking history in order to determine his/her preferences in food. In our previous study on the subject, we considered only the presence of certain ingredients in cooking recipes in order to determine user preferences. However, in order to ascertain them more accurately, we propose a scoring method for cooking recipes based on users' food preferences and the quantity of the ingredients used.

Keywords Cooking and browsing history · Cooking recipe recommendation · Preference extraction · Quantity of ingredient in recipe · Scoring method · User's culinary preferences

1 Introduction

Dietary habits are a subject of ever-growing interest in various areas of research because of the importance of healthy eating in preventing numerous diseases. Good eating habits are important for maintaining a healthy lifestyle. However, planning a

M. Ueda (✉)
University of Marketing and Distribution Sciences, 3-1 Gakuen Nishimachi,
Nishi-ku, Kobe 651-2188, Japan
e-mail: Mayumi_Ueda@red.umds.ac.jp

S. Nakajima
Kyoto Sangyo University, Motoyama, Kamigamo, Kita-ku, Kyoto 603-8555, Japan
e-mail: nakajima@cc.kyoto-su.ac.jp

© Springer Science+Business Media Dordrecht 2015
G.-C. Yang et al. (eds.), *Transactions on Engineering Technologies*,
DOI 10.1007/978-94-017-9588-3_29

healthy menu requires considering various factors, such as the nutritional values of the ingredients, the food in stock, food preferences, and cost. Thus, people need to expend considerable efforts to plan their daily meals. Against this backdrop, numerous cooking websites, such as Cookpad [1] and Rakuten Recipe [2], which recommend various food recipes, have been launched. Cookpad contains 1.5 million recipes and 20 million users [3], which reflects the high demand for recipe-providing services. However, these websites do not reflect users' specific preferences and circumstances. These two factors need to be considered in order to recommend recipes that users find highly satisfactory.

Considerable research has been carried out in the past on cooking recipe recommendation methods for menu planning support. Mino and Kobayashi proposed a method that considers the user's schedule [4], and assesses either the intake or consumption calories that are assigned to each event in the user's schedules. Karikome and Fujii [5] proposed a system that helps users plan nutritionally balanced menus and visualize their dietary habits. Their system calculates the nutritional value of each dish and records this information in a dietary log. The system then recommends recipes that foster nutrition. Freyne and Berkovsky [6] compared three recommendation strategies: content-based, collaborative, and hybrid.

In this paper, we propose a recipe recommendation method based on the user's culinary preferences [7]. Our method breaks recipes down into their ingredients, and scores them based on the frequency of their use and the specificity of their ingredients. Furthermore, our proposed system does not recommend foods that are similar to the dish the user has recently had on the ground that people do not want to repeatedly eat similar foods. Moreover, our system does not require any action from the user to reflect his/her food preferences, and instead estimates them automatically through his/her recipe browsing and cooking history. Furthermore, we think that our previous work in the area does not completely reflect user preferences, and thus propose the new recipe recommendation method based on users' food preferences according to the quantity of ingredients used in a dish. This paper is revised version of the conference paper that we gave a presentation at IMECS2014 [8].

This paper is structured as follows: in Sect. 2, we describe our method of scoring recipes and extracting user preferences. Section 3 contains our experimental results, and we offer our conclusions in Sect. 4.

2 Scoring Recipes and Extracting User Preferences

It has been noted that picky eating is one of the main reasons causing diet-related health issues. People do not want to eat food that they dislike, even if it perfectly addresses their nutritional needs. They hope to derive essential nutrition solely from their favorite foods. Therefore, we try to extract users' culinary preferences.

2.1 Preference for Ingredients

We express the user's food preferences I_k as follows:

$$I_k = I_k^+ - I_k^- \qquad (1)$$

2.1.1 User's Favorite Ingredients

Figure 1 shows the key idea underlying the estimation of a user's favorite ingredients through his/her cooking history. Our method identifies the ingredient that the user repeatedly consumes as his/her favorite ingredients. It resolves recipes down to their ingredients at the outset, and calculates *the score I_k^+ of ingredients* by incorporating the frequency of use of each ingredient in the dishes that the target user has consumed (*FF_k: foodstuff frequency*), as well as the specificity of the ingredients (*IRF_k: inverse recipe frequency*), into Eq. (2), which is based on the idea of term frequency-inverse document frequency (tf-idf):

$$I_k^+ = FF_k \times IRF_k \qquad (2)$$

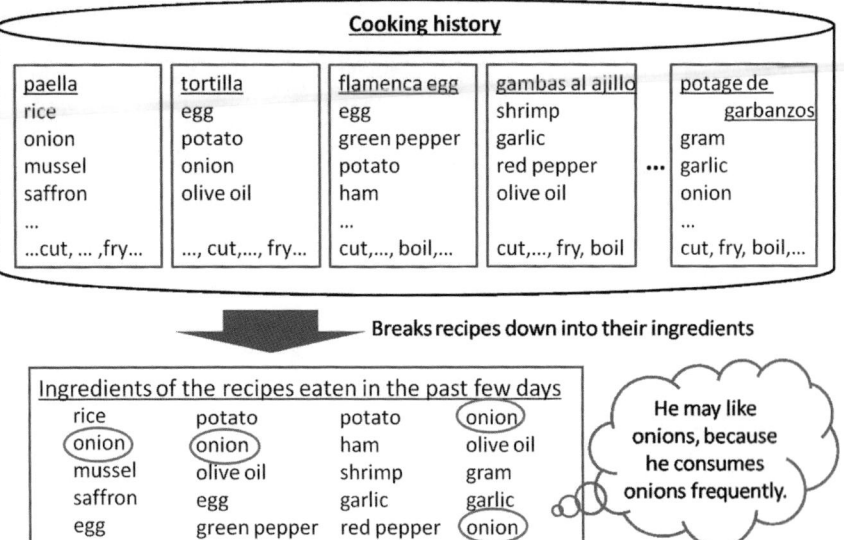

Fig. 1 Extracting favorite ingredients using cooking history

To determine the user's favorite ingredients by using *the frequency of use of ingredient k* (F_k), we utilize *the simple frequency of use of ingredient k* (F_k) during a certain period D, as shown in Eq. (3):

$$FF_k = \frac{F_k}{D} \tag{3}$$

We then calculate *the specificity of ingredient k* (IRF_k) using *the total number of recipes* (M) and *the number of recipes that contain ingredient k* (M_k), as shown in Eq. (4).

$$IRF_k = \log \frac{M}{M_k} \tag{4}$$

2.1.2 The Ingredients the User Dislikes

We believe that the user's food preferences are also influenced by his/her dislike for certain ingredients. We estimated the ingredients that a user dislikes by considering ingredients in recipes for food that he/she has never cooked, even if he/she has browsed the recipe details. Figure 2 shows the method to estimate the ingredients that a user dislikes using his/her recipe browsing history and cooking history. In the figure, N corresponds to the set of ingredients in recipes that the user has not browsed. C corresponds to the set of ingredients in the recipes that the user has cooked in the last few days, and U represents the set of ingredients in recipes that the user has not cooked, even if he/she has browsed them. For example, "shrimp" in

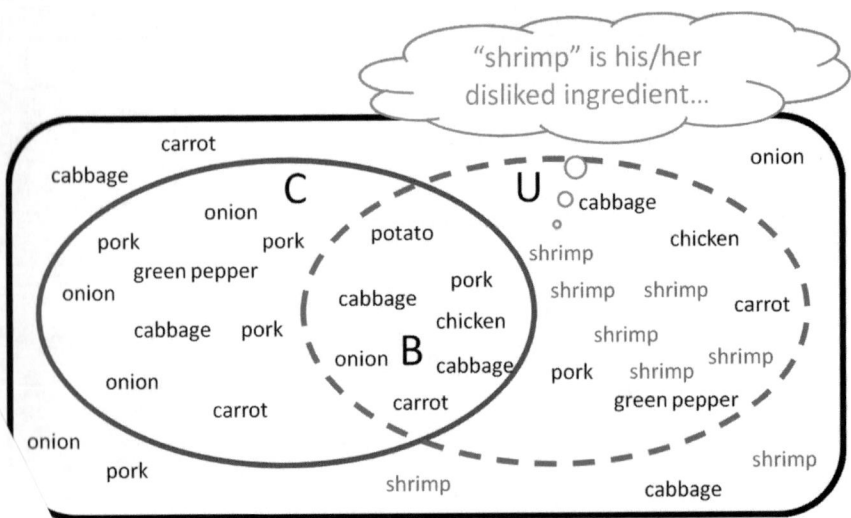

Extracting disliked ingredients using browsing and cooking history

Fig. 2 corresponds to an ingredient that the relevant user dislikes. We calculate *the score of the disliked ingredient k (I_k^-)* in Eq. (5):

$$I_k^-(x) = \begin{cases} 0 \ (0 < \frac{2|U_k|}{|A_k|} \le 0.5) \\ (\frac{2|U_k|}{|A_k|} - 1)^x (0.5 < \frac{2|U_k|}{|A_k|} \le 1) \end{cases} \tag{5}$$

where $|U_k|$ denotes the presence of ingredient k in U, and A_k denotes the presence of ingredient k in a recipe that the user has browsed. We calculate the ratio or frequency of the user's avoidance of the ingredient that he/she dislikes, because he/she will use the ingredients that he/she does not like. x in the above equation denotes the ratio or frequency of avoiding the ingredients, and we plan to calculate the value of x through preliminary experiments.

2.2 Quantity of Ingredient in Recipe

Figure 3 shows the basic concept of our recipe recommendation method. The figure on the left shows the concept underlying our previous work on users' culinary preferences, which estimates a user's preferences from his/her past actions, such as their recipe browsing history and cooking history. The figure on the right shows the new method based on the user's food preferences and the number of ingredients in the recipe. Consider *Recipe A* and *Recipe B* in Fig. 3. Recipe A contains 300 g of pork, 60 g of cabbage, and 100 g of potato. Recipe B contains 50 g of pork, 400 g of cabbage, and 20 g of potato.

Our previous work proposed a method to score recipes based on a user's food preferences based on his/her past actions, such as recipe browsing history and cooking history. To extract food preferences, our previous method breaks recipes down into their ingredients. That is, the score of recipes are determined by the contain or not contain of the ingredients. Because both recipes contain the same ingredients, they have the same score, even if the user dislikes potato, say left side of Fig. 3. This idea does not suit the eating habits.

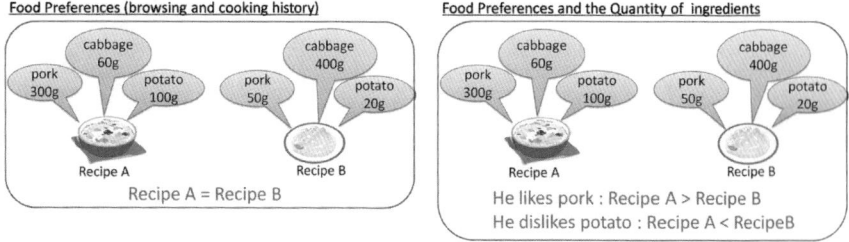

Fig. 3 The basic concept of our previous scoring method (*left*) and our new scoring method (*right*)

Our new method scores recipes based on a user's food preference as well as the amount of each ingredient used in the recipe. That is, the scores of a recipe that contains larger amounts of an ingredient that has previously been identified as the user's favorite will be higher than one that contains a smaller amount of it, even if both recipes contain the same ingredients. As shown by the right side of Fig. 3, even if both recipes contain the same ingredients, their scores are different. If the user likes pork, Recipe A has a higher score than Recipe B, even though both recipes contain the same ingredients.

Furthermore, we think that should not score recipes based on a uniform measure of quantity for all ingredients. Consider 100 g of an ingredient, for instance. We think that, in the context of cooking, 100 g of pepper is relatively large quantity, but 100 g of potatoes is not. Hence, we propose a scoring method based on a standard quantity and a dispersion quantity for each ingredient.

Figure 4 shows the basic idea behind our scoring method for three ingredients, according to the average and the dispersion quantity of each ingredient. The meaning of "30 g" varies across the three ingredients. As shown in the central figure of Fig. 4, "30 g of ingredient p" is a quite small quantity, and this, ingredient p has a relatively minor effect on the recipe. On the other hand, "30 g of ingredient q" is quite a large quantity, because of which ingredient q has a significant effect on the recipe.

Therefore, we use the dispersion quantity of ingredients to score recipes that is calculated by the positioning of each ingredient in the all recipes.

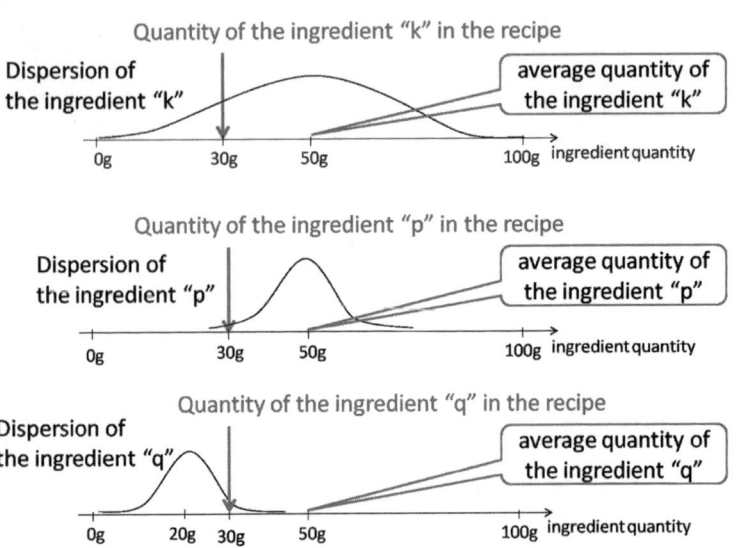

The basic idea of the scoring method using the average and dispersion quantity of each
nt

2.3 Recipe Scoring

2.3.1 User's Food Preferences

Our method scores cooking recipes in accordance with the results regarding a user's favorite/disliked ingredients, and then lists the recipes in decreasing order of scores. In general, people do not like eating dishes similar to ones they have eaten in the recent past. Therefore, our method weights recipes to avoid the repetition of similar dishes in its recommendations. *The score of cooking recipes* is defined in Eq. (6).

$$Score(R) = \sum_{k \in R} I_k - \alpha \sum_{d=1} (w_d \cdot sim(R, R_d)) \tag{6}$$

where d denotes the weight of avoiding the repetition of similar dishes, $sim(R, R_d)$ represents the similarities between the considered recipe R and the recipe R_d of the dish eaten d days ago. The weight w_d of avoiding similar dishes eaten d days ago is defined as:

$$w_d = 1 - \frac{d-1}{7} \quad (1 \leq d \leq 7) \tag{7}$$

2.3.2 The Quantity of Ingredients

Our method calculates the weights of ingredients k using the average and the standard deviation of the quantity of each ingredient. The average quantity of each ingredient is calculated from all recipes in the database. We calculate the standard deviation as:

$$\sigma_k = \sqrt{\frac{1}{n} \sum_{i=1}^{n} (g_{k(i)} - \overline{g_k})^2} \tag{8}$$

where n denotes the number of recipes that contain ingredient k, $g_{k(i)}$ represents the quantity of ingredient k in recipe i, and $\overline{g_k}$ is the average of $g_{k(i)}$.

We calculate the weights using the standard deviation and the distribution of the quantity of each ingredient. The ingredient with a deviation score of 50 ranks in the top half of all ingredients (Fig. 5), one with a deviation score of 60 ranks in the top 10 %, and the ingredient with a deviation score of 80 ranks in the top 0.15 % of all ingredients. We distribute the weights W_k between 0 and 2, according to the deviation scores. Cooking recipes are scored as follows:

$$Score(R) = \sum_{k \in R} (I_k \cdot W_k) \tag{9}$$

Fig. 5 The distribution of the
deviation value for the weight

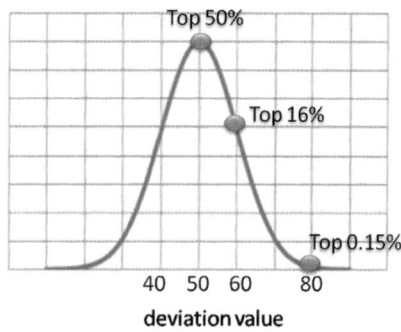

where W_k denotes the weights of ingredient k in the target recipe. For example, if the deviation score is 50, the weights of the ingredient is 1.0, and if the deviation score is 60, its weight is 1.68.

3 Accuracy Evaluation

3.1 Experimental Condition

In order to assess the appropriateness of the scoring method, we conducted simple experiments. We used 8,050 recipes extracted from a popular recipe search website in Japan [9], and used randomly selected recipes as the main dish.

We conducted our experiment as follows:

1. We presented a list of five recipes to the subjects of our experiment.
2. Each subject manually ranks recipes according to his/her preferences.
3. Repeat step 1 and 2 five times.

At beginning of the experiments, each subject enters his/her culinary history for the last 30 days. Our system calculates each subject's food preferences using this history.

We compared four ranking criteria: (1) ranking according to a user's subjective view (correct ranking), (2) ranking according to popular recipe search website, (3) ranking by user's food preferences (browsing and cooking history), and (4) ranking by the proposed method (user's food preference and the quantity of each ingredient).

2 Evaluation Results

ssessed the accuracy of our proposed method, the results of which are shown
'e 1. Table 2 shows the corresponding normalized discounted cumulative
CG) values. nDCG is a measure of similarity to the correct ranking. In this

Table 1 Ranking results of 4 methods

		(1) Correct ranking	(2) Ranking by popular website	(3) Browsing and cooking history	(4) Proposal method
1st	Recipe A	1	2	1	2
	Recipe B	2	1	5	5
	Recipe C	3	4	2	3
	Recipe D	4	5	3	4
	Recipe E	5	3	4	1
2nd	Recipe F	1	5	2	1
	Recipe G	2	4	1	2
	Recipe H	3	3	3	3
	Recipe I	4	2	5	5
	Recipe J	5	1	4	4
3rd	Recipe K	1	1	2	1
	Recipe L	2	2	3	2
	Recipe M	3	5	1	3
	Recipe N	4	4	5	5
	Recipe O	5	3	4	4
4th	Recipe P	1	1	2	1
	Recipe Q	2	2	3	2
	Recipe R	3	5	1	3
	Recipe S	4	4	5	5
	Recipe T	5	3	4	4
5th	Recipe U	1	2	1	1
	Recipe V	2	3	5	5
	Recipe W	3	2	2	4
	Recipe X	4	1	3	2
	Recipe Y	5	5	4	3

Table 2 Ranking results by nDCG value

	(1) Correct ranking	(2) Ranking by popular website	(3) Browsing and cooking history	(4) Proposal method
nDCG	1	0.8996	0.9072	0.9381

assessment, the ranking of recipes according to a user's subjective view is regarded as correct ranking. Namely, nDCG value of 1) ranking by user's subjective view (correct ranking) is 1.0.

The results in Table 2 indicates that our proposed method (ranking method 4 above) obtained the highest nDCG value—excluding, of cause, ranking method 1—ahead of ranking method 3—ranking according to browsing and cooking history. We thus see that our proposed method provides a ranking that is more similar to users' preferences than a conventional method, even though the nDCG value of ranking method 2 (ranking by popular website) is still quite high. Thus, it we conclude that our method, which considers users' food preferences and the quantity of each ingredient in a recipe, is effective for a personalized recipe recommendation system.

In our experiments, subjects sometimes took into account the category in which a recipe falls and the combination of ingredients in ranking food. We will try to consider these aspects in calculating recipe scores in future work.

4 Conclusion

In this paper, we presented a scoring method for a cooking recipe recommendation system by using users' food preferences and the quantity of each ingredient in a recipe. Our previously proposed method estimated a users' food preferences from his/her past actions, such as through their recipe browsing and cooking history. This method breaks recipes down into their ingredients and scores them according to the frequency of their use and the specificity of the ingredients. However, our previous method does not completely reflect users' preferences, because of which we proposed here a new recipe recommendation method based on users' culinary preferences and the quantity of each ingredient used in recipes. Since our method can estimate user preferences through their browsing and cooking history, the user does not need to do anything for his/her preferences to be ascertained by the system.

In order to verify the accuracy of our proposed method, we conducted simple experiments. We compared four ranking criteria: (1) correct ranking, (2) ranking by a popular website, (3) ranking by the user's preferences (browsing and cooking history), and (4) ranking by the proposed method (user's food preferences and the quantity of each ingredient in a recipe). The *nDCG* values of our proposed method is 0.9381, which shows that our method most accurately the user's preferences in food.

Our method recognizes the ingredients using the name of the ingredients. We will try to consider the preference of the processed food, for example tomato and ketchup, in the future work.

Acknowledgments This work was supported in part by the MEXT Grant-in-Aid for Scientific Research(C) (#23500140, #26330351).

References

okpad, http://cookpad.com/. Accessed 8 Jan 2014
uten Recipe, http://recipe.rakuten.co.jp/. Accessed 8 Jan 2014
ne of Service (Cookpad Inc. Corporate Information), https://info.cookpad.com/outline_of_
  ./. Accessed 8 Jan 2014

4. Y. Mino, I. Kobayashi, Recipe recommendation for a diet considering a user's schedule and the balance of nourishment, in *Proceedings of IEEE International Conference on Intelligent Computing and Intelligent Systems 2009*, pp. 383–391 (2009)
5. S. Karikome, A. Fujii, A system for supporting dietary habits: planning menus and visualizing nutritional intake balance, in *Proceedings of the 4th International Conference on Ubiquitous Information Management and Communication*, pp. 386–391 (2009)
6. J. Freyne, S. Berkovsky, Intelligent food planning: personalized recipe recommendation, in *Proceedings of the 14th International Conference on Intelligent User Interfaces* (2010)
7. M. Ueda, M. Takahata, S. Nakajima, Recipe recommendation method based on user's food preferences, in *Proceedings of the IADIS International Conference on e-Society 2011*, pp. 591–594 (2011)
8. M. Ueda, S. Asanuma, Y. Miyawaki, S. Nakajima, Recipe recommendation method by considering the user's preference and ingredient quantity of target recipe, in *Proceedings of the International Multiconference of Engineers and Computer Scientists 2014, IMECS 2014*, 12–14 Mar 2014, Hong Kong. Lecture Notes in Engineering and Computer Science, pp. 519–523
9. Ajinomoto Park, http://park.ajinomoto.co.jp/. Accessed 8 Jan 2014

Fast and Memory Efficient 3D-DWT Based Video Encoding Techniques with EZW Based Video Compression Mechanism

Vishal R. Satpute, Ch. Naveen, Kishore D. Kulat
and Avinash G. Keskar

Abstract This chapter deals with the video encoding techniques using Spatial and Temporal Discrete Wavelet transform (DWT). It discusses about two video encoding mechanisms and their performance at different levels of DWT. Memory is the major criteria for any video processing algorithm, so in this chapter focus on the efficient utilization of the system memory at increased level of spatial and temporal DWT is presented. Out of these two mechanisms, one of the mechanism implements multi resolution analysis in temporal axis. Here the chapter also discuss about implementing the different DWT level in spatial and temporal domain. In this chapter, the Haar wavelet is taken as the reference. Finally, the compression of the videos is achieved by using the standard embedded zero wavelet tree (EZW) mechanism. The performance of the EZW based compression in terms of the PSNR and compression ratio is shown for various videos in this chapter.

Keywords DWT · EZW · Haar-wavelet · Multi-resolution-analysis (MRA) · Spatial DWT · Temporal DWT · Video encoding–decoding · Video compression · Wavelet

V.R. Satpute (✉) · Ch. Naveen · K.D. Kulat · A.G. Keskar
Department of Electronics and Communication Engineering,
Visvesvaraya National Institute of Technology, Nagpur 440010, India
e-mail: vrsatpute@ece.vnit.ac.in

Ch. Naveen
e-mail: naveench@ieee.org

K.D. Kulat
e-mail: kdkulat@ece.vnit.ac.in

A.G. Keskar
e-mail: agkeskar@ece.vnit.ac.in

© Springer Science+Business Media Dordrecht 2015
G.-C. Yang et al. (eds.), *Transactions on Engineering Technologies*,
DOI 10.1007/978-94-017-9588-3_30

1 Introduction

In general the term video can be defined as the, *"recording of the moving visual images"*. A digital video can be represented as the, *"collection of the moving visual digital images arranged in a specific order"* and can be considered as a series of 2D frames arranged on time axis also called temporal axis. Few basic terminologies related to video are; *Frame*: Each digital image in a video cam be called as a *"frame"*, *Frame Rate*: *"Number of frames per second in a video"*, *Spatial Axis*: *"The axis defines the spatial relation of a frame"*, *Temporal Axis*: *"The axis defines the time domain information of the video"*. In the present world, the need for efficient video processing mechanisms has become a major issue due to its important role in the security, entertainment etc. The applications like video surveillance need the video processing mechanisms which handle the video efficiently by utilizing the minimum memory space and in minimum time. Here, we are going to discuss about such two algorithms which handle the video efficiently for encoding with minimum memory requirement and in minimum amount of time. It is well known that spatial DWT [1], it can be applied only on 2 dimensions i.e., on x and y axis. Since, video is a 3-dimensional object spatial DWT cannot be applied to videos directly. But, it can be applied indirectly by considering each frame of video as an image which is memory inefficient and takes lot of time as it process each frame entirely. Hence, there is an urgent need of the mechanisms which can process the video efficiently and in less time. Such kind of mechanisms is 3D-DWT mechanisms which add the application of DWT on the temporal axis [2, 3]. In this chapter the application of such two mechanisms on video are presented in which for both spatial and temporal DWT, the filter masks used are, [1/2,1/2] and [1/2, −1/2] for forward DWT with [2] and [−1, 1] for reverse DWT as low pass and high pass masks respectively. Here Low pass filter is represented as 'h', and high pass filter as 'g' [4]. These filters helps us to get finer details of the given image or video. The outputs of DWT are arranged in specific order which helps to get details of the spatial as well as frequency components of the given image or video.

The application of 2D-DWT on images (as shown in Fig. 1a) will lead to four outputs i.e. LL1, LH1, HL1, and HH1 as shown in Fig. 1b for level-1 2D-DWT.

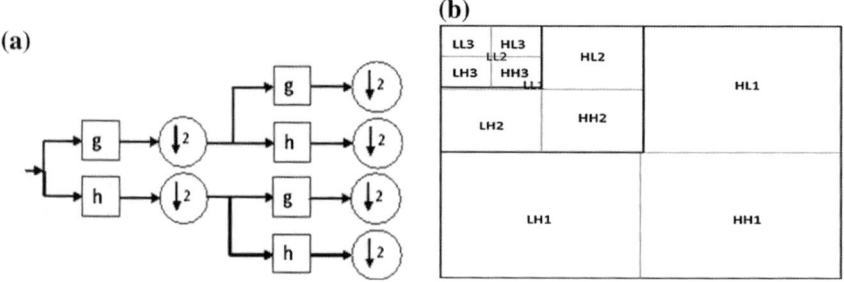

ıatial 2D-DWT of a given image; and **b** its higher level decomposition

(Here '1' indicates the level-1 output). For higher level of 2D-DWT, the low pass outputs i.e. LL is taken into consideration for further decomposing e.g. LL1 component of level-1 is used for level-2 2D-DWT and so on as shown in Fig. 1b. This technique is applied on the images and also on videos by considering each frame as an image. Since frame by frame operation is memory inefficient and time consuming we go for another technique in which DWT is applied on the third axis along with spatial DWT simultaneously and is called as Temporal DWT (3D-DWT) technique. After the encoding the video need to be compressed hence compression technique used is Embedded Zero Wavelet (EZW). The Embedded Zero Wavelet (EZW) is simple and remarkably effective algorithm for image compression which has a property of coding the bits in the order of their importance. Basic concept behind EZW is the concept of zero tree structure which occurs in the DWT applied image due to the spatial correlation of DWT [5–7]. To apply EZW on an image we need to follow three steps. The first step is to apply multi-level DWT on image. Second step consists of two passes namely Dominant pass and Subordinate pass. Dominant pass start with finding a threshold and modifying the image pixel values depending on threshold. Now the image is scanned in a special order and accordingly it will be assigned with codes POS, NEG, IZ, and ZTR for positive, negative, isolated zero and zero tree respectively [7, 8]. The pixels which are coded as POS, NEG are sent to sub-ordinate list. Now, elements in sub-ordinate list are processed in the entry order. In this pass we are going to refine the values for achieving loss less compression at the time of decoding [7]. The elements which are coded with POS and NEG are replaced by zeros in the image and the same process is repeated by making the threshold half of the previous, this process can be done up to the user defined threshold value or up to the threshold limit of 1. Finally, the generated string is coded using entropy coding algorithms such as Huffman coding, Arithmetic coding etc. but here we are using the simple coding by representing each symbol using two bits. One of the advantages of EZW is that we can stop decoding at any point of time to view the decoded image. Video encoding has certain advantages and disadvantages and is having some problems related with processing, these are discussed here. *Video Encoding*: Video encoding is defined as, *"Transformation of video from one domain to other domain or converting the video output into another digital form to make it compatible with certain codec's or mechanisms or algorithms"*. *Advantages of video encoding techniques*: One and the main advantage of the video encoding techniques is to make the video format compatible with the codec without disturbing its core part of operation. But it has its own disadvantages. *Disadvantages of Video encoding techniques*: The disadvantage of video encoding techniques is that it makes the computations complex, which is not good for a system with fewer amounts of memory and having the less computational capabilities. But one having the high computational capabilities and more memory to store then encoding may not be such a considerable problem. *Problems in video processing*: Due to the large memory and the time constraints video processing requires a high end processor and a huge amount of memory. When performing the video processing one should take these two parameters (memory and time) into account and design his/her system or algorithm.

The memory requirement is very high and is completely depends upon the three parameters i.e. *Video frame size, Video frame rate, and Video Duration.* Any one parameter can affect the performance of the algorithm drastically.

2 Temporal DWT Representation

The temporal DWT is nothing but the DWT applied to the frames and is called as 3D-DWT, if it is applied along with the spatial 2D-DWT [5, 9–13].

The application of 3D-DWT leads to two components in the temporal domain these are called as high pass (indicated by 1 for level-1 and 2 for level-2) and low pass (indicated by 1L for level-1 and 3 for level-2) frames and are indicated in Fig. 2a for temporal DWT. Component 1L can be further decomposed by using temporal DWT. One thing to be observed from Fig. 2b is that the number of frames for every increased level of DWT is becoming half. There arrangement for storing point of view or representation point of view can be done as shown in Fig. 2b [6]. In Fig. 2b, FP-1 to FP-N represents the frame pointer. Further decomposition can be processed by application of 3D-DWT on the low pass frames. *Reconstruction*: The reconstruction process of 3D-DWT involves the inverse DWT. This can be achieved by applying the IDWT on the components '2' and '3' initially and then with this result along with '1' the original video frames can be generated. Thus in general, to get back the original frames one need to apply the IDWT initially on to the latest decomposed outputs and along with this results one gets back the original video frames. This is very similar to the inverse DWT process applied to the images. Thus here one can perform the similar task of multilevel resolution process applied to the images on to the frames on temporal axis. Thus the frames will have multiple levels of processing. The process of reconstruction of IDWT is shown in Fig. 3a, b taken together. Now by using the result (2 3)' and '1' we can get back the original video.

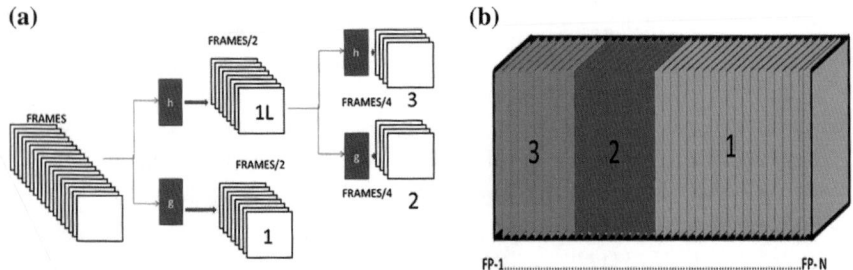

a Temporal DWT on video and **b** its representation

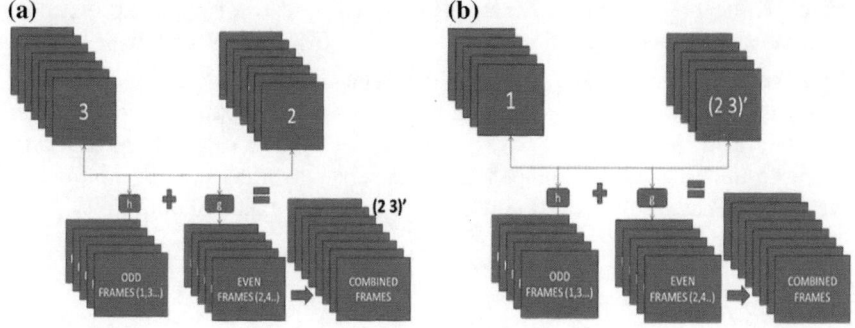

Fig. 3 a The first and **b** the last process of getting back the original video

3 Equations Related to Temporal DWT

Here discussion about the equations which help us in applying the K-level temporal DWT to the video is given. Here for explanation let us consider a video of size '**m**' by '**n**' by '**p**'.

3.1 Equations for Mechanism-1

Encoding and decoding equations for mechanism 1 are as follows:

$$
\left. \begin{aligned}
(LP_x)_K &= \sum_{i=1}^{\frac{m}{2K}} \sum_{j=1}^{\frac{n}{2K}} \frac{F(i,j,x) + F(i,j,x+1)}{2} \\
(HP_x)_K &= \sum_{i=1}^{\frac{m}{2K}} \sum_{j=1}^{\frac{n}{2K}} \frac{F(i,j,x) - F(i,j,x+1)}{2}
\end{aligned} \right\} \tag{1}
$$

where, 'K' denotes the level of temporal DWT applied, 'x' denotes the frame number and its maximum value is $p/(2^K)$, 'F' denotes the frames of the original video. The decoding equations are as follows:

$$
(FP_x)_{K-1} = \begin{cases}
\sum_{i=1}^{\frac{m}{2K}} \sum_{j=1}^{\frac{n}{2K}} (LP(i,j,\frac{x+1}{2}))_K + (HP(i,j,\frac{x+1}{2}))_K; & x \text{ is odd} \\
\sum_{i=1}^{\frac{m}{2K}} \sum_{j=1}^{\frac{n}{2K}} (LP(i,j,\frac{x}{2}))_K - (HP(i,j,\frac{x}{2}))_K; & x \text{ is even}
\end{cases} \tag{2}
$$

where, 'K' denotes the maximum level of the temporal DWT applied, 'x' denotes the frame number and its maximum value is $\frac{p}{(2^{k-1})}$ 'FP' denotes the low pass frames of the previous level. Now to get back the frames of previous temporal level i.e., $(K - 1)$th level, we need to know the high pass and low pass component of the present level i.e., Kth level. So, to get back the original video we have to start solving from Kth level, we cannot directly calculate the frames of any intermediate level without having the prior knowledge of the present temporal level.

3.2 Equations for Mechanism-2

Encoding and decoding equations for mechanism 2 are as follows:

$$(HP_x)_K = \sum_{i=1}^{m/(2^{K_s})} \sum_{j=1}^{n/(2^{K_s})} \frac{F(i,j,x) - F(i,j,x+1)}{2} \tag{3}$$

where, 'K_s' denotes the level of spatial DWT applied, 'x' denotes the frame number and its maximum value is $p/(2^{K_t})$, 'K_t' denotes the level of temporal DWT applied; 'F' denotes the frames of the original video. The decoding equations are as follows:

$$(FP_x)_{K_t-1} = \begin{cases} \sum_{i=1}^{\frac{m}{(2^\wedge K_s)}} \sum_{j=1}^{\frac{n}{(2^\wedge K_s)}} (LP(i,j,\frac{x+1}{2}))_{K_t} + (HP(i,j,\frac{x+1}{2}))_{K_t}, & x \text{ is odd} \\ \sum_{i=1}^{\frac{m}{(2^\wedge K_s)}} \sum_{j=1}^{\frac{n}{(2^\wedge K_s)}} (LP(i,j,\frac{x}{2}))_{K_t} - (HP(i,j,\frac{x}{2}))_{K_t}, & x \text{ is even} \end{cases} \tag{4}$$

where, 'K_t' denotes the maximum level of the temporal DWT applied, 'K_s' denotes the maximum level of the spatial DWT applied, 'x' denotes the frame number and its maximum value is $\frac{p}{(2^{K_t-1})}$ 'FP' denotes the low pass frames of the previous level. To get back the frames of previous temporal level i.e., $(K_t - 1)$th level, we need to know the high pass and low pass component of the present level i.e., K_tth level. So, to get back the original video we have to start solving from K_tth level, we cannot directly calculate the frames of any intermediate level without having the prior knowledge of the present temporal level. Equations (1)–(4) are applied only for the temporal DWT, for applying the spatial DWT the traditional method of DWT is followed [1]. If K-level temporal DWT is to be applied at a constant spatial DWT level then replace the corresponding level value in K_s in Eqs. (3) and (4) and solve the K level temporal DWT from Eqs. (3) and (4).

4 Video Encoding and Decoding

The two mechanisms presented are discussed here. The mechanism-1 can implement multi-level spatial 2D-DWT and the same level temporal DWT simultaneously [14]. The process for which is as follows:

1. Read the video and apply level-1 spatial 2D-DWT on each frame of the video.
2. Now on the output of step-1 apply the temporal DWT of level-1.
3. From the output in step-2 all low pass component is taken and spatial 2D-DWT is applied on it, this is level-2 spatial 2D-DWT.
4. Now, on the output of step-3 temporal DWT is applied which is called as level-2 temporal DWT. And for multi-level, this process is repeated until the number of frames or number of rows or number of columns in all low pass component reaches to an odd number.

Now, after applying the mechanism-1 on videos they are arranged in the order as shown in Fig. 6 [6]. Now, the steps explained above are shown in the diagrammatical form in Figs. 4 and 5. This arrangement will look like the multi resolution analysis applied on all three axes of a video [4]. So this mechanism can also be called as 3D multilevel pyramid decomposition [15].

The Reconstruction process of the mechanism-1 is similar to the reconstruction process explained in Sect. 2, but with slight changes. As mentioned in Sect. 2, reconstruction process starts with the all low pass component of the highest DWT level. For the ease of explanation let us consider the video which is used in encoding process on which level-3 spatial DWT and level-3 temporal DWT are applied. The steps for reconstruction are as follows: (refer Figs. 7 and 8). The process is as follows:

1. Consider all low pass component of the highest DWT level. By using the high pass components of that level get back the all low pass component of the previous level. From the Fig. 7a, all low pass components i.e. '9' which is passed though low pass filter. By using the two outputs of the temporal DWT low pass and high pass get back the level-3 of spatial DWT. And by using '9' which is low pass and by using '10', '11', '12' which are high pass components, we can get back '5'.
2. Now by using the output from step-1, apply the same procedure at the previous level to get the low pass component of its previous level. From the Fig. 7b, the output obtained will be '1', which is the low pass component of the previous level.
3. Process is continued until original video is reconstructed. (Refer Fig. 8)

As explained in the steps of mechanism-1 and its reconstruction, we are not using the entire video for the processing of the higher levels in spatial and temporal DWT. This helps us in utilizing the memory efficiently as we are not processing 1/4th part of the frame and 1/2th part of the frame (with respect to the current frame size and number of frames) for every increase in the DWT level at the time of

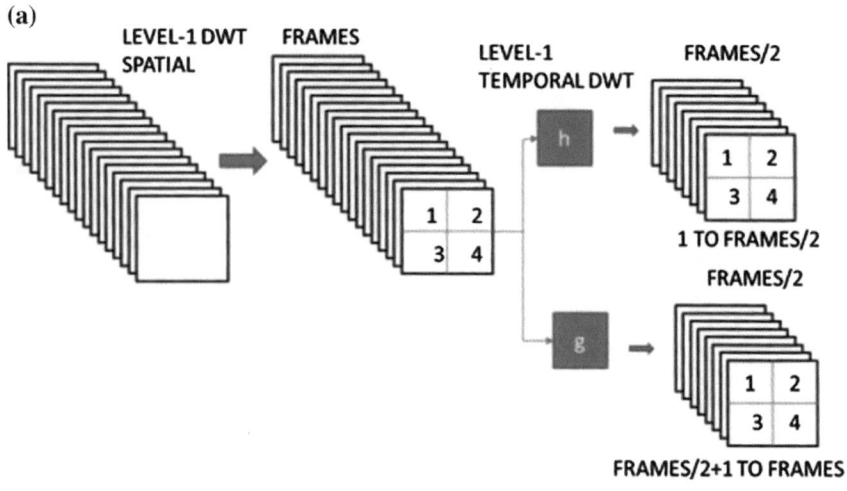

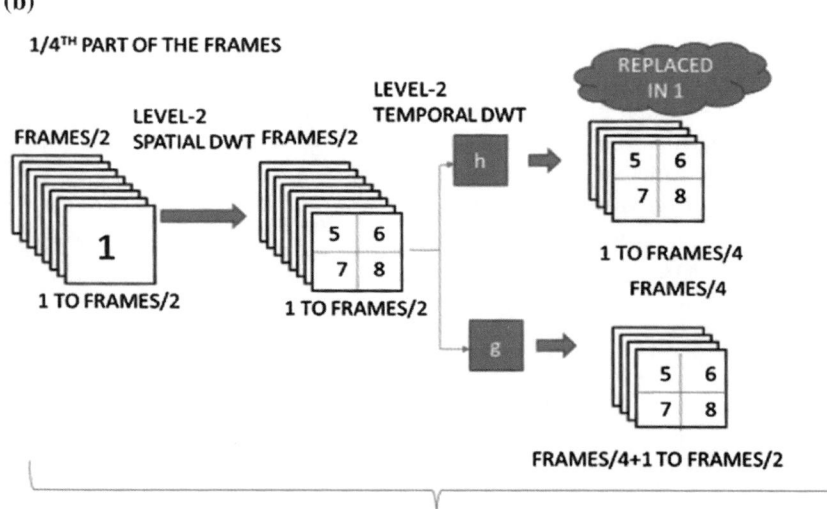

SECOND LEVEL SPATIAL AND TEMPORAL DWT RESPECTIVELY

Fig. 4 **a** Steps 1 and 2 to be followed; **b** steps 3 and 4 to be followed

ncoding. So the memory utilization will reduce by almost 3/4th when considered
th the spatial DWT applied frame by frame and time consumption also reduces as
are computing less number of pixels in the increased level. In this way we can
ve the efficient utilization of the system memory and time requirement.
g to the reconstruction, here also we are not utilizing the entire video at all

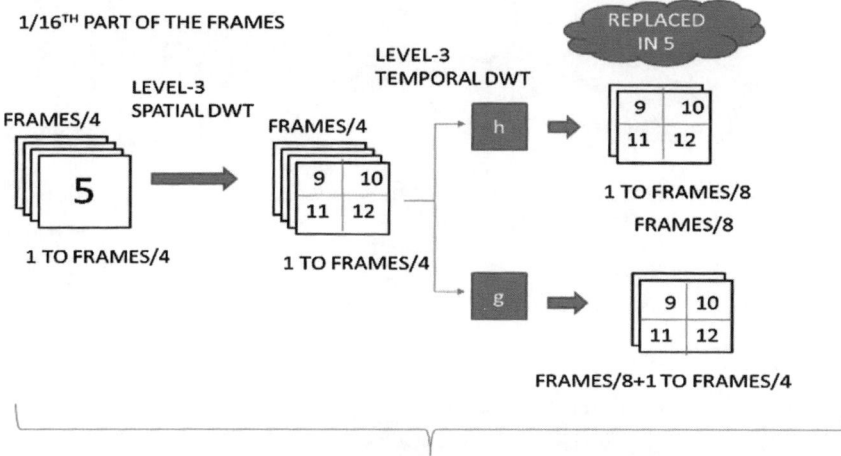

Fig. 5 Illustration of level-3 spatial DWT and level-3 temporal DWT

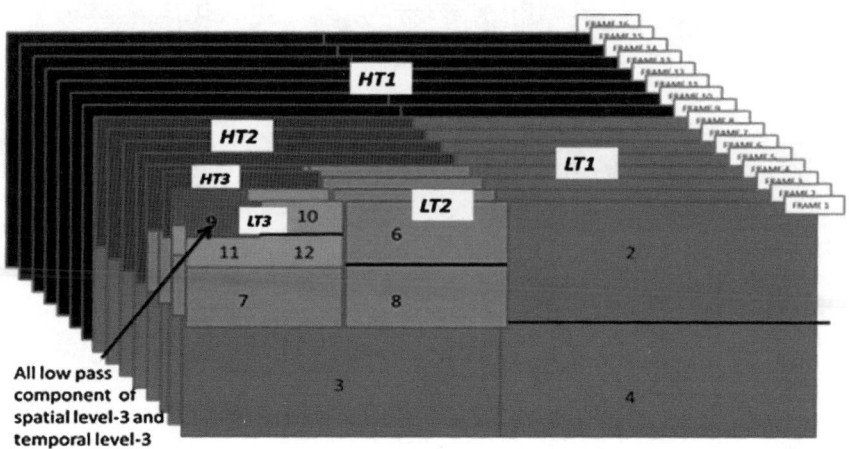

Fig. 6 Typical arrangement of the frames after the level-3 spatial DWT and level-3 temporal DWT

As we are approaching level-1 the usage of the system memory increases. So, at the time of reconstruction also one can achieve the efficient usage of system memory and time. For a video of size 'm' by 'n' and with 'p' number of frames. For the level-1 spatial and level-1 temporal DWT we require $m * n * p$ bytes of system memory for processing the video. Now, when we are going to level-2, high pass components are not considered so we require $(m/2) * (n/2) * (p/2)$ bytes of system memory for processing the video and if we go to next level further the

(b)

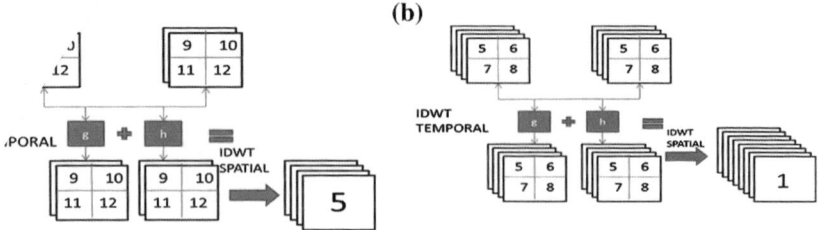

Fig. 7 a Step-1 and **b** step-2

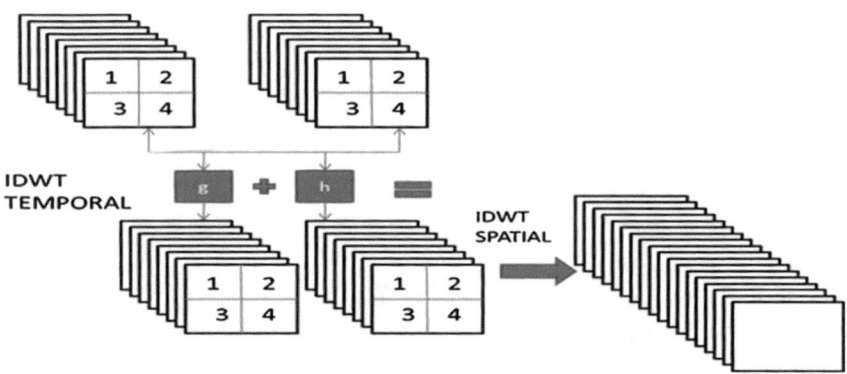

Fig. 8 Illustration of the next level of reconstruction

memory requirement will be reduced by a factor of 8 for every increment in the DWT level of spatial and temporal DWT with respect to the previous level.

The second mechanism i.e. mechanism-2 can implement multi-level spatial 2D-DWT and multi-level (may differ from spatial) temporal 3D-DWT. The process for which includes:

1. Read the video and apply multilevel spatial 2D-DWT on each frame of the video.
2. Now take only all low pass component from the output of step-1 and apply multi-level temporal DWT on all low pass component only.

The mechanism-2 is shown diagrammatically in Fig. 9.

Now, the output frames are arranged in the manner as shown in Fig. 10 [6]. In 10, SL represents corresponding spatial level indicated by number and TLH sents corresponding temporal level high pass component indicated by the 'r while TLL represents the all low pass component of the highest temporal vel applied. The reconstruction process steps for mechanism-2 are dis-
low:

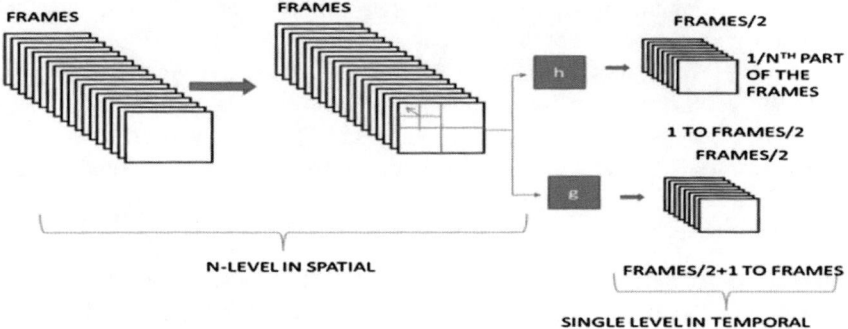

Fig. 9 Diagrammatical representation of mechanism-2

(a)

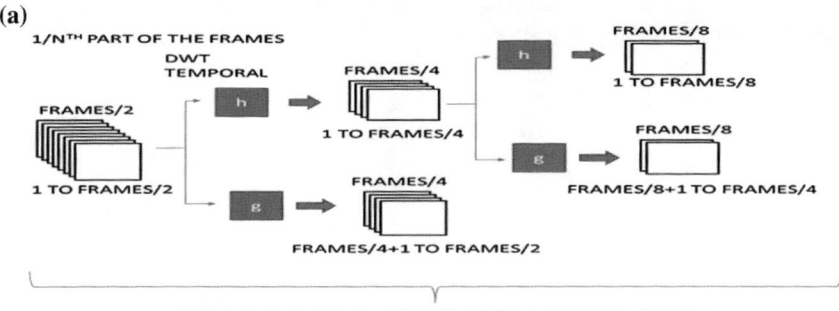

(b)

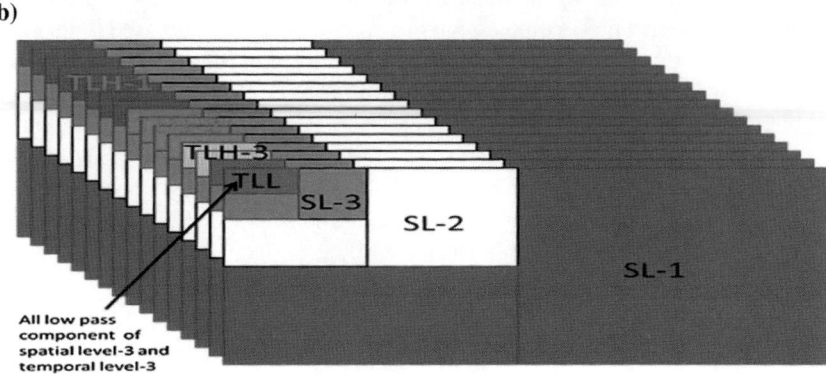

Fig. 10 Diagrammatical representation of mechanism-2 and arrangement of frames

1. Consider the all low pass component of the highest level of temporal DWT applied and by using the high pass component of the same level get back the low pass component of the previous level. From Fig. 10, using TLL and TLH-3 get back the all low pass component of the previous level using the reconstruction process of the temporal DWT explained in Sect. 2.

using the output in step-1 and high pass component of that level get the
pass component of the previous level using the reconstruction process of
oral DWT discussed in Sect. 2. From Fig. 10, using the output of step-1 and
H-2 get the all low pass component of the previous level.

repeat the steps 1 and 2 until all the temporal DWT levels applied on the video
are completed. So, now we are left with the video on which only spatial DWT is
applied. Using the 2D-IDWT on the output of step-3, one can get back our
original video.

In this mechanism also we are not using the full video when we are applying the
temporal DWT on video. This helps using efficient use of the system memory and
reduces the time for processing. For this mechanism, memory requirement go on
reducing with the increment in the spatial 2D-DWT level because we are processing
only the low pass component of the K-level spatial 2D-DWT output. As, we are
applying K-level 2D-DWT, the size of each frame at all low pass component will be
reduced by the factor of K in spatially. So, now we need only $1/(2 * K)$th part of
the system memory for processing the video. If we consider a video of size 'm' by
'n' with 'p' number of frames then we require 'm $*$ n $*$ p' number of bytes for
processing the K-spatial DWT and $((m/K) * (n/K))$ bytes to process the temporal
DWT.

4.1 Application of EZW on the Encoded Video

EZW is applied on the each frame of the encoded video for mechanism-1 and
mechanism-2. The typical compression or decompression system used in this work
is as shown in the Fig. 11. Encoded video sequence is given to the EZW encoder
[8]. Video sequence is compressed on frame by frame basis, each frame of the video
is given to the encoder and output of each frame is concatenated at the output. To
get back the original video, decoder must be in synchronization with the encoder.
This synchronization can be maintained by the header file which consists of frame
size, number of frames, level of 3D-DWT etc. The PSNR, MSE and compression
ratio of the reconstructed video is calculated by using equations [16].

$$MSE = \frac{1}{m * n}\sum_{x=1}^{m}\sum_{x=1}^{n}(I(x,y) - I'(x,y))^2 \qquad (5)$$

ame-by-frame compression of video

$$PSNR = 10 \log 10 \left(\frac{R^2}{MSE} \right) \tag{6}$$

$$Compression\ Ratio(CR) = \frac{size\ of\ compressed\ data}{size\ of\ original\ raw\ image} \tag{7}$$

5 Results and Comparison

The two proposed mechanisms are tested on various standard videos, videos taken from internet and videos captured in the laboratory under non-standard conditions as given in Table 1. Here, we are presenting the comparison between the mechanisms, based on the time taken to transform and reconstruct the video in Table 2 (for temporal and spatial level-1 and for level-2). Along with this, MSER [17] value and compression ratio is also provided in Table 3 (Fig. 12).

Table 1 Details of video database used

Video	Size	Frame rate	Duration (in s)	Total frames	Video format
Video 1	120 × 160	15	8	120	.avi
Video 2	120 × 160	15	5	86	.avi
Video 3	480 × 640	30	9	280	.avi
Video 4	240 × 320	30	4	140	.avi
Video 5	288 × 384	25	3	80	.avi
Video 6	240 × 320	8	5	46	.avi
Video 7	240 × 360	30	11	330	.avi
Video 8	480 × 640	10	14	140	.avi
Video 9	480 × 640	10	15	155	.avi
Video 10	240 × 320	8	5	40	.avi
Video 11	180 × 360	30	9	270	.avi
Video 12	240 × 320	29	18	540	.avi
Video 13	576 × 1024	25	10	256	.avi
Video 14	1,080 × 1,720	20	6	128	.avi
Video 15	144 × 176	30	13	392	.wmv

Table 2 Result of comparison of two mechanisms based on various parameters

Video	Time and MSER	Mechanism-1 (level-1)	Mechanism-2 (level-1)	Mechanism-1 (level-2)	Mechanism-2 (level-2)
Video-1	DWT	0.170923	0.161118	0.246506	0.173512
	IDWT	0.142283	0.133975	0.198421	0.136978
	MSER	3.529e−31	3.529e−31	8.025e−31	8.104e−31
Video-7	DWT	2.967789	2.578908	4.279791	3.054225
	IDWT	2.425089	2.214391	3.482500	2.487691
	MSER	3.479e−31	3.479e−31	9.340e−31	9.941e−31
Video-8	DWT	2.749909	2.422879	3.984854	2.969299
	IDWT	2.304075	2.043316	3.246693	2.289465
	MSER	3.413e−31	3.413e−31	9.336e−31	9.413e−31
Video-15	DWT	0.824465	0.669313	1.160209	0.777375
	IDWT	0.606323	0.549009	0.864553	0.618469
	MSER	1.525e−30	1.525e−30	3.474e−30	3.487e−30

Table 3 Comparative results of application of EZW on mechanism-1 and mechanism-2

Video	Outputs	Mechanism-1	Mechanism-2	.avi format
Video-1	MSE	4.606064	2.992383	35.2404
	PSNR	41.807160	43.536066	34.622226
	CR	0.291157	0.481138	0.2866
	Size occupied	670.8 KB	1.108 MB	665 KB
Video-2	MSE	3.126492	1.298460	87.416903
	PSNR	43.199504	47.009878	28.905921
	CR	0.668820	2.069733	0.677
	Size occupied	1.104 MB	3.416 MB	1.119 MB
Video-3	MSE	3.527473	2.516702	109.943574
	PSNR	42.900408	44.256265	30.269945
	CR	0.192577	0.349285	0.0351
	Size occupied	16.55 MB	30.04 MB	3.017 MB
Video-4	MSE	5.136806	2.210027	62.204
	PSNR	41.080496	44.725832	30.431
	CR	0.294607	0.693063	0.3051
	Size occupied	3.167 MB	7.451 MB	3.280 MB
Video-5	MSE	6.125640	3.271772	93.388483
	PSNR	40.339352	43.077129	28.677098
	CR	0.292897	0.434110	0.2753
	Size occupied	2.591 MB	3.840 MB	2.436 MB

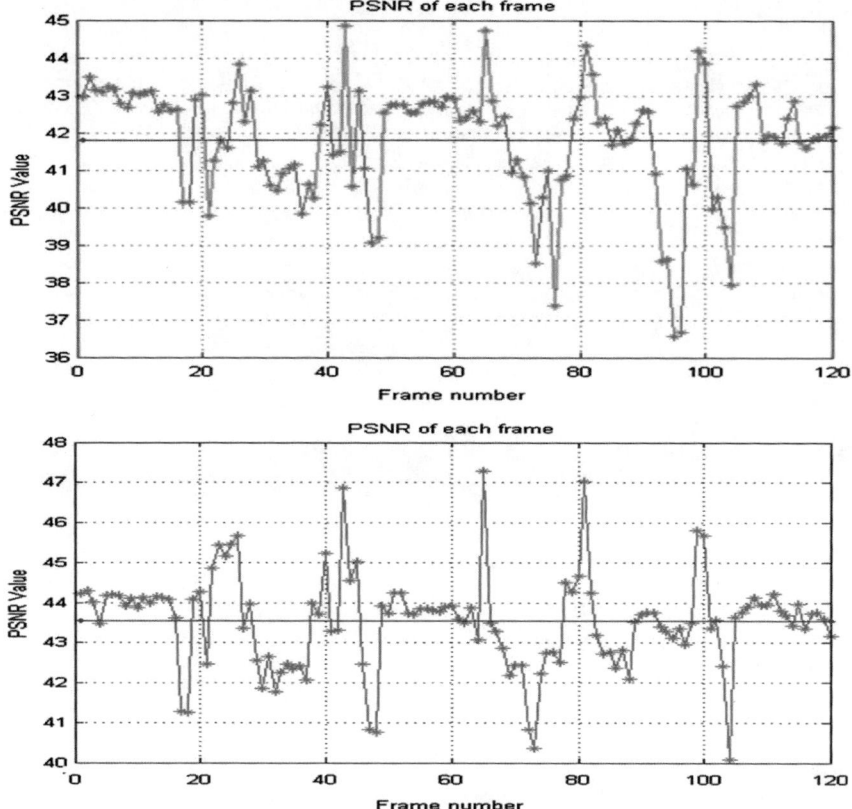

Fig. 12 PSNR versus frame number graph of mechanism-1 and 2 respectively for video-1

6 Conclusion

Thus, we conclude our chapter by explaining the two mechanisms which can be used for video encoding with efficient utilization of system memory and with minimal amount of time requirements. The standard EZW based video encoding helps us to compress the video data and the comparison is done with that of the standard AVI format. One important thing to be noted here is that, if we use 3D-EZW or 3D-SPIHT then better compression ratio can be achieved as it will enhance further temporal correlation of the video data and will provide more insight into the variations of the information available in the frames. The last but most important point to note here is that the time required to process this small amount of extracted information by the mechanism is very less and hence is best suitable for applications like video surveillance where time is an important parameter to be considered.

References

1. K. Sureshraju, V.R. Satpute, Dr. A.G. Keskar, Dr. K.D. Kulat, Image compression using wavelet transform compression ratio and PSNR calculations, in *Proceedings of the National Conference on Computer society and informatics NCCSI'12*, 23–24 Apr 2012
2. Ch. Naveen, V.R. Satpute, K.D. Kulat, A.G. Keskar, Fast and memory efficient 3D-DWT based video encoding techniques, in *Proceedings of the International Multi Conference of Engineers and Computer Scientists 2014 vol. I, IMECS 2014*, 12–14 Mar 2014, Hong Kong, pp. 427–433
3. N. Mehrseresht, D. Taubam, An efficient content adaptive motion-compensated 3D-DWT with enhanced spatial and temporal scalability. IEEE Trans. Image Process. **15**(6), 1397–1412 (2006)
4. S.A. Zu'bi, N. Islam, M. Abbod, 3D multi resolution analysis for reduced features segmentation of medical volumes using PCA, 978-1-4244-7456-1/10 © 2010 IEEE
5. M. Weeks, M. Bayoumi, 3-D Discrete wavelet transform architectures, 0-7803-4455-3/98©1998 IEEE
6. G. Liu, F. Zaho, Efficient compression algorithms for hyperspectral images based on correlation coefficients adaptive 3D zero tree coding. IET Image Process. (2008). doi:10.1049/ict-ipr:20070139
7. J.M. Shapiro, Embedded image coding using zero trees of wavelet coefficients. IEEE Trans. Signal Process. **41**(12), 3445–3462 (1993)
8. C. He, 1. Dong, Y.F. Zheng, Optimal 3-D coefficient tree structure for the 3-D wavelet video. Technical Report, HZ-2001, Wavelet Research Laboratory, The Department of Electrical Engineering, The Ohio State University, Aug 2001
9. A. Das, A. Hazra, S. Banerjee, An efficient architecture for 3-D discrete wavelet transform. IEEE Trans. Circ. Syst. Video Technol. 2010, pp. 286–296
10. Ch. Naveen, V.R. Satpute, K.D. Kulat, A.G. Keskar, Comparison of spatio-temporal 3D-DWT MRA with temporo-spatial 3D-DWT MRA, in *IEEE SCES-2014*, MNNIT Allahabad
11. A. Das, A. Hazra, S. Banerjee, An efficient architecture for 3-D discrete wavelet transform. IEEE Trans. Circ. Syst. Video Technol.
12. R.M. Jiang, D. Crookes, FPGA implementation of 3D discrete wavelet transform for real-time medical imaging, 1-4244-1342-7/07/$25.00 © 2007 IEEE
13. Ch. Naveen, V.R. Satpute, K.D. Kulat, A.G. Keskar, Video encoding techniques based on 3D-DWT, in *IEEE SCEECS 2014*, 1–2 Mar 2014, pp. 1–6
14. O. Lopez, M. Martinez-Rach, P. Pinol, M.P. Malumbers, J. Oliver, A fast 3D DWT video encoder with reduces memory usage suitable for IPTV, 978-1-4244-7493-6/10/© 2010 IEEE, 19–23 July 2010, pp. 1337–1341
15. R.M. Jiang, D. Crookes, FPGA implementation of 3D discrete wavelet transform for real-time medical imaging, 1-4244-1342-7© 2007 IEEE
16. H.L. Tan, Z. Li, Y.H. Tan, S. Rahardja, C. Yeo, A perceptually relevant MSE-based image quality metric. IEEE Trans. Image Process. **22**(11), 4447–4459 (2013)
17. Wikipedia: http://en.wikipedia.org/wiki/Mean_squared_error

PCA Based Extracting Feature Using Fast Fourier Transform for Facial Expression Recognition

Dehai Zhang, Da Ding, Jin Li and Qing Liu

Abstract Facial expression recognition is prevalent in research area, and Principal Component Analysis (PCA) is a very common method in use. Noticing that few researches focus on pre-processing of images, which also enhances the results of PCA algorithm, we propose an improved approach of PCA based on facial expression recognition algorithm using Fast Fourier Transform (FFT), which combines amplitude spectrum of one image with phase spectrum of another image as a mixed image. Our experiments are based on Yale database and self-made image database. Testing and evaluating in several ways, the experimental results indicate our approach is effective.

Keywords Amplitude spectrum · Facial expression recognition · FFT · PCA · Pre-processing · Phase spectrum · SVM

1 Introduction

Face recognition and facial expression recognition have been paid much attention in the past two to three decades, which play an important role in such areas as access control, human–computer interaction, production control, e-learning, fatigue driving detection, and emotional robot. There are ordinarily seven kinds of facial

D. Zhang (✉) · D. Ding · J. Li · Q. Liu
School of Software, Yunnan University, Kunming 650091, Yunnan, China
e-mail: dhzhang.ynu@gmail.com

D. Ding
e-mail: dingda6@126.com

J. Li
e-mail: lijin@ynu.edu.cn

Q. Liu
e-mail: Liuqing@ynu.edu.cn

© Springer Science+Business Media Dordrecht 2015
G.-C. Yang et al. (eds.), *Transactions on Engineering Technologies*,
DOI 10.1007/978-94-017-9588-3_31

413

expression to be classified: anger, disgust, fear, happiness, sadness, surprise and neutral [1].

However, as there are too many features in face or facial expression, most of these methods make trade-offs like hardware requirements, time to update image database, time for feature extraction, response time, etc. [2, 3]. So generally researchers would like to choose a special method that will reduce the amount of calculation and thus make the experiment much more efficient. Principal Component Analysis (PCA) is a main and long term studied method to extract feature sets by making sample projection from higher dimension to lower dimension [4]. Yet, for most occasions computers will run out of memory. Even with enough memory it will take much time to extract all eigenvalues [5]. Therefore, a better method called Fast Principal Component Analysis (FPCA), which will calculate in a much faster way, yet brings the same results as PCA does.

Besides PCA, many researches also proposed new algorithms to improve recognition rate [6, 7, 8]. Nevertheless, not like other researches that focus on improving algorithm in order to bring a better recognition rate, we are more interested in pre-processing of images, since it may also bring some positive effects on the final performance but in a more simple way. Although many researches also include pre-processing in their study, like adjusting face position, converting into grey-level images, or make some image corrections, most of them don't consider a way that enhancing features of a person [9]. Noticing this ignored aspect, we try to utilize a suitable method in pre-processing. And our method is Fast Fourier transform (FFT). By using FFT, we process original images and combine amplitude spectrum of one image with phase spectrum of another image to enhance features before extracting eigenvectors.

In this paper, a novel approach is proposed in order to improve the recognition rate of facial expression. We firstly discuss some basic algorithms to explain our method, Fast Fourier Transform. And then we evaluate our approach based on Yale database and self-made image database. Finally we present our results and compare to PCA only method.

2 Proposed Approaches

This section describes a set of approaches based on different fixed algorithms used in the paper. In the first step, we apply FFT for our training data to do pre-processing. The major task in this step is combining amplitude spectrum of one image with phase spectrum of another image into a mixed image. The next step is extracting feature by using FPCA. Finally, SVM is utilized for classification. Figure 1 shows the overall procedure of our work.

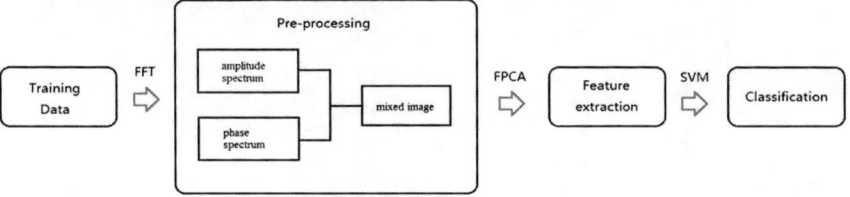

Fig. 1 Overall procedure of face recognition

2.1 Pre-processing

2.1.1 Fourier Transform

To discuss our method before, we firstly make a description about Fourier Transform. Virtually everything in the world can be described via a waveform. The Fourier Transform gives us a unique and powerful way of viewing these waveforms.

If we assume the period is T, Fourier series should be like:

$$f(x) = \frac{a_0}{2} + \sum_{k=1}^{+\infty}(a_k \cos n\, \omega_0 x + b_k \sin n\, \omega_0 x)$$ (1)

where $\omega_0 = \frac{2\pi}{T} = 2\pi u$, and $u = 1/T$, which is the frequency of $f(x)$.

It has already been proven that the first N terms of Fourier series is the best approximation of antiderivative $f(t)$ in given energy:

$$\lim_{N\to\infty} \int_0^T |f(t) - [\frac{a_0}{2} + \sum_{k=1}^{+\infty}(a_k \cos n\, \omega_0 x + b_k \sin n\, \omega_0 x)]|^2 dx = 0$$ (2)

In this paper, because we are dealing with images, so it is obvious for us to know two-dimension Discrete Fourier Transform (DFT):

$$F(u,v) = \sum_{X=0}^{M-1}\sum_{X=0}^{N-1} f(x,y)e^{-i2\pi(\frac{ux}{M}+\frac{vy}{N})}$$ (3)

where u and v are variables in domain.

The reason why we need Fast Fourier Transform (FFT) is that for a sequence of N, its definition of DFT transform and inverse transformation is:

$$\begin{cases} F(u) = \sum_{X=0}^{N-1} f(x) W_N^{ux}, & u = 0, 1, \ldots, N-1, W_N = e^{-j\frac{2\pi}{N}} \\ f(x) = \frac{1}{N} \sum_{u=0}^{N-1} F(u) W_N^{-ux}, & x = 0, 1, \ldots, N-1 \end{cases} \tag{4}$$

As we can see in Eq. (4), it is not hard to find that in order to calculate a sequence of N, it should do N^2 plural multiplications and $N(N-1)$ plural additions. Yet, the fundament of DFT is not so complex, W_N has its own periodicity, actually it only has N values. And also, these N values have symmetric relation in some degree. Because of periodicity and symmetry of W_N, we could make a conclusion as follows:

$$W_N^0 = 1, W_N^{\frac{N}{2}} = 1 \tag{5}$$

$$W_N^{N+r} = W_N^r, W_N^{\frac{N}{2}+r} = -W_N^r \tag{6}$$

where Eq. (5) represents some special values in matrix W, while Eq. (6) explains periodicity and symmetry of matrix W.

By utilizing periodicity of matrix W, some terms in DFT calculation can be combined; by utilizing its symmetry, we just need calculate half of W. Based on these two advantages, we could reduce a lot of operations, which is the basic idea of FFT.

2.1.2 Amplitude Spectrum and Phase Spectrum

In this paper, we apply FFT to combine amplitude spectrum of one image with phase spectrum of another image to enhance features of images.

The definition of amplitude spectrum:

$$|F(u,v)| = [Re(u,v)^2 + Im(u,v)^2]^{1/2} \tag{7}$$

Every amplitude spectrum $|F(u,v)|$ of point (u,v) in frequency domain could represent the ratio of sine (or cosine) plane wave to superimposition. Thus amplitude spectrum could reflect the frequency information and have a high practical value in filtering.

The definition of phase spectrum:

$$\varphi(u,v) = \arg \tan \frac{Im(u,v)}{Re(u,v)} \tag{8}$$

Though it is not so clear to figure out the information about phase spectrum as amplitude spectrum, it contains a kind of ratio relation between real part and imaginary part.

And in addition, we could restore $F(u, v)$ by amplitude spectrum and phase spectrum:

$$F(u, v) = |F(u, v)|e^{j\varphi(u,v)} \tag{9}$$

2.1.3 Images Combination

As we mentioned before, our focus is primary on pre-processing and we try to figure out a simple, an efficient or an elegant way to do this part well. And in this paper our answer is Fourier Transform. By combining amplitude spectrum of one image with phase spectrum of another image (both images belong to the same class), we got a mixed image that had main information of one image (the one with amplitude spectrum) and minor information of another image (the one with phase spectrum). Because our goal is to improve the recognition rate of face recognition, extracting eigenvectors from a mixed image would be better than extracting eigenvectors from an original image (as there were more features in the that image), therefore enhances the accuracy of SVM's results.

2.2 Extracting Feature

As we discussed before, our face recognition system is based on PCA. PCA is one of the oldest in analysing multiple variables, which is derived from Karhumen–Loeve (KL) transformation. It is Pearson who first proposed PCA in 1901, and in 1963 Karhumen–Loeve made a huge improvement about the original version of PCA [4].

If the image elements are considered to be random variables, then the image may be seen as a sample of a stochastic process. The PCA basis vectors are defined as the eigenvectors of covariance matrix S:

$$S = E[XX^T] \tag{10}$$

Since the eigenvectors associated with the largest eigenvalues have face-like images, they also are referred to as eigenfaces. If we suppose the eigenvectors of S are v_1, v_2, ..., v_n and are associated respectively with the eigenvalues $m_1 \geq m_2 \geq \cdots \geq m_n$. Then:

$$X = \sum_{i=1}^{n} \hat{x}_i v_i \tag{11}$$

However, if the number of dimension is very big, calculating matrix S will be a tough task that may even make computer gets stuck [9]. Fortunately, there is a good idea to fix it.

Assume that n is the number of samples, d is the number of dimension, and $Z_{n \times d}$ is the matrix that every sample in sample matrix X minus mean Eigenvalue $\bar{m}$, thus scatter matrix S should be $(ZZ^T)_{d \times d}$. Now consider the matrix $R = (ZZ^T)_{n \times n}$, for most occasion d is much greater than n, therefore the size of matrix R is much smaller than that of matrix S. So we have:

$$(ZZ^T)\vec{v} = \lambda \vec{v} \tag{12}$$

And from Eq. (12) we can get Eq. (13) by premultiplying Z^T:

$$(Z^T Z)(Z^T \vec{v}) = \lambda(Z^T \vec{v}) \tag{13}$$

Equation (13) shows that $Z^T \vec{v}$ is the eigenvalue of scatter matrix S, which means that we could calculate the smaller matrix R then calculate $Z^T \vec{v}$ instead of calculating $Z^T \vec{v}$ directly.

To harness PCA in our face recognition system, firstly our program read a sample of training database (m pixel * n pixel) and saved it in a sample matrix Face Container. Then by applying PCA, we converted m * n dimensions matrix into k dimensions matrix, and those k dimensions would represent all the dimensions in a training matrix, which would reduce a lot of operations in the SVM part.

2.3 Classification

In this paper, our classification builds on SVM. Unlike artificial neural network, by applying SVM we could calculate the optimum value based on small samples. The normal maximizing objective function of SVM is like:

$$L(\alpha) = \sum_{i=1}^{N} \alpha_i - \frac{1}{2} \sum_{i=1}^{N} \sum_{j=1}^{N} \alpha_i \alpha_j y_i y_j K(x_i, x_j) \tag{14}$$

where α_i is Lagrange multipliers.

In order to make a nonlinear projection, a kernel should be chosen [10]. In this study, we choose the radial basis kernel function as follow:

$$K(x, y) = \exp(-\gamma \|x - y\|^2) \tag{15}$$

In our method, we should determine both the error cost coefficient C and γ in radial basis kernel function. However, it is very hard using C and γ to make an expression by normal optimizing strategy. Fortunately, Doctor Chih-Jen Lin provides a very good way to solve this problem.

The main task in classification by using SVM is to determine C and γ, as normal optimizing strategies usually doesn't work well. Yet, in this aspect, Doctor Chih-Jen Lin proposed a program called grid.py to fix it. By using this program, we finally determine that the optimum value of C is 128, and the optimum value of γ is 0.0078125.

3 Test Results

In this section, we will evaluate the performance of proposed approaches for face recognition. We first did FFT for the entire training database, extracted PCA feature vectors and then applied each feature extraction method with SVM. All the works were done in Matlab2011b in a PC with 2.5 GHz Intel Core i5 and 4 GB 1,600 MHz DDR3 RAM.

3.1 Experiments on Yale Database and Discussion

The Yale database contains 165 images of 15 individuals. Each individual has 11 images with different facial expressions or configurations: normal, happy, sad, sleepy, surprised, and wink, center-light, left-light, and right-light, with and without glasses.

In the experiment, we firstly used FFT to combine amplitude spectrum of one image with phase spectrum of the next image (the last one will combine with the first one). All the mixed images we got are shown in Fig. 2 compared with original images.

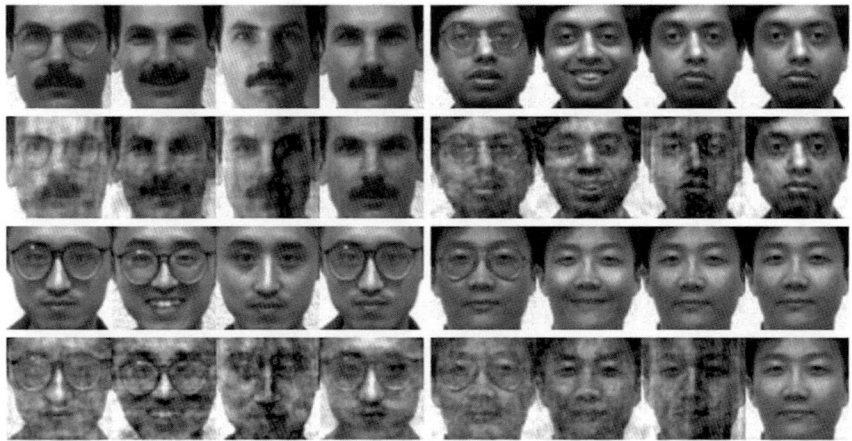

Fig. 2 Images after FFT processing (*above*) and original version (*below*)

Table 1 Recognition rate comparison of different approaches on Yale database

Training samples per individual	Value of k	PCA + SVM recognition rate (%)	FFT + PCA + SVM recognition rate (%)	Improvement rate (%)
3	20	81.905	83.81	2.326
	30	78.095	80	2.439
	40	81.905	83.81	2.326
4	20	84.127	85.714	1.886
	30	84.127	85.714	1.886
	40	82.540	84.127	1.923
5	20	84.211	90.79	7.813
	30	85.526	89.474	4.616
	40	84.211	93.421	10.937

To evaluate our method, we compare and contrast the results of PCA + SVM and that of FFT + PCA + SVM. And k (dimensions of Eigenvector) is set to 20, 30, and 40 respectively. The results are shown in Table 1.

As we can see from Table 1, the results produced by FFT + PCA + SVM are better than the results produced by FFT + PCA + SVM. And we can learn that when training samples of one individual is 5, the peak recognition rate is 93.421 %. And in [11], the peak recognition rate is 86.67 %, whose experiment was also based on Yale database and PCA and used 6 samples for training.

We also compare our results with DCV and KDDA, as shown in Fig. 3.

From Fig. 3, we learn that when the number of training samples is small, recognition rates of three approaches are close. However, when the number of training becomes greater, our method is better than DCV and KDDA.

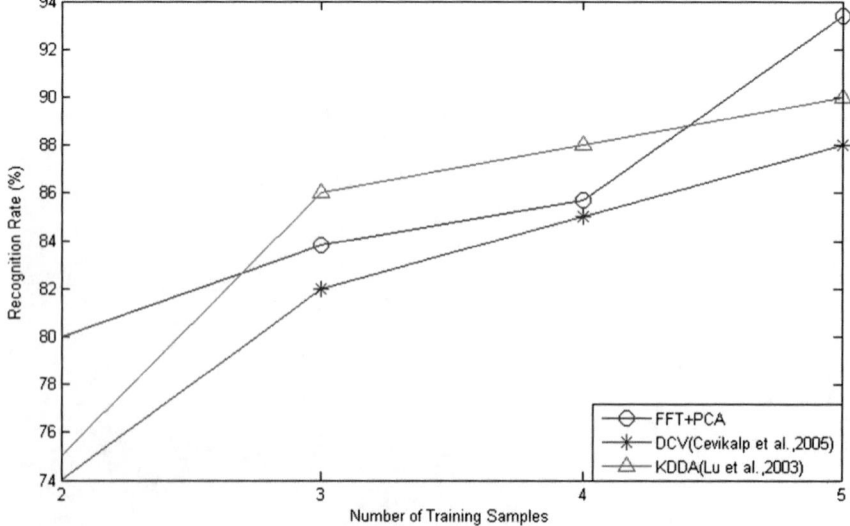

Fig. 3 Experiment comparison on Yale database

Table 2 Recognition rate based on FFT-Yale database as training database and original Yale database as testing database

Value of k	Recognition rate (%)
20	96.364
30	98.7879
40	97.576
50	98.7879
100	98.1818

Since we have an original Yale database and FFT-YALE database, we did another experiment that we chose FFT-Yale database as training database and chose original Yale database as testing database. The results are shown in Table 2.

From Table 2, we can see that recognition rate is very high. And it indicates that our proposed approach is encouraging.

3.2 Experiments on Self-made Image Database and Discussion

Another database in this paper is self-made image database, which contains 252 images of 7 facial expression classes (anger, disgust, fear, happiness, sadness, surprise and neutral) of 3 individuals (Da Ding, Ruxin Du, Hao Zhang). Each individual includes 84 images, and facial expression class has 12 images. In the experiment we pick half of the database (126 images) randomly as train images, and the rest half is used as test images. Figure 4 shows our image database.

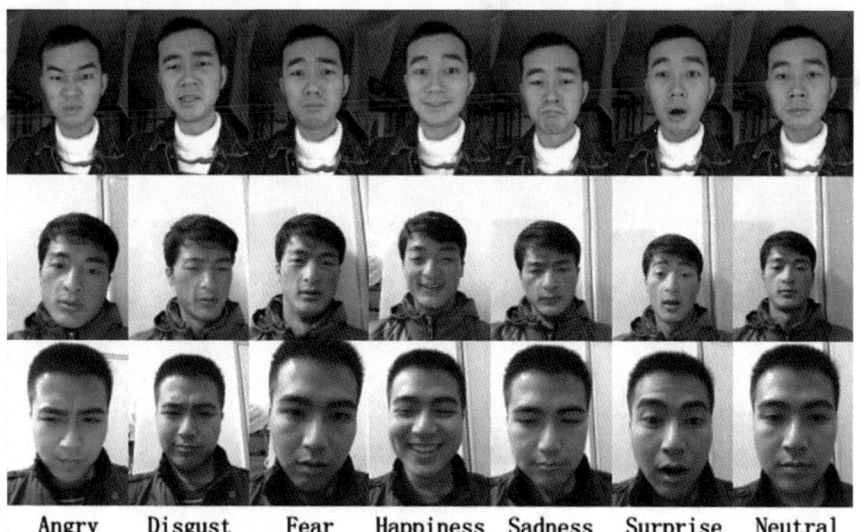

Angry   Disgust   Fear   Happiness   Sadness   Surprise   Neutral

Fig. 4 Our database including 7 facial expression of 3 individuals

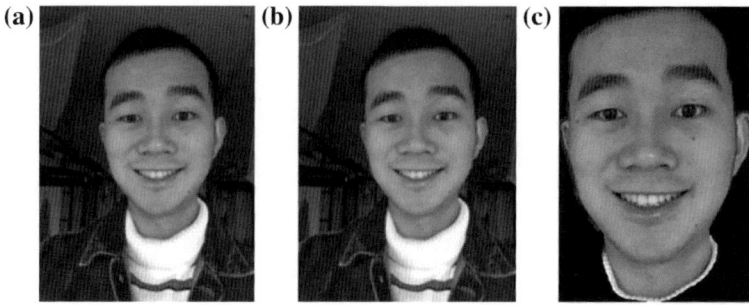

Fig. 5 Example of detecting face. **a** Original image. **b** After lighting compensation. **c** Identified face

Firstly, our purpose is to detect and extract individual's face from the background. After lighting compensation, extracting skin, and find skin color blocks, we find the identified face which has been converted to grey images, which is showed in Fig. 5. One example is displayed below.

Since we have detect faces (which assures that there is actually one face in the image), we used FFT to combine amplitude spectrum of one image with phase spectrum of the next image (the last one will combine with the first one). All the mixed images we got are shown in Fig. 6 compared with original images (Fig. 4).

After applying FFT, we then use PCA to extract facial expression feature in order to put the train images into right class. Our final results are displayed as Table 3.

Fig. 6 Original version (*above*) and images after FFT processing (*below*)

Table 3 Results of recognition rate using FFT + PCA

Facial expression	Individual		
	Da Ding (%)	Ruxin Du (%)	Hao Zhang (%)
Angry	100	100	100
Disgust	100	83.3	100
Fear	100	100	71.4
Happiness	83.3	100	83.3
Sadness	100	100	83.3
Surprise	100	100	100
Neutral	100	100	100
Average	97.6	97.6	90.4

Table 4 Comparison to PCA only method

Value of k	FFT + PCA (%)	PCA only (%)	Improvement (%)
10	96.0	90.5	6.14
20	94.4	89.7	5.31
30	95.2	92.1	3.45
40	97.6	93.7	4.24

And then we compare our method (FFT + PCA) to normal method (PCA only), and set k (dimensions of Eigenvector) to 10, 20, 30, 40 respectively. The results are shown in Table 4.

From Tables 3 and 4, we can clear see that our method (FFT + PCA) is more reasonable than PCA only, and the results indicate that our approach is satisfactory.

4 Conclusion

We present a full system of face recognition (FFT + PCA + SVM). Noticing that few researches focus on pre-processing of images, which will also improve the performance of classification, we apply Fourier Transform as a way to do pre-processing. Via FFT, we combine amplitude spectrum of one image with phase spectrum of another image as a mixed image. Then we harness PCA to extract Eigenvectors and use SVM as a classifier.

Our experiments are based on Yale database and self-made image database, and we compare our approach with others. The results indicate that our approach is effective. Since PCA only method is not as good as FFT + PCA in this paper, it is reasonable to conclude that our method is effective.

Acknowledgment This work was supported by the National Natural Science Foundation of China (Grant No. 61263043), the Foundation of Key Program of Department of Education of Yunnan Province (Grant No. 2011Z020, 2013Z049), the Key Discipline Foundation of School of

Software of Yunnan University (Grant No. 2012SE103), the Natural Science Foundation of Yunnan Province (2011FB020), and the Foundation of the Key Laboratory of Software Engineering of Yunnan Province (2012SE303).

References

1. D. Zhang, D. Ding, J. Li, Q. Liu, A novel way to improve facial expression recognition by applying fast fourier transform, in *Proceedings of the International MultiConference of Engineers and Computer Scientists 2014, IMECS 2014*, 12–14 Mar 2014, Hong Kong. Lecture Notes in Engineering and Computer Science, pp. 423–426
2. W. Gu, C. Xiang, Y.V. Venkatesh, D. Huang, H. Lin, Facial expression recognition using radial encoding of local gabor features and classifier synthesis. Pattern Recognit. **45**, 80–91 (2012)
3. B. Smith, An approach to graphs of linear forms (unpublished work style). Unpublished
4. J. Bekio-Calfa, J.M. Buenaposada, L. Baumela, Robust gender recognition by exploiting facial attributes dependencies. Pattern Recogn. Lett. **36**, 228–234 (2014)
5. F. Dornaika, A. Moujahid, B. Raducanu, Facial expression recognition using tracked facial actions: classifier performance analysis. Eng. Appl. Artif. Intell. **26**, 467–477 (2013)
6. L.J. Cao, K.S. Chua, W.K. Chong et al., A comparison of PCA, KPCA, and ICA for dimensionality reduction in support vector machine. Neurocomputing **55**, 321–336 (2003)
7. C. Liu, H. Wechsler, Gabor feature based classification using the enhanced fisher linear discriminant model for face recognition. IEEE Trans. Image Process. **11**, 467–476 (2002)
8. L.M.A. Levada, D.C. Correa, D.H.P. Salvadeo, J.H. Saito, D.A. Nelson, Novel approaches for face recognition: Template-matching using dynamic time warping and LSTM neural network supervised classification, in *International Conference on Systems, Signals and Image Processing*, pp. 241–244 (2008)
9. Y. Adini, Y. Moses, S. Ullman, Face recognition: the problem of compensating for changes illumination direction. IEEE Trans. Pattern Anal. Mach. Intell. **19**, 721–732 (1997)
10. H Wang, S.Z. Li, Y. Wang, Face recognition under varying lighting conditions using self-quotient image, in *Proceedings of the IEEE International Conference on Automatic Face and Gesture Recognition*, pp. 819–824 (2004)
11. A. Eftekhari, M. Forouzanfar, H.A. Moghaddam, J. Alirezaie, Block-wise 2D kenel PCA/LDA for face recognition. Inf. Process. Lett. **110**, 761–766 (2010)

Support System Using Microsoft Kinect and Mobile Phone for Daily Activity of Visually Impaired

Mohammad M. Rahman, Bruce Poon, Md. Ashraful Amin
and Hong Yan

Abstract The aim of this paper is to outline a system based on Microsoft Kinect and mobile devices that will provide assistant to visually impaired people. Our primary goal is to provide navigation aid that will help visually impaired to navigate. This includes detection and identification of face, texts and chairs. This is implemented using Microsoft Kinect and machine learning methods are used for this process as it requires rough identification of object. For data acquisition and processing, OpenCV, OpenKinect, Tesseract and Espeak are used. Features that have been incorporated for building this aiding tool are object detection and recognition, face detection and recognition, object location determination, optical character recognition and audio feedback. The face recognition system showed an accuracy of 90 %, the text recognition yielded an accuracy of 65 % and the chairs are recognized with more than 74 % accuracy. To identify denominations of bank notes, more accurate recognition is required. Mobile phone is used to identify bank note denomination. The proposed system can recognize Bangladeshi paper currency notes with 89.4 % accuracy on plain paper background and with 78.4 % accuracy tested on a complex background.

Keywords 3D camera · Human computer interaction (HCI) · Kinect · Mobile computing · Navigational aid · ORB · SIFT · SURF · Visual impairment

---

M.M. Rahman · Md.A. Amin
Computer Vision and Cybernetics Group, Computer Science and Engineering,
Independent University, Bangladesh, Dhaka 1229, Bangladesh
e-mail: motiur.rahman0@gmail.com

Md.A. Amin
e-mail: aminmdashraful@gmail.com

B. Poon (✉)
School of Electrical and Information Engineering, University of Sydney,
Sydney, NSW 2006, Australia
e-mail: bruce.poon@ieee.org

H. Yan
Department of Electronic Engineering, City University of Hong Kong, Hong Kong, China
e-mail: h.yan@cityu.edu.hk

© Springer Science+Business Media Dordrecht 2015
G.-C. Yang et al. (eds.), *Transactions on Engineering Technologies*,
DOI 10.1007/978-94-017-9588-3_32

1 Introduction

The proposed work here is motivated by the need of a navigational [1] and daily activity aid system for visually impaired people. Computer vision and mobile computing are powerful tools with great potential to enable a range of assistive technologies for the growing population of visually impaired. We intend to develop a system that will be able to detect objects (chairs), detect and recognize faces and texts, identify denominations of Bangladeshi note currency.

Although a number of researchers have proposed useful approaches to object, face and text detection [2–11], work related to detecting an object and to determine its location in a 3D environment is scarce. Moreover, the primary input devices used in our system, Microsoft Kinect and mobile phones, are relatively new and have yet to receive the required attention.

Object detection [2–5] is usually heavily concerned about detecting objects with accuracy being the most important factor of the system performance. Method proposed by Viola and Jones [6] overcame this challenge. The main strength of this algorithm is that the classifier cascade used could be resized itself to match the size of the sliding window in every iteration. Several methods on shape based recognition can be found in the work of Sclaroff and Pentland in [7]. They suggested an eigenvector or modal matching based approaches [8, 9]. In this approach, sample points in the image were casted into a finite element spring-mass model and correspondences are found by comparing modes of vibration.

Eigenface method for face recognition [10] identifies faces faster which makes it more suitable for real time applications. A very robust text processing system called Tesseract [11] was first designed by HP Labs and was later developed by Google.

Unlike the typical object, face or text processing system, mobile phone based currency note identification system requires more accurate processing and the result should be available in real-time. Most popular approaches to perform this task are Scale Invariant Feature Transform (SIFT) [12], Speeded up Robust Features (SURF) [13] and Oriented FAST and Rotated BRIEF (ORB) [14], a method developed in the lab of OpenCV [15].

2 Methodology

Two separate system flow charts are given in Fig. 1. Left one is to show the schematic view of the system developed using Kinect to detect and recognize objects (chair), face and English text. The right flow chart shows components of the system developed using mobile phone to detect Bangladeshi note currency and important signs.

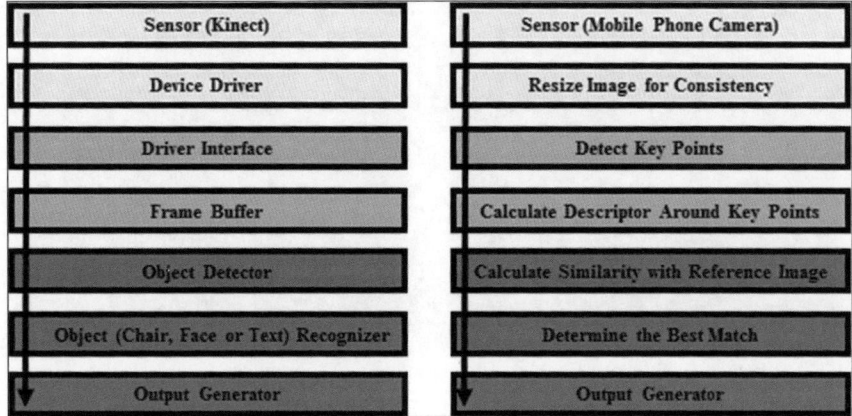

Fig. 1 System flow chart of the components of the Kinect (*left*) and mobile phone (*right*) based visual aid system

2.1 The Kinect Based System

2.1.1 Data Acquisition

For the Kinect based system the data is acquired using Microsoft Kinect. There is array of device driver to choose from to acquire data from the Kinect sensor and we chose an open source called OpenKinect [16]. This driver helped us to collect data from the device sensor without any dependency. To be able to adapt to various hardware without overwhelming the underlying system, we used an event driven mechanism. Instead of putting a pressure on the system to deliver frames of data as fast as possible, each component would run independently and react to events that would occur throughout the system. The frame-buffer component's sole responsibility is to always keep a frame of data at hand and always keep it up to date. Each frame actually contains a 2D color image and for each pixel in the image a relative depth map (Fig. 2).

2.1.2 Object Detection and Recognition

Each 2D color image is passed through a few cascades of object recognizer. In the training phase, each of these cascades are trained using images that contains objects of our interest (chairs and faces). After a cycle of Haar-training [15] the cascades were ready to be used. During detection of the objects, the classifier module tries to mark regions of interest on the image and tag them with the names of possible class

Depth View: VGA View:

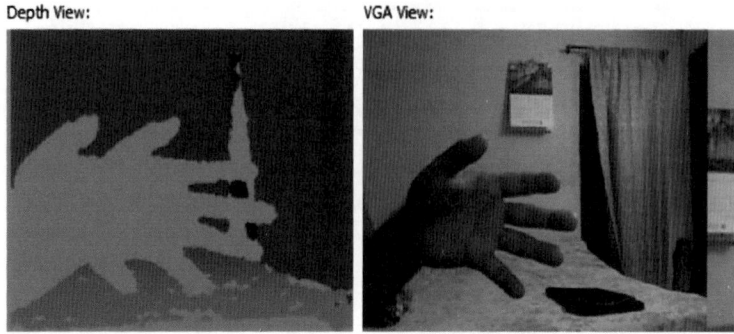

Fig. 2 A sample depth image (*left*) and corresponding 2D color image (*right*)

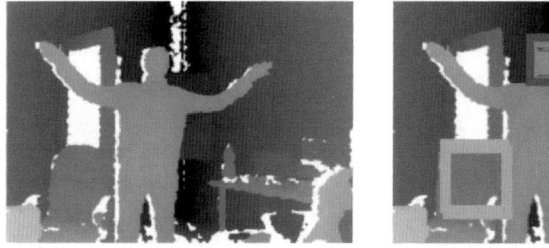

Fig. 3 Depth frame (*left* raw; *right* regions of interests marked)

of objects found in those regions. The depth frame, as received from the frame-buffer simultaneously, is also marked with appropriate tags so that the next step, where more detailed information is determined can use those data quickly to find the regions of interest and their properties. In Fig. 3, one frame containing a human face and a chair detected correctly is depicted.

Every detected object in the system is annotated with a tuple of two values: approximate distance between the object and the sensor, and the angle the object makes with the z-axis of the sensor. These values are calculated using positional information provided by the object detector and the depth data obtained from Kinect. However, there is a problem of using distance from the depth value provided by Kinect as some times the value provided are noisy due to environmental fluctuations as the measurement is performed using infrared camera. Instead of pixel by pixel distance measurement, we used a median based approach.

Determining Distance: To calculate distance between the object and camera following (left) median algorithm can be used. However, a faster median calculation that we implemented is given below (right).

```
1   def median(arr):                 1   def median(arr):
2     arr.sort()                     2     counts = [0] * 256
3     if len(arr) % 2 == 0:          3     for v in arr:
4       return (arr[len(a)/2-1] +    4       ++counts[v]
5   arr[len(a)/2])/2                 5     tmp = 0
6     else:                          6     for I, v in enumerate(counts):
        return arr[len(a)/2]         7       tmp += v
                                     8       if tmp >= len(arr)/2:
                                     9         return v
```

Determining Angular Displacement: The object detector is capable of providing coordinates in terms of pixels, considering the center of the image to be the origin. These coordinates in pixels were then converted to absolute distance in meters using a linear scaling followed by the use of Pythagoras formula. The coefficients of linear scaling formula were determined experimentally: a long wooden stick, 1 m in length was placed 3 m away from the camera. The number of pixel it covered in the image was measured. The coefficients were then computed from the available information. This process was repeated by keeping the stick at 4, 5 and 6 m distances to ensure that the linear scaling was effective enough and that such subtle change in the distance did not have any significant effect on accuracy of scaling process. This gave us distance of the object on xy-plane.

The sensor provides depth information in a non-standard unit. According to its specification, Kinect is capable of determining distance ranging between 2 and 7 m. The depth value, as provided by the device drivers, ranges between 0 and 255. The distance was calculated by multiplying a constant factor with the depth value provided by the driver. The constant factor was initially set to 0.027 which maps a value of 0–2 m and a value of 256–7 m. However, the final value of the constant factor was determined through trial and error. Ones the object the objects are detected and distance and orientation of the detected objects are known, next task is to identify face or text detected.

2.1.3 Face Recognition

Face recognition itself is a challenging problem to solve as variations caused by change in lighting, pose, age, sex is very hard so handle [17, 18]. Principle Component based Eigen face system developed by Turk and Pentland [10] is employed to solve the face recognition part of our system.

2.1.4 Text Recognition

For text recognition, we rely on the open source tools called Tesseract [11]. However, the obstacle that we really have to overcome to ensure a good text extraction mechanism involved the fact that the images containing text being fed to Tesseract will be completely arbitrary. Tesseract is optimized to extract text from well format and well scanned document images. The system, on the other hand, will only receive images from a Kinect device and those images may contain text in virtually any form.

During the preprocessing stage, our goal was to eliminate the effects of the background and noises on the image as much as possible. This involved applying a series of filters (sharpening, monochrome etc) to the image to reduce such unwanted elements and made the text as vivid as possible. The end result would be an image where the background was almost eliminated and the text would remain as a thin skeletal wireframe. Since this would work only for large texts in the image and that would cover basically all signboards and markers, the compromise was worthwhile.

Once the image was passed through the Tesseract, the recognized text was then passed through another small snippet of code which resolved some unexpected character issues. For example, it replaced all '$' signs appearing next to characters (other than digits) into the English character 's'. The text was further spell checked and common errors were fixed. This made the end result much more readable and hence easier for the text to audio engine to processes it better. Here note that this software is suitable for English. For other languages, we have to take different approach [19]. In Fig. 4, we provide a sample notice at the university library processed by our system.

2.2 Feedback

The output unit is the least complex unit of the whole system. It is comprised of a python program that actually works on the text that is supplied by our central processing unit. As we know earlier, the central processing unit actually uses the object classifiers and marks the object in the image frame. After that, the processing unit performs other calculations, sorts out the distance and location of the desired object and formulates a text describing the information. This text works like the input of the output unit. The output unit processes this text using a python package

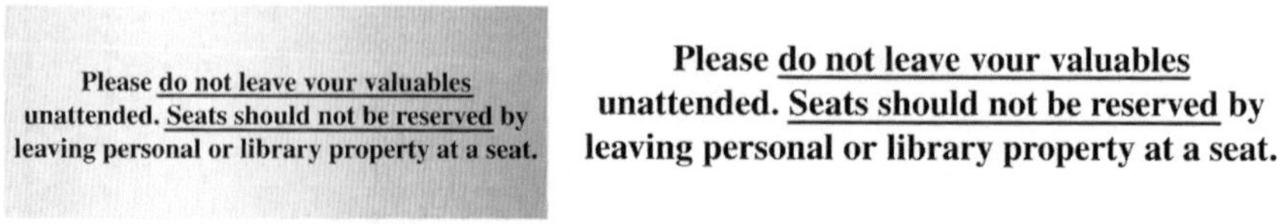

Fig. 4 Image preprocessing for text recognition

named "ESpeak" [20]. "ESpeak" is a voice synthesizer. It converts the text to speech. Therefore, the output unit gets the formulated text after object detection and location determination from the central processing unit which then converts the text to speech using "ESpeak" and deliver to the user.

2.3 The Mobile Based System

2.3.1 Data Acquisition

There are two parts in the data collection process. For the reference database, high quality images of the Bangladesh note currency and common signs are collected from reliable sources in the internet [21, 22]. Samples of such images are given in Fig. 5.

The second phase in data collection is for testing the system. Indeed, the system we built using mobile phone is a live system. It determines denomination of bank note and signs in real-time in every frame of image taken by a mobile phone. After an image is taken, it is preprocessed following standard sharpening. We then crop the image to 200 × 180 pixel and extract feature.

2.3.2 Feature Extraction

Feature extraction is performed in two steps. First, we determine the key location in an image where the real feature extractor should be applied. These locations are called key points. Features are acquired surrounding these key points. The descriptor and detector of the notes and signs are found and stored in a table. Both the detectors and the descriptors are found using Oriented BRIEF and Rotated

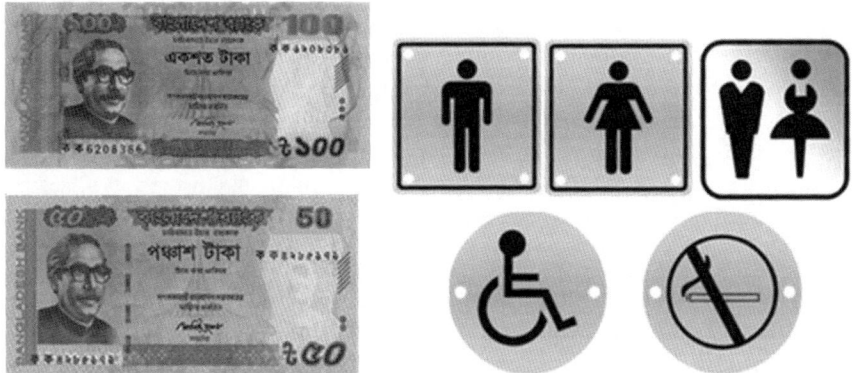

Fig. 5 Sample images from the reference database

Fig. 6 Key points identified on a 50 taka banknote

FAST (ORB) [14]. The number of considered features was 500, the number of pyramid level that was used was 8 and the edge threshold was 31 [23]. Key points marked in one banknote image is provided in Fig. 6. Ones the features are extracted, they are listed in a reference table for each of the reference images. We call this the reference object database. When the real-time system is deployed, the system determines the key points in each test image frame, extracts the features around the key points and then determines the similarity with each of the reference images.

2.3.3 Similarity Matching

Similarity matching has two issues. One is thresholding that is required due to the characteristics differences in the key point detection and description change because of background change. That is done through homography calculation.

Thresholding: The content of the input image often determines the quality of the descriptor that is chosen to represent a good match. In settings where there is plain background, the distance between the descriptors will be different from when there is complex background. In Figs. 7 and 8, we provide images (top) with plain and complex backgrounds, and descriptor (bottom) information for this two images. We can notice a significant difference in the description.

In normal situation where there is the presence of background, only descriptors whose distance is less than twice the distance of the minimum distance between the descriptors [24] are used.

$$\textbf{match\_desc\_dist} < 2 * \textbf{min(match\_desc\_dist)}$$

The frequency of the matched descriptors is analyzed. The highest frequency indicates the presence of a given bank note.

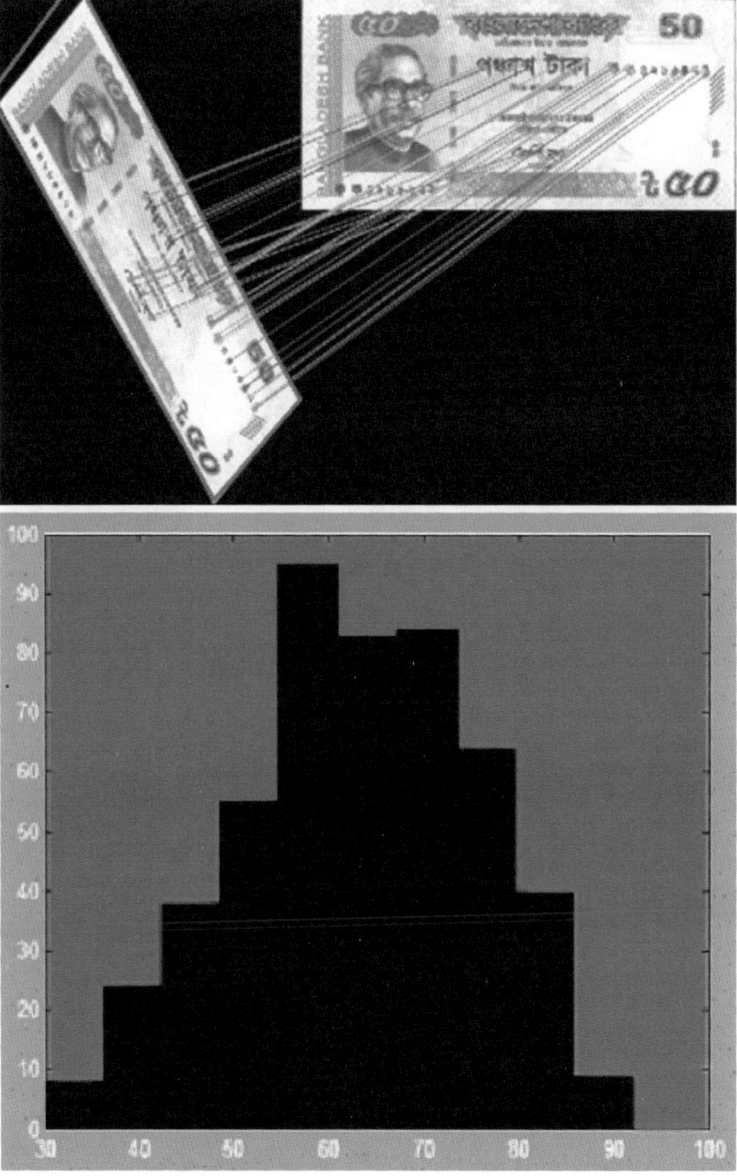

Fig. 7 Money with plain background (*top*) and the distance of the descriptors from training and testing samples (*bottom*)

Homography Calculation: A training image of the database has to undergo rotation along with translation to match with the image in real time. The phenomenon is known as homography and is calculated using:

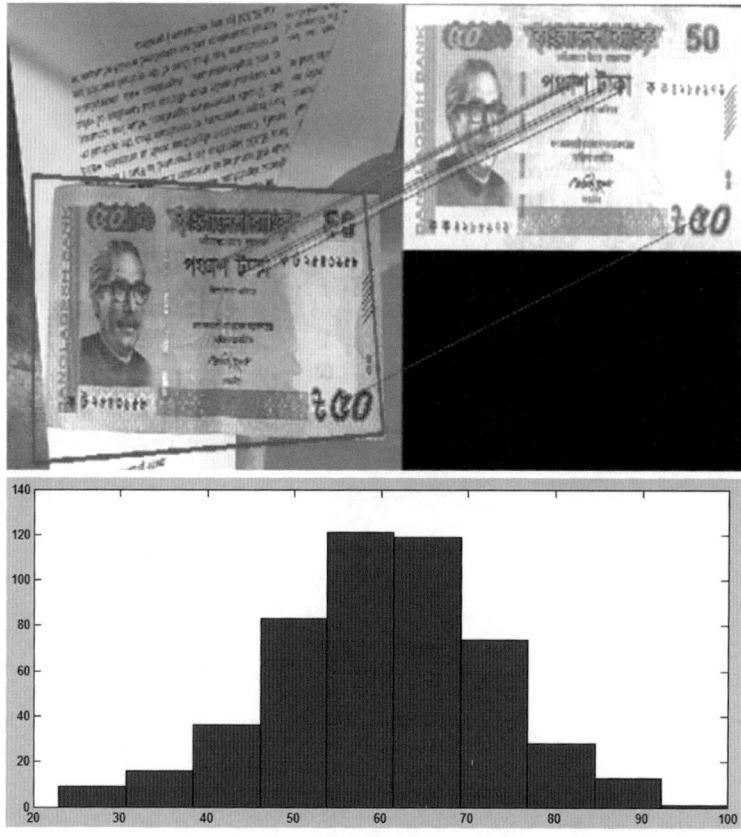

Fig. 8 Money with complex background (*top*) and the distance of the descriptors from training and testing samples (*bottom*)

$$x' = \frac{a_1 x + a_2 y + a_3}{a_7 x + a_8 y + 1} \quad \text{and} \quad y' = \frac{a_4 x + a_5 y + a_6}{a_7 x + a_8 y + 1}$$

x' and y' are coordinates in the test scene, whereas x and y are coordinates in the reference image. They are bounded to one another in value via the constants a_1–a_8. The above two equations are examples of over constrained system which can be solved using either Least Squared Error or via Random Sampling Consensus (RANSAC).

In Least Squared Error, the error between the estimated and actual output is minimized by a certain iterative procedure. In RANSAC, two points are randomly chosen to fit a line. The error between the estimated and the actual output is found out. If that error is above a threshold then two random points are again chosen. If the error is lower than the threshold, the loop stops.

3 Performance Evaluation

Performance of the proposed system is evaluated from two viewpoints: offline performance and online performance (real-time performance).

3.1 Offline Performance Evaluation

3.1.1 Offline Face Detection Accuracy

The performance analysis of offline face detection was evaluated using three different datasets: PIE [25], UMIST [26] and CBCL [27]. Each dataset contains around a thousand facial images, cropped to facial region only and there are slight variations in angle of image acquiring in those images. Figure 9 shows the number of true positives, false negatives and false positives for each dataset. The total height represents the actual size of the each dataset.

3.1.2 Offline Face Recognition Accuracy

Face recognition was performed using a small number of subjects. For each subject, around 50 images were captured. The number of images used to train was varied and the performance was measured. Figure 10 shows the face recognition accuracies with various training data sizes.

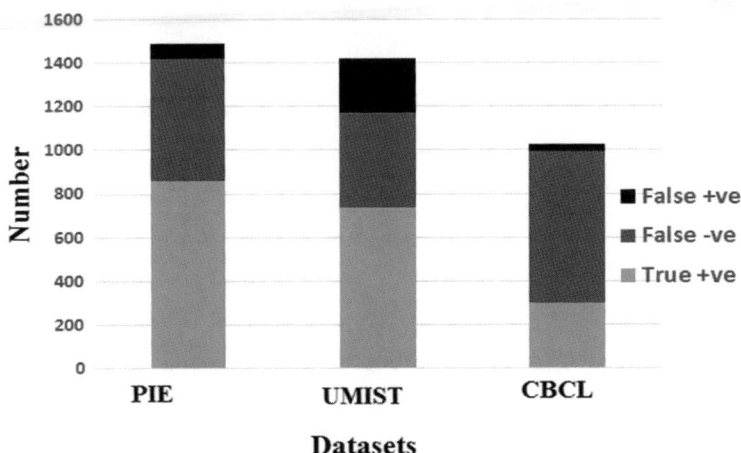

Fig. 9 Performance with various datasets for offline face detection

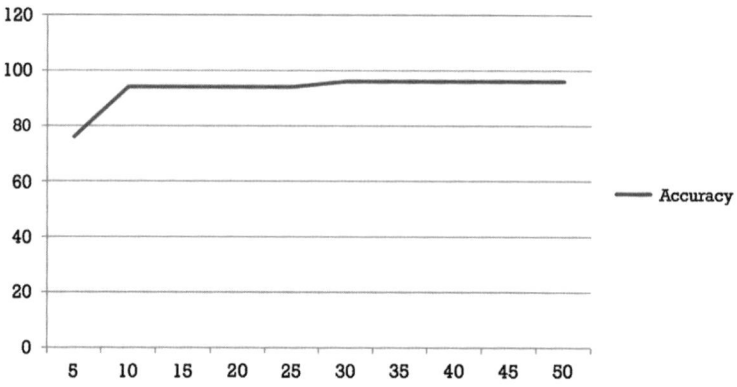

Fig. 10 Face recognition accuracy with varying size of training data

3.1.3 Offline Distance and Angle Calculation Accuracy

The distance measured used data obtained using the depth map of Kinect. Since the distance calculation formula was designed to account for trigonometric approximations, the expected error of the formula was supposed to be constant. However, due to the error inherent in the sensor itself and its depth measuring capabilities, there was an increase in error with increase in distance. Moreover, since the sensor was designed to work only within a short range of distance, distances calculated for objects within the first 2 m and the distance beyond 9 m were completely invalid and ignored. Figure 11 shows the average error percentages with various distances. In our experiment, the values for Mean and Median are the same. As such, the lines plotted for Mean and Median overlap each other.

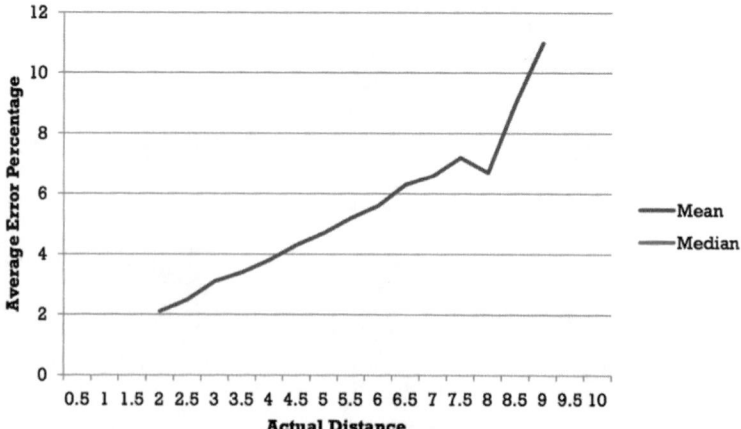

Fig. 11 Error in distance calculation using data from depth map (lower is better)

3.1.4 Bank Note Denomination Identification

In this experiment, 8 different bank currencies were used. Two sides, the front side as well as the backside of the currencies were matched against each other, forming a 16 × 16 confusion matrix. In confusion matrix 1 (left hand side of Fig. 12), note that

Confusion matrix for plain background (left)

	2 front	2 back	5 front	5 back	10 front	10 back	20 front	20 back	50 front	50 back	100 front	100 back	500 front	500 back	1000 front	1000 back	Sem
2 front	91.9	0.4	0.4	0.4	0.4	0.4	0.8	0.4	0.4	0.4	0.4	0.4	0.4	0.4	0.4	0.4	98.1
2 back	0.4	93.3	0.4	0.4	0.4	0.4	0.4	0.4	0.4	0.4	0.4	0.4	0.4	0.4	0.4	0.4	99.9
5 front	0.4	0.4	93.7	0.4	0.4	0.4	0.4	0.4	0.4	0.4	0.4	0.4	0.4	0.4	0.4	0.4	99.9
5 back	0.4	0.4	0.4	93.9	0.4	0.4	0.4	0.4	0.4	0.4	0.4	0.4	0.4	0.4	0.4	0.4	100.1
10 front	0.8	0.4	0.4	0.4	93.7	0.4	22.2	0.4	3.3	0.4	0.4	0.4	0.4	0.4	0.4	0.4	125.1
10 back	0.4	0.4	0.4	0.4	0.5	94.0	0.4	0.4	0.4	0.4	0.4	0.4	0.4	0.4	0.4	0.4	100.3
20 front	0.4	0.4	0.4	0.4	0.4	0.4	56.4	0.4	0.4	0.4	0.4	0.4	0.4	0.4	0.4	0.4	63.0
20 back	0.4	0.4	0.4	0.4	0.4	0.4	0.4	94.0	0.4	0.4	0.8	0.4	0.4	0.4	0.4	0.4	100.2
50 front	2.0	0.4	0.4	0.4	0.4	0.4	13.6	0.4	90.0	0.4	0.4	0.4	0.4	0.4	0.4	0.4	111.0
50 back	0.4	0.4	0.4	0.4	0.4	0.4	0.4	0.4	0.4	93.9	0.4	0.4	0.4	0.4	0.4	0.4	100.2
100 front	0.4	0.4	0.4	0.4	0.4	0.4	0.4	0.4	0.4	0.4	67.5	0.4	0.4	0.4	0.4	0.4	74.9
100 back	0.4	0.4	0.4	0.4	0.4	0.4	0.4	0.4	0.4	0.4	0.4	93.7	0.4	0.4	0.4	0.4	99.9
500 front	0.4	0.4	0.4	0.4	0.4	0.4	0.4	0.4	0.4	0.4	0.4	0.4	93.6	0.4	0.4	0.4	99.8
500 back	0.4	0.4	0.4	0.4	0.4	0.4	0.4	0.4	0.4	0.4	0.4	0.4	0.4	93.3	0.4	0.4	99.5
1000 front	0.4	0.4	0.4	0.4	0.4	0.4	1.6	0.4	0.8	0.4	26.3	0.4	0.4	0.4	93.8	0.4	127.6
1000 back	0.4	0.4	0.4	0.4	0.4	0.4	0.4	0.4	0.4	0.4	0.4	0.4	0.4	0.4	0.4	94.0	100.3
Sum	100.0	100.0	100.0	100.0	100.0	100.0	100.0	100.0	100.0	100.0	100.0	100.0	100.0	100.0	100.0	100.0	1599.8

Confusion matrix for complex background (right)

	2 front	2 back	5 front	5 back	10 front	10 back	20 front	20 back	50 front	50 back	100 front	100 back	500 front	500 back	1000 front	1000 back	sum
2 front	93.9	3.1	0.4	0.4	1.2	0.7	0.4	0.4	0.4	1.7	0.4	13.9	1.6	0.4	0.4	0.4	119.3
2 back	0.4	66.8	0.4	0.4	5.3	0.7	1.2	0.4	0.4	19.6	1.7	13.9	0.4	0.4	0.4	0.4	115.3
5 front	0.4	0.4	93.8	0.4	0.4	0.7	0.8	0.4	0.4	0.9	4.2	1.0	0.4	0.4	0.4	0.4	105.6
5 back	0.4	0.4	0.4	87.2	0.4	0.7	0.4	0.4	0.4	5.2	1.0	1.0	0.4	0.4	0.4	0.4	93.1
10 front	0.4	0.4	0.4	0.4	82.0	2.2	11.8	0.4	0.4	15.7	8.9	6.9	0.4	0.4	0.9	0.4	134.0
10 back	0.4	0.4	0.4	0.4	0.4	84.6	0.4	0.4	0.4	0.4	0.4	1.0	0.4	0.4	0.4	0.4	91.4
20 front	0.4	0.4	0.4	0.4	0.8	0.7	53.2	0.4	0.8	0.9	1.3	2.0	0.4	0.4	2.2	0.4	65.2
20 back	0.4	0.4	0.4	0.4	0.4	0.7	0.4	93.9	0.4	0.4	0.4	1.0	0.4	0.4	0.4	0.4	101.0
50 front	0.4	22.6	0.4	0.4	5.3	0.7	28.0	0.4	93.3	17.8	14.8	5.9	1.2	0.4	2.2	0.4	138.4
50 back	0.4	0.4	0.4	0.4	0.4	0.7	0.4	0.4	0.4	33.5	0.4	1.0	0.4	0.4	0.4	0.4	40.6
100 front	0.4	0.4	0.4	0.4	0.8	1.5	0.4	0.4	0.4	0.4	62.8	1.0	0.4	0.4	0.4	0.4	71.1
100 back	0.4	0.4	0.4	0.4	0.4	0.7	0.4	0.4	0.4	0.4	0.4	41.6	0.4	0.4	0.4	0.4	48.2
500 front	0.4	0.9	0.4	0.4	0.4	0.7	0.4	0.4	0.4	0.4	2.5	1.0	90.5	0.4	0.4	0.4	100.2
500 back	0.4	0.4	0.4	0.4	0.4	0.7	0.4	0.4	0.4	0.4	0.4	1.0	0.4	93.8	0.4	0.4	101.0
1000 front	0.4	1.8	0.4	0.4	0.8	2.9	0.8	0.4	0.4	0.9	0.4	5.0	1.6	0.4	83.6	0.4	107.2
1000 back	0.4	0.4	0.4	0.4	0.4	0.7	0.4	0.4	0.4	1.3	0.4	1.0	0.4	0.4	0.4	93.7	101.7
sum	100.0	100.0	100.0	100.0	100.0	100.0	100.0	100.0	100.0	100.0	100.0	100.0	100.0	100.0	100.0	100.0	1600.0

Fig. 12 Confusion matrix for plain background (*left*) and complex background (*right*) experiment on Bangladesh bank note denomination identification

Table 1 Execution time for each component

Configuration	Speed (average)
All components (kinect)	2.25 s per frame
Face recognition	1.7 s per frame
Text recognition	0.8 s per frame
Face and text recognition	1.9 s per frame
Object detection	2 s per frame
Face recognition and object detection	2.2 s per frame
Bank note and sign recognition (mobile)	0.25 s per frame

the most distinct notes are 10 and 1,000 taka and the most misclassified notes are those of 20 and 100 denominations. In confusion matrix 2 (right hand side of Fig. 12), note the recognition accuracy is not as good as those shown in confusion matrix 1. The system is most confused if a flipped 100 taka note is presented. From the confusion matrix, it can be seen that 50 taka front note is the most distinct note while 50 taka flipped note is the most misclassified note. Note the differences in the recognition accuracy for plain and complex background experiments.

3.2 Online or Real-Time Performance Evaluation

Since the system performed fairly well in offline tests, we used the same algorithms and approaches in our real-time use with slight optimizations. The performance measure then all came down to how quickly it processed every frame. In Table 1, the time required for each major component processing for in a frame is provided.

4 Conclusion and Future Work

In conclusion, we come to understand that it is possible to build a system that can come in aid to those who really need it with the available technologies all around us. People with visual impairment, those who are deprived of this very important sensory system, always tend to lead a difficult life. However, with a system such as this, their life can only get better. Our experiment shows that a complex system such as this can be built from a number of existing, stable components. At the same time, we can make it run efficiently using limited resources.

In our experiment, the whole system had been built upon several very simple principles. It was structured in a modular way for greater extensibility. Each chosen component was lightweight and open source for better availability, or was implemented customized to suit our needs. It was independent which reduced the number of articulation points of failure. Overall, the whole system was just one simple pipeline of processing techniques that started from two simple frames of data (rgb and depth) and ended at the generation of audio feedback.

At the end of the experiment, although we had a stable and efficient system, there was a considerable number of areas where it could be improved even more especially in terms of accuracy of some of the components. For example, object detection could be improved by employing a better training dataset and spending much more time in the training phase, tweaking the system to reach optimal performance quality.

This experiment shows that it is possible to detect and enumerate values of an unknown bank notes, even if they are very close to another in composition. Signs in images can also be determined. In future, we need to integrate both the systems together and preferably implement the whole system on a mobile device.

Acknowledgment This work is jointly supported by Independent University, Bangladesh and University Grants Commission of Bangladesh under Higher Education Quality Enhancement Project (HEQEP) Number: CP-3359.

References

1. M. Ridwan, E. Choudhury, B. Poon, M.A. Amin, H. Yan, A navigational aid system for visually impaired using microsoft kinect, in *Proceedings of the International MultiConference of Engineers and Computer Scientists, IMECS 2014*, 12–14 Mar 2014, Hong Kong. Lecture Notes in Engineering and Computer Science, pp. 417–422
2. A. Opelt, M. Fussenegger, A. Pinz, P. Auer, Weak hypotheses and boosting for generic object detection and recognition, in *Proceedings of the 8th European Conference on Computer Vision*. Lecture Notes in Computer Science, vol. 3022, pp. 71–84 (2004)
3. D.G. Lowe, Object recognition from local scale-invariant features. IEEE Trans. Pattern Anal. Mach. Intell. **2**, 1150–1157 (1999)
4. R. Fergus, P. Perona, A. Zisserman, Object class recognition by unsupervised scale-invariant learning. Proc. Comput. Vis. Pattern Recogn. **2**, 264–271 (2003)
5. S. Mahamud, M. Hebert, J. Shi, Object recognition using boosted discriminants. Proc. Comput. Vis. Pattern Recogn. **1**, 551–558 (2001)
6. P. Viola, M. Jones, Rapid object detection using boosted cascade of simple features, in *Proceedings of the Conference on Computer Vision and Pattern Recognition* (2009)
7. S. Sclaroff, A. Pentland, Modal matching for correspondence and recognition. IEEE Trans. Pattern Anal. Mach. Intell. **17**(6), 545–561 (1995)
8. L.S. Shapiro, J.M. Brady, Feature-based correspondence: an eigenvector approach. Image Vis. Comput. **10**(5), 283–288 (1992)
9. S. Umeyama, An eigen decomposition approach to weighted graph matching problems. IEEE Trans. Pattern Anal. Mach. Intell. **10**(1), 71–96 (1991)
10. M.A. Turk, A.P. Pentland, Eigenface for recognition. J. Cogn. Neurosci. **3**, 71–86 (1991)
11. Tesseract Open Source OCR Engine: Available http://code.google.com/p/tesseract-ocr/
12. D.G. Lowe, Distinctive image features from scale-invariant keypoints. Int. J. Comput. Vis. **60**, 91–110 (2004). http://www.cs.ubc.ca/~lowe/papers/ijcv04.pdf
13. H. Bay, A. Ess, T. Tuytelaars, L.V. Gool, Speeded-up robust features (SURF). Comput. Vis. Image Underst. **110**, 346–359 (2008). ftp://vision.ee.ethz.ch/publications/articles/eth_biwi_00517.pdf

14. E. Rublee, V. Rabaud, K. Konolige, G. Bradski, ORB: An efficient alternative to SIFT or SURF, in *Proceedings of the International Conference on Computer Vision*, pp. 2564–2571 (2011)
15. Open Source Computer Vision: Available http://opencv.org/
16. Open source Open Kinect project: Available http://openkinect.org/
17. M.A. Amin, H. Yan, An empirical study on the characteristics of gabor representations for face recognition. Int. J. Pattern Recognit Artif Intell. **23**(3), 401–431 (2009)
18. B. Poon, M.A. Amin, H. Yan, Performance evaluation and comparison of PCA based human face recognition methods for distorted images. Int. J. Mach. Learn. Cybernet. **2**(4), 245–259 (2011)
19. M.Z. Hossain, M.A. Amin, H. Yan, Rapid feature extraction for Bangla handwritten digit recognition, in *Proceedings of the International Conference of Machine Learning and Cybernetics*, pp. 1832–1837 (2011)
20. ESpeak—A Voice Synthesizer: Available http://espeak.sourceforge.net/docindex.html
21. Central Bank of Bangledesh: Available http://www.bangladesh-bank.org/currency/note.php
22. Samples of Sign Images: Available http://image.made-in-china.com/4f0j00NBJamowFULbd/Toilet-Sign.jpg
23. Open CV: Feature detection and description, Available http://docs.opencv.org/modules/features2d/doc/feature_detection_and_description.html#orb-orb
24. Issues with imgldx in Descriptor Matcher mexopencv: Stack overflow. Available http://stackoverflow.com/questions/20717025/issues-with-imgidx-in-descriptormatcher-mexopencv
25. The CMU Multi-PIE Face Database: Available http://www.multipie.org/
26. The Shefield (previously UMIST) Face Database: Available http://www.shef.ac.uk/eee/research/iel/research/face
27. MIT Center for Biological and Computational Learning Face Database: Available http://cbcl.mit.edu/software-datasets/FaceData2.html

Change Patterns Detection and Traceability Impact Analysis of Business Process Models

Watcharin Uronkarn and Twittie Senivongse

Abstract Business analysts define business process models for describing a series of tasks to produce services or products to serve business goals. Hence business process models represent business requirements for development of the software that enables automation of the business processes. When tasks in a business process are changed, such changes also trigger changes in the artifacts that have been produced during development of the related software. Analysis of an impact a business process change has on the software is useful for the software project manager and the system analyst to plan the effort to change the artifacts, including the software itself, accordingly. This paper presents a traceability impact analysis using traceability information of the old version of the business process model and its related software artifacts. The impact analysis is based on different patterns of business process changes that are made to the old business process model to create a new version of the model. Detection algorithms are used to identify various patterns of process structure changes so that business process tasks that are changed and software artifacts that would be impacted by the changes can be reported. The paper also discusses a supporting tool and its evaluation.

Keywords Business process · Business process model · Change detection · Change impact analysis · Change patterns · Traceability

W. Uronkarn · T. Senivongse (✉)
Department of Computer Engineering, Faculty of Engineering, Chulalongkorn University, Phyathai Road, Pathumwan, Bangkok 10330, Thailand
e-mail: twittie.s@chula.ac.th

W. Uronkarn
e-mail: watcharin.u@student.chula.ac.th

© Springer Science+Business Media Dordrecht 2015
G.-C. Yang et al. (eds.), *Transactions on Engineering Technologies*,
DOI 10.1007/978-94-017-9588-3_33

441

1 Introduction

Business process modeling is an activity to capture processes of business applications into business process models. From a technical perspective of a software project team, a business process model, such as one described in the Business Process Model and Notation (BPMN) [1], captures business requirements for the development of software that enables automation of the business process. That is, a business process model leads to development of several software artifacts such as requirement specifications, analysis and design models, implementation components, tests, and other project documents [2]. When tasks in a business process model are changed, such changes also trigger changes in the artifacts related to those tasks. Analysis of an impact a business process change has on the software is useful for the software project manager and the system analyst to plan the effort to change the artifacts, including the software itself, accordingly. For example, if the change has a small impact, change to software documents and code is preferable. On the other hand, it might be better to develop new software if the change affects the existing software to a large extent in such a way that only a small part can be reused. As a result, this paper addresses three requirements to manage business process changes:

Requirement#1: To determine the impact of a change on an old business process, we need to make the old business process traceable. As software requirements traceability refers to the ability to describe and follow the life of a requirement in both forward and backward directions, information about the relationships between software requirements and many kinds of associated artifacts have to be maintained [2] so that the scope of the initiating change can be analyzed. This is called *traceability impact analysis* [3]. In our case, we need to extend traceability information to also document the relationships between tasks in the old business process model and other kinds of software artifacts.

Requirement#2: To determine the impact of a business process change, we need to locate the change made to the old business process, detect the type of change, and identify the affected tasks.

Requirement#3: As the business process and the change may be complex and there may be several associated software artifacts, informative information regarding the types of change and parts of the existing software that are impacted should be provided to the project manager and the system analyst so that they can determine subsequent changes that are to be made to complete the change in the business process. This information should be provided in a manner that is as automated as possible.

To answer these requirements, this paper presents business process change detection for traceability impact analysis. We associate tasks (i.e., atomic business process activities) in the old business process model with traceability information that links them to other related software artifacts, i.e., requirements, use cases, design classes, and programs. When a new version of the business process model is designed to incorporate changes in business requirements, the two versions are compared to detect the business process tasks that are changed and the patterns (or types) of changes. Given the traceability information of the old model version, we can subsequently report the software artifacts that are associated with the changed tasks and would be affected by the changes. This paper is an extended version of the initial report of the approach [4]. In particular, this paper revisits the methodology and presents a complete set of algorithms to detect business process change patterns. We present traceability impact analysis of a real-world case of business process changes and discuss an evaluation of the approach.

The rest of this paper is organized as follows. Section 2 discusses related work and Sect. 3 revisits the approach. The algorithms for business process change patterns detection are listed in Sect. 4 followed by traceability impact analysis in Sect. 5. Section 6 discusses an evaluation and concludes the paper.

2 Related Work

Traceability impact analysis is a widely addressed issue in software requirements management. Literature has reported different approaches to enable traceability and the use of traceability information to analyze the scope and degree of requirement change impact at both code and design level of the software, by tracing the relationships between software artifacts. Here we focus on the impact of change that is not initiated at the software but at the business process level.

Piprani et al. [5] argue that, in reality, business requirements are generally surfaced over several years in memos, e-mails, meeting minutes, consultant reports etc. They propose an Object Role Modeling (ORM) based metamodel for modeling traceability information across these documents and across development phases. Their approach is seen as an effort to extend traceability information to include documents other than typical software artifacts. Similarly, we extend traceability to the business process level and allow tracing between a business process model and an analysis model of the software by documenting a relationship between a task in a business process model and a use case in a use case diagram. This is the approach taken by IBM's Rational System Architect for transforming a BPMN diagram to a UML use case diagram [6].

A number of researchers have tackled the problem of change impact analysis for business processes. For example, Wang et al. [7] and Xiao et al. [8] define change types and a change impact set for a business process that involves process changes (such as insert/remove/move a task) and external service changes that are propagated to the business processes. Unlike these approaches, ours targets business process changes at the modeling level, not the execution level, and our impact set will comprise the software artifacts that relate to the changed process model. In addition, we note that change types such as those by Refs. [7, 8] are rather primitive and not so informative for the system analyst to reason about business requirements change. We choose to adopt Dijkman's classification of structural differences between business processes [9] as the patterns of changes between two versions of a business process model. For the detail of this classification, refer to [9].

3 Overview of Change Impact Analysis of Business Process

The change impact analysis of a business process is based on comparison of the
model of the old business process and the new model which incorporates the
required process changes. The important tasks of the analysis, as shown in Fig. 1
and supported by our tool, involve (1) maintenance of traceability information
specifying the links between the old business process model and the software
artifacts that are generated in different development phases to automate the process

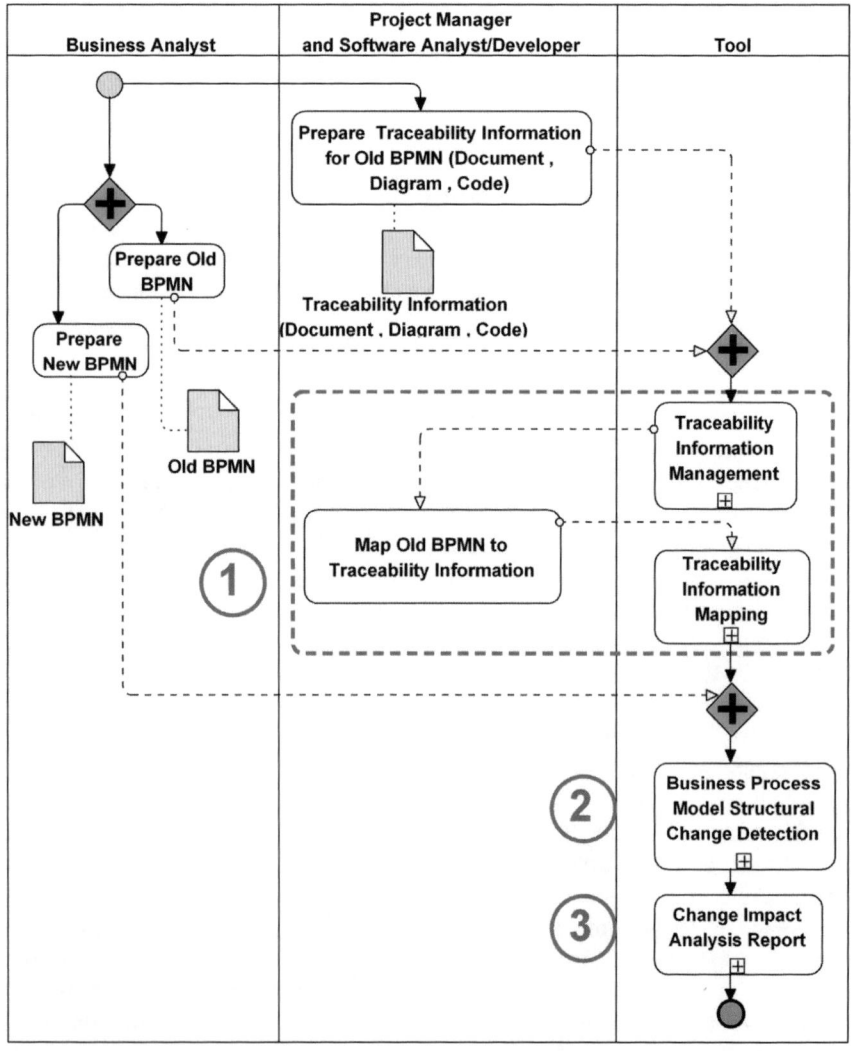

Fig. 1 Change impact analysis of business process

(i.e., response to *Requirement#1*); (2) detection of change patterns to reason about how the old and new business process models are different and to locate parts of the old model that are changed (i.e., response to *Requirement#2*); and (3) traceability impact analysis which analyze and report the software artifacts that are impacted by process changes (i.e., response to *Requirement#3*). Note that our tool analyzes the business process model in the BPMN notation.

Initially there were business requirements and a business analyst designed a business process model (i.e., old BPMN) whose tasks answered the requirements. Based on this old BPMN, the development team produced software artifacts during different software development phases, including the application program. We extend traceability to the business process level and allow tracing between the old business process model and an analysis model of the software by documenting a relationship between a task in the old BPMN and a use case in a use case diagram of

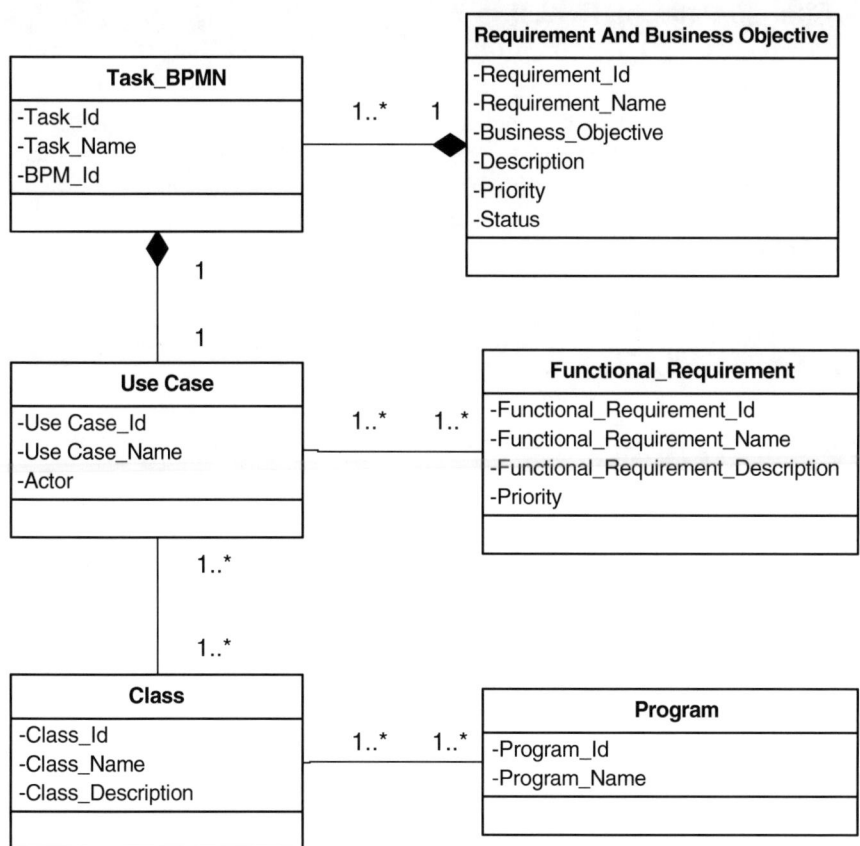

Fig. 2 Metamodel of traceability information mapping

the implemented software. The task then can be associated indirectly with other software artifacts that are linked further from the use case. Using the tool, the project team has to record the artifacts that are related to the old BPMN (i.e., the business requirements, use cases, software functional requirements, design classes, and programs) and define mapping between each task in the old BPMN and its related artifacts. The mapping is modeled by Fig. 2.

Later when there are changes in the business process, the business analyst designs a new version of the business process model (i.e., new BPMN) to reflect the changes. The question raised in the software project is, to what extent these changes will affect the associated application. Hence, the two versions of the BPMN will be compared to detect change patterns and report the tasks of the old BPMN and affected artifacts that may need modification.

4 Change Patterns Detection

To demonstrate the approach, we use the two versions of the business process of debt setting for spare parts and services of an agricultural machinery company in Thailand. We cover twelve patterns of process structure changes under three categories as classified by Dijkman [9]. The following describes the algorithms we use to detect if any of these change patterns is applied to a task in the old BPMN to create a new BPMN version.

4.1 Authorization Differences

In this category, a task in the old BPMN is assigned to different role(s) in the new BPMN. Each Pattern is depicted in Fig. 3.

4.1.1 Different Roles

In this pattern, a task is assigned to one role in the old BPMN and to another role in the new BPMN. The detection algorithm is as follows.

1. Check if a task exists in both versions of the BPMN.
2. Check if such a task is performed by a single role in the old BPMN and also by a single role in the new BPMN.
3. If the performing roles in both BPMNs are different, it is the case of the different roles pattern.

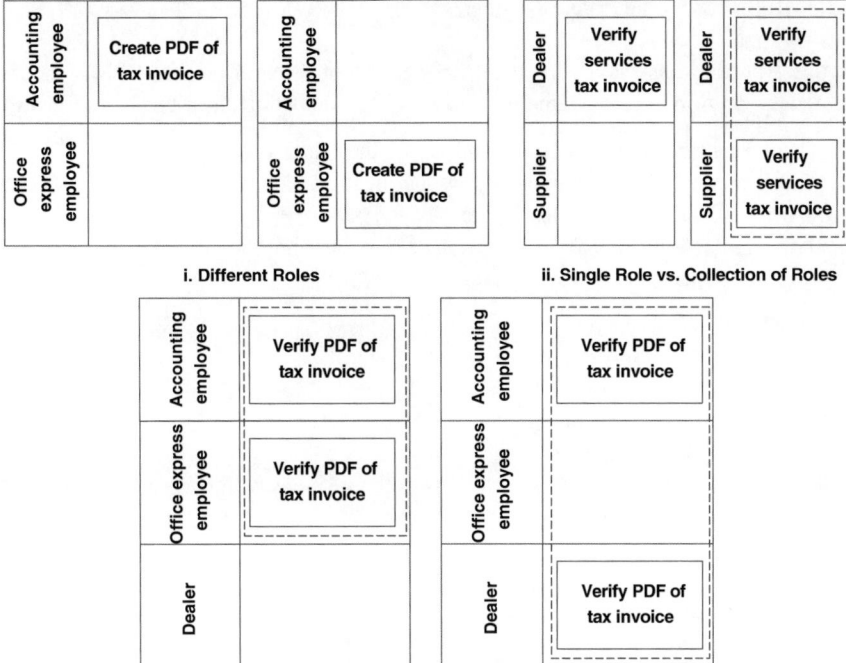

i. Different Roles   ii. Single Role vs. Collection of Roles

iii. Different Collections of Roles

Fig. 3 Change patterns: authorization differences

4.1.2 Single Role Versus Collection of Roles

In this pattern, a task is assigned to one role in one BPMN and to multiple roles in the other BPMN. The detection algorithm is as follows.

1. Check if a task exists in both versions of the BPMN.
2. Check if it is either of the following:

   (a) If such a task is performed by a single role in the old BPMN and by multiple roles in the new BPMN, it is the case of the single role versus collection of roles pattern; or

   (b) If such a task is performed by multiple roles in the old BPMN and by a single role in the new BPMN, it is the case of the single role versus collection of roles pattern.

4.1.3 Different Collections of Roles

In this pattern, a task is assigned to a collection of roles in the old BPMN and to another collection of roles in the new BPMN. The detection algorithm is as follows.

1. Check if a task exists in both versions of the BPMN.
2. Check if such a task is performed by a collection of roles in the old BPMN and also by a collection of roles in the new BPMN.
3. If the collections of roles in both BPMNs are different, it is the case of the different collections of roles pattern.

4.2 Control Flow Differences

In this category, the control-flow relations that a task has with other tasks in the old BPMN are different in the new BPMN. Each Pattern is depicted in Fig. 4.

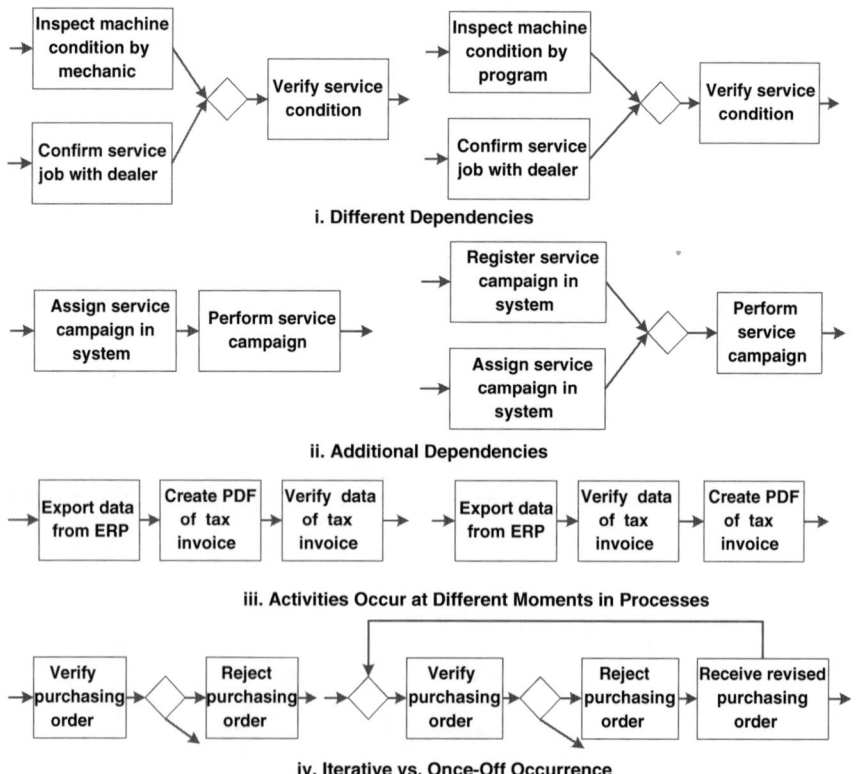

Fig. 4 Change patterns: control flow differences

4.2.1 Different Dependencies

In this pattern, a set of preceding tasks on which a task depends in the old BPMN is different in the new BPMN. The detection algorithm is as follows.

1. Check if a task exists in both versions of the BPMN.
2. Check if the preceding model element of such a task in the old BPMN is a gateway or a task:

   (a) In the case of gateway in the old BPMN, check if the preceding element of such a task in the new BPMN is a gateway or a task:

   - In the case of gateway in the new BPMN, if the set of tasks that precedes the gateway in the old BPMN is different from that in the new BPMN, it is the case of the different dependencies pattern.
   - In the case of task in the new BPMN, if the set of tasks that precedes the task in the old BPMN is different from that in the new BPMN, it is the case of the different dependencies pattern.

   (b) In the case of task in the old BPMN, check if the preceding element of such a task in the new BPMN is a gateway or a task:

   - In the case of gateway in the new BPMN, it is the case of the different dependencies pattern.
   - In the case of task in the new BPMN, if the set of tasks that precedes the task in the old BPMN is different from that in the new BPMN, it is the case of the different dependencies pattern.

4.2.2 Additional Dependencies

In this pattern, a set of preceding tasks on which a task depends in the old BPMN is a subset of the set on which such a task depends in the new BPMN. The detection algorithm is as follows.

1. Check if a task exists in both versions of the BPMN.
2. Check if the preceding model element of such a task in the old BPMN is a gateway or a task:

   (a) In the case of gateway in the old BPMN, check if the preceding element of such a task in the new BPMN is a gateway or a task:

   - In the case of gateway in the new BPMN, if the set of tasks that precedes the gateway in the old BPMN is a subset of that in the new BPMN, it is the case of the additional dependencies pattern.
   - In the case of task in the new BPMN, it is the case of the additional dependencies pattern.

(b) In the case of task in the old BPMN, check if the preceding element of such a task in the new BPMN is a gateway or a task:

- In the case of gateway in the new BPMN, it is the case of the additional dependencies pattern.
- In the case of task in the new BPMN, it is the case of the additional dependencies pattern.

4.2.3 Activities Occur at Different Moments in Processes

In this pattern, a set of preceding tasks and a set of succeeding tasks on which a task depends in the old BPMN are disjoint with their corresponding sets in the new BPMN. The detection algorithm is as follows.

1. Check if a task exists in both versions of the BPMN.
2. Check if the set of preceding tasks and the set of succeeding tasks of such a task are disjoint with their corresponding sets in the new BPMN, it is the case of the activities occur at different moments in processes pattern.

4.2.4 Iterative Versus Once-Off Occurrence

In this pattern, a task is part of a loop in one version of the BPMN while it is not in the other version. The detection algorithm is as follows.

1. Check if a task exists in both versions of the BPMN.
2. Check if it is either of the following:

(a) If there is no loop from any of the succeeding tasks back to the preceding tasks of such a task in the old BPMN but there is such a loop in the new BPMN, it is the case of the iterative versus once-off occurrence pattern; or
(b) If there is a loop from any of the succeeding tasks back to the preceding tasks of such a task in the old BPMN but there is not such a loop in the new BPMN, it is the case of the iterative versus once-off occurrence pattern.

4.3 Activity Differences

In this category, a unit of work that is performed by a collection of tasks in the old BPMN is performed by a different collection of tasks, or not at all, in the new BPMN. Each Pattern is depicted in Fig. 5.

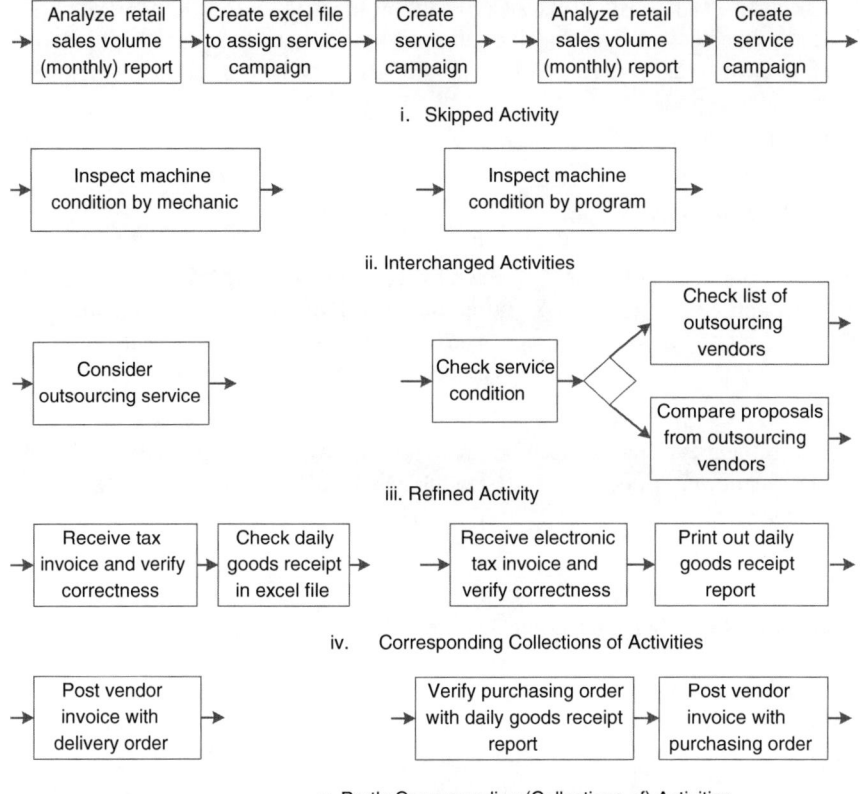

Fig. 5 Change patterns: activity differences

4.3.1 Skipped Activity

In this pattern, a task exists in the old BPMN but there is neither such a task nor an equivalent task in the new BPMN. The detection algorithm is as follows.

1. Check if a task exists in the old BPMN but does not in the new BPMN.
2. If an equivalent task of such a task is not specified in the new BPMN, it is the case of the skipped activity pattern.

4.3.2 Interchanged Activities

In this pattern, a unit of work performed by a task in the old BPMN is performed by a different but equivalent task in the new BPMN. Note that the equivalent task cannot be detected automatically by the algorithm and has to be specified by the business analyst. The detection algorithm is as follows.

1. Check if a task exists in the old BPMN but does not in the new BPMN.
2. If the business analyst specifies that such a task in the old BPMN is fully equivalent to another task in the new BPMN, it is the case of the interchanged activities pattern.

4.3.3 Refined Activity

In this pattern, a unit of work performed by a task in the old BPMN is performed by an equivalent collection of tasks in the new BPMN. Note that the equivalent collection of tasks cannot be detected automatically by the algorithm and has to be specified by the business analyst. The detection algorithm is as follows.

1. Check if a task exists in the old BPMN but does not in the new BPMN.
2. If the business analyst specifies that such a task in the old BPMN is fully equivalent to a collection of tasks in the new BPMN, it is the case of the refined activities pattern.

4.3.4 Corresponding Collections of Activities

In this pattern, a unit of work performed by a collection of tasks in the old BPMN is performed by an equivalent collection of tasks in the new BPMN. Note that the equivalent collection of tasks cannot be detected automatically by the algorithm and has to be specified by the business analyst. The detection algorithm is as follows.

1. Check if a collection of tasks exists in the old BPMN but does not in the new BPMN.
2. If the business analyst specifies that such a collection of tasks in the old BPMN is fully equivalent to another collection of tasks in the new BPMN, it is the case of the corresponding collections of activities pattern.

4.3.5 Partly Corresponding (Collections of) Activities

In this pattern, a unit of work performed by a (collection of) task(s) in the old BPMN and that performed by a (collection) of task(s) in the new BPMN are partly equivalent and partly different. Note that the corresponding collection of tasks cannot be detected automatically by the algorithm and has to be specified by the business analyst. The detection algorithm is as follows.

1. Check if a collection of tasks exists in the old BPMN but does not in the new BPMN.

2. If the business analyst specifies that such a collection of tasks in the old BPMN is partly equivalent to another collection of tasks in the new BPMN, it is the case of the partly corresponding (collections of) activities pattern.

5 Traceability Impact Analysis

An impact analysis of changes that are made to the old BPMN is possible by traceability information of the tasks in the old BPMN and related software artifacts. When the change pattern detection algorithms in Sect. 4 are used to identify changed tasks, impacted software artifacts are analyzed.

5.1 Traceability Information

The traceability information can be managed by the project team via our tool. Information about the artifacts of the software that automates the old BPMN can be input to the tool. The old BPMN can then be uploaded to the tool and its tasks are extracted. Then the system analyst can select each of the extracted tasks as in Fig. 6a to specify software artifacts that are related to it as in Fig. 6b. The artifacts include business requirements, use cases in a use case diagram, software requirements in a software requirement specification, classes in a class diagram, and programs. As mentioned in Sect. 4, detection of four change patterns cannot be completely automated. Therefore, when the new BPMN is uploaded to the tool and its tasks are extracted, the business analyst assists pattern detection by selecting which (collections of) tasks in the old and new BPMNs are equivalent and whether they are fully or partly equivalent such as in Fig. 7.

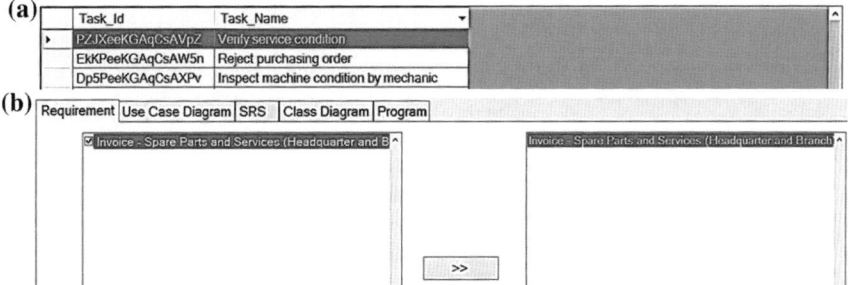

Fig. 6 Traceability information management **a** selecting task extracted from old BPMN, **b** associating artifacts with extracted task

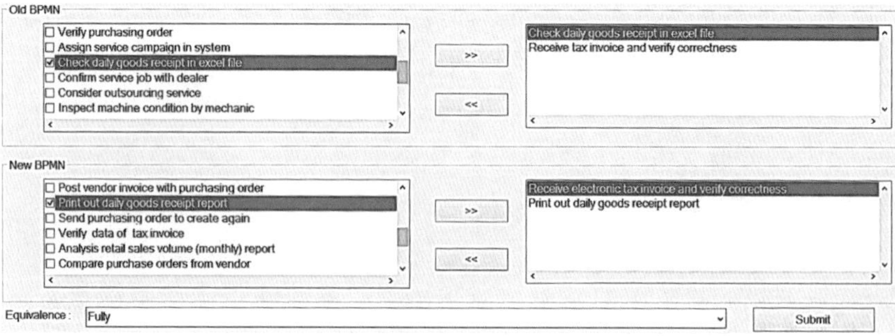

Fig. 7 Specifying equivalent tasks between old and new BPMNs

Authorization Differences Type

No.	Name of Changed Task	Differences Types	Cause Of differences
1	Create PDF of tax invoice	Different roles	Change role from 'Accounting employee' to 'Office express employee'
2	Print out original tax invoice and send to dealer	Different roles	Change role from 'Accounting employee' to 'Office express employee'

Fig. 8 Example of changed tasks in the old BPMN and relevant change patterns

5.2 Change Impact Analysis

When change patterns are detected, the tasks to which the change patterns are applied form an impact set. For each impacted task in the set, the tool traces its traceability information mapping to its associated artifacts to analyze further impact. The tool reports the impacted tasks in the old BPMN and change patterns that are applied to them such as in Fig. 8. In addition, the tool reports each kind of software artifacts that are affected; an example of business requirements affected by changes in several tasks is shown in Fig. 9. The software project team can use these analysis results to guide them through evaluation of the impact scope and planning for the change.

Requirement and Business Objective Information

No.	Name of Changed Task	Differences Types	Name of impacted Requirement
1	Create PDF of tax invoice	Different roles	Invoice - Spare Parts and Services (Headquarter and Branch)
2	Inspect machine condition by mechanic	Interchanged activities	Invoice - Spare Parts and Services (Headquarter and Branch)
3	Verify service condition	Different dependencies	Invoice - Spare Parts and Services (Headquarter and Branch)

Fig. 9 Example of impacted artifacts

6 Discussion and Conclusion

The presented approach uses the business process change patterns to drive the traceability impact analysis. Traceability information is extended to the business process level to record the relationships between business process tasks and software artifacts that they originate.

In an evaluation, we use ten cases of real-word business processes that altogether incorporate all change patterns. The tool can detect change patterns and analyze changed tasks and impacted artifacts correctly. However, the performance of the approach and the tool also depends on the correctness of the task equivalence information specified by the business analyst since four change patterns can be detected only semi-automatically. To improve the situation, techniques to determine semantic similarity between tasks can be adopted to analyze semantic knowledge about tasks. We can also have the impact analysis results visualized to make the analysis report more intuitive.

References

1. Object Management Group, Business process model and notation (BPMN) (Online). Available: http://www.omg.org/spec/BPMN/2.0/PDF
2. F.A.C. Pinheiro, J.A. Goguen, An object-oriented tool for tracing requirements, in *IEEE Software*, Mar 1996, pp. 52–64
3. S.A. Bohner, R.S. Arnold, *Software Change Impact Analysis* (IEEE Computer Society Press, Los Alamitos, 1996)
4. W. Uronkarn, T. Senivongse, Change pattern-driven traceability of business processes, in *Proceedings of the International MultiConference of Engineers and Computer Scientists 2014, IMECS 2014*, 12–14 Mar 2014, Hong Kong. Lecture Notes in Engineering and Computer Science, pp. 601–606
5. B. Piprani, M. Borg, J. Chabot, É. Chartrand, An adaptable ORM metamodel to support traceability of business requirements across system development life cycle phases, in *Proceedings of on the Move to Meaningful Internet Systems: OTM 2008 Workshops*. Lecture Notes in Computer Science, vol. 5333, pp. 728–737 (2008)
6. IBM: Mapping business process diagrams to UML use case diagrams (online). Available: http://pic.dhe.ibm.com/infocenter/rsysarch/v11/index.jsp?topic=/com.ibm.sa.bpr.doc/topics/t_ovwmapbp2uml.html
7. Y. Wang, J. Yang, W. Zhao, Change impact analysis for service based business processes, in *Proceedings of IEEE International Conference on Service-Oriented Computing and Applications, SOCA 2010*, 13–15 Dec 2010, Australia, pp. 1–8
8. H. Xiao, J. Guo, Y. Zou, Supporting change impact analysis for service oriented business applications, in *Proceedings of International Workshop on Systems Development in SOA Environments, SDSOA 2007*, 20–26 May 2007, Minneapolis, 6 pp
9. R. Dijkman, A classification of differences between similar business processes, in *Proceedings of 11th IEEE International Conference on Enterprise Distributed Object Computing Conference, EDOC 2007*, 15–19 Oct 2007, Annapolis, pp. 37–47

Author Index

© Springer Science+Business Media Dordrecht 2015
G.-C. Yang et al. (eds.), *Transactions on Engineering Technologies*,
DOI 10.1007/978-94-017-9588-3

457

Subject Index

Note: Page numbers followed by "f" and "t" indicate figures and tables respectively

© Springer Science+Business Media Dordrecht 2015
G.-C. Yang et al. (eds.), *Transactions on Engineering Technologies*,
DOI 10.1007/978-94-017-9588-3

Printed by Printforce, the Netherlands